CNC Programming Techniques

An Insider's Guide to Effective Methods and Applications

First Edition

CNC Programming Techniques

An Insider's Guide to Effective Methods and Applications

Peter Smid

Industrial Press, Inc.
200 Madison Avenue
New York, New York 10016-4078, USA
http://www.industrialpress.com

Library of Congress Cataloging-in-Publication Data

Smid, Peter.
CNC programming techniques:
Peter Smid.
p. cm.
ISBN 0-8311-3185-3

First Edition

CNC Programming Techniques

Industrial Press Inc.
200 Madison Avenue
New York, NY 10016-4078

Editor: John Carleo
Cover Design: Janet Romano

1 2 3 4 5 6 7 8 9 10

Cover photo provided by GE Fanuc Automation.
For more information, visit www.gefanuc.com or call 1-800-GE Fanuc.

Acknowledgments

To

John Carleo

Janet Romano

and Patrick Hansard

... without you, this book would not happen ...

About the Author

Peter Smid is a professional consultant, educator and speaker, with many years of practical, hands-on experience, in the industrial and educational fields. During his career, he has gathered an extensive experience with CNC and CAD/CAM applications on all levels. He consults to manufacturing industry and educational institutions on practical use of Computerized Numerical Control technology, part programming, CAD/CAM, advanced machining, tooling, setup, and many other related fields. His comprehensive industrial background in CNC programming, machining and company oriented training has assisted several hundred companies to benefit from his wide-ranging knowledge.

Mr. Smid's long time association with advanced manufacturing companies and CNC machinery vendors, as well as his affiliation with a number of Community and Technical College industrial technology programs and machine shop skills training, have enabled him to broaden his professional and consulting skills in the areas of CNC and CAD/CAM training, computer applications and needs analysis, software evaluation, system bench marking, programming, hardware selection, software customization, and operations management.

Over the years, Mr. Smid has developed and delivered hundreds of customized educational programs to thousands of instructors and students at colleges and universities across United States, Canada and Europe, as well as to a large number of manufacturing companies and private sector organizations and individuals.

He has actively participated in many industrial trade shows, conferences, workshops and various seminars, including submission of papers, delivering presentations and a number of speaking engagements to professional organizations. He is also the author of articles, monthly magazine columns, and many in-house publications on the subject of CNC and CAD/CAM. During his many years as a professional in the CNC industrial and educational field, he has developed tens of thousands of pages of high quality training materials.

Peter Smid is also the author of

CNC Programming Handbook, A Comprehensive Guide to CNC Programming

Fanuc CNC Custom Macros: Practical Resources for Fanuc Custom Macro B Users

Both hardcover books have been published by Industrial Press. Inc.

The author always welcomes comments, suggestions and other input from educators, students and industrial users.

You can send e-mail to the author from the ***CNC Programming Techniques*** page at

www.industrialpress.com

Preface

I would like to express my most sincere thanks to all programmers, machinists, operators, engineers, students and many other readers and users who made my two previous books - also CNC oriented - such a great success. Both were published by *Industrial Press, Inc.* (New York, NY, USA):

- **CNC Programming Handbook, Second Edition with CD-ROM**
  A Comprehensive Guide to CNC Programming
  ISBN: (0-8311-) 3158-6

  and

- **Fanuc CNC Custom Macros, with CD-ROM**
  Practical Resources for Fanuc Custom Macro B Users
  ISBN: (0-8311-) 3157-8

This third handbook also relates to the subject of CNC programming, this time from a somewhat different angle. First, there several programming subjects that are virtually impossible to find anywhere else, for example, how to program cams or tapered end mills. Other, more common, subjects are covered in a great depth, such as the coverage of cutter radius offset or thread milling.

As in my previous publications, I have included many overall and detailed drawings, to help visualize the subject or procedures covered. Where applicable, a complete programming example is provided, or - at least the most significant part is shown.

In view of the recent, and rather significant, emergence of metric system in many North American industries, particularly in the USA, I have focused on more examples presented in metric units than those in imperial units. Working on the premise that a professional CNC programmer should have no problem working with either units selection (after all, number are numbers), many examples in this handbook emphasize the metric system. For balance, a significant number of examples using imperial units are also included. Speaking of *imperial* units - in my previous books, I had used the term *English* units instead. It may seems frivolous, but the fact is that modern Great Britain is now a metric country and the so called *English units* are the thing of the past - of the *imperial* era in British history.

I also feel that I should mention the relationship of this book to the *CNC Programming Handbook*. In terms of focus, these are very different publications. *CNC Programming Techniques* is a book that does not replace my previous books, but complements them in a special way. In terms of subjects covered, there are minor similarities in some chapters, but the coverage of each subject is fresh, and with much more detail provided. At the end of the book, I had included references to subjects covered in the *CNC Programming Handbook*. My feeling was that those readers who may need some additional background will benefit from these references. On the other hand, those, who do not need the background can safely ignore those few pages and explore the subjects covered in this book only.

I sincerely hope that this book will help you become even a better CNC programmer (or even a better CNC Operator) by understanding not just the 'hows' but also the 'whys' of many programming techniques. Thanks you for your continuing interest.

Peter Smid
November 2005

TABLE OF CONTENTS

1 PART PROGRAM DEVELOPMENT

This introductory chapter to the *CNC Programming Techniques* presents the most basic technique of all - *how to develop a part program* in an organized way. Its purpose is to present an engineering drawing, evaluate it and develop all procedures required to write the final program.

In strict terms, program development using a step by step procedure is not the most appropriate approach, because it suggests *'finish step 1 before starting step 2'*. That is never the case in CNC programming, as many 'steps' interact with each other. For example, a change in setup may influence the tooling selection, width or depth of cut, etc. Keep this in mind when studying this chapter.

Program Development Drawing

The drawing below will be used throughout the chapter, including required details, calculations, all complete with explanations of individual steps required in the CNC program development. The drawing includes some of the most common machining operations - face cutting, machining holes, contouring, circular pocket, and a slot milling. The design has been kept intentionally simple.

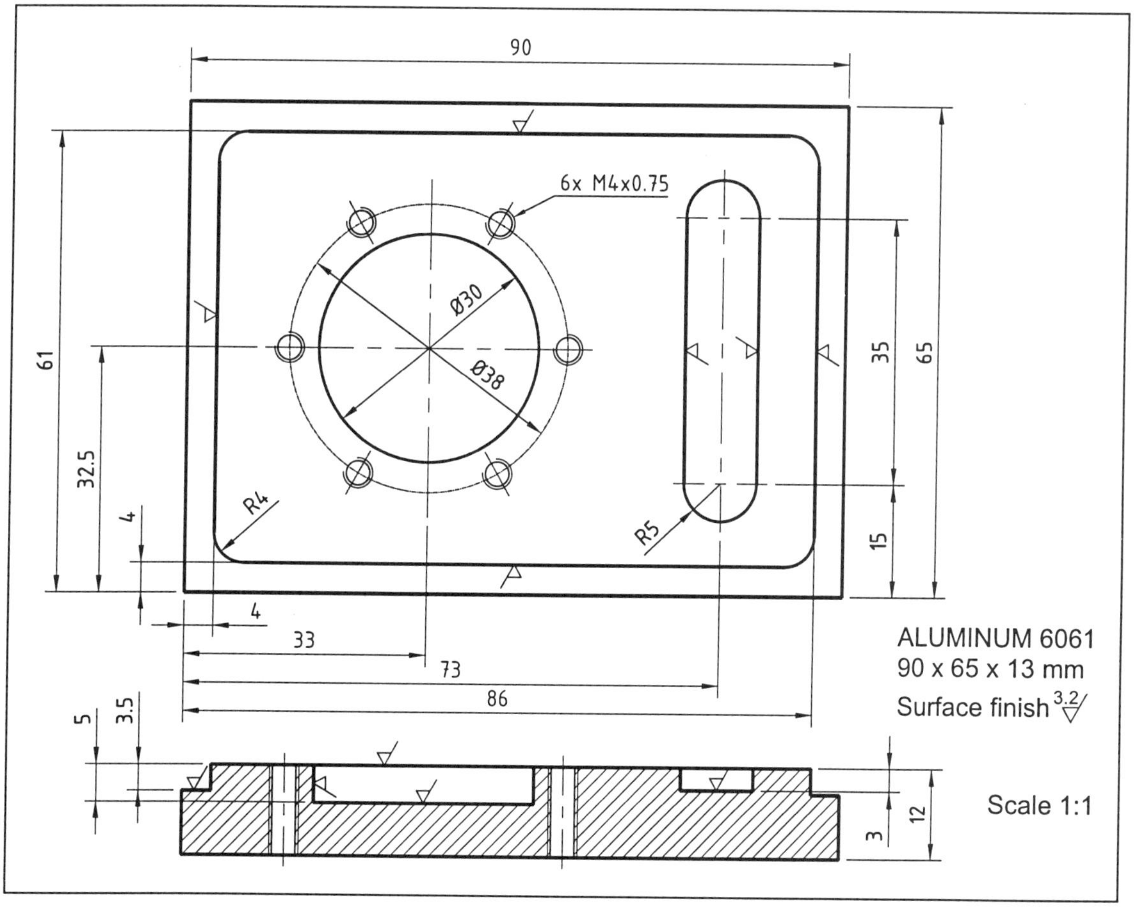

Drawing Evaluation

The first thing a CNC programmer should always do before writing the program, is to evaluate the drawing in order to get a general idea about the part. Such an evaluation includes several observations that can be summed up:

- Drawing units and scale
- Dimensioning method
- Tolerances
- Material type, size, shape and condition
- Surface finish requirements
- Title block information
- Drawing revisions
- Bill of Materials (BOM) - if available
- Omissions and other errors

In the enclosed drawing the ***drawing units*** are not specified directly, but it is obvious from the drawing that it uses metric dimensions in millimeters. ***Scale*** is not always specified in the drawing, often because of various forms of copying may not match the original scale. The drawing for this chapter shows a full size drawing and is specified as 1:1.

Drawing ***dimensions*** are always important for the CNC programmer, at least for two reasons. One, they establish the important features of the part, two, they serve as the most important resource in determine the part zero (part origin). In the drawing, all dimensions originate from the lower left corner of the part - in this case, that will also be the most suitable location of the part zero. Keep in mind that this is not always the case.

Tolerances are closely related to dimensions. This particular drawing does not contain any tolerances, so the aim of the programmer (and operator) will be to adhere to general company standards.

Not all drawings describe every aspect of the ***material*** the part is made of, but for programming purposes, the type, size, shape and condition of the blank material are the most important. The sample drawing specifies the material type and its size. *Aluminum 6061* is easy to machine, and fairly high speeds and feeds can be used for efficient machining. The material size is specified in the drawing as 90 × 65 × 13 mm (L × W × D). Here comes the first item that will have a direct relationship with the tooling selection and machining operations. Although the length and width of the material are the same as the final length and width of the part, that is not the case with the material depth (thickness). There is 1 mm difference that has be accounted for during programming and even setup.

The drawing also specifies the overall ***surface finish*** of 3.2 for all marked surfaces. Not all drawings specify individual surfaces. The amount of 3.2 is a statistical deviation from the ideal profile and is measured in μm (millionth of a meter = 0.000 001 m = 10^{-6}). In practical terms, the required finish of 3.2 μm is attainable with standard milling cutters at relatively fast spindle speeds and moderate feedrates, assuming proper setup and quality tooling.

Small and simple drawings seldom have an elaborate ***title block***. Title block is usually a rectangular area in a corner of the drawing that contains text data, such as drawing name, part number, draftsperson, dates, revisions, material, etc. ***Revisions*** are changes made to the drawing from its original version - they are very important to the CNC programmer. Always make sure you develop the program using the latest drawing version. Keeping a copy of the drawing is also a good idea.

Bill of Materials (abbreviated as BOM) is a special list that contains individual items requ manufacture the part. Typical items included in BOM are stock items, purchased items and v other parts required for assembly. Large complex drawing are more likely to have BOM than small simple drawings.

Even the best engineers and draftspersons make a mistake. One important part of drawing evaluation is to look for ***errors, omissions*** and other discrepancies. It is most frustrating to go through many calculations just to find out that a critical dimension was missing. The programmer should also look for double dimensioning, where one dimension is in conflict with another dimension. Finally, when evaluating a drawing, never try to find dimensions by scaling the lines or arcs, never assume a dimension or a feature. *If in doubt, always ask.*

Material and Stock

Although the enclosed sample drawing does include material the part is made of, there are many issues related to the material before machining (and programming). Apart of size of the material, its ***shape and condition*** are equally important. Shape could be a simple block or cylinder, it could be cored or solid, it could be a complex casting or forging, and so on. Shape of the material is most important in program development during setup and tooling decisions as well as toolpath. The condition of the material includes the overall quality - such as burrs, flakes, premachining, hardness, etc. It also includes the accuracy of the shape. For example, it is important that the material supplied for the sample drawing is exactly 90 × 65 mm in length and width, square at corners, as these sizes are already final and need no machining. There is a little bit more flexibility in the material thickness, because the top will have to be faced off to suit drawing dimension of the part height.

Part Setup

A high quality vise, specially designed for CNC work is the most common fixture (part holding devise) for small to medium size parts on vertical machining centers.

Part Reference Point

Part reference point is another common name for *part zero* or *part origin.* Before any toolpath can be developed in the form of coordinates, the CNC programmer has to select *part zero.* As a general rule, the part zero for setups in a vise should always be located on a non moving jaw (fixed jaw), and a part stopper or similar device is also recommended for repeatability.

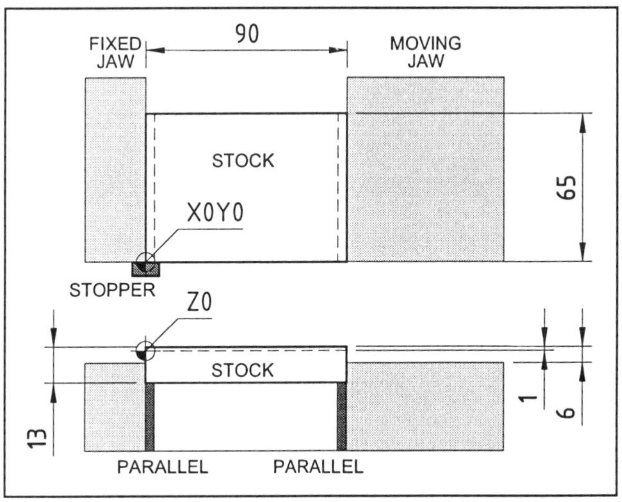

Part Orientation

How the part blank (stock) is oriented in the vise often influences the method of machining. Looking at the sample drawing, the part blank can be oriented either horizontally or vertically, looking from the CNC operator's direction. Horizontal orientation has the advantage that the machined part will match the drawing. Another benefit is that the lower left corner of the part will be located at the intersection of the vise non moving jaw and a stopper. The only benefit of vertical orientation would be the width of grip being 65 mm rather than 90 mm in the horizontal orientation, minimizing any 'bending' effect caused by pressure of the jaws. For this job, we select the horizontal orientation, as the 25 mm difference between widths has no practical disadvantage.

Selecting Part Zero

Based on the previous considerations in this section, the selection of part zero for the X and Y axes presents no problem. The lower left corner of the part will be the part zero. This is also the coordinate location that will be used to set the work offset G54.

Setting the part zero for the Z-axis requires some thinking as well as evaluation of several possibilities. Unless the machine shop uses off-machine presetter to set the tool length, the most common method of tool length setting is the *touch-off* method. Selecting Z0 is important part of the setup. The most common method is the *top of the finished part*, but bottom of the part or some other fixture location may also be considered. In this example, the top of the *finished* part will be Z0, which brings up a question - what to do with the 1 mm extra height? The top face will be face milled, and the 1 mm extra thickness will be removed by that operation. All tool lengths will be set on the finished face. Later section of this chapter will provide more details. Look at the illustration shown on the previous page.

Tooling Selection

Selecting tool holders and cutting tools is another very critical part of part program development process. Tool holders are generally used with different cutting tools and remain unchanged for extended periods of time. Tools, such as drills, reamers, taps, end mills, carbide inserts, etc. are perishable tools. Some toolholders are used for a specific tool group, such as a collet holder is a better choice for end mills, Weldon™ type toolholders prevent tool spinning, Jacobs™ chucks are used for drills, etc. Some tools, such as taps, require special holders, designed for a single purpose.

Tooling selection is always closely related to the setup and cutting conditions. When selecting tools, always keep in mind these other two considerations. When the part setup method has been established, the tools can be selected on the basis of drawing dimensions and machining operations required. Tools are always selected on the basis of machining operations required.

Identifying Machining Operations

From the sample drawing, even a casual look will identify the types of operations required to machine the part - all are very common and can be adapted to other jobs:

- **Face milling**
- **Contour milling**
- **Circular pocket milling**
- **Slot milling**
- **Spot drilling**
- **Drilling**
- **Tapping**

One of the most important machining rules is that heavier operations are always done before lighter operations. That does not just mean roughing before finishing, which is a common sense, it also means milling before drilling, for example. Milling has the tendency to shift the material in the XY axes, whereby drilling pushes the material towards the fixture (Z-axis). The above operations list is suitable to be used as the order of operations.

Face Milling

As the part is 1 mm thicker than the final thickness, a face milling tool has to be used to machine the top face to provide final thickness of 12 mm, as per drawing. The amount of one millimeter represents a small depth for facing, so only one depth cut will be necessary to face the part. In order to select the best face mill diameter, the part length and/or width have to be considered as well. In many cases, the length or width of the part also determines the *direction* of the cut. The length of the part is 90 mm, its width is 65 mm. If either direction is acceptable, a standard ∅100 mm face mill would be the best choice. A ∅75 mm face mill will only be able to face along the X-axis. If both face mills are available, which one is a better choice? Number of flutes (cutting inserts) in the tool should also be considered, to employ as many edges at the same time. If the ∅100 mm face mill has more cutting edges, it will be a better choice. It will also be the choice for this job.

Cutting direction is very important as well. Although the ∅100 mm face mill can cut in four directions along the two axes, choice of the X-direction, from the X+ to the X- (right to left) is recommended. The reason is that the cutter will be 'pushing' against the vise fixed jaw, creating favorable cutting conditions. Cutting along the Y-axis can cause the face cutter to pull the part up. Last, and equally important consideration is the location of the face mill center relative to the part. It is not recommended to locate the cutter center at the middle of the part, but slightly off the middle. Such tool position will better control of chips during the entry and exit, to and from the material. It also minimizes chatter.

The Y-axis location of the face mill center must still guarantee that the full part width of 65 mm can be machined. To calculate the maximum shift possible, take one half of the difference between the face mill diameter (100 mm) and the part width (65 mm):

Maximum shift from part middle = (100 - 65) / 2 = 17.5 mm

Select the Y-location a little less, to make the tool overhang the part. For this example, 15 mm is a reasonable shift, with 2.5 mm left for edge overlap - see illustration.

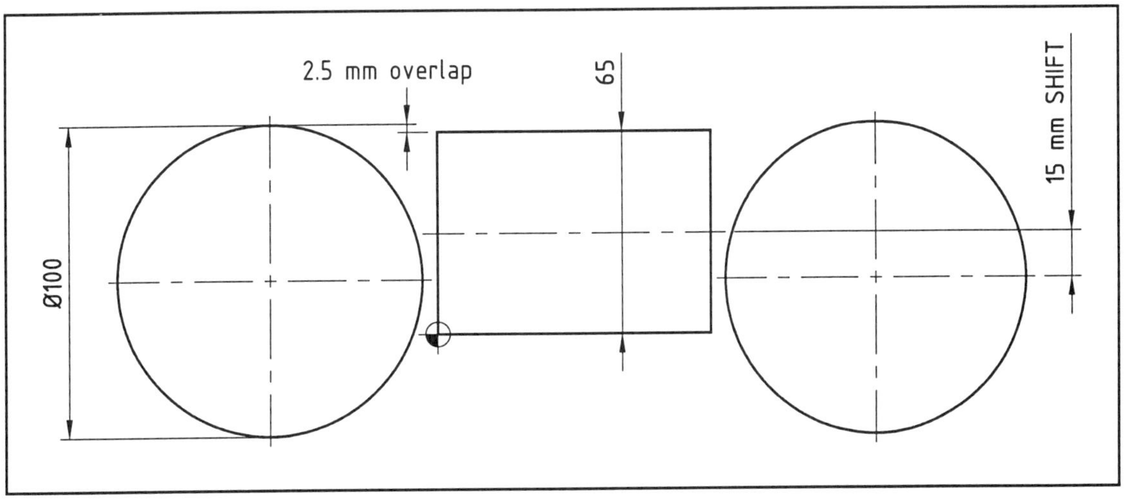

Contour Milling

The filleted rectangular shape is located four millimeters from all edges, except at the corners, where the distance is much greater. This corner distance is very important when selecting cutter diameter. If only a single contouring toolpath is required, the cutter diameter should be larger than the corner distance *C* - *see illustration*. If the corner distance is too large for any cutter size, two or more passes may be required, or some other method of machining needs to be selected.

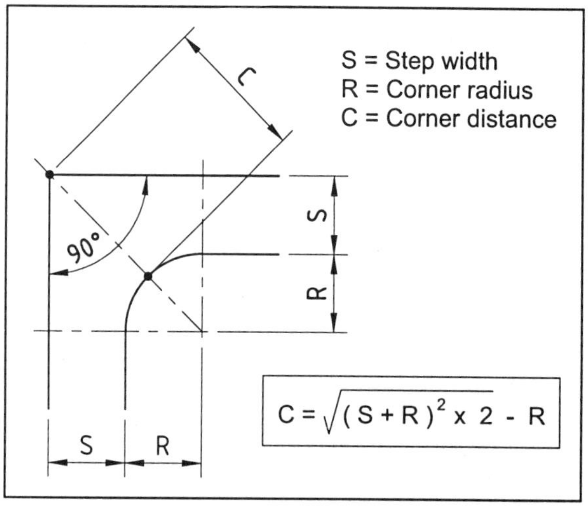

The step depth along the edges is 3.5 mm and should also be considered when the cutting tools are selected. For this job, the depth does not present any special difficulties, and can be machined in one pass.

In our case, the step width *S* is 4 mm, the corner radius *R* is also 4 mm. Using the above formula, the *C* dimension (maximum corner distance) is 7.31371 mm. In practice, any end mill 8 mm in diameter or larger can be used to clean the corners in a single pass. This conclusion is correct, but actual cutting conditions have to be considered as well. Selecting the ∅8 mm end mill will not be a good choice in this case, because the radius of the end mill is the same as the step width (4 mm). The center of the cutter will follow the exact edge of the part, which is not the best machining condition. In addition, a larger tool diameter will provide more tool rigidity and prevent deflection. A two-flute or a three-flute end mill, especially designed for cutting aluminum will be the best choice. As for the diameter, the selection should not always be related to the current operation only, but other operations as well. Either a ∅10 mm or ∅12 mm end mill will be a good choice - the final decision will depend on the tool requirements for the circular pocket. The slot width is 10 mm, so the tool diameter must be smaller than that for optimum cutting conditions.

Circular Pocket Milling

The pocket in the drawing is 30 mm in diameter and 5 mm in depth. Unlike the end mill selected for contouring, the end mill for the pocket must be *center-cutting*, in order to plunge into solid material. Center-cutting end mills are also called slot drills, because they were originally designed for milling of standard slots. High helix, three-flute precision ground center-cutting end mills, are the best overall solution to cutting aluminum. Three flute end mills are often frowned upon - unfairly, because they cannot be measured with standard verniers or micrometers. Yet, they offer the strength, the chip flow and the surface finish that many aluminum parts require. Of course, the more common two-flute end mills are also a suitable choice.

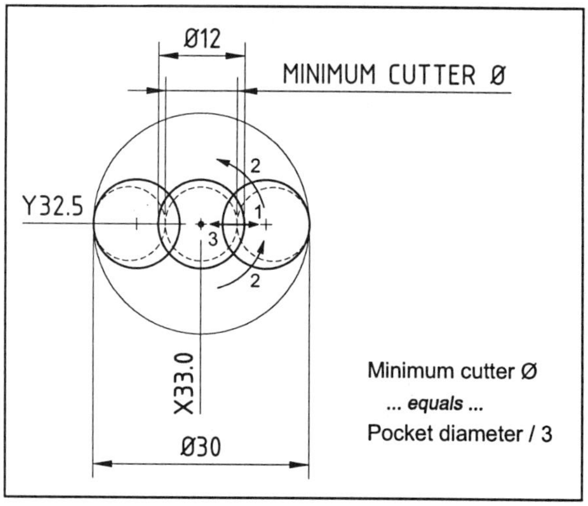

Circular pockets can be machined many different ways, but the most economical method is the single pass method, particularly for small pockets. As shown in the illustration, the minimum diameter of the selected cutter must be one third of the pocket diameter. The reason for the one third is that such tool selection guarantees the complete clean-up of the pocket bottom. The circular pocket in the drawing is ∅30 mm, so the minimum cutter diameter required is 30/3 = ∅10 mm. Although correct mathematically, in practice it is always better to use a slightly larger cutter diameter, to overlap all cuts at the pocket bottom.

In the description of contouring (see above), the choice of tools was narrowed down to two end mills - ∅10 mm end mill or ∅12 mm end mill. Either tool will work well for the contouring, but only the ∅12 mm end mill is also suitable for machining the circular pocket. The selection of ∅12 mm end mill for *both* operations will also offer benefits in a shorter setup time, tool maintenance, and even inventory. The final selection is the 12 mm end mill, used for both - the contour and the pocket.

There is one final condition that has to be considered. In order to machine the pocket. a center-cutting end mill has to be used. That is not required to machine the contour. If only one tool is used for both operations, it has to be a center-cutting end mill for this job. It is perfectly suitable for the contour, and its center-cutting design is mandatory for cutting the circular pocket.

Slot Milling

Unless there are some very tight tolerances required, a standard slot can be machined with a single tool, usually a center-cutting end mill. There is an established - *and very specific* - procedure to cut standard slots (described later). It includes a single pass between centers, followed by the internal contour to make the slot according to the engineering drawing. The main consideration in tooling selection is the *slot width*.

In the drawing, the slot radius is 5 mm, so its width is the double the radius, 10 mm. Whatever your programming style may be, resist the temptation to use a ∅10 mm center cutting end mill between slot end centers only - the quality of the slot will be very poor and so will its dimensions. Much better choice is to select a center-cutting end mill that is a bit smaller than the slot width. Is the size of such an end mill important? If so, why? YES - the size of the selected end mill is very important, because it will control the *amount of stock left* on the slot walls for finishing. For example, if we choose a ∅7 mm end mill, it will leave 1.5 mm per side for finishing; ∅9 mm end mill will leave 0.5 mm per side for finishing, and ∅8 mm end mill will leave 1 mm per side for finishing.

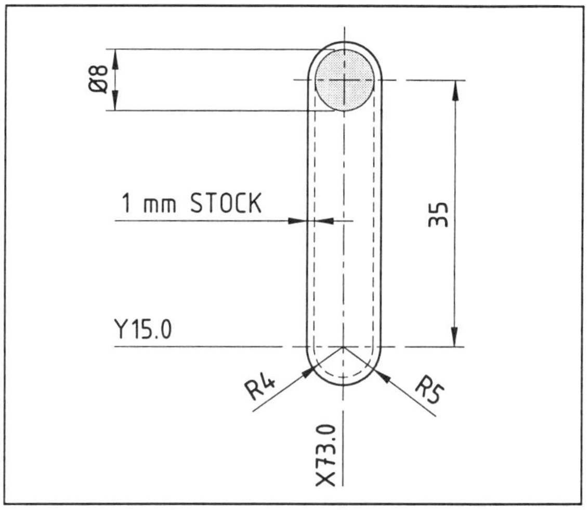

All of the three basic selections are correct, but the choice has to be made for this particular part, as represented by the drawing. Leaving 1 mm per side for finishing is reasonable, so the cutting tool selected for the slot will be an ∅8 mm center-cutting end mill (two or three flutes) - *see illustration*.

If the slot is dimensioned with tight tolerances, it may be a better choice to select *two* suitable tools - one that would make the roughing cut, the other the finishing along the slot walls. Whether one or two tools are used, the finishing cut should be made with cutter radius offset in effect, so the final slot dimensions can be fine-tuned at the machine, via the control system.

Spot Drilling

When drilling holes, a small chamfer - or even a small corner break - is very desirable at the top of the hole. A chamfer will eliminate burrs or sharp edges that are the natural result of drilling, allowing a smoother entry of a tap, or just eliminating the burrs for easier handling. Spot drill is a tool that is used for this purpose; it resembles a regular drill and has two main purposes:

- **To start-up a hole with a small dimple at its exact location ... control of hole location**
- **To machine a chamfer on a hole, by controlling the depth of cut ... control of chamfer size**

In the example, no chamfering or corner breaking of the hole is specified. In CNC work, it is quite customary to break sharp corners, unless the drawing instructions specifically prohibits it.

Unlike any drill selection, spot drills are offered only in about three to six diameters, depending on the choice of individual tool manufacturers. One of the most common spot drills used is a very versatile diameter of 10 mm (or ∅12.7 mm = ∅0.5 inch in Imperial units). If you are not familiar with spot drills, bear in mind that only the *angular portion* of the drill is used, never its own full diameter. Also keep in mind that the majority of spot drills have a 90° included angle at the tool tip. This is a very practical angle, as it allows to make 45° chamfers on small and medium size holes.

One purpose of spot drilling is to make a small dimple at the XY hole location - the exact size of the dimple is not important, so 2-3 mm depth will usually be sufficient, depending on the hole diameter. On the other hand, many holes are not just spot drilled for the purpose of maintaining their XY location - they are also *chamfered* with same tool, to a particular chamfer specification, typically at 45°.

The six holes in the drawing have to be spot drilled, drilled and tapped. The tap diameter is 4 mm, which is the largest size of the hole. Any functional chamfer for a given hole diameter must be bigger than the largest hole. The drawing specifications may include the chamfer size - otherwise the decision is in the hands of the programmer. For the drawing in the given example, no chamfers are given, so an arbitrary decision has to be made - by the CNC part programmer. Chamfers are usually very small, typically within the range of 0.125 to 0.5 mm (0.005 to 0.02 inches) at 45 degrees. For small holes, the chamfer can also be quite small. As this is an arbitrary decision for the example provided, the six holes in the drawing will be 0.35 x 45°.

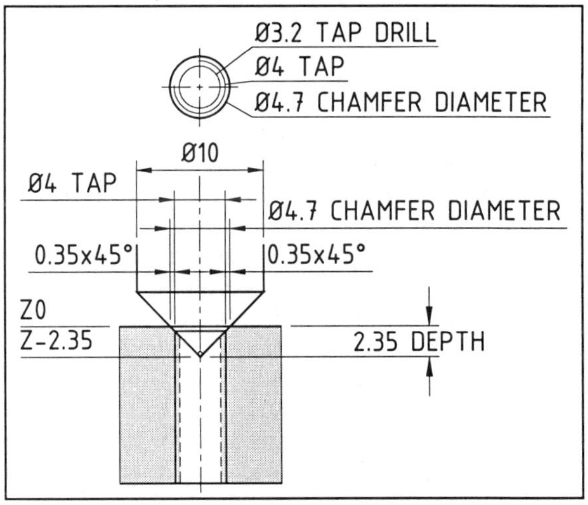

When using spot drills, always consider the fact that only a portion to the angular tool tip is used. That means the programmed depth controls the chamfer size - *or* - the chamfer size is controlled by the programmed depth. If we choose the chamfer size to be 0.35, that means the chamfer is applied on both sides of the hole centerline. The ∅4 mm hole with 0.35 mm per side chamfer will have a chamfer diameter of 0.35 + 4 + 0.35 = 4.7 mm. The depth is calculated directly from the chamfer diameter and the tool point angle. Since the spot drill point angle is 90°, it means the programmed depth *must be* one half of the chamfer diameter:

DEPTH OF SPOT DRILL = (2 × chamfer size + hole diameter) / 2 = (2 × 0.35 + 4) / 2 = 2.35 mm

Drilling

To drill the six holes, the drill selection must be related to the tapping operation that follows. There is a correlation between the tap size and the pre-drill hole size. The drilled hole must be smaller that the nominal tap size, in order to provide material to be cut by the tapping operation. The best source of information about the tap drill selection can be found in various charts from tooling manufacturers, and also in the *Machinery's Handbook*, published by Industrial Press, Inc. In either source, the suggested drill selection - called the *tap drill* - can be found. For an M4x0.75 metric tap, the common selection of the tap drill size is ∅3.2 mm (or ∅3.25 mm).

The drawing calls for a through hole, drilled and tapped as per given drawing dimensions. The thickness of the part *T* at this stage is 12 mm (already faced), which means the drill has to penetrate the part thickness with its full body diameter. It is never a good idea to line the end of the full drill diameter *D* with the bottom of the part. In practice, the tap drill should penetrate not only the thickness of the part, but also provide additional 1-2 mm of breakthrough clearance. In addition to the part thickness *T* and the breakthrough clearance *C*, we also have to consider the drill point length (shown as *P* in the illustration). In the programming section, the drill data will be calculated, resulting in the Z-depth of the drill (its final position below the part).

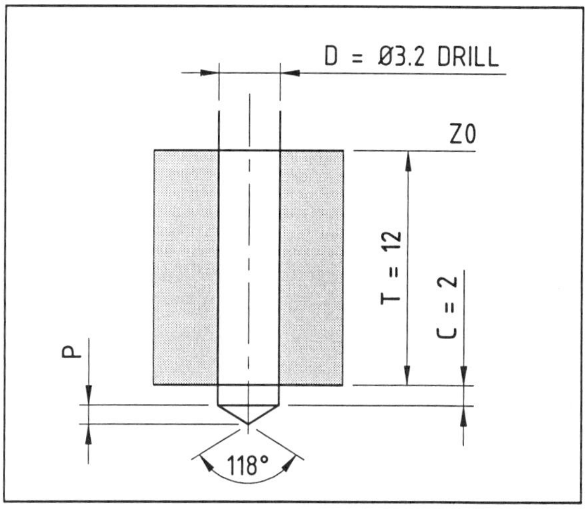

As a conclusion to the tap drill selection, standard ∅3.2 metric drill will be used for the sample part.

Tapping

Once the drilling depth is established (through the part in this example), the tapping depth presents no problems, particularly for through holes. In fact, the final Z-depth for the drill may be used as the final tapping depth, eliminating additional calculations.

Tapping operations have other issues to consider, particularly the relationship between the spindle speed and the tap pitch. These two items will also influence the starting position before tapping begins, as well as the feedrate calculation. This subject will be discussed later in this chapter. In this example, a standard tapping head (tension/compression type) will be assumed. Another name for this type of tapping head is a *floating tap holder*. Its purpose is to prevent tap breakage, when the tap has reached the final depth, but the spindle still decelerates to a full stop. It also work in the opposite direction, when the return feedrate has already started, but the spindle has not completed its acceleration to the programmed spindle aped (r/min).

Summary of Tools Used

Tools used for this job have been carefully selected, based on several important considerations. Check again the details associated with each operation. There are other ways to machine this part - consider the presented method as only one of several possibilities. One of the programmers responsibilities is to assign a tool number to each tool. Keep in mind, that ascending order from tool number 1 is not always practical, as many frequently used tools loaded in the machine tool magazine may use the same number from one job to another.

For this drawing example and the program development, the cutting tools will be numbered in their order of use, as shown in the following table, with particular descriptions:

Tool Number	Description	Size in mm	Type	Comments
T01	Face mill	⌀100	5-6 cutting edges	Top of the face - one pass only
T02	Center-cutting end mill	⌀12	3-flutes	Contour and circular pocket
T03	Center-cutting end mill	⌀8	2 or 3 flutes	Slot - complete
T04	Spot drill	⌀10	90° point angle	Chamfer 0.35 x 45°
T05	Tap drill	⌀3.2	118° point angle	Through the part thickness
T06	Plug tap	M4×0.75	High spiral flutes	Through the part thickness

Needless to say, a change in the drawing specifications may have a profound effect on the tooling selection. Changes in the setup will also have to be considered for the newly selected cutting tool, or adjusted as necessary. *RANDOM MEMORY TOOL CHANGE METHOD WILL USED.*

Machining Data

Machining data considerations cover a large and important area of program development. What is commonly referred to as selecting 'speeds and feeds' is much more than that. When the programmer reaches this stage, the following decisions have to be made - *for each tool* - if applicable:

- Spindle speed in r/min
- Cutting feedrate per minute
- Depth of cut
- Width of cut

There may be many other related decisions to be made, depending on the complexity of the part.

Spindle Speed

Most speeds are calculated from a standard machining formula, based on a peripheral surface speed per minute. For metric units, the surface speed is in *m/min*, for Imperial units it is in *ft/min*:

Metric	Imperial
$r/min = \dfrac{m/min \times 1000}{\pi \times D}$	$r/min = \dfrac{ft/min \times 12}{\pi \times D}$

In both formulas, the parameter *D* refers to the diameter of the tool in milling, or the diameter of the part in turning, in millimeters or in inches. The surface speed for many materials can be found in various tooling catalogues and technical publications.

Cutting Feedrate

Feedrate can be calculated from the spindle speed, chipload per tooth and the number of cutting edges (teeth or flutes):

Feedrate calculation - Metric or Imperial
Feedrate = r/min × C × N

☞ ... where *C* is chipload per tooth in either inches or millimeters and *N* is the number of flutes (teeth)
or the number of cutting edges

Tooling Data

For the purpose of this example, the spindle speed and feedrate calculations will be made based on the following surface speeds and chipload for each machining group (higher rates are possible):

Tool Number	Description	Size in mm	Surface speed in m/min	Chipload per tooth in mm	Spindle r/min	Feedrate mm/min
T01	Face mill	∅100	150	0.35	477	501.0
T02	Center-cutting end mill	∅12	55	0.06	1459	175.0
T03	Center-cutting end mill	∅8	55	0.06	2188	263.0
T04	Spot drill ∅10 mm	∅4.7	25	0.08	1693	135.0
T05	Tap drill	∅3.2	25	0.07	2487	174.0
T06	Plug tap	M4×0.75	10	N/A	796	597.0

The feedrate for the face mill was calculated for three cutting edges in the material, and the feedrate for both end mills, for two flutes cutting. Tapping speed was calculated by multiplying the spindle speed by the thread pitch.

The cutting data presented here are only examples - always evaluate the actual work conditions

Details of Operations

In this section, each operation associated with a particular tool will be described in sufficient detail. Although all individual procedures are correct in principle, there are many other ways to produce the desired results. Hopefully, with the presentation of one method, you can adapt the acquired knowledge and develop another method. Actual program code will be generated for each tool.

The main goal of this chapter is not only to develop a working CNC program, but also describe and explain the steps required to complete this important task.

Tool 1 - Face Milling

The first part of this operation has been decided earlier, in the tooling selection. A ∅100 mm face mill with 5-6 cutting edges has been selected. In order to provide the best cutting conditions, the center of the face mill has been shifted by 15 mm, still leaving a 2.5 mm edge overlap.

In order to develop a part program for this tool, the start and end points of the facing cut must also be selected - or calculated. As a good machining practice, the top face surface finish will be of better quality, if the cutter starts and ends 'in the air' - *away from the part*. Selecting a 5 mm clearance from the two opposite edges is arbitrary, but reasonable. The X-coordinate at the start point (P1) will be:

X-coordinate of P1 = 90 + 5 + 50 = X145.0 (part length + clearance + cutter radius)

At the end point (P2), the calculation is similar, but does not include the part length:

X-coordinate of P2 = - 5 + 50 = X-45.0 (clearance + cutter radius)

The Y-coordinate is the same for both points. Based on the 15 mm shift and 32.5 mm half width of the part, the Y-coordinate is:

Y-coordinate (P1 and P2) = 65/2 -15 = Y17.5

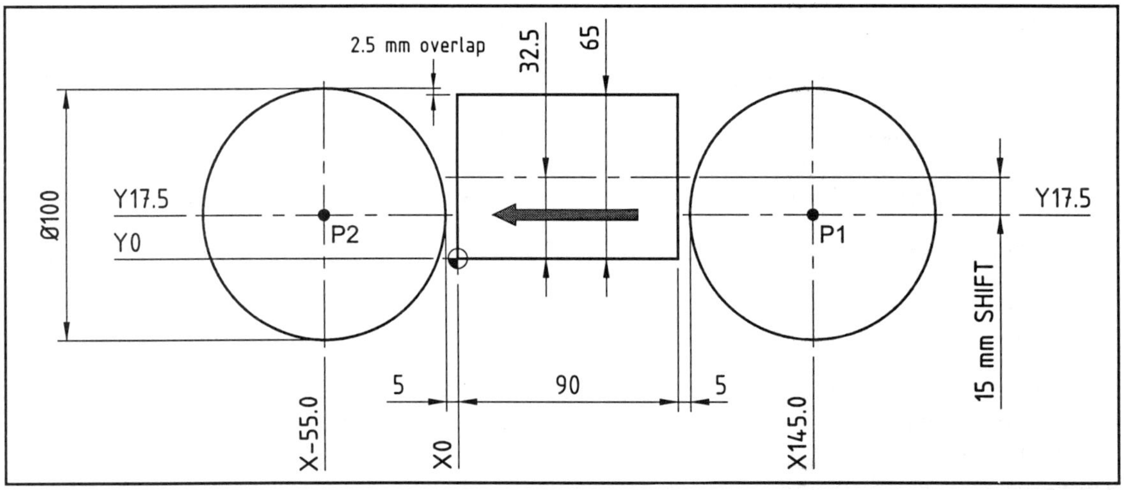

The program for this tool can now be written. This is the first tool of the program, and the format structure will reflect that:

```
(T01 - 100 MM FACE MILL - 1 MM OFF THE TOP FACE)
N1 G21
N2 G17 G40 G80 T01
N3 M06
N4 G90 G54 G00 X145.0 Y17.5 S477 M03 T02
N5 G43 Z10.0 H01 M08
N6 Z0
N7 G01 X-55.0 F501.0
N8 G00 Z10.0 M09
N9 G28 Z10.0 M05
N10 M01
```

Tool 2 - Outside Contour

Tool 2 is a ∅12 mm center-cutting end mill. It will be used for two operations - the contour and the pocket. The first activity of Tool 2 is to machine the outside contour.

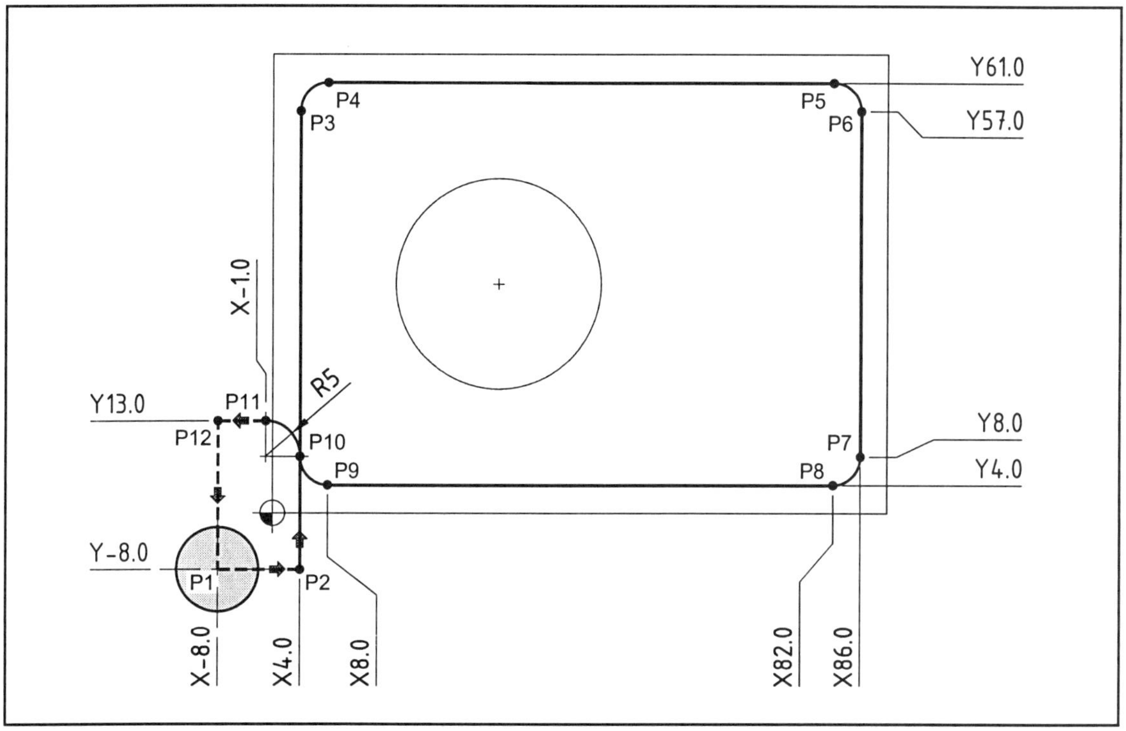

The calculation of individual end points on the contour itself should present no problems - the 4 mm radius is constant around the contour. Study the lower left corner area of the illustration, and you will see that some heavy activity takes place right there. This is the area where the cutter is first positioned, where it approaches the part contour (a motion called *lead-in*) and also where it leaves the contour (a motion called *lead-out*). Because the contour is closed, the lead-in and lead-out motions are close together. In an open contour, they may be much further apart from each other, although the basic principles will not change.

When a large number of points exists in a single contour, making a chart or a table of points and their coordinates will make the program development much easier and better organized. It is always a good idea to define all points in the order of machining, so the coordinates can be easily transferred from the table to the program:

Pt	X	Y	Pt	X	Y	Pt	X	Y
P1	X-8.0	Y-8.0						
P2	X4.0	Y-8.0	P6	X86.0	Y57.0	P10	X4.0	Y8.0
P3	X4.0	Y57.0	P7	X86.0	Y8.0	P11	X-1.0	Y13.0
P4	X8.0	Y61.0	P8	X82.0	Y4.0	P12	X-8.0	Y13.0
P5	X82.0	Y61.0	P9	X8.0	Y4.0	P1	X-8.0	Y-8.0

Once the coordinate points and their machining order has been established, the program for the contouring operation can be written (*e/mill* is the same as *end mill*):

```
(T02 - 12 MM CENTER-CUTTING E/MILL)
(OUTSIDE CONTOUR CUTTING - D52 = 6.000)
N11 T02
N12 M06
N13 G90 G54 G00 X-8.0 Y-8.0 S1459 M03    (P1)
N14 G43 Z10.0 H02 M08
N15 Z-3.5
N16 G41 G01 X4.0 D52 F175.0              (P2)
N17 Y57.0                                (P3)
N18 G02 X8.0 Y61.0 I4.0 J0               (P4)
N19 G01 X82.0                            (P5)
N20 G02 X86.0 Y57.0 I0 J-4.0             (P6)
N21 G01 Y8.0                             (P7)
N22 G02 X82.0 Y4.0 I-4.0 J0              (P8)
N23 G01 X8.0                             (P9)
N24 G02 X4.0 Y8.0 I0 J4.0                (P10)
N25 G03 X-1.0 Y13.0 I-5.0 J0             (P11)
N26 G00 X-8.0                            (P12)
N27 G40 Y-8.0                            (P1)
N28 Z2.0
<machining will continue for circular pocket>
```

The important programming features have been used in the above program section:

- **Cutter radius offset**
- **Numbering of offsets**

The most important programming feature for this toolpath was the use of drawing dimensions for the programmed contour and combining it with the cutter radius offset (G41 ... D52). When a part program contains cutter radius offset, the CNC operator enters the tool radius amount into the control registry and lets the computer do all calculations. There are several do's and don'ts, but overall it is quite an easy way to develop a part program. The operator must know how the toolpath was generated - in typical manual programming, the drawing dimensions are used. In this case, the nominal amount for the offset is the cutter radius (D52 = 6.000). Typical to many CAD/CAM systems, the program output may be to the center of the cutter. In this case, the nominal amount of the offset will be zero (D52 = 0.000). It is always a good idea to include the *suggested* offset amount in the program itself - see the above example.

This part of the program uses tool 2 (T02). As each tool also requires tool length offset, it makes sense to assign the tool length offset the same number the tool number (H02). On the other hand, not all tools in the program use the cutter radius offset. If the tool does require cutter radius offset, the address *D* must be programmed, also with the offset number. That presents a difficulty if the control system has only a single memory bank for *both* types of offsets, so called *shared memory*. In this case, the programmer usually shifts the number by an arbitrary amount (such as 50 for D52). The shift amount can be different, but is should always be higher than the number of tools that can be stored in the magazine. The amount of 50 is also about half way between the range of offsets, usually available in the range of 01-99.

The program also includes a lead-in motion (block N16) and lead-out motions (N25-N27). G41 specifies that radius offset will be to the left of cutting direction, providing a climb milling action.

Cutter radius offset cannot be started or canceled while an arc motion is in effect

Tool 2 - Circular Pocket

The second activity of Tool 2 is to machine the circular pocket. From its last position in block N28, the tool will move towards the pocket center. Circular and other symmetrical pockets are much easier to program when the first cut starts at the pocket center. The cutter diameter of 12 mm has been selected for one important reason - it can cut the pocket in just a single cut around and still maintain a clean bottom. This cut will be preceded by a lead-in, and followed by a lead-out. Cutter radius offset will be applied and canceled during a linear motion.

When machining a circular pocket, the relationship between the pocket size, tool size, and lead-in/out arc is very important. In the illustration at right, the relationship is shown.

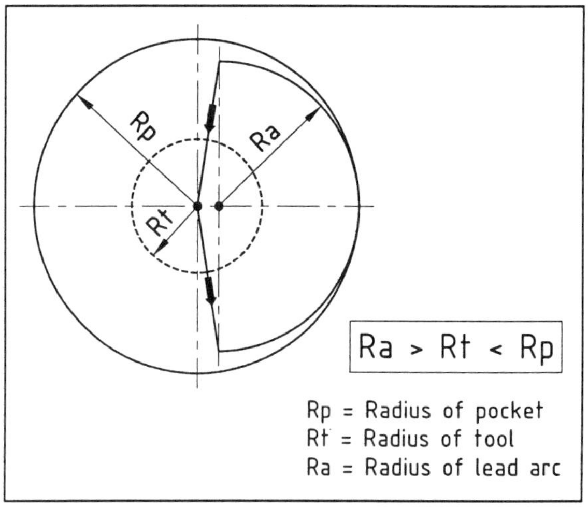

Also shown is the direction of the cut, starting from the pocket center. The toolpath is simple, starting at the center:

☞ **1 - Lead-in line - G41 G01 in effect**
2 - Lead-in arc - G03 with 90° sweep
3 - Full circle to cut the pocket - G03
4 - Lead-out arc - G03 with 90° sweep
5 - Lead-out line - G40 G01 in effect

This procedure will be applied to the actual program (continued):

```
(POCKET CUTTING - D62 = 6.000)
N29 X33.0 Y32.5                  (A)
N30 G01 Z-5.0 Z100.0
N31 G41 X35.0 Y19.5 D62 F175.0  (B)
N32 G03 X48.0 Y32.5 I0 J13.0    ©)
N33 I-15.0                       (D)
N34 X35.0 Y45.5 I-13.0 J0        (E)
N35 G40 G01 X33.0 Y32.5          (A)
N36 G00 Z10.0 M09
N37 G28 Z10.0 M05
N38 M01
```

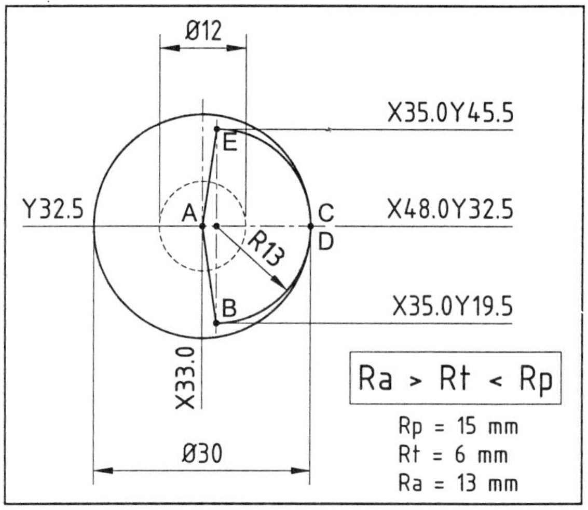

Note that the cutter radius offset has been changed - *for the same tool!* Instead of D52 that controls the outside contour, D62 controls the final size of the pocket. The radius amount stored in the control could be the same for both applications, but the operator can fine tune one without affecting the other.

When selecting the lead-in/out arc, first watch the rule of relationship as shown in the illustrations. For the example in this chapter, the *Ra* (lead arc radius) must be larger than *Rt* (tool radius). Based on the pocket radius *Rp*, the range should be between 6 mm and 15 mm. In order to achieve better machining results, selecting a larger radius provides better results, as the tangential entry into the pocket diameter is much smoother. If a tolerance is given on the pocket diameter and/or depth, two cuts would be necessary, perhaps even two tools. However, the programming procedure described here will remain exactly the same.

Tool 3 - Slot Milling

Tool 3 is an ∅8 mm center-cutting end mill. It will be used for roughing and finishing of the vertical slot. The approach towards the finish pass is very much similar to the approach described for the circular pocket, also in climb milling mode. The major difference is that the radius of the tool and the lead arc radius will be much closer together, details are described for the finishing cut.

First, the roughing toolpath - this one cannot be any simpler. The end mill will make a rapid move to the center of one slot radius in XY axes, feeds to the full depth of 3 mm and makes a straight cut to the center of the opposite radius.

The question *'which slot end should I start from?'* is unnecessary - start from either end, it makes no difference in machining at all. In the example, the tool will rough the slot from its upper position to its lower position, as shown at right:

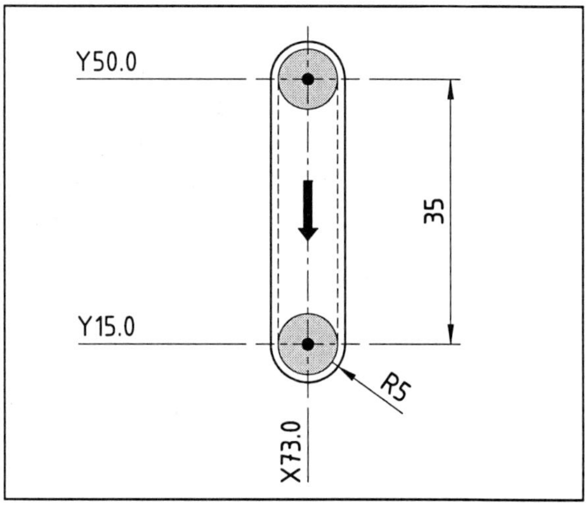


```
(T03 - 8 MM CENTER-CUTTING E/MILL)
(D53 = 4.000)
N39 T03
N40 M06
N41 G90 G54 G00 X73.0 Y50.0 S2188 M03 T04
N42 G43 Z10.0 H03 M08
N43 Z2.0
N44 G01 Z-3.0 F100.0
N45 Y15.0 F263.0
```

... will continue for slot finishing

Radius of the tool *Rt* is 4 mm, radius of the slot *Rs* is 5 mm. As the lead arc *Ra* must be somewhere between these two values, there is not much flexibility here. For the example in this chapter, the lead arc radius *Ra* will be 4.5 mm. The programmed toolpath will follow the same process already established for the circular pocket toolpath, but instead of the pocket being machined, it will be the slot:

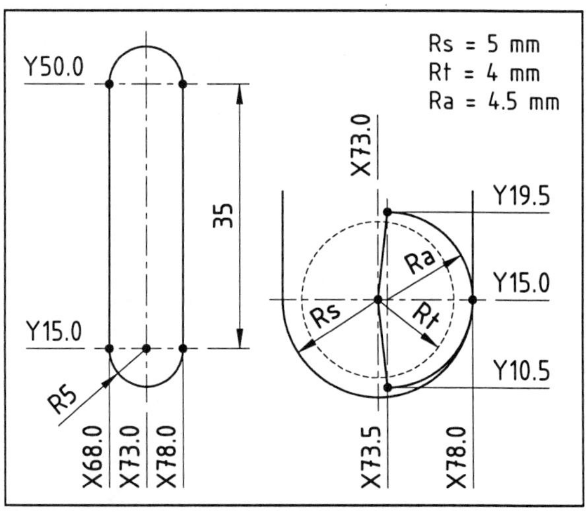


```
(SLOT FINISHING)
N46 G41 X73.5 Y10.5 D53
N47 G03 X78.0 Y15.0 I0 J4.5
N48 G01 Y50.0
N49 G03 X68.0 I-5.0 J0
N50 G01 Y15.0
N51 G03 X78.0 I5.0 J0
N52 X73.5 Y19.5 I-4.5 J0
N53 G40 G01 X73.0 Y15.0
N54 G00 Z10.0 M09
N55 G28 Z10.0 M05
N56 M01
```

As for the pocket cutting, two tools may be used for controlling the precision of the slot, particularly if tight tolerances are required. The programming technique will not change.

Tool 4 - Spot Drilling

There are two major differences between a standard drill and a spot drill. One is in the tool design, the other in the way the tool is used. The design between the two drill types affects the flutes, the web thickness, the overall length, and the tool point angle. The way how the two types of drill are used is a major consideration when programming. In the earlier section covering the tool selection, the details for using the spot drill for this example has already been covered, and the illustration is presented as a reference.

In a summary, the spot drill depth of each hole spotted will be Z-2.35, at the calculated XY location of each hole.

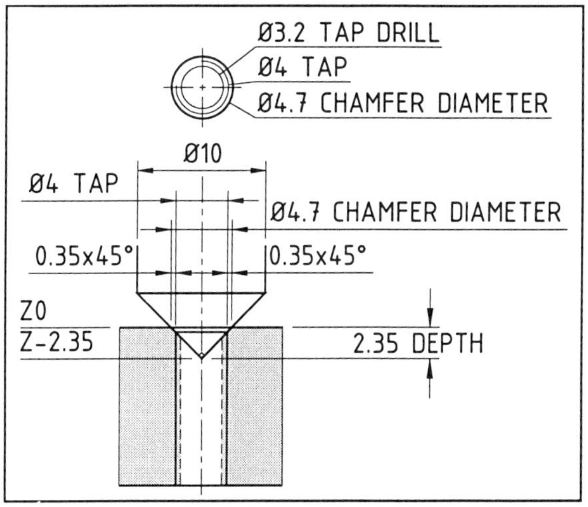

Apart from the depth of cut, another critical part of programming a spot drill is to calculate the XY coordinates of the six holes, which means a good calculator will be needed. Once these coordinates are established, they will also be used for the drilling and tapping of this bolt hole pattern.

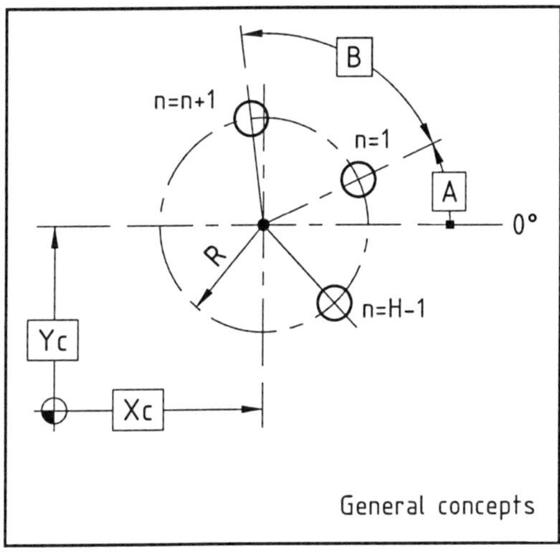

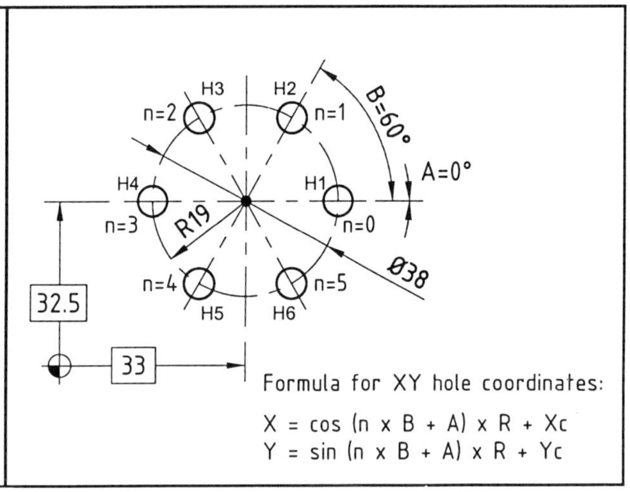

On the left of the above illustration is a general concept of calculating bolt circle hole locations. On the right, the concept is applied for this particular example. Using the formula shown, the XY coordinates of all six holes can be defined:

```
H1(X)=cos(0×60+0)×19+33 = X52.0      H1(Y)=sin(0×60+0)×19+32.5 = X32.5
H2(X)=cos(1×60+0)×19+33 = X42.5      H2(Y)=sin(1×60+0)×19+32.5 = X48.954
H3(X)=cos(2×60+0)×19+33 = X23.5      H3(Y)=sin(2×60+0)×19+32.5 = X48.954
H4(X)=cos(3×60+0)×19+33 = X14.0      H4(Y)=sin(3×60+0)×19+32.5 = X32.5
H5(X)=cos(4×60+0)×19+33 = X23.5      H5(Y)=sin(4×60+0)×19+32.5 = X16.046
H6(X)=cos(5×60+0)×19+33 = X42.5      H6(Y)=sin(5×60+0)×19+32.5 = X16.046
```

Once the coordinates are known (depth is known already), the spot drilling program can be written:

```
(T04 - 10 MM SPOT DRILL - CHAMFER DIAMETER = 4.7)
N57 T04
N58 M06
N59 G90 G54 G00 X52.0 Y32.5 S1693 M03 T05
N60 G43 Z10.0 H04 M08
N61 G99 G82 R2.0 Z-2.35 P200 F135.0        (H1)
N62 X42.5 Y48.954                          (H2)
N63 X23.5                                  (H3)
N64 X14.0 Y32.5                            (H4)
N65 X23.5 Y16.046                          (H5)
N66 X42.4                                  (H6)
N67 G80 G00 Z10.0 M09
N68 G28 Z10.0 M05
N69 M01
```

G82 fixed cycle has been used for the spot drill operation. This cycle is very similar to the G81 drilling cycle (see next operation), but it requires a dwell. The purpose of dwell is to pause at the bottom of the hole, before retracting to the clear position - *the reason?* In order to make sure the surface of the spot drilled hole is smooth, the tool has to rotate at least one spindle revolution, to make it clean. To achieve this goal, the following formula calculates the minimum dwell for fixed cycles - units are in milliseconds and there are no decimal places when milliseconds are used (1 sec = 1000 ms):

$$\text{Minimum dwell (ms)} = (60 \times 1000) / \text{r/min}$$

In the program, the spindle speed is 1693 r/min, so the minimum dwell in milliseconds is 35.44 ms. Although that is the minimum dwell mathematically, practically we have to look at another possible situation, and that is the status of the spindle override switch, located at the control panel. On the majority of CNC machines, this switch has the range of 50-120%. For the dwell calculation, the concern should be with the lowest setting. In order to guarantee at least one full revolution of the spot drill, even at 50% setting, the minimum dwell has to be doubled. In the example, it would be 35.44 ms × 2, which is 70.88 ms. Doubling or even tripling this time will result in more revolutions at the bottom of the cut. The 200 ms in the program will guarantee almost three revolutions at 50% spindle override.

Tool 5 - Drilling

Most of the drilling parameters have been establish earlier, in the tool selection section of this chapter. Additional programming decisions focus on the method of drilling the hole - whether a single cut is sufficient for the depth required, or several interrupted cuts are needed. Interrupted drilling - also called *peck drilling* - is not needed in this example. For holes that penetrate the material - so called through holes - the program must provide a breakthrough clearance *C* (2 mm in the example). In addition, the tool point length *P* (also known as the drill pont length), has to be calculated. Constant of 0.3 is used for standard twist drills, with 118° point angle. Adding the breakthrough clearance *C* and the drill point length *P* to the thickness of the part *T*, the final drilling depth can be calculated:

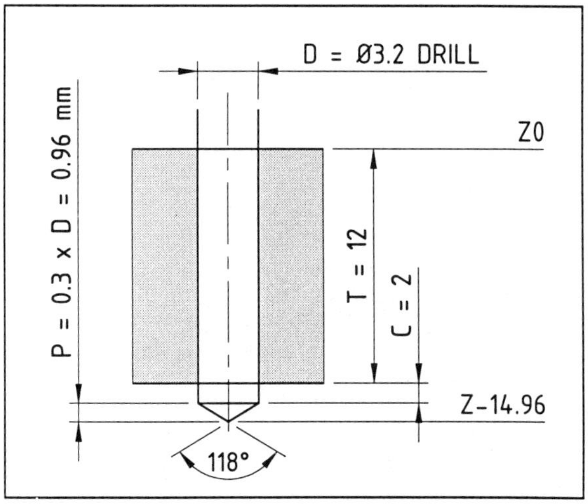

Z-depth = T + C + P = 12 + 2 + 0.96 = 14.96 = Z-14.96 in the program

The drilling program can be written, using the calculated values and the previous XY coordinates:

```
(T05 - 3.2 MM TAP DRILL - THROUGH)
N70 T05
N71 M06
N72 G90 G54 G00 X52.0 Y32.5 S2487 M03 T06
N73 G43 Z10.0 H05 M08
N74 G99 G81 R2.0 Z-14.96 F174.0          (H1)
N75 X42.5 Y48.954                        (H2)
N76 X23.5                                (H3)
N77 X14.0 Y32.5                          (H4)
N78 X23.5 Y16.046                        (H5)
N79 X42.4                                (H6)
N80 G80 G00 Z10.0 M09
N81 G28 Z10.0 M05
N82 M01
```

At the machine, the CNC operator can make some changes to the program, as needed - for example, the G81 cycle can be changed to G83 or G73 peck drilling cycle, just by changing the cycle number and adding the Q-amount of each peck. Here is a variation on the above program (cycle call only):

```
N74 G99 G83 R2.0 Z-14.96 Q5.0 F174.0     (H1)
```

The depth of each peck will be 5 mm. Working with fixed cycles for machining holes offers a generous amount of flexibility - in programming as well as at the machine.

Tool 6 - Tapping

Once the spot drill and drill calculations are out of the way, the tapping is quite simple. There is no need to calculate the XY hole locations (already done for the spot drill). There is also no need to calculate the final tapping depth - the depth already calculated for the drill will be suitable for tapping as well. So - what considerations are unique to tapping? *Feed level clearance and the feedrate!*

```
(T06 - M4x0.75 TAP DRILL - THROUGH)
N83 T06
N84 M06
N85 G90 G54 G00 X52.0 Y32.5 S796 M03 T01
N86 G43 Z10.0 H06 M08
N87 G99 G84 R5.0 Z-14.96 F597.0          (H1)
N88 X42.5 Y48.954                        (H2)
N89 X23.5                                (H3)
N90 X14.0 Y32.5                          (H4)
N91 X23.5 Y16.046                        (H5)
N92 X42.4                                (H6)
N93 G80 G00 Z10.0 M09
N94 G28 Z10.0 M05
N95 G28 X42.4 Y16.046
N96 M30
%
```

Note the increased feed level clearance (R-level) - the increase is an adjustment needed to absorb feed acceleration before the tool touches the part, due to heavy feedrate. The feedrate itself is always a combination of two related items - spindle speed and tap pitch. The tapping feedrate is:

Tapping feedrate = r/min × tap pitch = 796 × 0.75 = 597.0 = F597.0 (underfeeding may be applicable)

That concludes the chapter on part program development. Complete program is listed next.

Complete Program

```
(T01 - 100 MM FACE MILL - 1 MM OFF THE TOP FACE)
N1 G21
N2 G17 G40 G80 T01
N3 M06
N4 G90 G54 G00 X145.0 Y17.5 S477 M03 T02
N5 G43 Z10.0 H01 M08
N6 Z0
N7 G01 X-55.0 F501.0
N8 G00 Z10.0 M09
N9 G28 Z10.0 M05
N10 M01

(T02 - 12 MM CENTER-CUTTING E/MILL)
(OUTSIDE CONTOUR CUTTING - D52 = 6.000)
N11 T02
N12 M06
N13 G90 G54 G00 X-8.0 Y-8.0 S1459 M03          (P1)
N14 G43 Z10.0 H02 M08
N15 Z-3.5
N16 G41 G01 X4.0 D52 F175.0                    (P2)
N17 Y57.0                                      (P3)
N18 G02 X8.0 Y61.0 I4.0 J0                     (P4)
N19 G01 X82.0                                  (P5)
N20 G02 X86.0 Y57.0 I0 J-4.0                   (P6)
N21 G01 Y8.0                                   (P7)
N22 G02 X82.0 Y4.0 I-4.0 J0                    (P8)
N23 G01 X8.0                                   (P9)
N24 G02 X4.0 Y8.0 I0 J4.0                      (P10)
N25 G03 X-1.0 Y13.0 I-5.0 J0                   (P11)
N26 G00 X-8.0                                  (P12)
N27 G40 Y-8.0                                  (P1)
N28 Z2.0
(POCKET CUTTING - D62 = 6.000)
N29 X33.0 Y32.5                                ( A )
N30 G01 Z-5.0 Z100.0
N31 G41 X35.0 Y19.5 D62 F175.0                 ( B )
N32 G03 X48.0 Y32.5 I0 J13.0                   ( C )
N33 I-15.0                                     ( D )
N34 X35.0 Y45.5 I-13.0 J0                      ( E )
N35 G40 G01 X33.0 Y32.5                        ( A )
N36 G00 Z10.0 M09
N37 G28 Z10.0 M05
N38 M01

(T03 - 8 MM CENTER-CUTTING E/MILL)
(D53 = 4.000)
N39 T03
N40 M06
N41 G90 G54 G00 X73.0 Y50.0 S2188 M03 T04
N42 G43 Z10.0 H03 M08
N43 Z2.0
N44 G01 Z-3.0 F100.0
N45 Y15.0 F263.0
(SLOT FINISHING)
N46 G41 X73.5 Y10.5 D53
N47 G03 X78.0 Y15.0 I0 J4.5
N48 G01 Y50.0
N49 G03 X68.0 I-5.0 J0
N50 G01 Y15.0
N51 G03 X78.0 I5.0 J0
N52 X73.5 Y19.5 I-4.5 J0
N53 G40 G01 X73.0 Y15.0
```

```
N54 G00 Z10.0 M09
N55 G28 Z10.0 M05
N56 M01

(T04 - 10 MM SPOT DRILL - CHAMFER DIAMETER = 4.7)
N57 T04
N58 M06
N59 G90 G54 G00 X52.0 Y32.5 S1693 M03 T05
N60 G43 Z10.0 H04 M08
N61 G99 G82 R2.0 Z-2.35 P200 F135.0               (H1)
N62 X42.5 Y48.954                                  (H2)
N63 X23.5                                          (H3)
N64 X14.0 Y32.5                                    (H4)
N65 X23.5 Y16.046                                  (H5)
N66 X42.4                                          (H6)
N67 G80 G00 Z10.0 M09
N68 G28 Z10.0 M05
N69 M01

(T05 - 3.2 MM TAP DRILL - THROUGH)
N70 T05
N71 M06
N72 G90 G54 G00 X52.0 Y32.5 S2487 M03 T06
N73 G43 Z10.0 H05 M08
N74 G99 G81 R2.0 Z-14.96 F174.0                    (H1)
N75 X42.5 Y48.954                                  (H2)
N76 X23.5                                          (H3)
N77 X14.0 Y32.5                                    (H4)
N78 X23.5 Y16.046                                  (H5)
N79 X42.4                                          (H6)
N80 G80 G00 Z10.0 M09
N81 G28 Z10.0 M05
N82 M01

(T06 - M4x0.75 TAP DRILL - THROUGH)
N83 T06
N84 M06
N85 G90 G54 G00 X52.0 Y32.5 S796 M03 T01
N86 G43 Z10.0 H06 M08
N87 G99 G84 R5.0 Z-14.96 F597.0                    (H1)
N88 X42.5 Y48.954                                  (H2)
N89 X23.5                                          (H3)
N90 X14.0 Y32.5                                    (H4)
N91 X23.5 Y16.046                                  (H5)
N92 X42.4                                          (H6)
N93 G80 G00 Z10.0 M09
N94 G28 Z10.0 M05
N95 G28 X42.4 Y16.046
N96 M30
%
```

In this introductory chapter, you have learned many CNC programming techniques. They all can be easily adapted to a large number of programs. It is always difficult to put something in print, when so many variations exist. For example, the speeds and feeds used in this chapter may prove to be too low for a production of a large volume of parts. You may also find that your way of machining may be better than the one shown here. That is all to be expected. After all, CNC programming is almost like following a recipe - the ingredients are there, even the process - but it still needs the skilled hand of the cook - *the CNC programmer* - to make all the elements work well together.

2 CALCULATING CONTOUR POINTS

It is much easier to calculate the absolute XY coordinates for a simple contour, such as a rectangle, but it is much harder to calculate points where the contour includes angles and partial radiuses. These parts are usually programmed with the assistance of a CAD/CAM system (CAM), but if such system is not available or in other situations, the CNC programmer must resort to the old fashion way, by using a pocket calculator. Most calculations will be using trigonometric functions, but knowing the basic arithmetic and algebraic operations, knowing formulas, being familiar with solving triangles is still the major requirement. This chapter will present some techniques that have proven to be suitable for solving a majority of problems associated with calculation of more difficult contour points.

Tools and Knowledge

Any tool can only be used correctly if the user has a sufficient amount of knowledge about the purpose of the tool and how to use such a tool. In CNC manual programming, we are talking about three major tools - pencil, paper, and a calculator. An old cartoon has also shown a fourth tool - a very large eraser. Of course, in these days, pencil is most likely replaced by a text editor (even *Windows Notepad®* will do in an emergency), and actual printing on paper is not always necessary, as the program can be transferred to the control system via a cable, using DNC software. Eraser is part of the editor, and Windows even provide a simple calculator. In practice, a physical calculator is much better choice, and it does not have to be expensive. The most suitable calculator is the scientific type (it must have *sin*, *cos*, and *tan* functions) and it should have more than one memory. This is a personal recommendation - calculated data are often used later, for subsequent calculations. Having more than one memory enables the programmer to store initial calculations with much higher precision; it also eliminates punching in many rounded numbers.

Even the most sophisticated calculator will not solve a problem on its own - it is up to the user to provide proper entry method of various calculations and formulas. Mathematical calculations are vital part of programming for CNC machine tools. Geometrically more complex parts will require more calculations, but not necessarily more complex calculations. Common arithmetical and algebraic calculations, combined with the solution of right angle triangles, are virtually all the tools of mathematics you need.

Mathematical Knowledge

Arithmetic is the basic knowledge of mathematics and includes addition, subtraction, multiplication and division function while working with integers and real numbers. Algebra is a step higher, as it includes working with both known and unknown values (symbols, such as letters, for example), mainly in the form of equations and formulas. Trigonometry stands on its own - it is the most important part of geometry for CNC programming. Trigonometry covers the relationship between angles and sides of a triangle, particularly the area of solving *right angle triangles* (also known as just *right triangles*). A right angle triangle has three sides and three angles. Only two sides may have the same length, and one angle must be 90 degrees, which is an angle *never* used in calculations. Often the word 'trigonometry' is treated with a negative attitude by some programmers. Yet, solving trigonometric problems can be quite a simple and routine work. Various charts, tables and formulas can come handy, and many of them can be easily found in publically available technical books, such as the *Machinery's Handbook*. One such trigonometric table is presented on the next page.

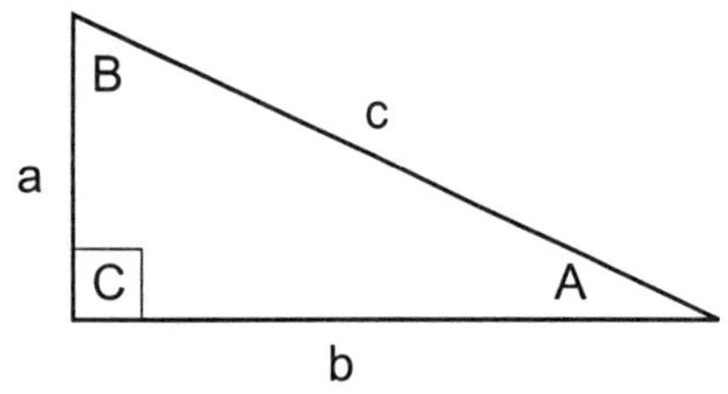

$a / c = \sin A = \cos B$		
$b / c = \cos A = \sin B$		
$a / b = \tan A = \cot B$		
$b / a = \cot A = \tan B$		
$a = c \times \sin A$	$b = c \times \cos A$	$c = a / \sin A$
$a = c \times \cos B$	$b = c \times \sin B$	$c = a / \cos B$
$a = b \times \tan A$	$b = a \times \tan B$	$c = b / \sin B$
$a = b / \tan B$	$b = a / \tan A$	$c = b / \cos A$
$a = \sqrt{c^2 - b^2}$	$b = \sqrt{c^2 - a^2}$	$c = \sqrt{a^2 + b^2}$
$A = 90° - B$	$B = 90° - A$	$C = 90°$

Table of equations for solving common right angle triangles

In order to solve a right angle triangle, there are some very basic mathematical rules you have to understand. The most important rule deals with the known dimensions of a triangle:

Two sides of a right angle triangle must be known

One side and one angle of the right angle triangle must be known

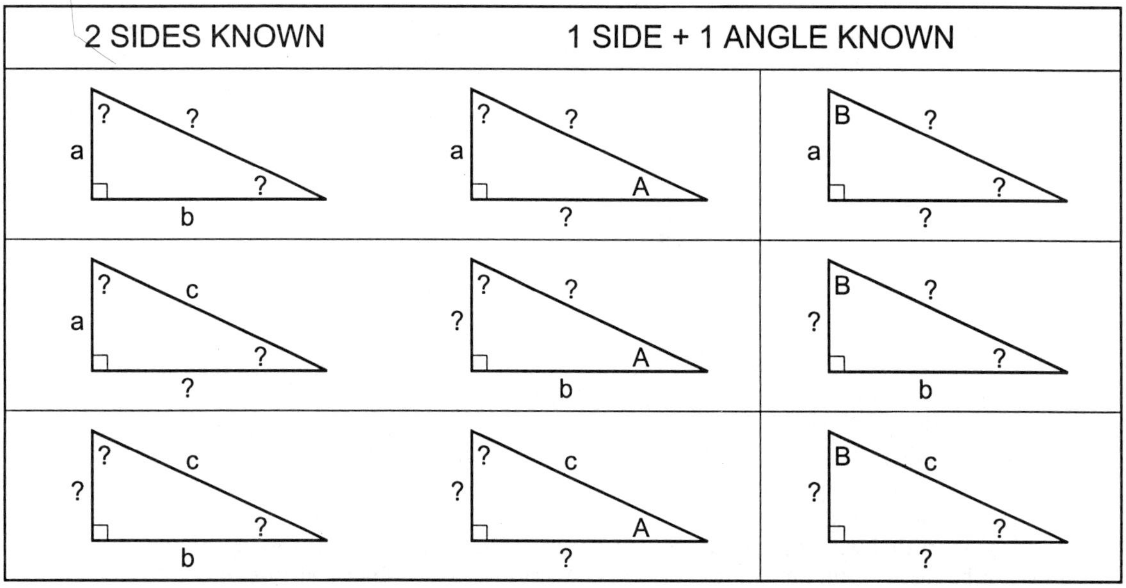

The above illustration shows known dimensions (shown as *a, b, c* and *A, B*) and dimensions that can be calculated (shown with *question mark*). Angle *C* is not specified, as it is always 90°.

Other mathematical rules are equally important, such as similar triangles, as well as rules of angles:

The sum of angles in a triangle is always 360°

The sum of complementary angles A and B in a right angle triangle equal to 90°

Organized Approach

What is the most important issue in solving a trigonometric problem - or any problem - is to approach the attempt at solution in an organized way. Solving triangles for a CNC program is no different. The most important key to solving trigonometric problems is the ***ability to visualize the triangle*** you want to solve. Such a triangle is not always obvious in the drawing or a work sketch, and must be identified first. In this respect, the following suggestions may be very useful.

Process of Calculating XY Coordinates

To illustrate the various techniques that can be used to calculate coordinate points, the drawing provided below will be used. The contour is still relatively simple, but it includes some major elements found on much more complex contours. Keep in mind that the purpose of this chapter is to provide a specific process that results in a useful solution. For that reason, the part zero will be the center of R16 (lower left). This may not be the most suitable origin for setup, but not an impossible one either.

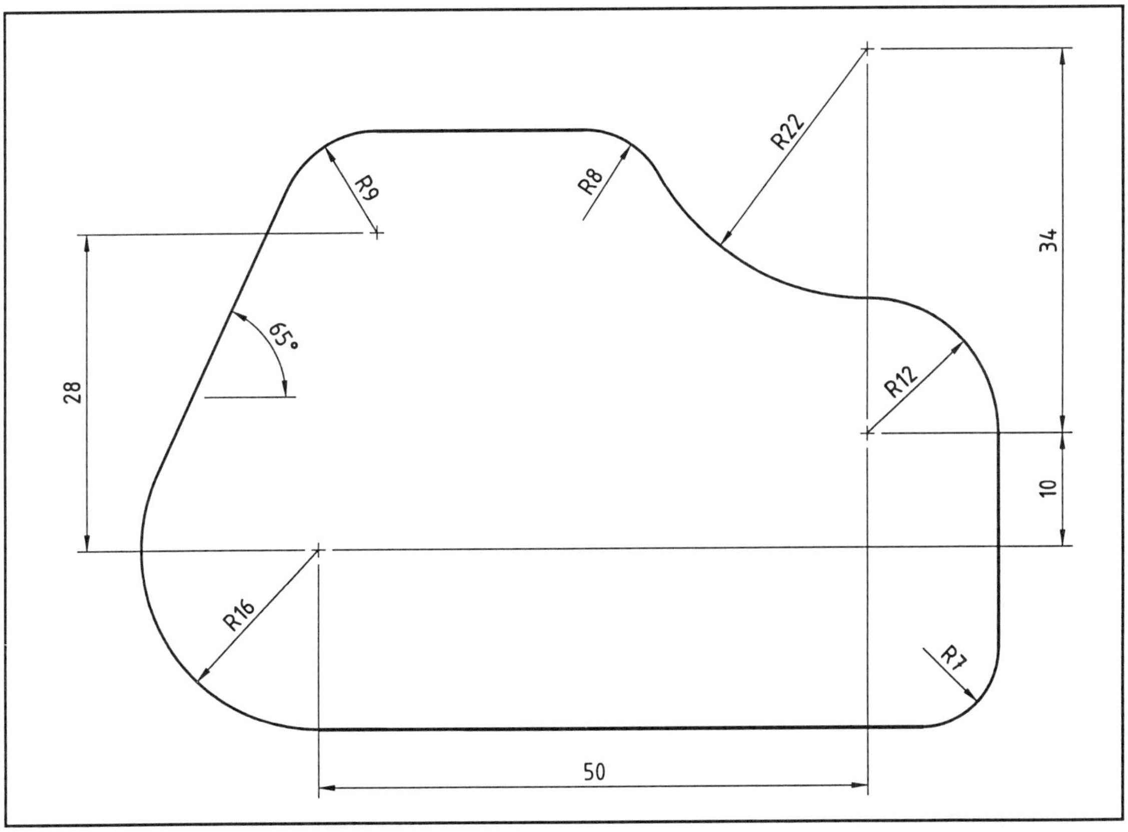

Based on these common rules, you can approach all calculations in a planned and organized manner. If possible, use the drawing itself or make an approximate work sketch of the detail. When making a sketch, try to make it in scale and close to the original. Using color pencils may also improve the visibility of important features.

Step 1 - Establish the Main Contour Points

First, establish the point of reference, such as the part zero (X0Y0) - this point may be identified in the sketch as *P0* or just with the origin symbol.

Follow the contour carefully, and identify each contour point by a reference number (for example, *P1, P2, ...*). If possible, identify these points in the *order of machining* - it will be that much easier later to write the program.

Identify all quadrant points that exist on the contour but are *not* endpoints (for example, *Q1, Q2, ...*)

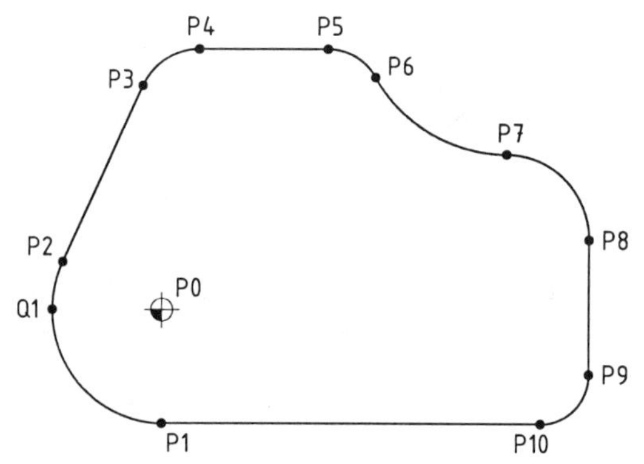

The points numbering shown would suit well an external contour cutting in climb milling mode.

Step 2 - Fill-in the Coordinate Sheet

Once the points are identified, carefully evaluate the drawing dimensions - it is quite common to find that a large number of points do *not* need any special calculations, at least not beyond some basic arithmetic. There are ten points (*P1* to *P10*) in the drawing that are endpoints of the contour, which may mean up to twenty coordinate values that will have to be calculated. Although that seems like a lot, if you eliminate all endpoints that are already known, the number of calculations will drop, often quite significantly.

A very useful method is to use a simple table or a form often known as the *coordinate sheet*. A typical coordinate sheet will have a narrow column for the points and two or three major columns for the coordinates. The illustration shows a coordinate sheet that contains all contour points (endpoints), where the contour coordinates are *known*.

Pt	X-COORDINATE	Y-COORDINATE
P0	X0	Y0
P1	X0	Y-16.0
P2		
P3		
P4		Y37.0
P5		Y37.0
P6		
P7	X50.0	Y22.0
P8	X62.0	Y10.0
P9	X62.0	Y-9.0
P10	X55.0	Y-16.0
Q1	X-16.0	Y0

Note that *all known coordinates* have been entered, even if they were the same as for the previous point. It is a good idea to fill the coordinate sheet completely, but unnecessary in the CNC program.

When the actual calculations are made for the unknown points, the arc center often becomes part of the calculation. In the next step or two, the critical center point locations will also be added to the coordinate sheet, at least as a reference.

Step 3 - Identify Calculation Zones

Any complex contour should never be considered as a single entity for making calculations. Much better approach is to separate the contour into *logical areas* or *zones*. What is a logical zone? In order to establish coordinate points, you can only calculate a *new* value based on an *existing* value. That means an *existing known point* will become the start of calculating the next point, which is so far unknown, either partially or completely. A new zone is defined, if there is no logical connection between known points. Keep in mind that the *'zone system'* is just a suggested method of making the calculations easier and organized. Feel free to express and develop your own personal approach.

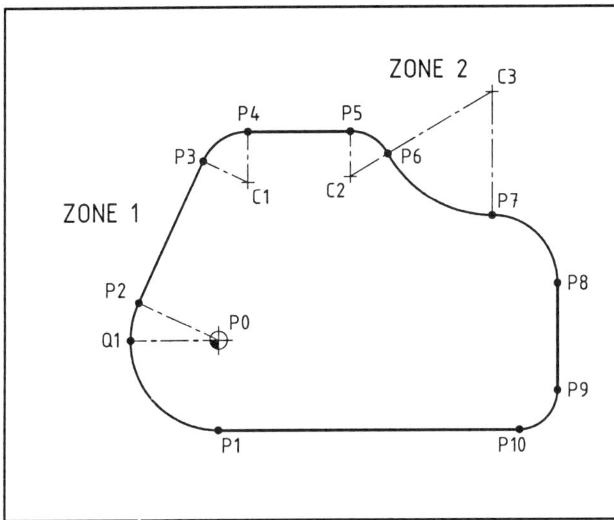

Pt	X-COORDINATE	Y-COORDINATE
P0	X0	Y0
P1	X0	Y-16.0
P2		
P3		
P4		Y37.0
P5		Y37.0
P6		
P7	X50.0	Y22.0
P8	X62.0	Y10.0
P9	X62.0	Y-9.0
P10	X55.0	Y-16.0
Q1	X-16.0	Y0
C1		Y28.0
C2		Y29.0
C3	X50.0	Y44.0

In the drawing and the table, contour points *P1, P7, P8, P9,* and *P10* have all *known* XY coordinates. What is important to understand is that the unknown point *P2* depends on the known point *P0.* Point *P3* is unknown and related to *P4*, which is known only partially (Y-coordinate). There is a connection, however, between *P2* and *P4* - *the 65° angle*. Logically, it follows that points *P2*, *P3*, and *P4* will somehow be related to each, at least in some calculations. These points were placed in the *Zone 1*.

Continuing with the contour evaluation, there is no direct dimensional relationship between points *P4* and *P5* whatsoever. However, *P5* is closely tied with *P6* and *P7*. Point *P5* has its Y-coordinate known, *P7* has both coordinates known. That makes points *P5, P6,* and *P7* connected, so they were placed in *Zone 2*.

When evaluating a particular contour for the purpose of making the calculations better organized, also keep in mind that the completely or even partially known center points are also great help.

Step 4 - Helpful Ideas for Calculations

Virtually all calculations of this kind will include trigonometric functions. Since trigonometric functions relate to right angle triangles, it makes sense to create vertical and horizontal lines (also known as *orthogonal* lines), *in addition* to the lines connecting the arc center with the point to be calculated. In the following examples, you will see that the combination of two orthogonal and one angular line create a triangle - *a right angle triangle*. This triangle can be solved by using the table provided at the beginning of this chapter, resulting in the required XY coordinates. Once these coordinates are known, there is enough information (data) collected to write the CNC program.

Once each triangle is identified, fill-in all known dimensions, such as the length of sides, radius and angles. The next section, focusing on *Zone 1* (as shown above), will contain several interim calculations - these will also be used to establish the final XY points, to be used in the CNC program.

The main emphasis for calculating the coordinates in both zones will be on the identification, maintenance and preservation of individual calculations. Preservation of calculations means not falling into the trap of accumulated rounding error. Here are some suggestions to make better calculations:

- **Use a calculator that has more than one permanent memory - three memories can help a lot**
- **Always work with numbers that are at least one decimal place more precise than required by CNC**
- **Keep the last set of calculations - they may come handy when you expect it the least**

The second item in the above list may benefit from a more detailed explanation. For solving a triangle, you will use a scientific calculator, one that supports *sin, cos,* and *tan* functions. When a formula is entered, for example, `16 × cos25`, the result will be displayed with fewer decimal places than the calculator actually stores in its internal memory, but still with much more precision than actually needed. The result of the example may have eight decimal places, showing a rather large number of `14.50092459`. This result is just too large to handle on daily basis. Writing the number rounded to 3 decimal places (as `14.501`) will introduce a rounding error when used with repeated calculations. The best approach is to store the calculated result with its highest accuracy in the calculator memory and recall the stored value for any subsequent calculations.

In reality, there are several occasions when the described and preferred method cannot be used. One common problem is the lack of multiple memory storage inexpensive calculators have. Yet, even this 'cheap' calculator has at least one memory. Unfortunately, for CNC work, one memory is usually not enough - the previous calculations will compete with newer calculations for a single storage. So, what is the best solution to achieve better precision?

The solution described next may not qualify to be the best solution for all possible applications, but it does provide accurate results most of the time. The solution is to *write down* the result of each calculation - as provided by the calculator - with the accuracy of at least one decimal place higher than required by the CNC program. For example, the above calculation of `16 × cos25` will be used:

`16 × cos25` = **14.50092459** **... typical calculator display (8 to 12 decimal places)**

`16 × cos25` = **14.5009** **... suggested rounding of metric is 4 decimal places (3+1)**

`16 × cos25` = **14.501** **... final value used (3 decimal places maximum output)**

In conclusion, the main emphasis of this section is to promote the idea that rounding of a calculation should always be done very carefully - *and always as the last activity*. That way the programmed toolpath will use the most precise coordinate points possible, *for the selected units*.

Selected units of measurement are *very* important in the part program. The short example above has evaluated a single metric dimension as well as the calculations surrounding this type of dimensioning. In metric system, the smallest programmable input is 0.001 mm (one micron), so the initial dimension of 14.50092459 mm will result in 14.501 mm when the final rounding is done. Stored (interim) calculations should have more decimal places, at least four. In Imperial units, the logical approach will be exactly the same, except the smallest programmable input is 0.0001 inch - five decimal places recommended as minimum for hand calculations.

Other ideas could be presented in this section, and on the same subject. The beauty of CNC programming is that there are no exact *recipes* to follow, no *'one-or-the-other'* selection to be made, and not too many *'step-by-step'* procedures. This working situation leaves the CNC programmer the freedom to approach the job with no preconceived ideas. These comments are applicable not just in this chapter, but for every step of the programming process.

Step 5 - Calculations for Zone 1

The purpose of defining the *Zone 1* is to create favorable conditions to calculate points *P2*, *P3,* and *P4*. Make a sketch of the relevant details within *Zone 1*, including all known dimensions - see illustration. Using some of the helpful ideas above, first triangle will be identified, to calculate *P2*.

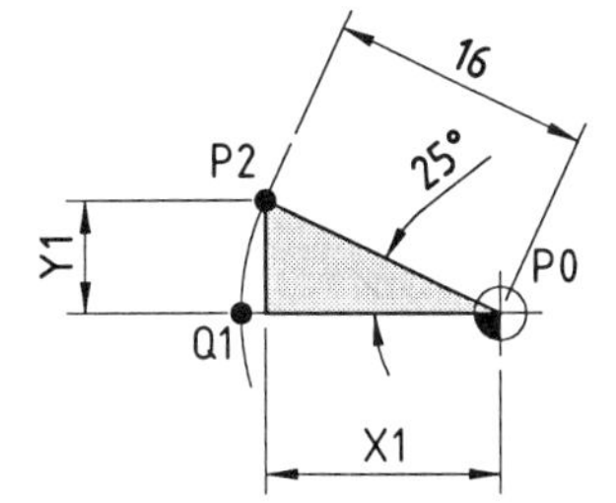

ZONE 1 - P2 CALCULATION

Both dimensions *X1* and *Y1* can be calculated, using sine and cosine functions:

```
X1 = 16 × cos25 = 14.50092
Y1 = 16 × sin25 = 6.76189
```

As both of these dimensions originate from part zero, they will be used directly to define P2. Final rounding will be to three decimal places (metric):

```
X-14.501 Y6.762     (P2)
```

Calculation of *P3* and eventually *P4* is dependent on the previous calculation, because the 65° line is common to both - it connects *P2* with *P3*. The trigonometric calculation is similar to the previous calculation:

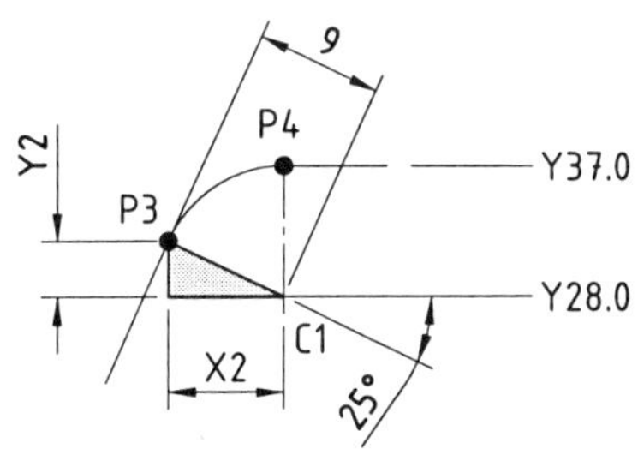

ZONE 1 - P3 CALCULATION (PARTIAL)

```
X2 = 9 × cos25 = 8.15677
Y2 = 9 × sin25 = 3.80356
```

Although both sides of the triangle are known, only the Y-coordinate can be calculated for *P3* at this time:

```
P3 Y-coordinate = 28 + Y2 = 28 + 3.80356 = 31.80356 = Y31.804
```

The only known coordinate for *P4* is the Y-coordinate. Based on direct drawing dimensions, the Y-coordinate for *P4* will be:

```
P4 Y-coordinate = 28 + 9 = Y37.0
```

The next step is to use the connecting line between *P2* and *P3* and calculate the missing dimensions - the X-coordinates for *P3* and *P4*.

This is step could be considered the most difficult in the whole project. Not because of the calculations themselves, but because of the triangle to solve is not readily available. In cases such as this, the best approach - in fact, the only approach - is to make a sketch connecting *P2* with *P3*. As this is a line at an angle, it will most likely become the hypotenuse of the triangle sought. Once the hypotenuse is known, a horizontal line from P3 and vertical line from P2 will complete the triangle. There are enough data collected, so this triangle is easy to solve.

Detailed illustration on the next page shows the necessary features.

The main key to successful calculations is to connect features that are known in order to find feature that are unknown

In addition to the provided vertical dimension of 28 mm, all remaining data have already been collected - *X1* and *X2*, as well as *Y1* and *Y2*. Ideally, they should be stored in individual memories of your calculator. If that is not possible, they should be written down with at least a four decimal places precision.

Dimensions *X3* and *Y3* have to be calculated next. The *Y3* dimension does not require trigonometric solution, only simple arithmetic functions:

```
Y3 = 28 - Y1 +Y2
Y3 = 28 - 6.76189 + 3.80356
Y3 = 25.04167
```

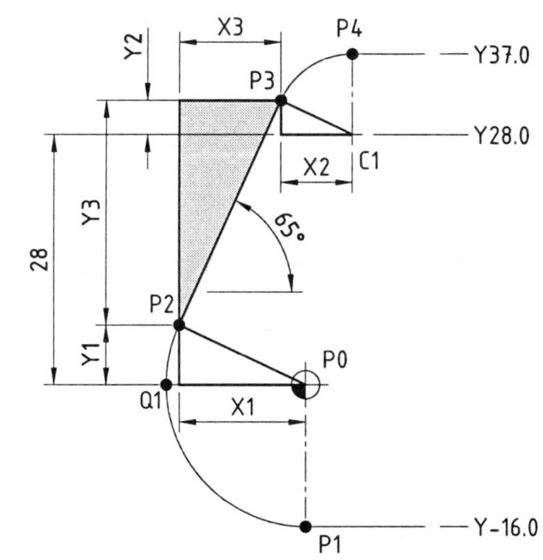

ZONE 1 - P3 and P4 CALCULATION

Dimension *X3* can now be calculated, using the new and previous calculations:

```
X3 = Y3 × tan25 = 25.04167 × tan25 = 11.67712
```

Finally, the point *P3* can be identified in both coordinates:

P3 X-coordinate = X3 - X1 = 11.67712 - 14.50092 = -2.82380 = X-2.824
P3 Y-coordinate = Y31.804 *(previous result)*

The Y-coordinate for point *P4* is already known to be Y37.0, based on the drawing dimensions. To calculate the X-coordinate for point *P4*, another arithmetic function will be used:

```
P4 X-coordinate = X2 + X3 - X1 = 8.15677 + 11.67712 - 14.50092 = 5.33297
P4 X-coordinate = X5.333
```

Rounded to three decimal places, the points P3 and P4 are completely known:

X-2.824 Y31.804	(P3)
X5.333 Y37.0	(P4)

Once the points *P2*, *P3*, and *P4* are known, all calculations within *Zone 1* have been completed. If you are using a coordinate sheet or a similar table to keep track of points, this is a good time to update it. Final coordinate table will be shown later, when all point coordinates are known, rounded to three decimal place accuracy for metric units or four decimal place accuracy for Imperial units.

Rounding should always be done on the LAST calculation

Step 6 - Calculations for Zone 2

The next step - *Step 6* - will organize calculations of coordinates for points *P5, P6* and *P7*. From the method of organization, there is no direct mathematical connection between points in *Zone 1*, so a new zone - *Zone 2* - will be established.

The purpose of defining the *Zone 2* is to create favorable conditions to calculate points *P5, P6* and *P7*. Make a sketch of the relevant details within *Zone 2*, including all known dimensions - as shown in the illustration.

Note that center *C2* has a known Y-coordinate, and center *C3* has both *X* and *Y* coordinates known.

When two arcs are tangent to each other, the line between their centers crosses at the arcs endpoints

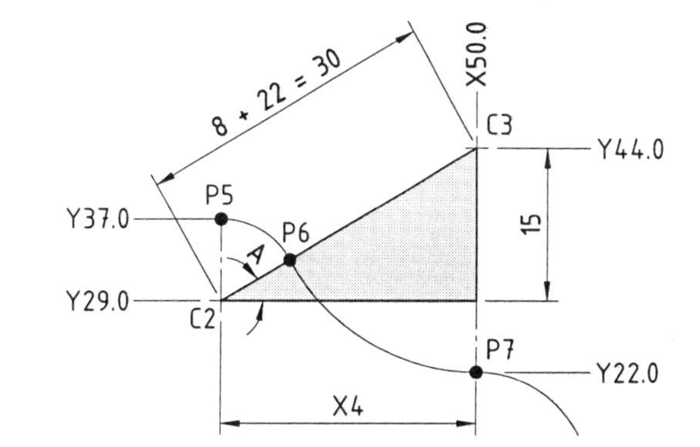

ZONE 2 - P5 CALCULATION

Point *P6* fits the above definition perfectly. The triangle shown in the illustration is the first step in the *Zone 2* - the reason? Two sides of the triangle are known, therefore the currently unknown angle *A* can be established for further calculations:

Angle A = unknown, to be calculated
Side opposite = 15 mm
Hypotenuse = sum of the two radiuses 8 + 22 = 30 mm

To calculate the angle *A*, the relationship between the opposite side and the hypotenuse is the *sine function* of the angle. From the trigonometric table at the beginning of this chapter, you will see that:

`a / c = sinA ==> 15 / 30 = sinA = 0.5` *0.5 is the sine of angle A*

Using the inverse trigonometric function ($\sin^{-1}$), the angle whose sine is 0.5 will be found to be exactly 30°, therefore, angle **A = 30°**. Using this newly established angle, the currently unknown dimension *X4* can be calculated:

```
X4 = 30 × cos30 = 30 × 0.86603 = 25.98076
```

Once the dimension *X4* is known, the X-coordinate of *P5* can be established:

```
P5 X-coordinate = 50 - X4 = 50 - 25.98076 = 24.01924 = X24.019
```

With the Y37.0 coordinate already known from the drawing, the *P5* coordinates are:

```
X24.019 Y37.0     (P5)
```

Once the *P6* is known, all trigonometric calculations will be completed. All coordinates of the remaining points can be easily calculated from the drawing (see the initial table of coordinates). As the point *P6* is part of the triangle shown above, the calculation will be virtually the same as for *P5*, just using different data.

The angle *A* is now known to be 30°, so to calculate the *X5* and *Y5* dimensions is a fairly straightforward process, using the hypotenuse of 22 mm, which is the radius of the larger arc:

```
X5 = 22 × cos30 = 22 × 0.86603
X5 = 19.05256

Y5 = 22 × sin30 = 22 × 0.5
Y5 = 11.0
```

As in the previous *P5* calculation, the known arc center *C3* will be used to calculate the XY coordinates for *P6*:

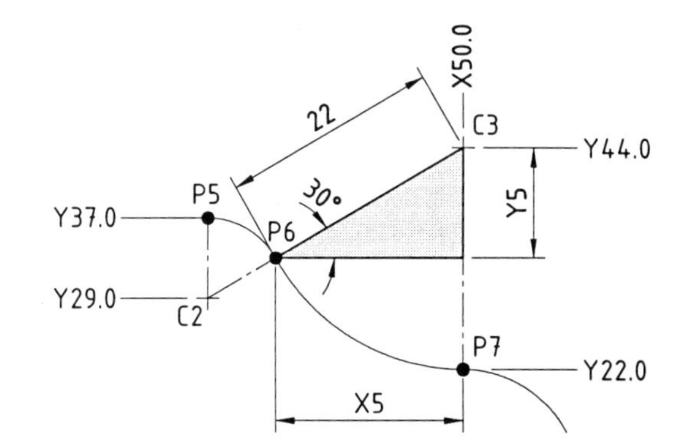

ZONE 2 - P6 CALCULATION

```
P6 X-coordinate = 50 - X5 = 50 - 19.05256 = 30.94744 = X30.947
P6 Y-coordinate = 44 - Y5 = 44 - 11 = 33 = Y33.0
```

With both coordinates calculated, the XY coordinates of *P6* are:

```
X30.947 Y33.0     (P6)
```

This is the last trigonometric calculation - all required points (XY coordinates) to program the contour are known, based on the initial drawing.

Updating Coordinate Sheet

Once all the XY coordinates are established, based on the step for individual calculations, the coordinate sheet should be updated. Although not strictly necessary, the idea is that there is a concise record available in one place.

If the points in the coordinate sheet are in the order of machining, the program is so much easier to write.

Pt	X-COORDINATE	Y-COORDINATE
P1	X0	Y-16.0
P2	X-14.501	Y6.762
P3	X-2.824	Y31.804
P4	X5.333	Y37.0
P5	X24.019	Y37.0
P6	X30.947	Y33.0
P7	X50.0	Y22.0
P8	X62.0	Y10.0
P9	X62.0	Y-9.0
P10	X55.0	Y-16.0

Writing the CNC Program

The sole purpose of this chapter was to provide means and ideas in handling contours with angular lines, tangent lines, tangent arcs, unknown dimensions, etc. The calculations described have provided all necessary XY coordinates for the ten contour points required to write the CNC program. The remainder of the process is relatively simple - just use these points as part of the program structure, provide the beginning, the end, the arc vectors, the motion commands - you get the idea.

3 FORMULAS FOR CONTOURING

In the previous chapter, the main focus was on evaluating a part contour and establishing contour points necessary to develop a suitable toolpath. This chapter continues that subject, with additional geometry solutions and with the focus on calculations using various shortcuts by employing *formulas*.

During toolpath development, it is quite common to define a typical contour as a combination of lines and arcs meeting at their respective endpoints. Some contours will have more lines or more arcs - in some cases, they may have only lines or only arcs. In CNC programming, the approach to finding coordinate points (contour points) is generally the same, regardless of the line/arc combination within a contour. There are three main possibilities that will each require a little different approach by the CNC programmer, when developing the desired toolpath:

- **Intersecting contour point between a LINE element and another LINE element**
- **Intersecting or tangent contour point between a LINE element and an ARC element**
- **Intersecting or tangent contour point between an ARC element and another ARC element**

Regardless of the endpoint combinations, the tool cutting direction will be either in the *climb* milling mode or the *conventional* milling mode (predominantly climb milling mode).

In the majority of programming applications, the CNC programmer has to use some mathematical methods to establish the XY coordinates of the desired contour point.

Contour Point Between Two Lines (Lathe)

A contour point between two linear motions will always be an *intersection*. From the three available possibilities of contour transitions, this is usually the one that can be resolved (that is, *calculated*) in the easiest way. The two examples shown here represent the typical applications of this kind of contour points used on a chamfer of a lathe part.

- **Typical lathe chamfer - no lead-in**
- **Typical lathe chamfer - with lead-in**

For example, if diameter *D1* is ∅32 mm, the chamfer length *L* is 5 mm, the clearance *C* is 3 mm and the angle *A* is 28°, then:

D2 for the chamfer *without* lead-in will be:

D2 = 32 - 2 × 5 × tan28 = 26.683

D2 for the chamfer *with* lead-in will be:

L1 = 5 + 3 = 8
D2 = 32 - 2 × 8 × tan28 = 23.493

For milling, the formula is almost the same, but does *not* use the multiplier of 2.

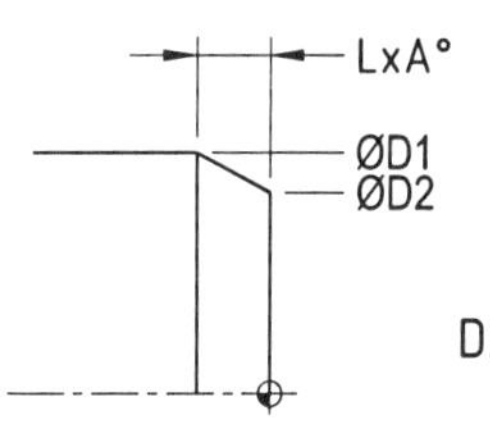

Dimensions L, A, and ØD1 must be known

D2 = D1 - 2 x L x tanA

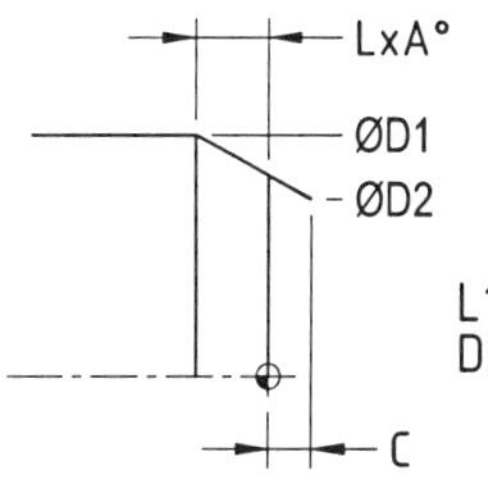

Dimensions L, A, C, and ØD1 must be known

L1 = L + C
D2 = D1 - 2 x L1 x tanA

Contour Point Between Line and Arc

Whenever a linear motion changes into an arc motion (or an arc motion changes into a linear motion), the contour point can be either an intersection or a tangency. While an *intersection* will create two possible points between the line and the arc, *tangency* will always create a single point of contact.

Intersecting Contour Point

The illustration below shows two typical intersections between a line and an arc. In both cases, the absolute coordinates for points *A* and *B* must be calculated. The upper example is very simple:

X-coordinate for A = X + R
Y-coordinate for A = Y

X-coordinate for B = X
Y-coordinate for B = Y + R

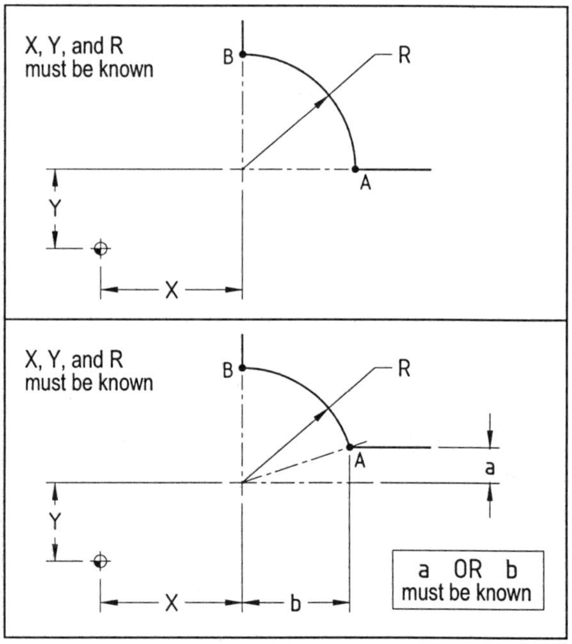

The lower example is only marginally more difficult. One of the dimensions *a* or *b* must be known, the other will be calculated:

☞ **... If dimension *a* is known (more typical):**

X-coordinate for A $= X + \sqrt{R^2 - a^2}$
Y-coordinate for A = Y + a

X-coordinate for B = X
Y-coordinate for B = Y + R

☞ **... If dimension *b* is known (less typical):**

X-coordinate for A = X + b
Y-coordinate for A $= Y + \sqrt{R^2 - b^2}$

X-coordinate for B = X
Y-coordinate for B = Y + R

The *Pythagorean Theorem* method has been used, as two sides of a triangle are known.

Example 1 - if the radius *R* is 15 mm and the dimension *b* is 12 mm, then the dimension *a* will be:

$$a = \sqrt{R^2 - b^2} = \sqrt{15^2 - 12^2} = 9.000$$

Example 2 - if the radius *R* is 18.5 mm and the dimension *a* is 5 mm, then the dimension *b* will be:

$$b = \sqrt{R^2 - a^2} = \sqrt{18.5^2 - 5^2} = 17.812$$

In both cases, the XY dimensions will have to be included for the program input. There are other variations of intersecting lines, and their solution always depends on the method of dimensioning used in the engineering drawing.

Tangent Contour Point

The procedure used in the previous chapter described the calculations between lines and arcs in a methodical way. This section will address those tangent arcs that can be calculated using a formula.

Arcs programmed within a contour are sometimes called the *partial arcs*. In fact, all arcs in the previous chapter were partial arcs. The term *partial arcs* is not really an accurate description for the subject of programming arcs, but it is used quite frequently in many articles and books. Generally, it refers to an arc that is an inherent portion of a specific contouring toolpath. As the word *arc* literally means *the part of a circle,* the term *partial arc* is redundant and makes no sense.

In CNC programming, one of the most common application of an arc tangent between two lines is at 90 degrees - between a horizontal *and* a vertical line. The illustration below shows four such arcs. In order to calculate the contour change point (tangency points), only additions and subtractions will be required, rather than trigonometric functions.

For example, if the program zero (X0Y0) is at the lower left corner of the part shown, the coordinates of the eight points will be:

```
A = X0    Y6.5
B = X0    Y21.5
C = X6.5  Y28.0
D = X38.5 Y28.0
E = X45.0 Y21.5
F = X45.0 Y6.5
G = X38.5 Y0
H = X6.5  Y0
```

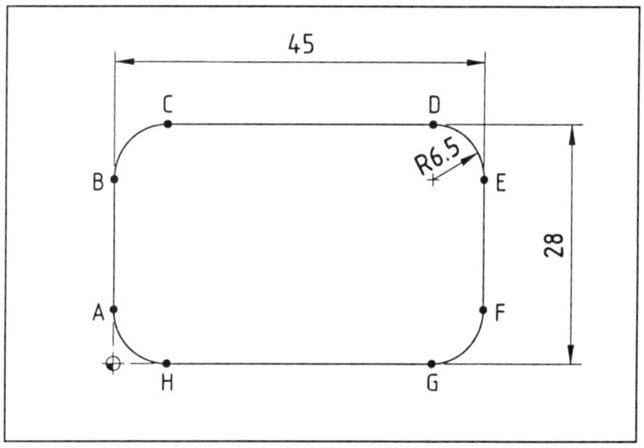

Another very common occurrence of two tangent lines in part programming is when one of the lines is at an angle, and the other is orthogonal (*i.e.,* vertical or horizontal). Although such calculations may present a few difficulties to the beginner, using a formula greatly simplifies the task at hand.

The illustration at right shows two methods of orientations used for drawing dimensions. Keep in mind that each orientation has four positions - they will be explained later in this section.

In both views, the main goal is to calculate the angle *B*, so it can be used later, within the formulas. Radius *R*, angle *A* and sharp point *P* are known. The *P* point is only used for determining the program coordinates and is *not* required for the angle *B* calculation.

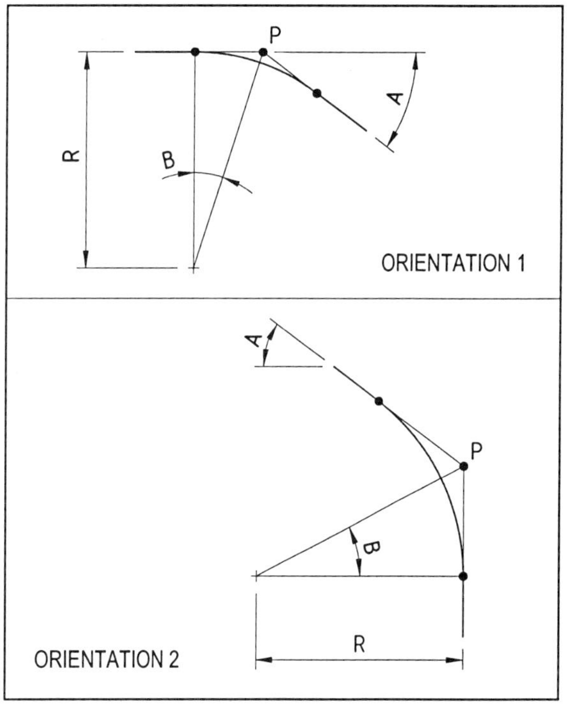

The greatest challenge is solving triangles is to know which triangle to solve

To calculate the angle *B*, you have to know something about angles and right triangles. Fortunately, the knowledge can be narrowly focused. In the illustration below, the main triangle is shaded and other relationships are identified as well. Each shaded triangle can be 'flipped' about the line drawn between the arc center and the point *P*. This 'other' triangle is *identical* to the shaded triangle. At the moment, it may seem a trivial observation, but it will come very handy for the actual calculations of XY coordinates used in the part program.

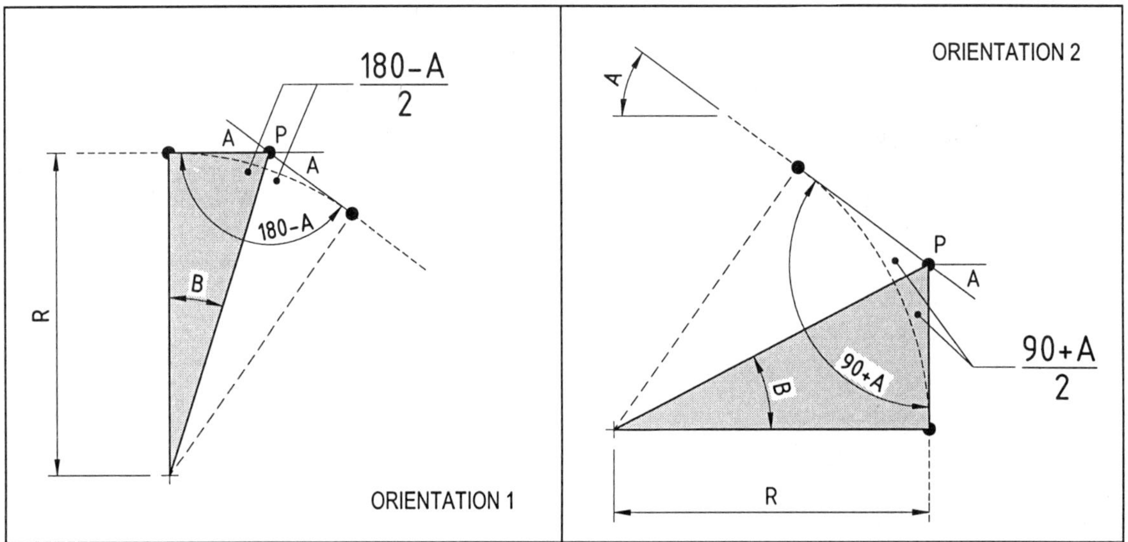

Orientation 1	Orientation 2
$B = 90 - \frac{180-A}{2} = A / 2$	$B = 90 - \frac{90+A}{2} = (90 - A) / 2$
B = A / 2	**B = (90 - A) / 2**

As the dimensions *R* and *A* must be known to calculate the triangle, and the sharp point *P* must be known to calculate the XY coordinates, two examples can be given for reference. Only the angle *B* is calculated at this time:

Example for angle B calculation - *Orientation 1*

Given data: *R = 27.0,* **A = 17°**, *P = X260.0 Y185.0*

B = 17 / 2 = 8.5°

Example for angle B calculation - *Orientation 2*

Given data: *R = 14.5,* **A = 23.78°**, *P = X343.0 Y212.75*

B = (90 - 23.78) / 2 = 33.11°

Note that the non-listed data - *although given* - have not been used in the calculation of angle *B*.

When developing the toolpath geometry (*i.e.*, individual contour change points), the calculation of the angle *B* is just the beginning. The main purpose of the remaining calculations is to find the XY coordinates of all tangent points within a given contour. The formulas shown on the next two pages cover *all* possibilities of a contour change point, where two lines are tangent to an arc - one line is either horizontal or vertical, the other line is defined at an angle.

Because of the dimensioning method and shared data in the drawing, it is possible to develop a formula that can be used instead of long hand calculations. The illustration below shows *Orientation 1* that was established earlier in this section. Rather than using reference to X and Y, the letter W has been chosen as a neutral way to identify that particular side. Looking at the illustration, it should become soon apparent that all four parts are identical - only their orientation is different. That should help to identify the proper corner as it appears in the engineering drawing. If these and the following formulas are used on a CNC lathe, some results will have to be doubled when used as X-diameter.

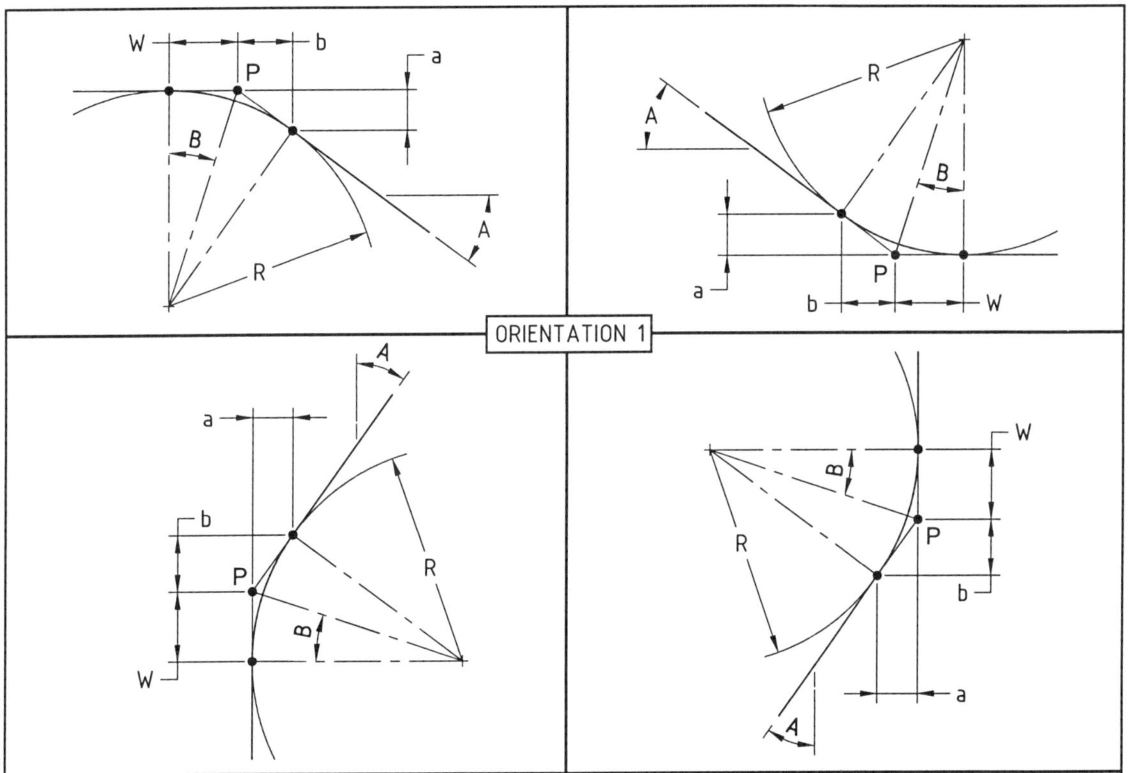

Known data: radius R, angle A, and point P	W = R x tanB	
B = A / 2 Common Formulas ►	a = W x sinA	b = W x cosA

Example - *Orientation 1*:

R = 30.0
A = 35°
P = X27.0 Y40.0

Calculations - see results at right:

B = 35 / 2 = 17.5°
W = 30 × tan17.5 = 9.459
a = 9.459 × sin35 = 5.425
b = 9.459 × cos35 = 7.748

Make sure to identify the angle A correctly

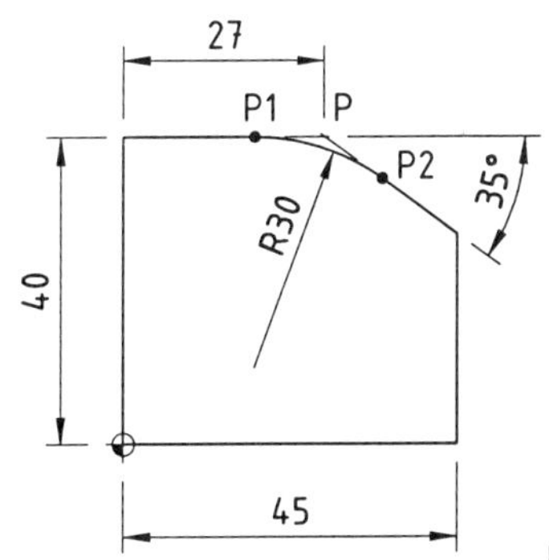

Point P:
X27.0 Y40.0

Point P1:
X = 27 - 9.459
Y = 40.0

Point P2:
X = 27 + 7.748
Y = 40 - 5.425

P1: X17.541 Y40.0
P2: X34.748 Y34.575

Calculations for the *Orientation 2* are identical - the formulas have not changed. What *did* change was the calculation of angle *B* (which is based on the angle *A* definition in the drawing).

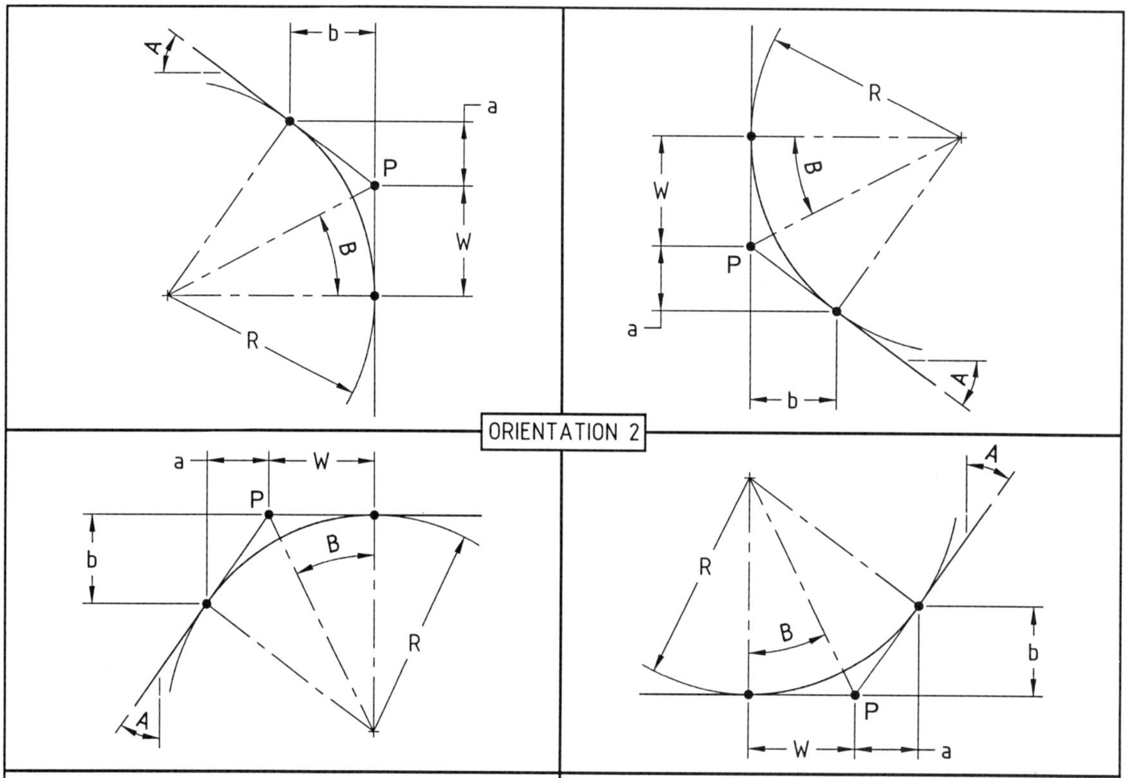

Known data: radius R, angle A, and point P	W = R x tanB	
B = (90 - A) / 2 Common Formulas ►	a = W x sinA	b = W x cosA

Example - *Orientation 2*:

R = 27.0
A = 90 - 36 = 54°   **** RECALCULATED****
P = X48.0 Y???   **Y = UNKNOWN**

Calculations - see results at right:

B = (90 - 54) / 2 = 18°
W = 27 × tan18 = 8.773
a = 8.773 × sin54 = 7.097
b = 8.773 × cos54 = 5.157

Make sure to identify the angle A correctly

The above message is very important, especially in this example. Also note that if the point *P* is not provided in the drawing, it must be calculated first.

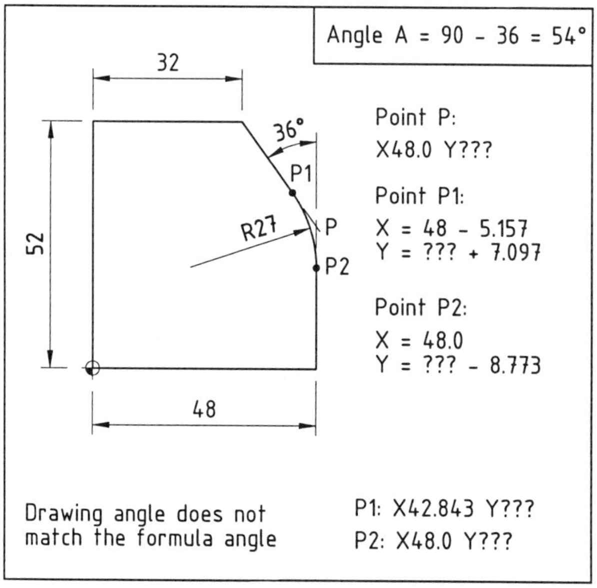

Calculating the Sharp Point

Calculation of the XY coordinates needed in the above example cannot be completed, if the sharp point *P* is not dimensioned in the drawing or is dimensioned only partially. In the examples presented so far, the assumption was that the sharp corner *P* is defined in the drawing directly. Depending on the engineering intentions, that is not always the case. As the presented formulas depend on the point *P*, if its coordinates are unknown, they have to be *calculated.* Such calculations are quite straightforward.

Using the previous drawing, there is only one triangle to solve.

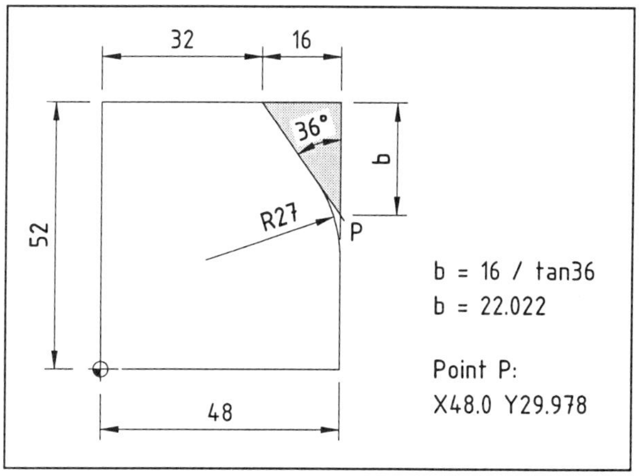

Completing the Calculations

The unfinished version of the calculations can now be completed:

Example - *Orientation 2 - UPDATED*:

R = 27.0
A = 90 - 36 = 54° **** RECALCULATED****
P = X48.0 Y29.978

Calculations - see results at right:

B = (90 - 54) / 2 = 18°
W = 27 × tan18 = 8.773
a = 8.773 × sin54 = 7.097
b = 8.773 × cos54 = 5.157

Make sure to identify the angle A correctly

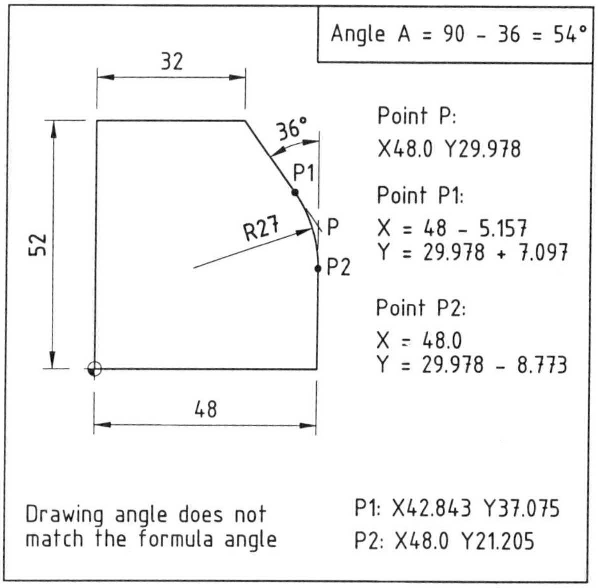

In the summary: calculations of coordinate points for blend arcs when one line is orthogonal and the other at an angle can be formulated into a few short steps. The most important is to find the orientation of the corner listed here and match it with the corner in the drawing. Equally important is to use the angle as shown in the formulas, even if the drawing shows the opposite angle. If the sharp point *P* is not known by its coordinates, a simple calculation will provide the missing data.

There is another configuration of an arc that is tangent to two lines, and that is when both lines are at an angle. Undoubtedly, this situation presents much more complex situation, where simple formulas may be hard to develop. Drawings that include this type of corner blend use several methods of dimensioning and in order to create a formula, all methods would have to be known. While it is certainly possible, the formula would be too complex to present any real benefits. Even without a formula, such drawing configurations can be solved step by step using trigonometric functions.

Contour Point Between Two Arcs

Whenever an arc motion changes into another arc motion, the contour point can be either an *intersection* or a *tangency*. The common rule specified in the previous section applies here as well - while an intersection will create *two* possible points between the line and the arc, tangency will always create a *single* point of contact between the adjacent entities. Out of the three examples shown in this chapter, the tangency between two arcs is also fairly common, and as such deserves more detailed evaluation. Two intersecting arcs are far less common, but will be covered here as well.

Intersecting Arcs

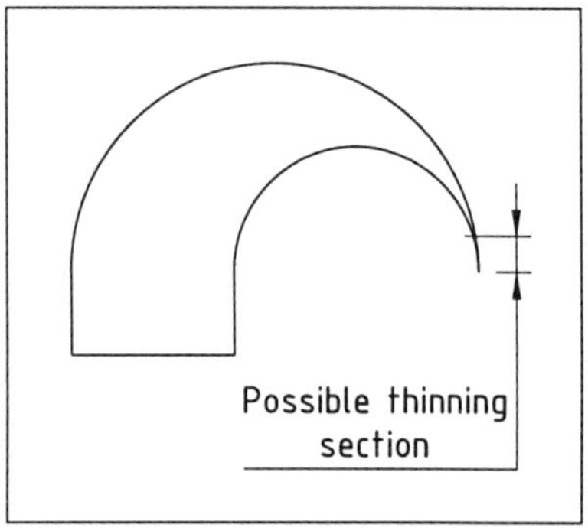

Carefully evaluate the drawing at right - it shows a fairly possible situation when two arcs intersect. Although this drawing shows a part that can be machined in theory, this combination of intersecting arcs will likely cause a thinning section that can be easily bent or pried away.

In order to illustrate the calculations for intersecting arcs, a more reasonable drawing will be used. The drawing below is very simple - the three intersecting points at the top only need some subtraction and addition. The problem is that the radius is not given and must be calculated for the arc toolpath.

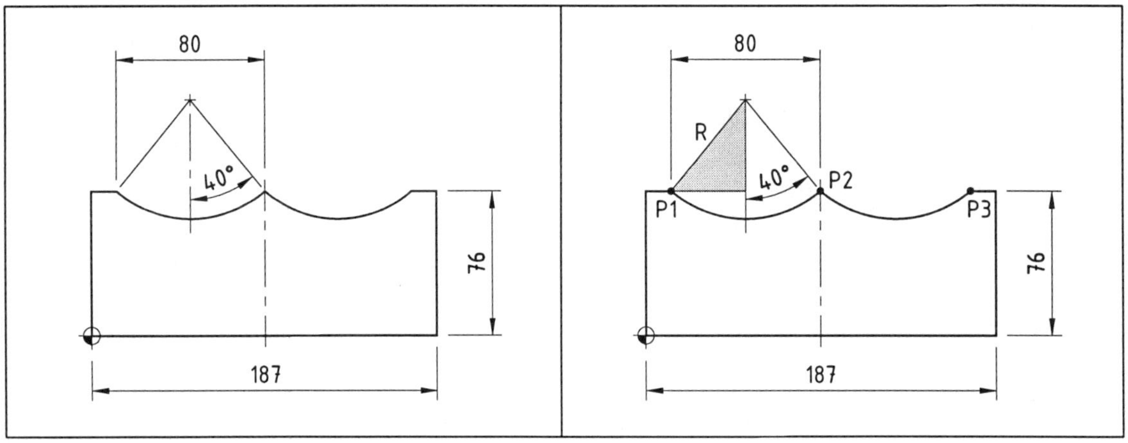

The three points P1, P2, and P3 are easy:

X-coordinate for P1 = 187 / 2 - 80 = X13.5 Y-coordinate for P1 = Y76.0
X-coordinate for P2 = 187 / 2 = X93.5 Y-coordinate for P2 = Y76.0
X-coordinate for P3 = 187 / 2 + 80 = X173.5 Y-coordinate for P2 = Y76.0

Even the calculation of the radius R is not difficult:

R = (80 / 2) / sin40 = 40 / sin40 = 62.229 mm

If *I* and *J* vectors are used for cutting the arcs in the order of P1 to P3, I will be 40.0 and J will have to be calculated:

J = 40 / tan40 = 47.670 ... I40.0 J47.67 in the program

Because of the easy transition between the two arcs, this part can be machined successfully.

Tangent Arcs

When one arc is tangent to another arc, there one important fact to remember - a ***line drawn between the arc centers creates the point of tangency***. This applies every time and at any orientation.

Several typical examples are shown at right. The highlights indicate a possible toolpath contour.

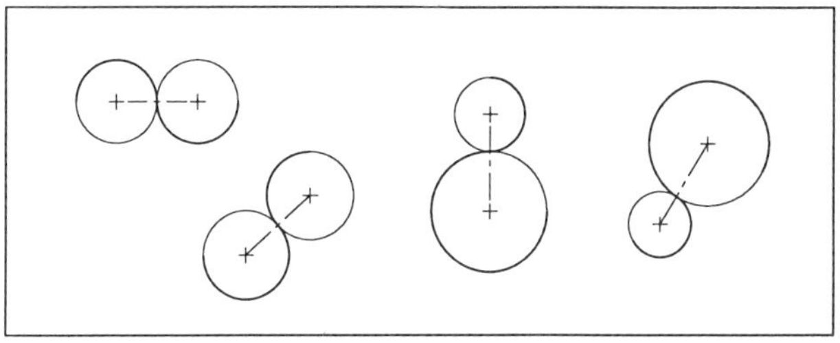

Transition between two arcs where the line between centers is either vertical or horizontal, is the easiest transition. In the example drawing, both possibilities are shown, and the XY coordinates for each arc point are listed here:

P1 = X0 Y100.0
P2 = X40.0 Y100.0
P3 = X100.0 Y100.0
P4 = X150.0 Y100.0
P5 = X150.0 Y50.0
P6 = X150.0 Y26.0

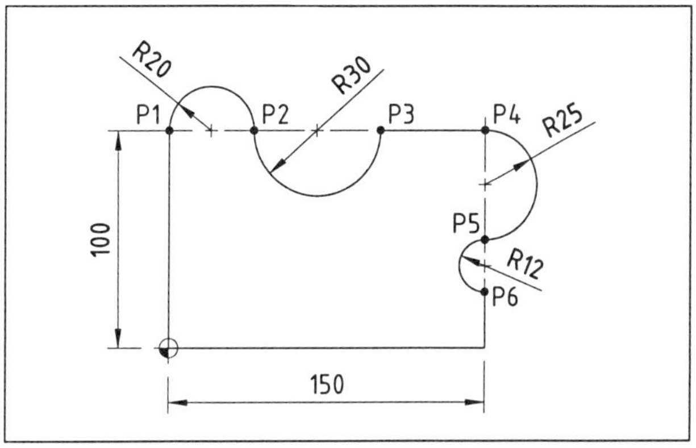

When the line between centers is at an angle, some trigonometric calculations are necessary, depending on how the drawing is dimensioned. As there are many ways to dimension such a drawing, creating a formula that would cover all possibilities will not be possible. However, solving the problem trigonometrically does not present any degree of complexity. General rule is that if the angle is given, the *sin*, *cos*, and *tan* functions will be used. On the other hand, if only two sides are known, the *Pythagorean Theorem* will be applied.

In the illustration shown at right is the part drawing. Note that the large radius is not defined. This is just one of the many variations that can be used by the designers to identify the important features of the part.

Just like in any drawing, the unknown point coordinates have to be calculated. The one coordinate that is most elusive is the tangency point between the two arcs. In situations like this, the first thing you should do is to *draw a line between the arc centers*. The point where the line crosses the arcs is the tangency point - it is also a point programmed by its coordinates. As this line will become the hypotenuse of the triangle to solve, the other two sides have to be drawn as well. That is done easily by drawing a vertical line from one arc center and a horizontal line from the other arc center.

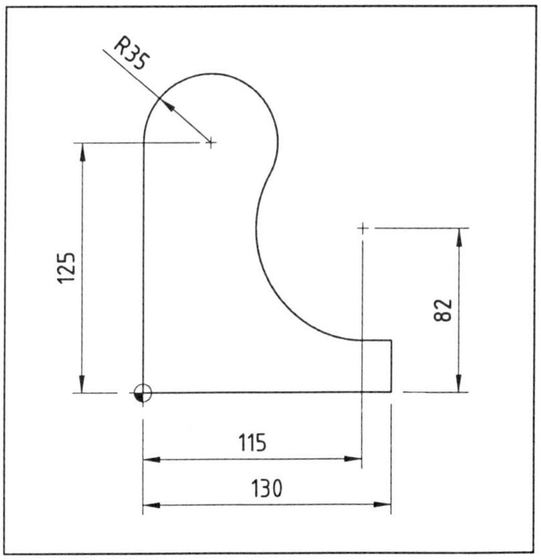

The complete calculations are shown on the next page, with the corresponding illustrations.

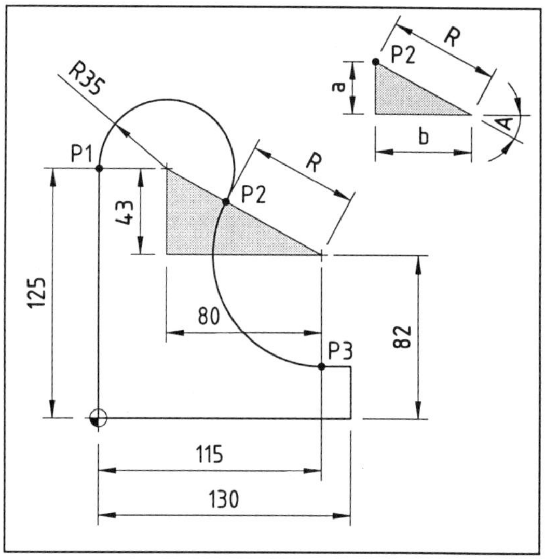

At right, the unknown dimensions must be calculated, particularly the unknown radius - they are identified. With the radius known, the points XY coordinates can be calculated.

The three lines originating at the center of the two arcs create a rectangle. By a simple subtraction, the two sides of this triangle are known - one side is 43 mm, the other is 80 mm. The trigonometric relationship between these two sides is the tangent of the smaller angle, but the angle is not needed for any calculations, so the solution will be to use the *Pythagorean Theorem* method to find the hypotenuse of this triangle:

Hypotenuse $= \sqrt{43^2 + 80^2} = 90.824$

The radius R is 35 mm smaller, therefore:

R = 90.824 - 35 = 55.824

At this stage, only the point P1 is known (no special calculations are necessary) but P3 must be calculated by subtracting the 55.824 radius from the given Y-dimension of 82 mm:

X-coordinate for P3 = X115.0
Y-coordinate for P3 = 82 - 55.824 = Y26.176

Point P2 is the trickiest of the three, but also not difficult. In the inset, the detail shows two unknown sides *a* and *b*, and the known hypotenuse R=55.824. While the Pythagorean solution worked before, it will not work now, and the angle *A* of the triangle *must* be known. Using known dimensions of 43 and 80, the relationship of 43 to 80 is the tangent of the unknown angle *A*:

tanA = 43 / 80 = 0.538
Angle A = 28.258°

The two sides can now be calculated:

a = 55.824 × sin28.258 = 26.429
b = 55.824 × cos28.258 = 49.171

With the little help of simple arithmetic, XY coordinates of the three points can now be listed:

X-coordinates for P1: X0
Y-coordinates for P1: Y125.0

X-coordinates for P2: X65.829
Y-coordinates for P2: Y108.429

X-coordinates for P3: X115.0
Y-coordinates for P3: Y26.176

These and other examples in this chapter have looked at the methods of developing a contouring toolpath for lines and arcs that intersect or are tangent to each other. While many shortcuts using formulas can be used in calculating the point coordinates, there are times when the plain step-by-step method is the only way. Fortunately for the CNC programmer, after a while, only a handful of methods will be required.

4 USING CUTTER RADIUS OFFSET

This chapter covers one of the most important - *and very often misunderstood* - subject used in modern CNC programming - the subject that is in reality very simple; the subject that deals with the methods how to handle the tool diameter (actually its radius) for any given contour toolpath. The subject of this chapter deals with the radius size of the selected cutter as it relates to the actual programmed toolpath, typical applied to contouring.

Incidentally, this subject has been first addressed by the high-level control systems since the early 1980's. During the time since its conception, the methods of programming has changed a few times. As might be expected, the most current method offers the best combination of programming power as well as programming ease, when this feature is used in program development. In spite of the numerous articles written on the subject over the years, many CNC programmers and CNC operators alike consider this subject as advanced or even very advanced. In spite of the apprehension of some users, the subject of cutter radius versus the contour toolpath can be very simple in everyday situations. Perhaps one of the reasons that gives credence to the impression of this subject as being 'difficult' are the several methods of approach that are available to the CNC programmer as well as some rules associated with these methods. The main purpose of this chapter is to provide explanations and solutions. The general rules apply equally to milling and turning, only the actual methods used to achieve the same goal differ somewhat. Both milling and turning applications will be covered, starting with typical usage in milling.

General Concepts

In every program, the first tool motion is always towards a specified location, relative to part zero. Such a motion usually originates at machine zero, is programmed as an *absolute* motion within a certain work offset setting, typically using the G54 designation. As the illustration shows, the G54 is defined from machine zero to part zero along each axis. Why is this statement important when the subject is the radius offset and not the work offset? The answer is that *they are both related.* Work offset is always defined to the *center* of the spindle - to the *centerline* of the cutting tool - *it completely disregards the radius of the cutter!*

When you want to cut a piece of wooden board to an exact length, you measure the length and with a pencil make the cutting line. Then you position the saw just *touching* the pencil line - *on the side you don't care about.* This is exactly what forms the basis of cutter radius offset.

When the cutting tool moves to a given location, its *center* reaches that location. This is perfectly normal for drilling, reaming, tapping and similar operations that machine holes. The centerline of the drill, for example, matches the centerline of the hole, and the hole diameter becomes the drill diameter after machining. For contouring, this situation is not suitable - it must relate to the *cutter edge.*

If the cutter is positioned above a location that represents a contour, a great part of the cutter will be *over* the contour, rather than *away from* the contour. What is expected in this case, is to reposition the cutting tool in such a way that its physical cutting edge is *touching* the contour - that it is *tangential* to the contour. In order to achieve this goal, the position of the cutter has to be adjusted by the cutter radius. Such adjustment is called *cutter radius compensation* or *cutter radius offset.*

Cutter radius offset allows programming to the drawing dimensions

Several terms have been used in machine shops - the afore mentioned *cutter radius compensation*, and the shorter *cutter radius offset*. In turning, the commonly used term is *tool nose radius compensation* or the shorter *tool nose radius offset*. As all these terms refer to the same subject, for the purposes of this book, it will be useful to work with a single term - a term that covers both CNC machining center applications as well as CNC turning applications - *radius offset* or *cutter radius offset*.

The cutter radius can be *adjusted* - or *offset* - right in the part program, by simply entering the individual locations for the cutter center rather than the contour points of the drawing. In fact, this was a method used for many years, in the early stages of numerical control the part programmer had to calculate the *equidistant* toolpath points. With the advancement of CNC computer technology, this method is no longer required, having ben replaced by the *drawing dimensions* method. Although the resulting toolpath will be the same in both cases, the method of achieving it make a big difference. A simple comparison of the manual (equidistant toolpath) and the computerized (drawing dimensions toolpath) method may clarify the concept. From the illustration it is clear that using the drawing dimensions method will save a lot of programming time.

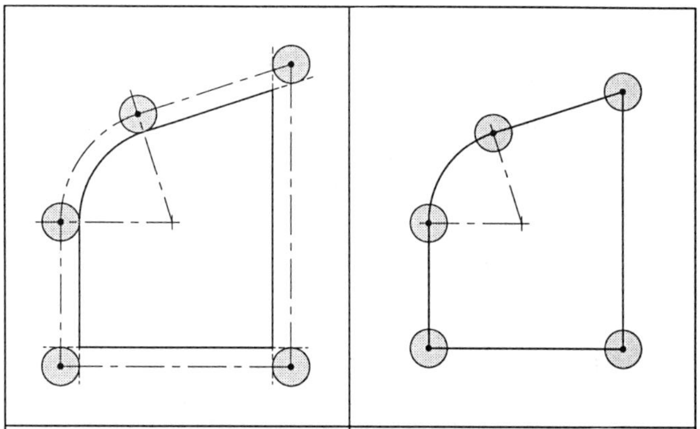

Equidistant toolpath | Drawing dimensions

Benefits Of Cutter Radius Offset

Before trying anything new, the CNC programmer has to feel comfortable that using the cutter radius offset feature will bring certain benefits to the programming process. Using cutter radius offset offers many important benefits, summed up in the following list:

- **Drawing dimensions can be used in the program by ignoring the cutter diameter (radius) in the program**
- **Fewer calculation are required during programming**
- **Allows tight control of part dimensions (tolerances) at the CNC machine**
- **Contouring can use the same coordinates for roughing and finishing**
- **Enables the CNC operator to perform a test cut or a trial cut for the first part**
- **Actual cutter size can be different than the suggested programmed size**

The benefits and basic principles applied for the machining centers are equally valid for CNC lathes as well. Specific lathe applications are covered elsewhere in this chapter.

Controlling Cutter Radius

In terms of machine tool controls and their handling of cutter radius, the same logic that is applied to one control type will also be applied to another control type. For example, a radius of an end mill will be treated the same way as a radius of a turning tool insert corner. The main differences are in the data registered at the control. For milling cutters, the cutter radius only is sufficient. For lathe tools, both the radius and the orientation tip must be registered. In programming, there is no difference in approach for these two types of controls.

Radius Offset Commands

Commands G40-G41-G42

Virtually every CNC system regardless of the make is designed to use the same preparatory commands - *G-codes* - to activate and cancel the cutter radius mode.

There are three such commands available:

G-code	Description	Comment
G40	Cutter radius offset CANCEL	Programming of actual contour points - no offset is in effect
G41	Cutter radius offset LEFT	LEFT and RIGHT is always relative to the tool motion, viewed into the cutting direction
G42	Cutter radius offset RIGHT	

Neither G41 nor G42 can be programmed by itself in a block. In order for the control system to perform the radius offset for a given cutter, the radius of the cutter must be stored in the control registry. As more than one tool can be used in the same program - and each tool can be of different size - the control registry has to accommodate settings for all tools, not just for one. That can be achieved by using the address *D* in the CNC program, as an *addition* to the G41 or G42 command:

`G41 X.. D05` ... ***the radius amount stored in the offset registry number 5 will be used***

For example, if the 05 registry number stores the offset amount of 6.000, the G41 command will force the cutting tool to follow the programmed contour as an *equidistant* contour 6 mm away from the actual contour. The word *equidistant* means that the cutter centerline will always be at the *same distance* from the defined contour.

Using the D-offset Number

Not all control units are created equal. Lower level controls do not have many features that the higher level controls have. One of the features that amplifies this difference is the availability of offset registers for *tool length* and *tool radius*. While the higher level controls have *separate* registers for each of tool length and tool radius, lower level controls have only a *single* register that is used to *share* both types of offsets. For the CNC programmer it means using *different* offset numbers for the tool length offset *H* and the cutter radius offset *D*.

A common practice is to add a certain arbitrary amount to the D-address. Normally, the tool length offset H-address is the same as the tool number *T*. For example, a tool programmed as T06 will use tool length offset H06. This makes sense, because *every tool* needs at least one tool length offset number. Some tools may also require a cutter radius offset number *D*, but not *all* tools. If the control system provides separate offset registers, the D-address number should be the same as the tool number and the tool length number. On the other hand, if the *H* and *D* offsets are *shared* in the same offset registry, their programmed numbers must be different. In such cases, the programmer *adds* arbitrary amounts of 30, 40, 50 or so, to the T/H number:

EXAMPLE - Cutting tool is identified as T06 is in the part program			
Type of offset registry	Tool Number	Length Offset Number	Radius Offset Number
Separate for H and D	T06	H06	D06
Shared by H and D	T06	H06	D56

The arbitrary offset number typically depends on the number of tools available in the tool magazine, and the number of offsets available in the shared registry. In the table example above, arbitrary number of 50 has been added to the T/H number. This is a suitable number for controls that have 99 offsets maximum. Feel free to choose any reasonable number, but once selected, use it consistently.

The arbitrary D-offset number should always be higher than the maximum number of tools in the magazine

In order to prevent confusion, develop a numbering system that never conflicts with other offsets.

Basic Programming Techniques

Every toolpath that uses cutter radius offset always has three distinct modes:

- **Cutter radius *activation*** ... start-up *Radius offset becomes ACTIVE*
- **Cutter radius *application*** ... usage while active *Radius offset is currently USED*
- **Cutter radius *cancellation*** ... cancellation *Radius offset becomes INACTIVE*

Certain specific rules apply to each mode and must be followed for a successful application of cutter radius offset in the part program as well as at the CNC machine. Before describing practical details of each mode, evaluate the main items listed in the summary of each mode:

Cutter Radius Activation

This section describes the common general rules that relate to the *start-up* of cutter radius offset in a typical CNC program for milling. In a list form, these are the main points:

- **Cutter radius offset should be turned ON (with G41/G42 + D-offset) *after* the tool has reached the Z-depth**
- **Start tool position must always be in G40 mode**
- **Position of the cutter center should always be *away* from the program contour**
- **Cutter start location must be greater than or equal to the cutter radius + stock + clearance**
- **Material area to be removed is determined by the CNC programmer**
- **G41/G42 should be programmed in a toolpath *towards* the part contour, not along the actual part edge**
- **G41/G42 cannot be programmed together with an arc motion - allowed only in G00 or G01 mode**
- **Dxx offset number should be programmed in the same block as G41/G42 or before, but never after**

Rather a disproportional number of errors (or alarms) issued by the control system relate to the improper use of cutter radius offset. When looking for solutions to the problem (alarm), start searching the program in the section immediately surrounding the blocks that relate to the tool approach (lead-in) towards the machined contour. The error may lie elsewhere, for example in the incorrect entry to the offset register. The subject of practical lead-in and lead-out tool motions is described later in this chapter, along with suitable illustrations and examples.

Pay an extra careful attention to the XY coordinates of the start position. The combination of cutter radius offset, stock allowance and some extra clearance define such position accurately, but programmers do not always follow this suggestion. For example, a 6 mm radius plus 0.5 mm stock plus additional 1.5 mm clearance means the start location should be 8 mm away from the machined contour.

Cutter Radius Application

This section list only one - *but very critical* - rule that relates to the usage of cutter radius offset in a CNC program *between* its activation (G41/G42) and its cancellation (G40).

- **Between the G41/G42 block and the G40 block, there must be no more than *two* consecutive blocks that do *not* contain an axis motion**

- **The stored D-offset amount must always be *smaller* than the *smallest inner arc radius* of the part, while in the cutter radius offset mode**

Errors caused by the improper use of cutter radius offset while it is activated (applied to the contour toolpath) are usually confined to the program itself. Assuming the operator's settings at the machine control are correct, one or both of the two items above will most likely be the cause of the error.

Cutter Radius Cancellation

This section describes the common general rules that relate to the *cancellation* of cutter radius offset in a CNC program. In many ways, the cutter radius cancellation is the exact opposite of the cutter radius activation (see above).

- **Cutter should end away from the part = tool edge away from the part contour**
- **End tool position clearance must be greater or equal to the cutter radius + stock + clearance**
- **End tool position must always be in G40 mode**
- **Offset should be turned OFF (G40) *before* the tool retracts from the Z-depth**
- **No D-address is required for G40 block - D00 is usually permissible (subject to verification)**
- **G40 cannot be programmed together with an arc motion - only in G00 or G01 mode**

Just because the contour has been completed successfully, it does not mean the cancellation of the cutter radius - when no longer needed - should be done without some precautions and good planning. The single most important item listed above is to cancel the G41/G42 mode before retracting to a clear tool position. Of course, all other conditions are important as well.

D-offset Stored Amount

At the machine, the CNC operator has many responsibilities - one being the setting and adjusting various offset settings, including the settings related o the active cutter radius. In CNC programming, there are four ways to handle the cutter radius:

1. **As an equidistant centerline of the cutter - no provision for adjustment at the machine (G40 mode)**
2. **As an equidistant centerline of the cutter - with provision for adjustment at the machine (G41/G42 mode)**
3. **As drawing based dimensions - no provision for adjustment at the machine (G40 mode)**
4. **As drawing based dimensions - with provision for adjustment at the machine (G41/G42 mode)**

The first two options require that the precise cutter radius is known during programming, the last two options ignore the radius, as it can be set at the control.

Equidistant Centerline - G40 Mode

In many CAM systems, this is the default mode. There is no easy way to adjust the cutter radius at the machine, if necessary. While quite suitable - *and actually preferred* - method for roughing operations, having no fine-tuning options for finishing cuts limits this method to roughing only.

Equidistant Centerline - G41/G42 Mode

The same CAM system that typically provides the default mode with G40, can also be preset to provide G41/G42 commands, still outputting the coordinates to the cutter center. This allows the CNC operator to fine-tune critical dimensions at the machine. While this program output is not practical for manual program development, it has distinct advantages when used by a CAM software.

The nominal offset amount will be zero - for example, D51 = 0.000

Drawing Dimensions - G40 Mode

Using drawing dimensions in G40 mode means the cutter radius is ignored. There are several possible example of using this method. They all relate on the actual tool diameter to make the desired profile. Probably the most common application is rough machining of grooves, where their centerline is defined in the drawing. For typical contour finishing, this is not a suitable method.

Drawing Dimensions - G41/G42 Mode

The most common method of programming the cutter radius manually (without the use of a CAM system) is to ignore it altogether and use the drawing based dimensions in the program directly. The program will contain the G41/G42 commands. This allows the CNC operator to fine-tune critical dimensions at the machine. While this program output is very practical for manual program development, the operator is responsible for the actual settings at the machine.

The nominal offset amount will be the cutter radius - for example, D51 = 6.000

Radius vs Diameter

Virtually every control unit comes with a system parameter setting that allows the user to enter the *radius* amount or the *diameter* amount in the D-offset register. Virtually every control has the same parameter set to the radius, *not* the diameter. Radius entry is far more popular (as well as more logical) and this setting should not be changed, unless there is an absolute need for the opposite. Compensated tool path is always the programmed toolpath extended or reduced by the tool *radius amount* stored in the D-offset. For example, on Fanuc FS16M, parameter #5004 controls whether the D-offset is input as a radius or as a diameter. #5004=0 selects a radius input, #5004=1 selects a diameter input. In both cases, the actual compensation is *always* by the radius of the cutter. Other controls follow the same pattern - check the control manual for details - and stay with the radius setting.

With only a few exceptions, the actual application of the cutter radius offset in the program is as important as correct settings at the CNC machine. Many radius offset related problems can be detected to the start or end of the contour, in the form of a *lead-in* and a *lead-out* tool motions.

Contour Lead-In and Lead-Out

The terms *lead-in* and *lead-out* refer to the programming method of moving the tool from a start position towards the contour (lead-in) and moving the tool from the contour back to the start position (lead-out). Start position is typically located *close to* the contour but *not touching* the contour.

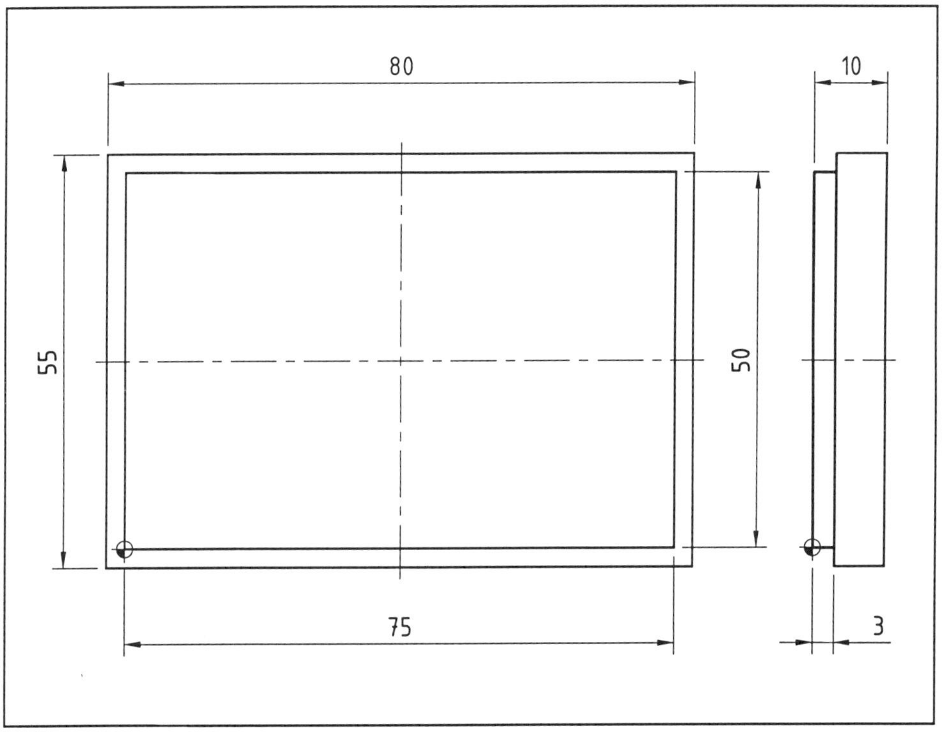

In order to illustrate the concepts of lead-in and lead-out motions, a simple rectangular contour will be used. All examples are based on external contouring of a 75×50 mm rectangle from 80×55×10 mm stock (2.5 mm wide edges). Climb milling mode is programmed with D51 offset number and the tool in the spindle. Program zero is at the lower left corner of the *contour* for the X and Y axes, and the top of the part for the Z-axis. Cutting tool used is a ∅12 mm end mill at 450 r/min and 225 mm/min feedrate. Start position is at X-20.0 Y-20.0, unless otherwise noted - it has been selected for clarity of illustrations only, and can be smaller for a more efficient program.

By definition, *lead-in* is the toolpath motion where the cutter radius offset is applied, *lead-out* is the toolpath motion where the cutter radius offset is canceled. In practical terms, one motion is the opposite of the other motion. Two methods will be shown - linear to linear and linear to arc.

Methods For Lead-In - Linear Motion

Since the first actual cutting will be towards the X0Y50.0 position, and the tool has to be on the left, there are three possible approaches:

Lead-in method 1: **Diagonally towards X0Y0 - then cutting towards Y50.0**

Lead-in method 2: **X+ direction toward X0 - then cutting towards Y50.0**

Lead-in method 3: **Y+ first towards Y0, then X+ towards X0 - then cutting towards Y50.0**

Selecting one method over another means understanding their respective advantages and disadvantages. All three lead-in methods shown are correct, so the decision which one to use will be based solely on the advantages over other methods.

Lead-in method 1: **Diagonally towards X0Y0 - then cutting towards Y50.0**

This approach is fairly efficient, but cutting feedrate must be used as stock can be cut during the tool approach. Another option would be to split the lead-in motion into two motions, but that is not really a practical alternative as it adds extra calculations.

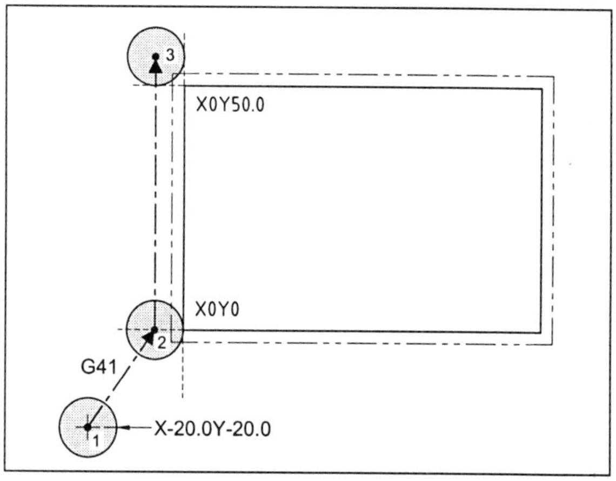

Block N6 is the lead-in block:

```
...
N3 G90 G00 G54 X-20.0 Y-20.0 S450 M03    (START POSITION)
N4 G43 Z5.0 H01 M08                      (TOOL LENGTH OFFSET AND COOLANT ON)
N5 Z-3.0                                 (FINAL Z-DEPTH)
N6 G01 G41 X0 Y0 D51 F225.0              (START RADIUS OFFSET)
N7 Y50.0                                 (CUT TO Y50.0 AT F225.0)
...
```

Lead-in method 2: **X+ direction toward X0 - then cutting towards Y50.0**

This lead-in is also fairly efficient because cutting feedrate can be increased as no stock is cut during the approach. Motion from start position 1 to position 2 can also be a rapid motion. This version is much better than the previous method, and can be made even more efficient by programming the start point closer to the part (see earlier note).

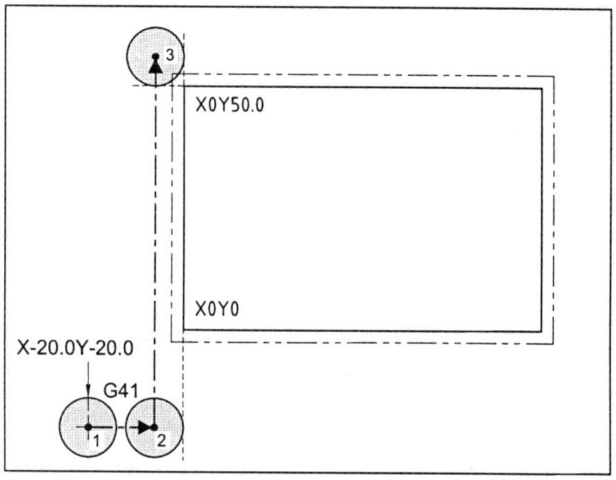

Block N6 is the lead-in block:

```
...
N3 G90 G00 G54 X-20.0 Y-20.0 S450 M03    (START POSITION)
N4 G43 Z5.0 H01 M08                      (TOOL LENGTH OFFSET AND COOLANT ON)
N5 Z-3.0                                 (FINAL Z-DEPTH)
N6 G01 G41 X0 D51 F500.0                 (START RADIUS OFFSET)
N7 Y50.0 F225.0                          (CUT TO Y50.0 AT F225.0)
...
```

Lead-in method 3: **Y+ first towards Y0, then X+ towards X0, then cutting towards Y50.0**

This is also a legitimate lead-in but rather inefficient, because it contains an unnecessary tool motion. We will reject it, but show the program segment anyway.

Block N7 is the lead-in block:

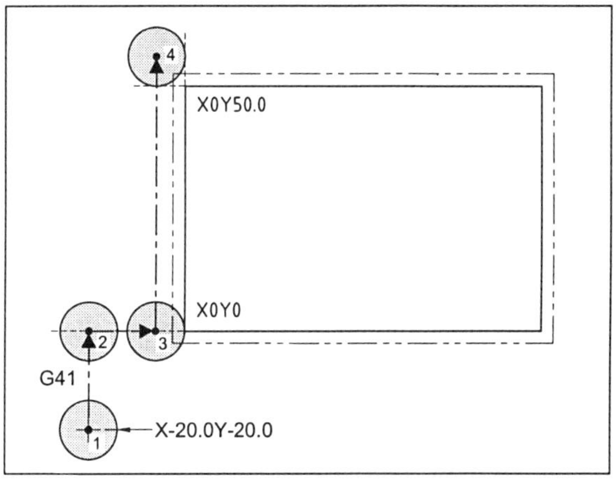


```
...
N3 G90 G00 G54 X-20.0 Y-20.0 S450 M03     (START POSITION)
N4 G43 Z5.0 H01 M08                        (TOOL LENGTH OFFSET AND COOLANT ON)
N5 Z-3.0                                   (FINAL Z-DEPTH)
N6 Y0
N7 G01 G41 X0 D51 F225.0                   (START RADIUS OFFSET)
N8 Y50.0                                   (CUT TO Y50.0 AT F225.0)
...
```

The second lead-in method is the best choice for most jobs and will be used in other examples.

What happens during tool lead-in ?

The one block `N6 G01 G41 X0 Y0 D51 F225.0` *or* `N6 G01 G41 X0 D51 F225.0` is not enough to complete the cutter radius offset. What is the offset to the left? What is the offset to the right? In terms of a toolpath, the left vs. right is *always determined by the* ***next*** *tool motion.* We know from the examples that the next motion is to Y50.0 position. How does the control system 'know' this location?

The illustration shows what happens if there is no 'next' tool motion programmed. The cutter starts from position X-20.0Y-20.0 with the programmed target of X0. While the G41 and D51 are programmed correctly, with no 'next' motion, the radius compensation will fail, as there are two possibilities. Of course the next motion direction may be at an angle, but the concept shown will remain the same.

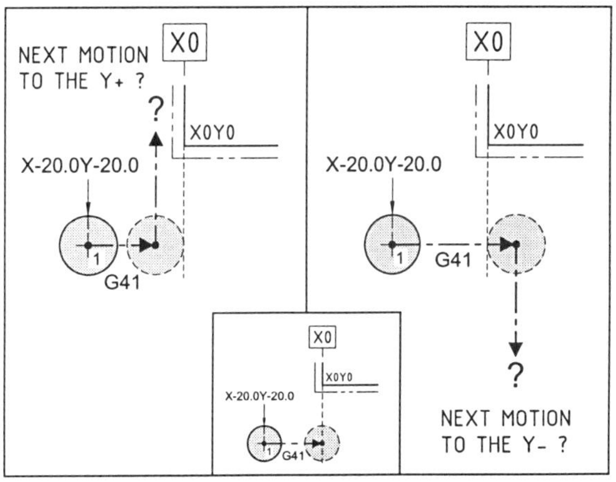

The result (shown in the inset) will be:

The center of the cutter will be at X0

Modern cutter radius compensation is based on the principle of reading the next block, without actually executing it. This type of compensation is called the 'look-ahead' type, and is the basis of cutter radius offset used by all modern CNC machines.

Axis motion missing

Another possible source of problems associated with cutter radius offset is missing motion block, while G41 or G42 are active. Since the control system is looking for an axis motion in order to determine what is a *left* and what is a *right* direction, what happens if there is *no axis motion* in the next block? For example:

```
N6 G01 G41 X0 D51 F500.0          (START RADIUS OFFSET)
N7 F225.0                         (** AXIS MOTION NOT PROGRAMMED **)
N8 Y50.0                          (CUT TO Y50.0 AT F225.0)
...
```

In this case, the control system will 'look' at the next block after the no-motion block, which will be N8 in the example. If it finds a axis motion, it compensates accurately, if not, no compensation takes place. Some controls may 'look-ahead' more than two blocks, but it much safer if you avoid any non-cutting motions while the in the compensation mode (G41 or G42), and definitely avoid more than one non-cutting block. For example, many controls will *not* compensate properly when the following sequence is programmed:

```
N6 G01 G41 X0 D51 F500.0          (START RADIUS OFFSET)
N7 F225.0                         (** AXIS MOTION NOT PROGRAMMED **)
N8 S600                           (** AXIS MOTION NOT PROGRAMMED **)
N9 Y50.0                          (CUT TO Y50.0 AT F225.0)
...
```

The non-cutting blocks are only examples - they show a poor program structure; the control will behave the same way, regardless what instructions the no-motion block or blocks contain.

Methods For Lead-In - Arc Motion

Not every first motion of a contour is a *linear* motion. Often it is necessary to start machining the contour on an arc, for example, when machining a circular pocket. Review the basic programming rules listed in the *Cutter Radius Activation* section of this chapter, and one rule should stand out:

G41/G42 cannot be programmed together with an arc motion - allowed only in G00 or G01 mode

The above rule is quite specific - the cutter radius offset will *only* be activated while in a rapid motion mode G00 or a linear cutting mode G01. It also means, that such a tool motion must be programmed *before* the arc motion of the contour (left) or another lead-in motion (right), as illustrated at right. Using a circular pocket as an example, two lead-in methods are shown - note the preceding linear motion before arc.

These methods work for any contour where the first motion is an arc.

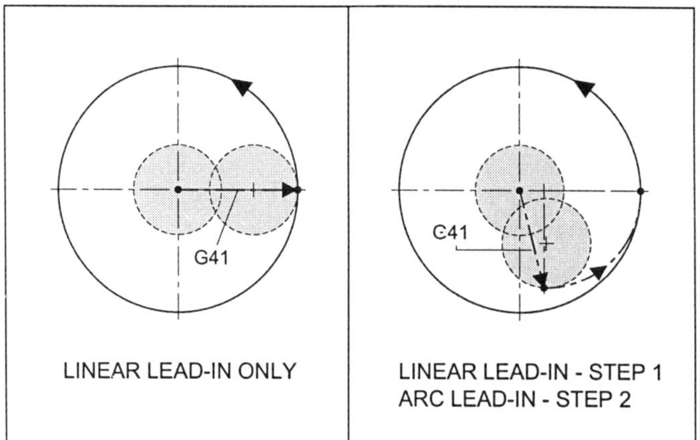

G41 IS ACTIVATED DURING LINEAR MOTION

Methods For Lead-Out - Linear Motion

Just like the purpose of the lead-in is to *activate* the cutter radius offset, the purpose of the lead-out is to *cancel* it. If the last contour entity is a linear motion, the lead-out motion will also be linear (or rapid). The most common - and *safest* - method is to complete the last entity of the contour, then make one more motion, in which the offset cancellation will take place. The cancellation should take place before retracting from the depth along Z-axis.

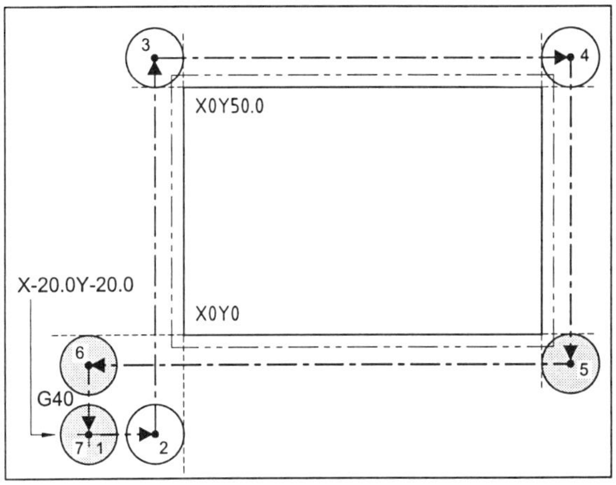

In the illustration at right, the last cutting motion of the contour is from location 5 to location 6 and is followed by the cancellation motion from location 6 to location 7 (1).

Methods For Lead-Out - Arc Motion

If the last contour entity is an arc motion, the lead-out motion *must be* linear (or rapid). All rules for a lead-out listed for the linear motion apply as well here. For machining purposes and proper function of the radius offset, it is customary to program an arc lead-out when the last contour entity is as arc. However, this motion cannot contain the offset cancellation, and must be followed by a linear motion in the G40 mode. In a way, the lead-out of this kind is a two-block program entry.

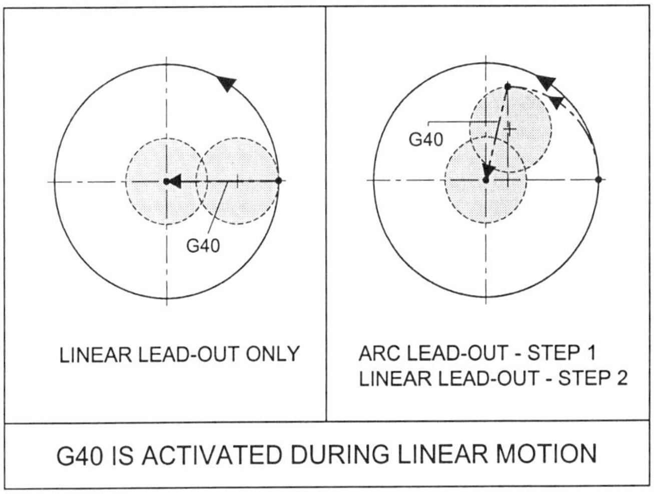

Program Example

The lead-in and lead-out motions are used to turn on and off the cutter radius offset. The method shown in the illustration at right should help in determining the proper start position of the cutter, once the Z-depth has been reached. It can be used in any orientation.

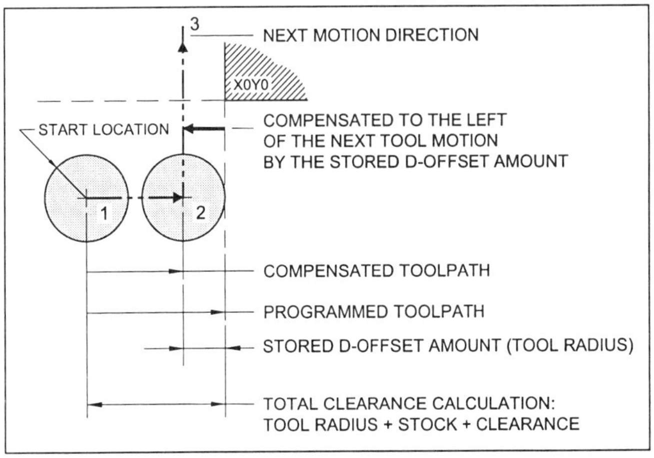

The following program uses the initial drawing and is a complete example of using the lead-in to activate G41 and lead-out to cancel cutter radius offset (G40). All blocks contain descriptive comments.

```
(T01 - 12 MM DIA END MILL - D51 = 6.0)
(CUTTER RADIUS OFFSET USED)
(DRAWING DIMENSIONS PROGRAMMED)
N1 G21                                  (METRIC UNITS)
N2 G17 G40 G80                          (START-UP BLOCK)
N3 G90 G54 G00 X-20.0 Y-20.0 S450 M03   (MOVE TO START POSITION)
N4 G43 Z5.0 H01 M08                     (CLEARANCE ABOVE PART)
N5 Z-3.0                                (DEPTH OF CUT)
N6 G01 G41 X0 D51 F400.0                (START RADIUS OFFSET)
N7 Y50.0 F225.0                         (NEXT MOTION IS Y+)
N8 X75.0                                (NEXT MOTION IS X+)
N9 Y0                                   (NEXT MOTION IS Y-)
N10 X-20.0                              (NEXT MOTION IS X-)
N11 G00 G40 Y-20.0 M09                  (CANCEL RADIUS OFFSET)
N12 Z5.0 M05                            (CLEAR ABOVE PART)
N13 G28 X-20.0 Y-20.0 Z5.0              (HOME IN XYZ)
N14 M30                                 (PROGRAM END)
%
```

The explanation provided so far has covered the most typical applications of lead-in and lead-out for cutter radius offset for external contour toolpath. Programming the lead-in and lead-out for an internal toolpath requires some additional attention.

Internal Contours

Programming internal contours shares the rules and precautions with the methods described for external contours. The major concern when approaching an internal contour is a possible lack of space. This is typical to programming a finishing contour in various slots and similar toolpaths.

Linear Slot Machining

In the illustration at right, a typical slot is shown. The program will be developed for one tool only - a separate tool can be used for roughing. Tool will start at the center of one end (P0)and rough the slot to the center at the other end (P1). Once this point is reached, the tool will make a standard *line-arc lead-in* towards the slot inner wall, contour the slot all the way around, back to the first point. From there, a standard *arc-line lead-out* will complete the job. Tool used will be a ∅10 mm center cutting end mill to machine the slot.

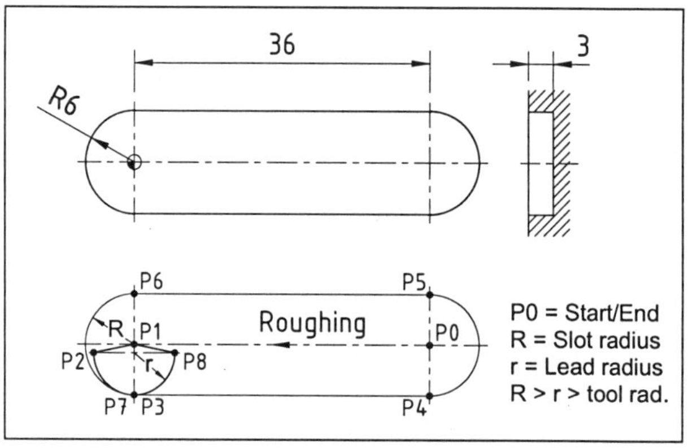

When selecting the end mill, keep in mind the relationships between the three radiuses. The end mill must have a smaller radius that the lead-in/lead-out arc, which has to be smaller than the slot radius. The order of cutting will be from P0 to P1, P2, P3, P4, P5, P6, P7, P8, and back to P1. To satisfy the conditions, the lead-in and lead-out radius must be greater than 5mm and smaller than 6 mm, with 5.5 mm selected for the program. The stock left for finishing will the difference between the tool radius and the slot radius: 6 - 10 /2 = 1 mm in this example.

Now, the program can be written, using the toolpath defined in the above illustration.

```
N1 G21                                      (T01 - 10 MM END MILL D51=5.000)
N2 G17 G40 G80
N3 G90 G54 G00 X36.0 Y0 S1200 M03           (P0)
N4 G43 Z2.0 H01 M08
N5 G01 Z-3.0 F100.0
N6 X0 F150.0                                (P1)
N7 G41 X-5.5 Y-0.5 D51                      (P2)
N8 G03 X0 Y-6.0 I5.5 J0                     (P3)
N9 G01 X36.0                                (P4)
N10 G03 Y6.0 I0 J6.0                        (P5)
N11 G01 X0                                  (P6)
N12 G03 Y-6.0 I0 J-6.0                      (P7)
N13 X5.5 Y-0.5 I0 J5.5                      (P8)
N14 G01 G40 X0 Y0                           (P1)
N15 G00 Z2.0 M09
N16 G28 X0 Y0 Z2.0 M05
N17 M30
%
```

Circular Slot Machining

Circular slot - or a circular groove - is similar to a straight slot. The cutting tool must be smaller than the slot (or groove) width and typically enters the slot at a quadrant position, roughs it out all the way around, than makes a line-arc lead-in in the climb milling mode, finishes one wall, and returns back to the starting point via arc-line lead-out. From the same position, the other wall is completed by repeating the process, but in the opposite direction, in order to maintain climb milling mode.

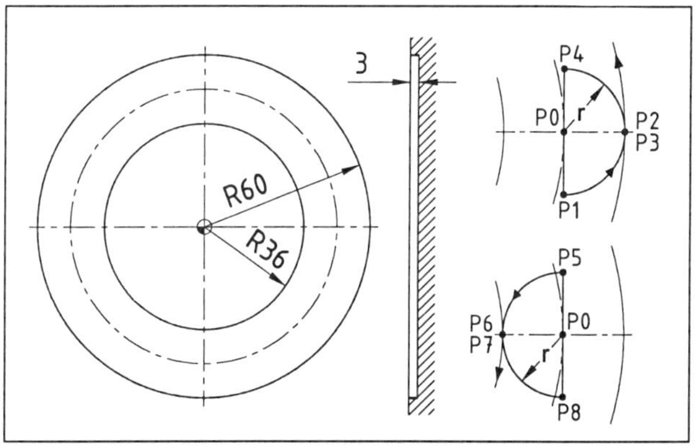

In order to select a suitable cutter diameter, the selection must guarantee to clean the slot bottom. Therefore, the cutter diameter must be larger than *one third of the slot width*:

Minimum cutter diameter > (60 - 36) / 3 = 8 mm

The lead radius *r* must be the same as *one half of the slot* width:

Lead radius *r* = (60 - 36) / 2 = 12 mm

The cutter diameter must be greater than 8 mm and smaller than 12 mm, so a ∅10 mm center cutting end mill will be selected for the example:

```
N1 G21                                      (T01 - 10 MM END MILL D51=5.000)
N2 G17 G40 G80
N3 G90 G54 G00 X48.0 Y0 S1200 M03           (P0)
N4 G43 Z2.0 H01 M08
N5 G01 Z-3.0 F100.0
(ROUGHING THE SLOT)
N6 G03 I-48.0 F150.0                        (FULL CIRCLE - ROUGHING)
(*** FINISHING RADIUS R60 ***)
N7 G01 G41 Y-12.0 D51                       (P1)
N8 G03 X60.0 Y0 I0 J12.0                    (P2)
N9 I-60.0                                   (P3)
N10 X48.0 Y12.0 I-12.0 J0                   (P4)
N11 G01 G40 Y0                              (P0)
```

```
(*** FINISHING RADIUS R36 ***)
N12 Y12.0                                 (P5)
N13 G03 X36.0 Y0 I0 J-12.0                (P6)
N14 X-36.0                                (P7)
N15 X48.0 Y-12.0 I12.0 J0                 (P8)
N16 G01 G40 Y0                            (P0)
N17 G00 Z2.0 M09
N18 G28 X48.0 Y0 Z2.0 M05
N19 M30
%
```

Strictly speaking, the cancellation of G41 in block N11 and its subsequent reinstallment in block N12 is not necessary, but it is a good programming practice to complete one contour before starting another. Also, it allows an easier way to change the D-offset number, in cases where controlling the slot *width* is important.

Finishing Internal Contour

A closed internal contour may be a pocket with a bottom or a similar design with an open core. Such a shape has several inside corners and is the subject for a fairly common error relating to cutter radius offset. Both the problem and the solution also applies to similar situations, for example, rectangular pockets. Although the drawing in the illustration shows sharp corners on the open core, the tool will form its own radius in the corners, which will have to be removed by other means, if necessary.

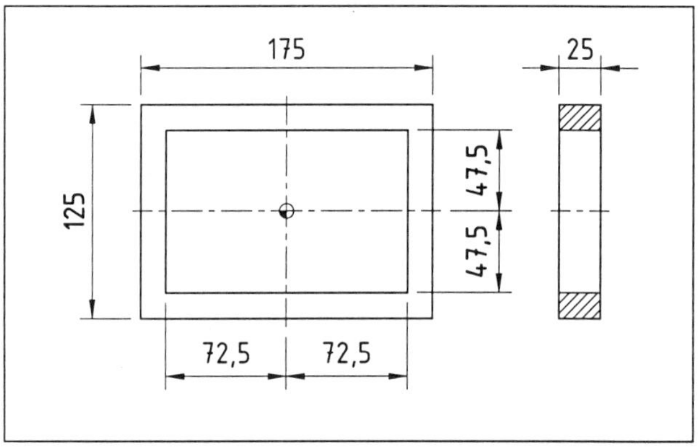

The following program contains a common error - study the program an try to find the error:

```
N1 G21                                    (T01 - 25 MM END MILL D51=12.500)
N2 G17 G40 G80
N3 G90 G54 G00 X35.0 Y0 S850 M03          (START POSITION)
N4 G43 Z2.0 H01 M08
N5 G01 Z-26.0 F100.0
N6 G41 X72.5 Y47.5 D51 F150.0
N7 X-72.5
N8 Y-47.5
N9 X72.5
N10 Y47.5
N11 G00 G40 X0 Y0                         (CENTER OF PART)
N12 Z2.0 M09
N13 G28 X0 Y0 Z2.0 M05
N14 M30
%
```

The lead-in linear motion in block N6 applies the cutter radius offset at a corner location of the contour (upper right corner), which is also the last tool location. This method of programming will cause an severe undercut that cannot be detected by the control system, therefore no alarm will be generated. On the next page, there is an illustration of the effect as well as the correct method of programming a closed internal contour with cutter radius offset in effect.

On the *left side* of the illustration is shown the overcut that happens when the radius offset is activated in a motion towards a corner. On the *right side* of the illustration is shown the overcut that happens when the radius offset is cancelled in a motion away from a corner.

In practice, it is not likely to actually see the error at the end.

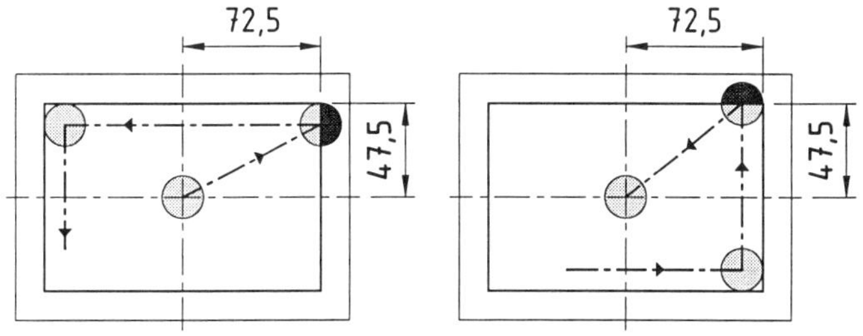

OVERCUT AT THE START OF CONTOUR — OVERCUT AT THE END OF CONTOUR

The correct method of programming a radius offset for a closed inside contour is to start in the middle of the selected contour element or in a similar position. The next illustration shows two *correct* methods - one with a linear lead-in/lead-out, the other with an line-arc lead-in/lead-out.

Middle position of the lower horizontal contour has been selected in the example. In both cases, the start and end is in the middle of the rectangle, however, this location can be different in order to save unnecessary cutting, particularly for large parts. The two program listings showing the correct approach in applying cutter radius offset:

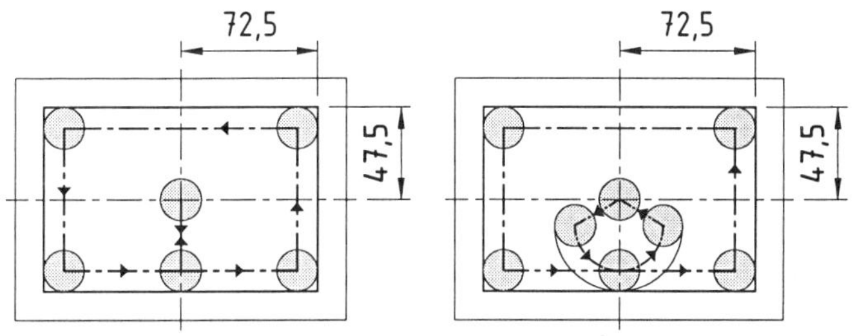

LINEAR LEAD-IN / LEAD-OUT — LINE-ARC LEAD-IN / LEAD-OUT

N1 G21 (*** LINEAR LEADS ***)	N1 G21 (*** LINE-ARC LEADS ***)
N2 G17 G40 G80	N2 G17 G40 G80
N3 G90 G54 G00 X0 Y0 S850 M03	N3 G90 G54 G00 X0 Y0 S850 M03
N4 G43 Z2.0 H01 M08	N4 G43 Z2.0 H01 M08
N5 G01 Z-26.0 F100.0	N5 G01 Z-26.0 F100.0
N6 G41 Y-47.5 D51 F150.0	N6 G41 X-30.0 Y-17.5 D51 F150.0
N7 X72.5	N7 G03 X0 Y-47.5 I30.0 J0
N8 Y47.5	N8 G01 X72.5
N9 X-72.5	N9 Y47.5
N10 Y-47.5	N10 X-72.5
N11 X0	N11 Y-47.5
N12 G00 G40 Y0	N12 X0
N13 Z2.0 M09	N13 G03 X30.0 Y-17.5 I0 J30.0
N14 G28 X0 Y0 Z2.0 M05	N14 G00 G40 X0 Y0
N15 M30	N15 Z2.0 M09
%	N16 G28 X0 Y0 Z2.0 M05
	N17 M30
	%

Maintaining Dimensional Sizes

This section seems to be aimed at the CNC operators rather than CNC programmers. While it is true that the operator is generally responsible for maintaining all offsets for a given program, it is the programmer that has to incorporate these offsets in the program. It is not enough to put a G41 D.. here or there and ignore its effect on the actual machining. Understanding the radius offset in order to maintain dimensional sizes is important in both programming and machining.

Basic Rule

Several different methods have developed over the years to explain the concept of adjusting the stored amount of cutter radius. Many of these methods used terms such as 'smaller tool', 'larger tool', 'increased travel', decreased travel', etc. In many cases, it was also important to specify whether the offset is applied on an external contour or an internal contour. There is much simpler way to make decision as how to change the current amount of the radius offset setting.

To update the stored D-offset amount, ignore the cutter size and concentrate instead on the *stock* to be added to the contour or removed from it. This method is much easier to remember and can be used equally for external and internal contouring, without any modifications:

To ADD stock to the measured size, use LARGER D-offset amount

To REMOVE stock from the measured size, use SMALLER D-offset amount

To reach a certain target position, tool with bigger D-offset amount travels a shorter distance than a tool with smaller D-offset amount - and vice versa.

Handling Dimensional Tolerances

Programming a cutter radius offset has many great advantages in maintaining the dimensional tolerances demanded by the part drawing. When it comes to handling tolerances *in the program*, there are two general methods programmers use:

1. **Programming the *nominal* size only - ignore tolerances - the operator sets required offsets**
2. **Programming the *nominal+tolerance* size - build-in tolerances - the operator has less work**

Before jumping to an uninformed conclusion and select one method over the other, consider a very important aspect of machining - ***the tool life***.

Tool Life Issues

Every cutter will wear out during machining, however slightly. Although the cutter may still be able to produce tens or even hundreds of parts more, it will not maintain the same part dimensions. On CNC mills and machining centers, the external width of parts will be increasing, their internal width will be decreasing. On CNC lathes, outside diameters will become larger, inside diameters smaller. As long as the effect of the cutter wear maintains the dimensional tolerance (and surface finish quality), there is nothing to worry. On the other hand, the programmer and/or the operator can greatly influence *how many parts the cutter will make*, before the need of cutter replacement. Consider the following example. The two drawings on the next page show two simple parts - both are identical, except for their tolerances - the specified width of 74 mm is called the *nominal dimension*.

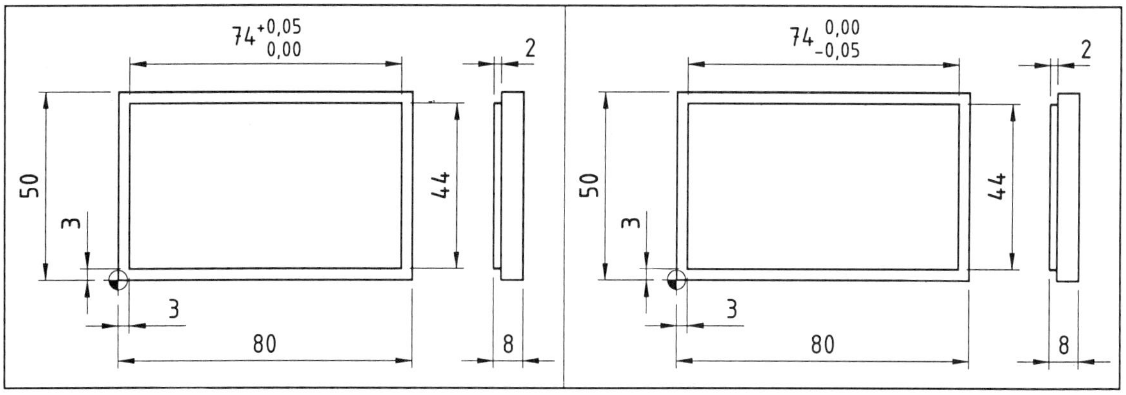

In order to increase the tool life, some programmers choose to include a certain amount of tolerance in the dimension, for example, programming X77.025 (left) or X76.975 (right) instead of X77.0. The idea is that if a higher tolerance is specified, it will take the tool longer to get outside the tolerance range. The idea is certainly a valid one, and every programmer and operator should be aware of it. What really matters is how this idea is implemented - *including* the tolerance in the program or *not including* the tolerance in the program.

Adherents to both methods have some valid points and a heated discussion could take place on the subject. Personally, I prefer to ignore tolerance in the program, for a couple of reasons. Changing and fine-tuning offsets is one of the main jobs of a CNC operator. That is the first reason. The second - *and more compelling* - reason is that tolerances are often the subject of engineering change, while the nominal size remains intact. There is also a third reason - in some instances, the inclusion of the tolerance in the programmed dimension may adversely affect neighboring dimensions, for example, the 3 mm edge width in the above drawing. All three reasons make the nominal dimensions method more flexible and practical.

Ignoring the tolerances in the program gives the operator important responsibilities but not *additional* responsibilities - a skilled operator already has them anyway. The CNC operator may still opt for the mean value offset setting, but at the machine - *not through the program*. The usual procedure to change offset setting to maintain drawing tolerances is:

1. **Establish the target size (can be the mean of the tolerance range or the high tolerance)**
2. **Find the difference between the measured size and the target size**
3. **Check if the difference is per side or per width (stock amount = per side)**
4. ***Increase* the D-offset amount to add stock, *decrease* the D-offset amount to remove stock**
5. **Both definitions in *Item 4* are equally applicable to external and internal cutting**

Item 4 is the key to manage offsets at the CNC machine - in order to *add* stock to the material, *add* to the stored D-amount; in order to take stock *away* from the material, *subtract* from the stored D-amount.

One final consideration regarding the drawing above. How does the increase or decrease by 0.025 mm affect the part sizes without a tolerance, particularly the 3 mm edge width and the 44 mm contour width? Yes, they will be affected. The 3 mm edge is a single dimension (measured per side), so it will become 0.025 mm smaller (left) or larger (right). The 44 mm width is measured over two edges, so the 0.025 mm will affect it on both sides equally, exactly the same way as the 74 mm contour length. Keep in mind that both the 3 mm edge and the 44 mm contour width are open sizes, where no tolerances are specified.

Handling Cutter Radius Offset Errors

One of the reasons many CNC programmers or operators shun away from using cutter radius offset is that they are not familiar with various solutions available to them in case of offset failure. The key to solving *any* problem is *knowing its cause* - the real cause must be known or found out before a problem can be resolved. This proven approach to problem solving is no different for a CNC program using cutter radius offset that fails. Unlike many other problems, if a cutter radius offset fails during setup or machining, the reason for the failure is always the same:

The stored cutter radius cannot fit into the space allocated

In case of radius offset error, this simple statement is always true. Modern control systems are using the 'look-ahead' type of cutter radius offset. That means any offset error that can be detected will be detected *before* it does any damage. Solving an offset related error should be always be done at the CNC machine, whether the problem is related to various settings or to the program itself.

Common Errors

The following list identifies the most common areas to look in case of a cutter radius error:

- **Check the actual offset setting - is the amount correct?**
- **Does the offset number match the program ?**
- **Check for decimal errors (missing decimal point, decimal point in the wrong place)**
- **Do the units match? (Metric tools vs Imperial tools)**
- **Check the actual tool travel as programmed (distance-to-go)**
- **Look for a registered radius amount that is greater than the smallest inside arc programmed (control setting)**

Keep in mind that the control system only interprets the program as stored in the control memory. The paper version of the program may be correct, yet not correspond to the stored version in the control. This problem arises mainly when the program is entered via keyboard at the machine or edited at the machine. Also keep in mind that the control does not distinguish between cutting actual part and what is commonly known as 'cutting air'. It only processes the motions programmed - the error statements such as *'... overcutting will occur in cutter radius compensation ...'* or *'CRC interference'* has to be interpreted in this spirit.

A common source of an offset problem is a cutter radius intentionally programmed *larger* to achieve extra stock for roughing. For example, a ∅10 mm end mill will cut a 5.5 mm inside corner radius correctly for finishing, where the D-offset is set to 5.000. However, the same tool using the D-offset of 6.000 for roughing with 0.5 mm stock will cause an error condition (overcutting alarm) - see sketch.

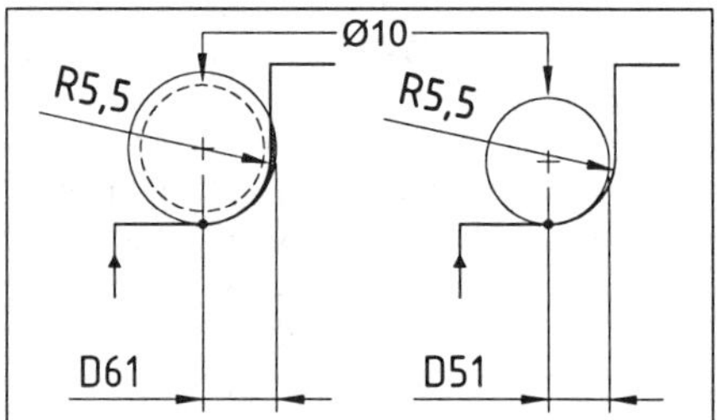

Cutter = Ø10 mm
Roughing with D61=6.000
Finishing with D51=5.000

Usually it is easier to find the causes for radius offset failure at the control than in the program. Because the control reads the blocks *ahead* of the current block, study the program listing from the current block to several blocks ahead.

Offset Programmed Too Late or Too Early

In terms of program input, it seems very easy to program the cutter radius offset. After all, it needs to be turned *ON* with the G41/G42 D.. command and turned *OFF* with the G40 command. All contouring motions between the two commands will follow the compensated toolpath. Simple? Yes, providing some basic precautions are observed.

Using the initial drawing in this chapter, two common errors and their effect will be illustrated:

- **Applying the G41/G42 Dxx command *TOO LATE***
- **Applying the G40 cancellation command *TOO EARLY***

If the cutter radius is started *after* the tool motion towards the first entity of a contour has been completed, the offset has been started *too late* and the toolpath will be incorrect

If the cutter radius is started *before* the motion of the last entity of a contour has been completed, the offset has been started *too early* and the toolpath will be incorrect

Example 1 - Cutter radius offset *start-up* is programmed *too late*:

```
...
N5 Z-12.0                (POS. 1)
N6 G01 X0 F400.0         (POS. 2)
N7 G41 Y50.0 D51 F225.0  (POS. 3)
N8 X100.0
...
```

A fairly innocent looking program segment - yet, it includes a very serious error. When the tool moves from position 1 to position 2, the radius offset is not in effect as yet. During the motion from position 2 to position 3, the radius offset is activated, resulting in a severe overcut.

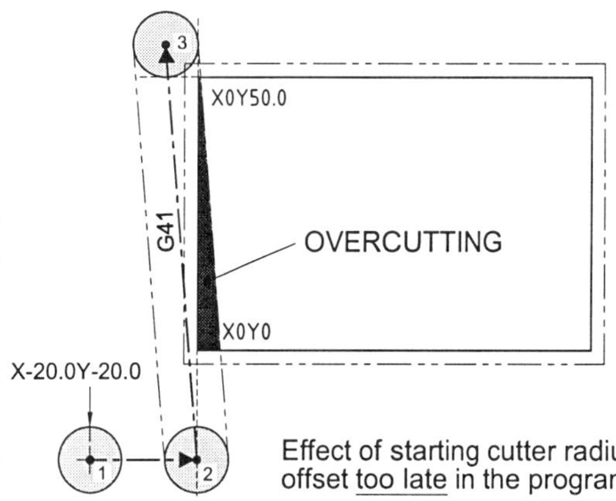

Effect of starting cutter radius offset too late in the program

Example 2 - Cutter radius offset *cancellation* is programmed *too early*:

```
...
N9 Y0                    (POS. 5)
N10 G40 X-20.0           (POS. 6)
N11 G00 Y-20.0 M09       (POS. 7)
N12 G28 ...
. . .
```

Again, the program segment looks fine. The tool was correctly compensated all the way to position 5. As the G40 cancellation command is programmed during the tool motion from position 5 to position 6, the result will also be a severe overcutting.

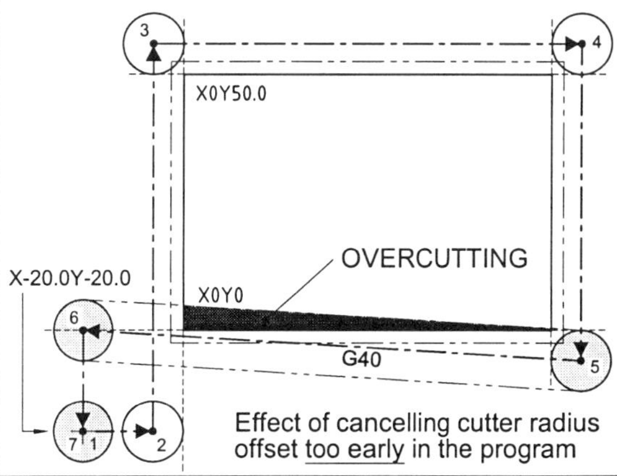

Effect of cancelling cutter radius offset too early in the program

Offset Start or End on an Arc

One of the conditions mentioned earlier in this chapter was that the cutter radius offset *cannot* be started or cancelled on an arc. For example, all these program statements are illegal and will result in an error condition (alarm):

```
G41 G02 X.. Y.. I.. J.. D..
G41 G03 X.. Y.. I.. J.. D..
G41 G02 X.. Y.. R.. D..
G41 G03 X.. Y.. R.. D..

G40 G02 X.. Y.. I.. J..
G40 G03 X.. Y.. I.. J..
G40 G02 X.. Y.. R..
G40 G03 X.. Y.. R..
```

Of course, if G42 is used instead of the more common G41, the result will be the same. This restriction only applies to the *actual start or cancellation* - it does not affect any arcs programmed between the G41/G42 D.. command and the G40 command.

Tool Nose Radius Offset

Using a cutter radius offset on CNC lathes is usually called *tool nose radius compensation* or *tool nose radius offset*. Although there are many similarities with cutter radius offset used for milling, turning and lathe boring tools are different - they cannot be represented just by a circle, because they have cutting edges oriented into different directions.

Command Point and Radius Center

As an example, take the two most common lathe tools - typical turning tool and typical boring bar - and focus on the way they are setup on the lathe. Particularly, focus on the way the insert is oriented in respect to the tool nose radius center. In the chapter ***Turning and Boring*** are additional details related to the tool nose radius. The illustration shows a typical tool orientation of the two common tools for CNC lathes of the *rear type*.

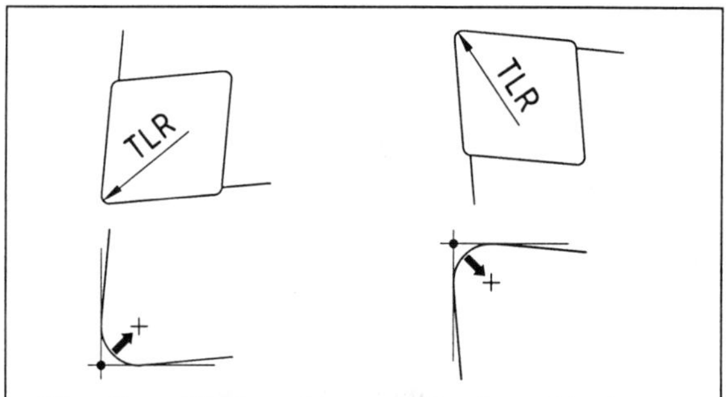

Direction of tool center from command point (external - turning)

Direction of tool center from command point (internal - boring)

Although the two tools selected for the example are common, they do not represent all options that exist for other tools used on CNC lathes. Fanuc provides a column *T* for the *tip orientation* number. Typical offset screen for either *Geometry* offset of *Wear* offset will have five columns:

No.	X	Z	R	T
01				
02				
03				
04				
...				

There are normally two screens, identical in layout. The offset number for *Geometry* offset is preceded by the letter *G*, the offset number for the *Wear* offset is preceded by the letter *W*, both omitted in the illustration. The total of ten entries, from 0 to 9, may be entered into the *T* column of the offset screen. Note that the tip number will always be the same for both geometry and wear offsets.

Tool Tip Orientation

The chart at right identifies the directions of tool radius center orientation, measured from the command point to the tool center for CNC lathes of the ***REAR*** type:

Tool tip orientation numbers 0 or 9 define the command point at the center of the tool nose radius.

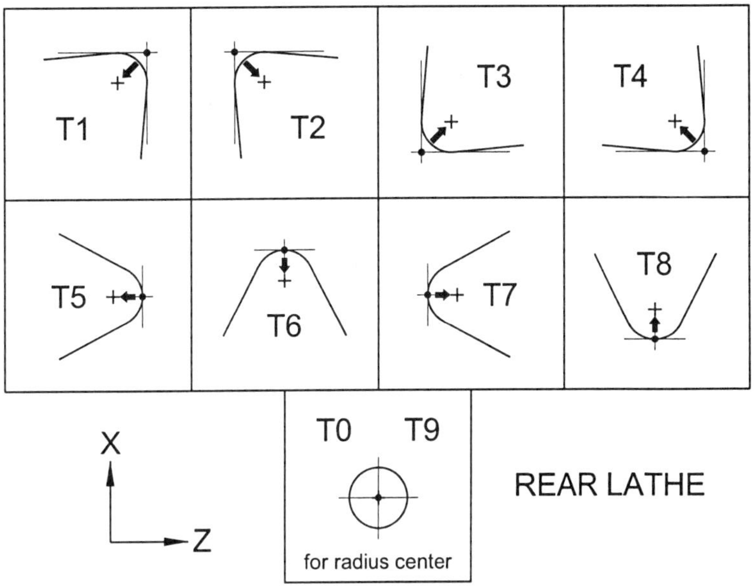

The chart at right identifies the directions of tool radius center orientation, measured from the command point to the tool center for CNC lathes of the ***FRONT*** type:

In Chapter 20, covering the subject of Four-axis Lathe, look for specific practical applications)

Tool tip orientation numbers 0 or 9 define the command point at the center of the tool nose radius.

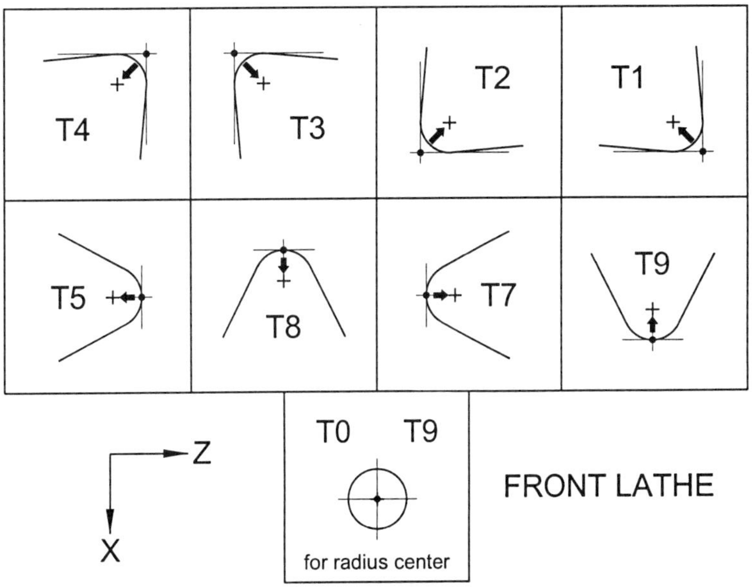

Tool tip orientation number *T* is only effective in G41/G42 mode and must be set at the control before the program is processed

Common Tool Nose Radius Errors

All common causes of errors listed for milling operations apply equally to lathes. However, one error - and probably the most common error - is *insufficient clearance*. There are two specific situations where this error happens more often. Look carefully at three typical examples (0.8 tool radius assumed - equivalent to 0.0313 inches).

Error in Facing

Although it is not necessary to program the radius offset for facing towards the part centerline, it is commonly done to be consistent with other contours. Once such a cut is completed, the tool typically enters machining along the Z-axis or XZ-axes. It is the *method of transition* from one direction to another where the error happens.

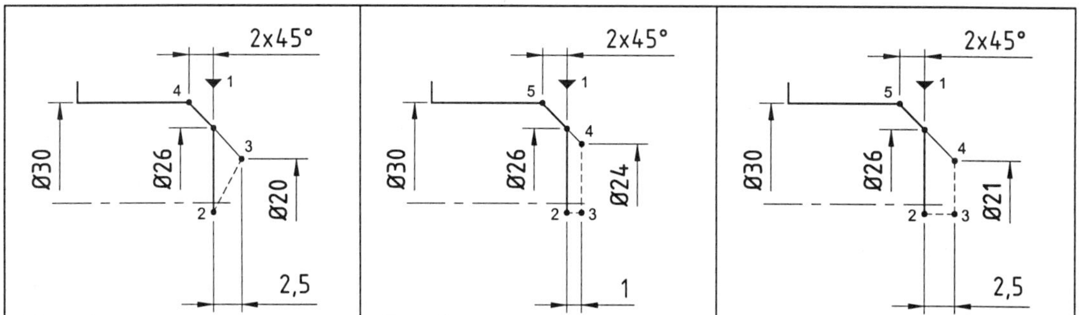

Error 1: Angular departure error | Error 2: Insufficient clearance | Correct method

ERROR 1 - shows an *2.5 mm angular* departure from the face end. If the program contains G41 for facing and G42 for chamfer and diameter turning, the face will *not* be completed:

```
(G21)
G41 G00 X35.0 Z0 T....
G01 X-1.8 F0.1
G42 G00 X20.0 Y2.5     (ANGULAR DEPARTURE WILL CAUSE INSUFFICIENT CUTTING)
G01 X30.0 Z-2.0
...
```

Incidentally, the clearance of 2.5 mm in front of the face *is* sufficient for all standard tool nose radiuses - 0.4 mm, 0.8 mm, and 1.2 mm (0.016", 0.032", and 0.047").

ERROR 2 - shows *1 mm Z-axis* departure from the face end. If the program contains G41 for facing and G42 for chamfer and diameter turning, overcutting alarm will be generated:

```
(G21)
G41 G00 X35.0 Z0 T....
G01 X-1.8 F0.1
G00 Z1.0               (1.0 MM CLEARANCE WILL CAUSE OVERCUTTING ALARM)
G42 X24.0
G01 X30.0 Z-2.0
...
```

As the next illustration shows, there will be insufficient cutting for *Error 1* or overcutting in the air for *Error 2*, which the control cannot detect. The illustration also shows the correct application.

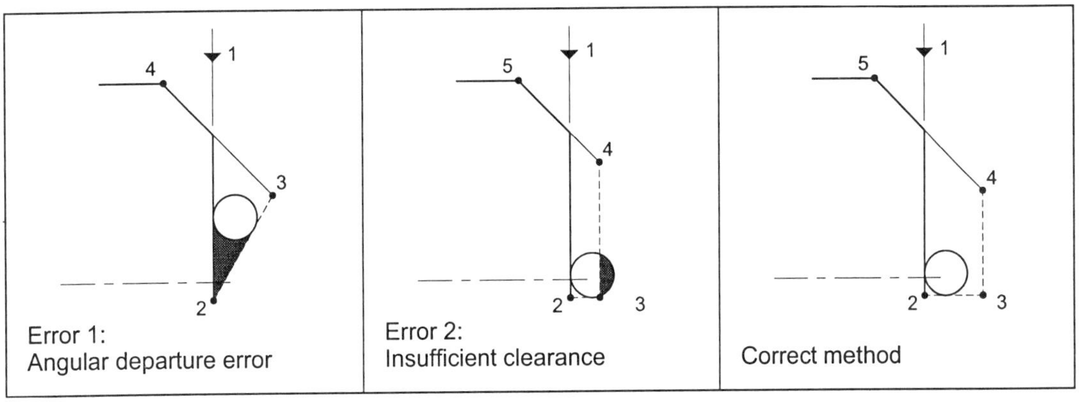

***CORRECT METHOD* - shows *2.5 mm Z-axis* departure from the face end. If the program contains G41 for facing and G42 for chamfer and diameter turning, *NO* alarm will be generated:**

```
G21
...
G41 G00 X35.0 Z0 T....
G01 X-1.8 F0.1
G00 Z2.5                    (2.5 MM CLEARANCE IS SUFFICIENT)
G42 X21.0
G01 X30.0 Z-2.0
...
```

The reason for either error is that the radius of the cutter could not fit in the space provided. In the table, all three common tool nose radiuses are provided with minimum clearances:

Metric			Imperial		
Nose Radius	Minimum X-clearance on DIA	Minimum Z-clearance	Nose Radius	Minimum X-clearance on DIA	Minimum Z-clearance
0.4 mm	1.6 mm	0.8 mm	0.0156 inch	0.0625 inch	0.0313 inch
0.8 mm	3.2 mm	1.6 mm	0.0313 inch	0.1250 inch	0.0625 inch
1.2 mm	4.8 mm	2.4 mm	0.0469 inch	0.1875 inch	0.0938 inch

Keep in mind these are minimum clearances only. It is always a good and practical idea to round up each clearance. Another, and often the best, method is to *always program the maximum rounded clearance for **any** tool nose radius*. If you program 5 mm on diameter and 2.5 mm (or even 3 mm) for Z-clearance in metric, and 0.2 on diameter and 0.1 Z-clearance in inches, no alarm will be generated.

Error in Boring

In boring operations, it is not unusual to have a boring bar with minimum radial clearance. Part programmers normally select boring bar diameter as large as possible for the work, in order to eliminate deflection and/or chatter. After the last cutting motion, an alarm will occur, if there is no sufficient clearance when the tool moves away from the bored contour in X-axis. To avoid problems demonstrated in the above situations, follow two simple rules (also see the table above):

Always program clearance in the X-direction that is greater than QUADRUPLE tool nose radius

Always program clearance in the Z-direction that is greater than DOUBLE tool nose radius

For example, evaluate the clearance X-axis motion in G41 offset mode for the example at right:

```
G21
...
G41 ... T....
...
G02 X28.0 Z-.. R..
G01 Z-34.0 F0.1
G00 x23.0  (5 MM CLEARANCE IS SUFFICIENT)
Z3.0       (RAPID TO Z-FRONT CLEARANCE)
...
```

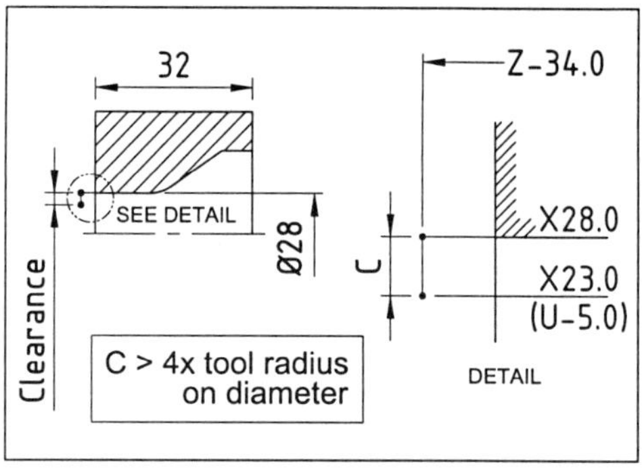

Question - What will be the actual X-motion *per side* (actual clearance) for 0.8 mm tool nose radius?

U-5.0 is 2.5 mm per side - into this clearance must fit 0.8 mm tool nose radius - *twice!*
That means 2.5 - 2 × 0.8 = 0.9 mm actual clearance (about 0.035 inches).

Error in Grooving

When machining large grooves, a special button tool (a grooving tool with a radius) is used to rough and finish the groove by contouring, rather than simple cutting on groove walls and bottom. While standard square grooves may require a small corner break, they *do not* usually require the tool radius offset to be programmed. Grooves containing large radiuses are a little different.

The illustration shows a typical groove shape that has to be contoured. It also shows two suitable methods of contouring, one with a command point set to the bottom of the grooving tool ('six o'clock position'), the other by programming to the center of the radius. This latter method is always guarantee to work, and still can use radius offset from the wear screen, providing G41 or G42 has been included in the program.

Command point set to a side of the tool is not suitable.

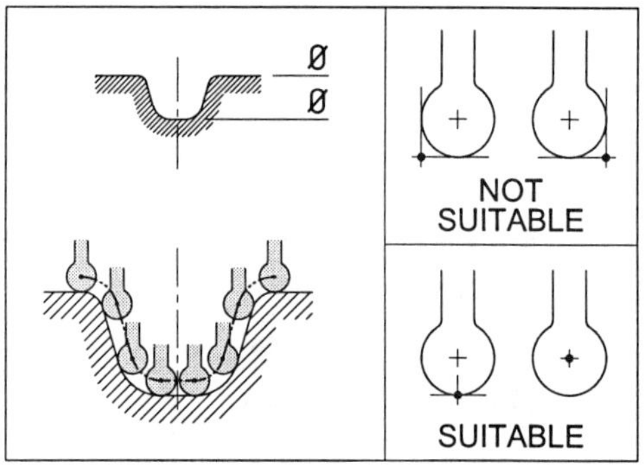

5 PART REVERSAL IN MILLING

Project Description

This special project illustrates a fairly common machining application in programming of CNC machining centers - *a part that has to be reversed within a **single** program*. Intentionally simple, yet very typical, example is shown here to illustrate the concept.

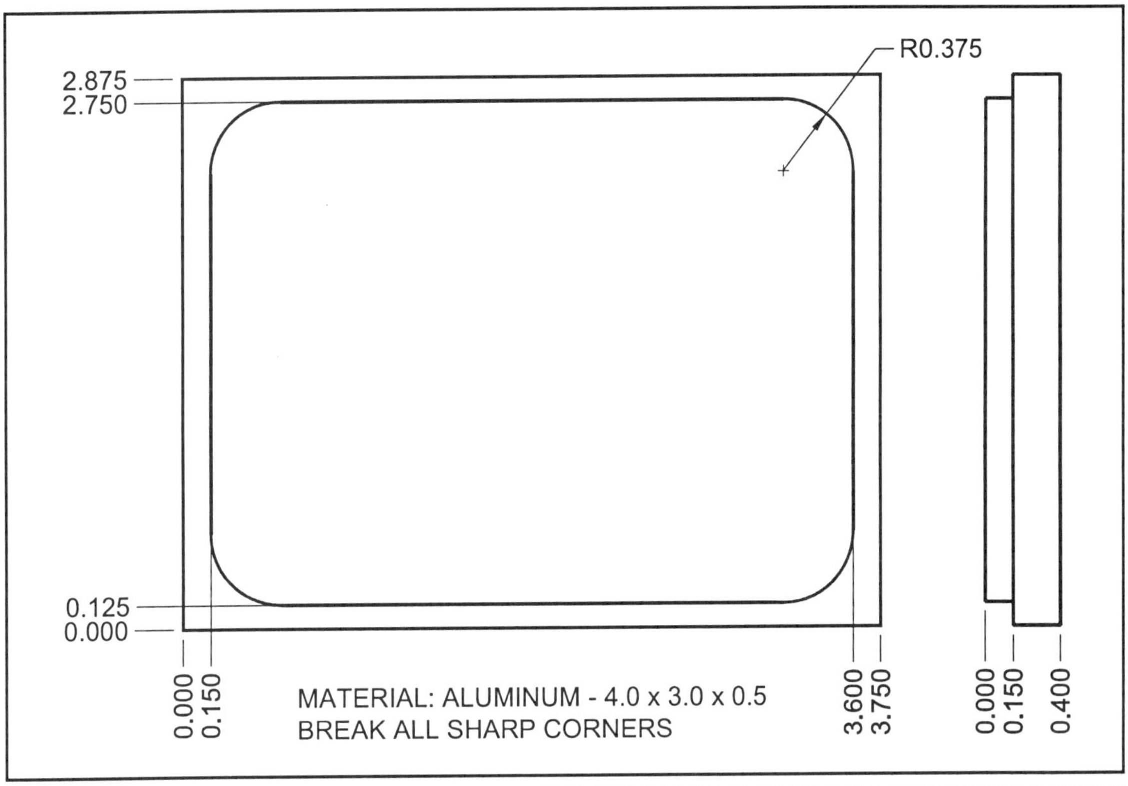

The objective of this project is simple:
All faces and contours of the above part have to be machined in a SINGLE program

Single program is a program that contains all operations required to complete a given part. Some parts require two or more operations, which also means two or more programs. A single part program may include a *part reversal* in the program body, typically associated with the *Program Stop* function M00. The program stop is applied *before* the program end function M30 is reached.

Material and Setup Conditions

◆ **Material used:**

Aluminum plate - 4 × 3 × 0.5 inches

- **Suggested setup:**

For the actual machine tool setup, standard machinist vise is suitable, with the aluminum stock mounted horizontally, resting on some solid support (such as parallels) that provides enough clearance for the machining depths.

The stock lower left corner is located against the lower left corner of the vise.

This lower left corner should be set against the fixed *(non-moving)* jaw, along the *X* axis and a stopper along the *Y* axis.

See illustration at right

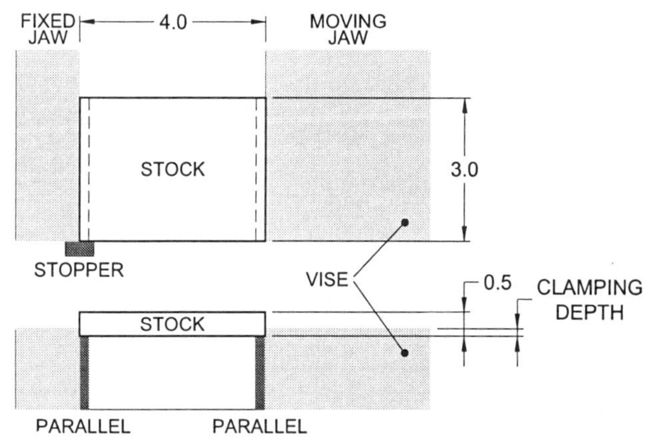

Schematic representation only

Cutting Tools

To illustrate the concept of part reversal, only three cutting tools will be used:

- **T01 = ∅2.5 Face mill** ***H01 tool length offset***
- **T02 = ∅0.5 End mill** ***H02 tool length offset*** *and* ***D52 cutter radius offset***
- **T03 = ∅0.375 Chamfer mill** ***H03 tool length offset*** *and* ***D53 cutter radius offset***

The tool T03 (chamfer mill) is not required for the purposes of this project, but it is a recommended tool to include, in order to break sharp corners wherever possible.

Material Removal

The next decision relates to the removal of excessive material. Since the machining takes place at *both* sides of the part length *and both* sides of the part width, how much material is to be removed from one side and how much from the other? Unless some special circumstances interfere, the common approach would be an *equal* removal of material along the contour. Consider the stock size and the overall finished sizes (length and width):

- **Stock length = 4.0 Part length = 3.750**
- **Stock width = 3.0 Part width = 2.875**

As the drawing indicates, the finished base rectangle of 3.75 × 2.875 × 0.4 is to be machined from existing material stock of 4 × 3 × 0.5 size. Let's look at the *XY* dimensions first. As all four sides have to be machined, it makes sense to remove equal amount of material from each side:

- **Material per side machined along the X axis: (4.0 - 3.750) / 2 = 0.1250**
- **Material per side machined along the Y axis: (3.0 - 2.875) / 2 = 0.0625**

Tool T02 (∅0.5 end mill) can easily handle 0.125 width of cut along the *X* axis and 0.0625 width of cut along the *Y* axis - in a single contour cut. There is also another consideration - whether the tool can clean the four corners in a single contour or not. Tool diameter 0.5 is sufficient for this contour.

To calculate the material removal for the *Z* axis, the methods are similar:

- **Stock height = 0.5**
- **Part height = 0.4**

There is 0.1 difference between the stock thickness and the finished part thickness. Following the decisions made for the *X* and *Y* axes earlier, 0.05 should be removed from each face of the stock. *However, there is small catch!* The vise holds a significant portion of the material that is not accessible by the tool without a collision. Equally important consideration is the maximum depth of cut for the first clamping.

The drawing shows that the bottom base height is 0.4-0.15=0.25. In order to clean up the sides (edges) of the base, the tool has to cut at the depth that is *greater* than 0.25 - such as 0.275 used in this project. That leaves 0.225 theoretical maximum for actual clamping depth, providing there is no other face cut required - *which is NOT the case here.*

By splitting the remaining stock equally, the maximum clamping depth in the first clamping will be 0.5-0.05-0.275=0.175. Depending on the actual height of parallels, it may be enough to hold the part securely. In many cases, much better method would be to take a minimum depth of face cut for the first clamping, and remove the rest of the material during the second clamping, even if it means more than one face cut. In this project, the first depth of cut will be 0.025, and the maximum clamping depth will be 0.5-0.025-0.275=0.200. Perhaps not too significant a change, but it provides more flexibility and robustness to the part holding. Two face cuts will be programmed after part reversal (second clamping), to provide a more versatile example.

In both clampings, the *Z* axis program zero will be at the top of the actual *finished* face!

Detailed illustrations are provided for each version of the setup.

Machining Process

As in most machining applications, there are more than one machining method available. In order to illustrate a certain programming concept, only the most suitable method can be selected. In this project, all six faces (four sides, the top and the bottom) of the stock have to be machined in a single setup - *i.e., in one program* - practically, it means the CNC program requires two sections (representing the *first* and the *second* clamping), with a program stop function M00 and related part reversal in between.

Decisions relating to the material removal have been made already, and the machining will be distributed for each clamping according to this procedure:

Clamping 1

1. **Face the part - minimum cleanup of 0.025 (two passes across at the same depth)**
2. **Finish rectangular contour to the depth of 0.275 - one peripheral cut only**
3. **Chamfer rectangular contour - minimum corner break (same contour as in *Item 2*)**
4. **Stop program (M00) and reverse the part (flip upside down)**

Clamping 2

5. **Face the part to final thickness of 0.4 - two face cuts (two passes across at each depth)**
6. **Finish contour to size - one peripheral cut only (includes 4 corner radii)**
7. **Chamfer finished contour - minimum corner break (same contour as in *Item 6*)**

This programming method is reasonable and also best suited to illustrate the part reversal concepts. Although only a milling example has been provided, the same concepts apply to turning as well.

Program Zero Selection

When a part is reversed in the middle of a single program, the major programming decision is to select the program zero (part origin) properly. Normally, the program zero is selected at a certain physical location of the finished part, such as a corner or a hole center - a location that does not change and can be easily measured by an edge finder, corner finder, a dial indicator or by similar means. The actual program zero location is established during setup at the CNC machine, and is stored in the offset registry. The program does not contain the actual measured distances, but a work offset reference, such as G54. By definition, the work offset is the *distance from the machine zero to the program zero, measured along the X and Y axes.*

If there is no part reversal in the program, no special consideration is required - the program zero corresponds to the actual measured location, often determined by the origin of drawing dimensions.

First Clamping

Now, the decision has to be made regarding the program zero location. Can the part zero be the lower left corner of the vise that is easy to reach with an edge finder or a corner finder? Or should it be the actual corner of the finished part, still hidden within the material? Consider the following comparison with programmed *XY* coordinates provided for each corner of the part base:

- ***Left image*** **... the work offset is set to the actual corner of the stock (vise), *NOT* the finished contour**
- ***Right image*** **... the work offset is set to the actual corner of the finished contour**

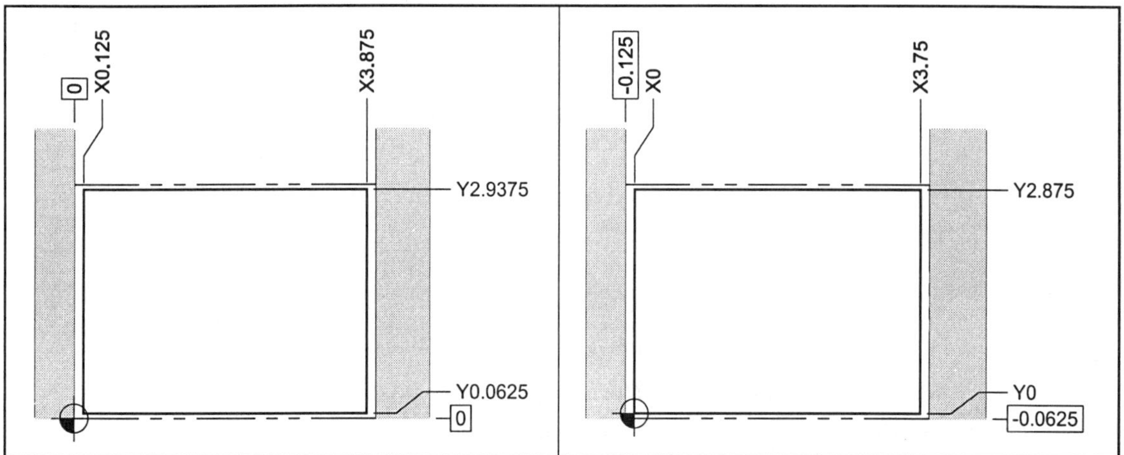

WORK OFFSET IS ALWAYS MEASURED FROM MACHINE ZERO TO PART ZERO

It is important to realize that both settings will produce a correctly machined part, providing both the program and the setup are also correct. That brings up an important question - *which setting is better?* Consider the left illustration on the previous page. There is nothing wrong with this method, but the program must reflect the setup. Here is a portion of the program, using tool T02:

```
(PROGRAM MATCHING THE LEFT ILLUSTRATION)
G90 G54 G00 X-0.6 Y-0.6 S1800 M03          (G54 TO THE LL CORNER OF THE VISE)
G43 Z1.0 H02 M08
G01 Z-0.275 F25.0
G41 X0.125 D52                              (SUGGESTED D52 = 0.25)
Y2.9375 F8.0
X3.875
Y0.0625
X-0.6
G40 G00 Y-0.6
Z1.0
```

Although the program is technically correct, it also offers a very impractical programming approach. The reason is that the *XY* coordinates are not calculated from the drawing origin, but from the *setup origin*. The programmer has to calculate each and every coordinate separately, always considering the amount of stock to be removed in each axis. If this reason appears trivial, consider situation where the toolpath is not just a simple rectangle, but a complex contour with many angles and radii. Compare the previous program with the next one, reflecting the setup shown in the right illustration on the previous page:

```
(PROGRAM MATCHING THE RIGHT ILLUSTRATION)
G90 G54 G00 X-0.6 Y-0.6 S1800 M03          (G54 TO THE LL CORNER OF THE VISE)
G43 Z1.0 H02 M08
G01 Z-0.275 F25.0
G41 X0 D52                                  (SUGGESTED D52 = 0.25)
Y2.875 F8.0
X3.75
Y0
X-0.6
G40 G00 Y-0.6
Z1.0
```

Since the common method is to program drawing dimensions at all times, the first clamping will use the concepts of the version shown on the right in the above illustration.

Second Clamping

Once the part is reversed, there is no problem to match the corner of the now finished part, with the lower left corner of the vise.

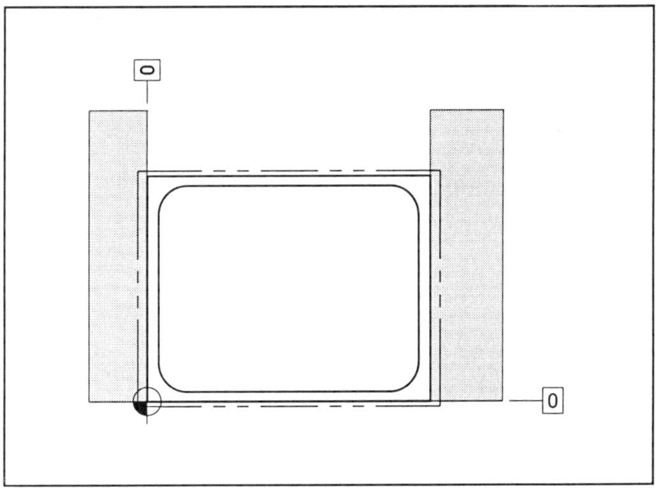

The illustration at right shows the setup for second clamping

So far, the focus was exclusively on the *XY* axes. A similar explanation is also needed for the tool length setting, along the Z-axis. Since this setting will be established by an offset measurement, let's look at all methods of programming a part reversal for milling. Three different methods will be presented. All three methods will be using the lower left corner of the vise and top face of the *finished* part as the main reference, regardless of the clamping.

Programming Methods

There are at least three programming methods available to handle such a situation. In all cases, the contours in both the first and second clamping will be programmed using the drawing dimensions:

- **Use two WORK OFFSETS - one for the first clamping, another for the second clamping**
- **Use one work offset with G52 LOCAL COORDINATE SYSTEM programmed**
- **Use G10 DATUM SHIFT to adjust the work offset through the program**

All three methods of programming will be discussed in detail after evaluating the part thickness and tool length settings. Note that some of these programming methods will require certain features to be installed in the control system.

Tool Length Settings

In many ways, the setting of the Z axis should be the easiest to understand. The part zero should always be *the top of the finished face*. As both clampings involve some face milling and a non-changing part support (such as parallels used in this example), the Z0 for the first clamping will be *higher* than Z0 for the second clamping. Study the two illustrations shown on this and the following pages:

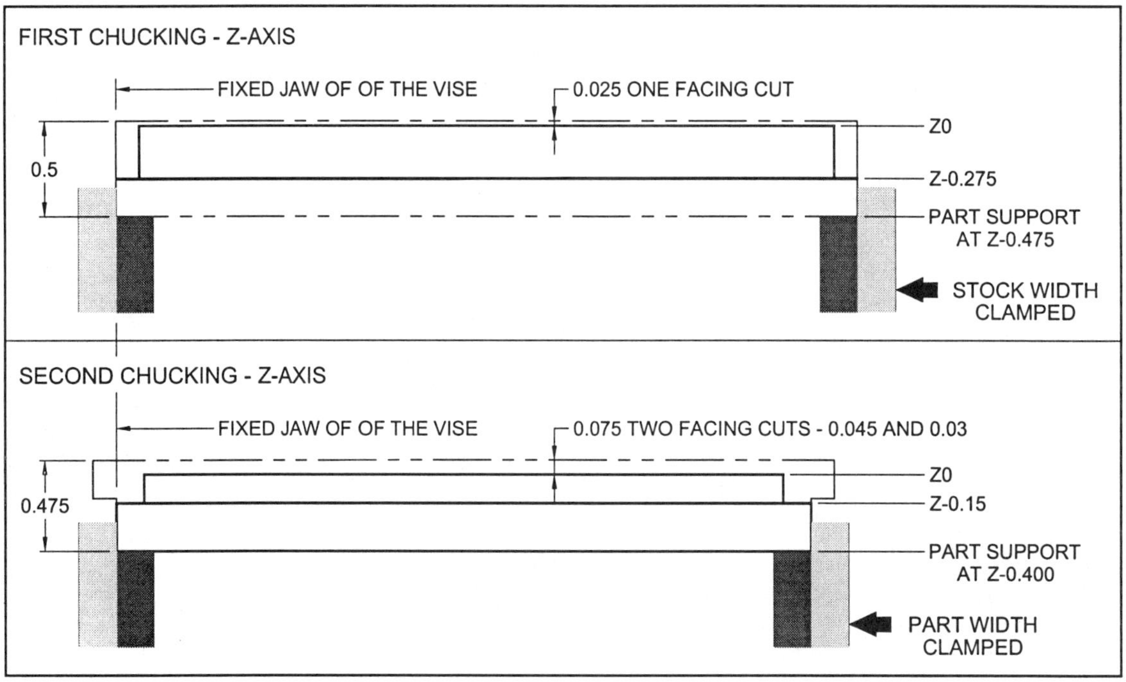

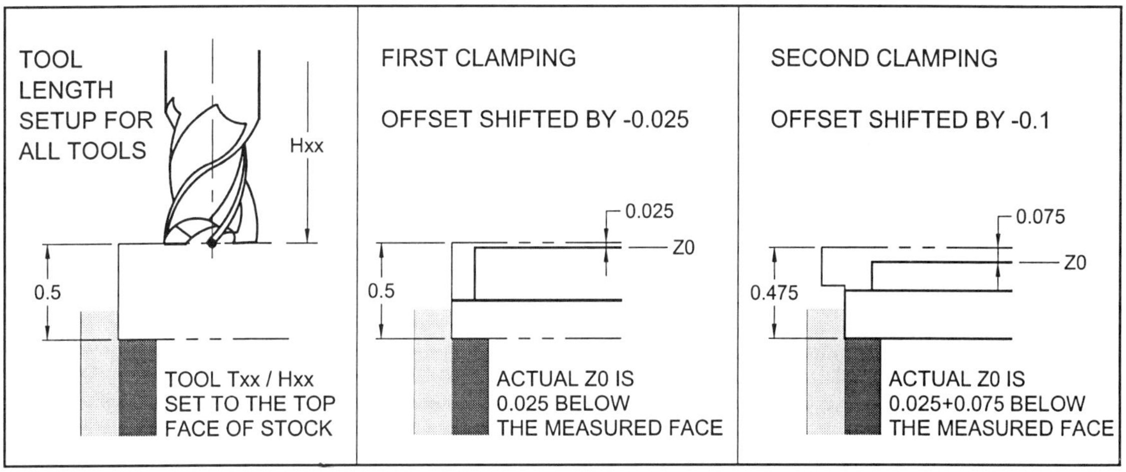

First Clamping

The most important item to remember at this stage is that the tool length offset for each tool is measured only *once* for the *whole* program (*both* clampings). This measurement for each tool is stored in the appropriate tool length offsets registry screens. The tool length measurement in these examples used a common *touch-off* method, that is measured from the Z-axis home position to the top of the stock when it was still 0.5 high (before any machining). As the above illustration demonstrates, the actual *Z0* (of the finished face) is *lower* by -0.025 for the first clamping and an *additional* -0.075 for the second clamping - for the total of -0.100.

This important reality has to be reflected in the entry data of each tool length offset in the registry. For this project, the three tools have touched-off the *original* stock face (before any machining) with these simplified sample measurements:

Tool T01 = -11.0 **Tool T02 = -12.0** **Tool T03 = -11.5**

As is customary, the tool length measurement for T01 will be stored in tool length offset H01 registry, the tool length measurement for T02 will be stored in the H02 registry, and so on for all tools. Since the touch-off method of measuring did *not* touch the real Z0, each tool length offset will have to be adjusted by the difference *between the measured face and the actual Z0 face*, which is 0.025 lower (for the first clamping). The tool length offsets could be changed to these settings:

H01 = -11.025 **H02 = -12.025** **H03 = -11.525**

Of course, every experienced CNC operator knows that there is a much simpler method to achieve the same goal. Rather than adjusting individual settings for each tool individually, only *a single* adjustment can be made for *all tools*, regardless of how many are used in the program. This method requires changing the Z setting in the *active work offset* (such as G54), while leaving the original tool length offsets intact:

G54		LENGTH OFFSET	
X	0.0000	H01	-11.0000
Y	0.0000	H02	-12.0000
Z	-0.0250	H03	-11.5000

☞ **Keep in mind that this solution covers only the *first* clamping and will *not* work for the second clamping**

Second Clamping

Once the part is reversed (flipped over), two *additional* facing cuts will be machined *above* the new Z0 - this Z0 will now be *another* 0.075 lower then for the first clamping, which presents a serious problem - where to register this additional 0.075 difference?

One possible solution would be to use different *Hxx* numbers for the three tools, for example H11, H22, H33. Although correct in principle, this method of programming is extremely impractical and could be very confusing to the CNC operator. A much improved method is to use ***two work offsets***! Incidentally, this method will also solve the problem of the *XY* settings, not yet discussed.

Using the WORK OFFSET Method - G54-G55

Since both the *XY* and the *Z* offset settings have to be considered together, let's start with a totally new illustration of the offsets screens - this time showing *two* work offsets registry windows and *one* tool length offset registry window (for three tools):

G54		G55		LENGTH OFFSET	
X	0.0000	X	0.0000	H01	-11.0000
Y	0.0000	Y	0.0000	H02	-12.0000
Z	0.0000	Z	0.0000	H03	-11.5000

It is worth noting that both work offsets G54 and G55 are empty at the moment - they have zero values - and there is a good reason for it. Not every programming job requires part reversal in the middle of the program. In fact, the majority of jobs are only a single clamping jobs. In many shops, depending on the type of work, it is quite common to keep a certain holding device (a vise, for example) mounted on the machine table for as long as it is used or is not in the way, in order to save setup time. In such situations, the lower left corner of the vise will be set only once and registered in - *yes, **in the work offset G54***! It makes sense and it makes the setup more efficient, if the type of work allows it. The next illustration show such a setting:

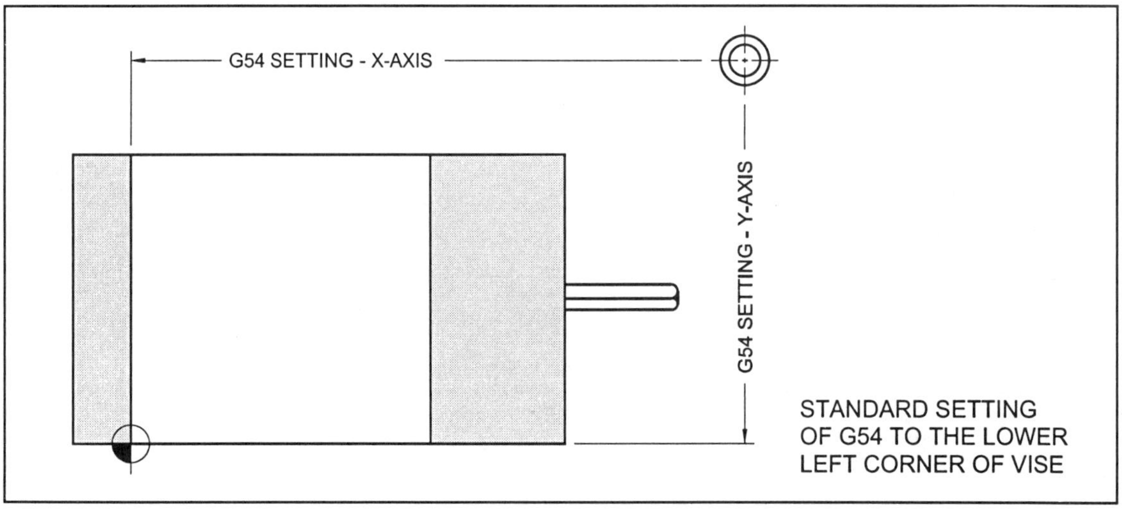

Considering this possibility (and actual reality in many CNC shops), it does not make much sense to change the work offset setting for the G54 values. In our example, the G54 work offset setting will only be used for the *second clamping* (measured to the lower left corner of the vise), and *another* work offset - G55 - will be used for the *first* clamping! The illustration shows the general concept:

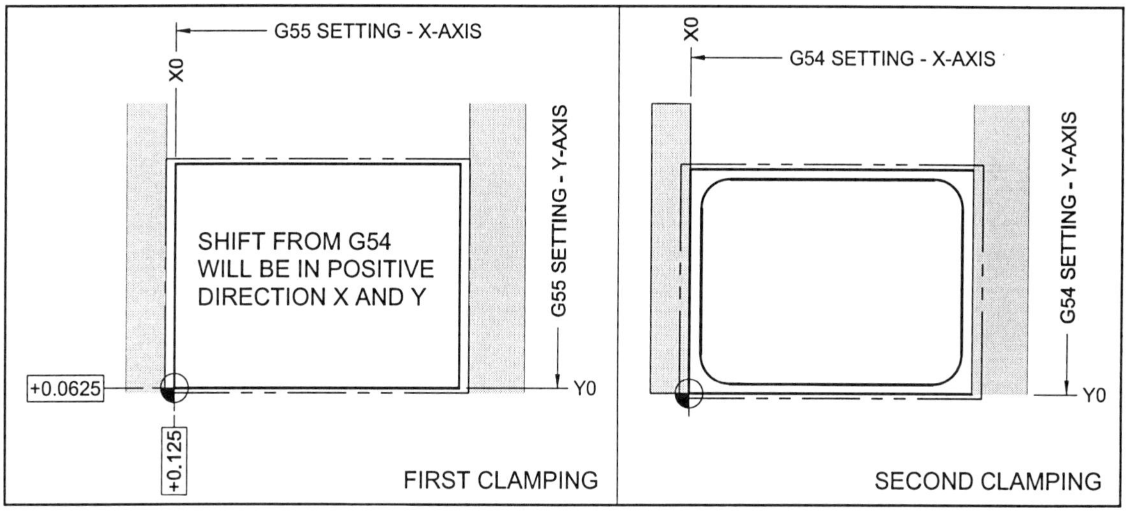

To conclude the subject of part reversal programming using G54 and G55 work offsets method, the sample offset settings will be as shown (the tool length offsets remain the same):

G54		G55		LENGTH OFFSET	
X	-13.6000	X	-13.4750	H01	-11.0000
Y	-11.4000	Y	-11.3375	H02	-12.0000
Z	-0.1000	Z	-0.0250	H03	-11.5000

G54 setting in the above example represents the ***actual measured distance from machine zero to part zero*** (lower left corner of the vise). *G55 **must** be calculated from the existing G54 setting!*

Note the *Z* axis settings within the G54 and G55 registers. Normally, the Z-setting in the work offset is not used - it is equal to zero. If there is a value for the Z-setting other than zero, it will be *added* to the length offsets of all tools that use the particular work offset. Think of this setting as applying to all tool lengths rather than any single tool length.

In this practical example, the actual tool length offset was measured to the top face of the *stock* (not to the finished face). The G55 Z-setting indicates how much material will be faced off during the first clamping. The Z-setting for G54 work offset must contain an *additional* 0.075 (see earlier calculations), which means the combined setting must be Z-0.100.

This concludes the section of the first method of programming a part reversal, using *work offsets*. Using this method, the CNC operator is required to do quite a bit of work and human errors are quite possible. If this program is to be used repeatedly, it is better to use one of the other two methods, where the part program itself will control all differences between the first and the second clamping.

Common Toolpath

Regardless which method of programming is used, the toolpath is the identical in all of them for easier comparison of the various methods. The following two illustrations show the toolpath for both clampings along *X* and *Y* axes (contouring only).

The first illustration shows the initial setup and toolpath for the *first* clamping of the part:

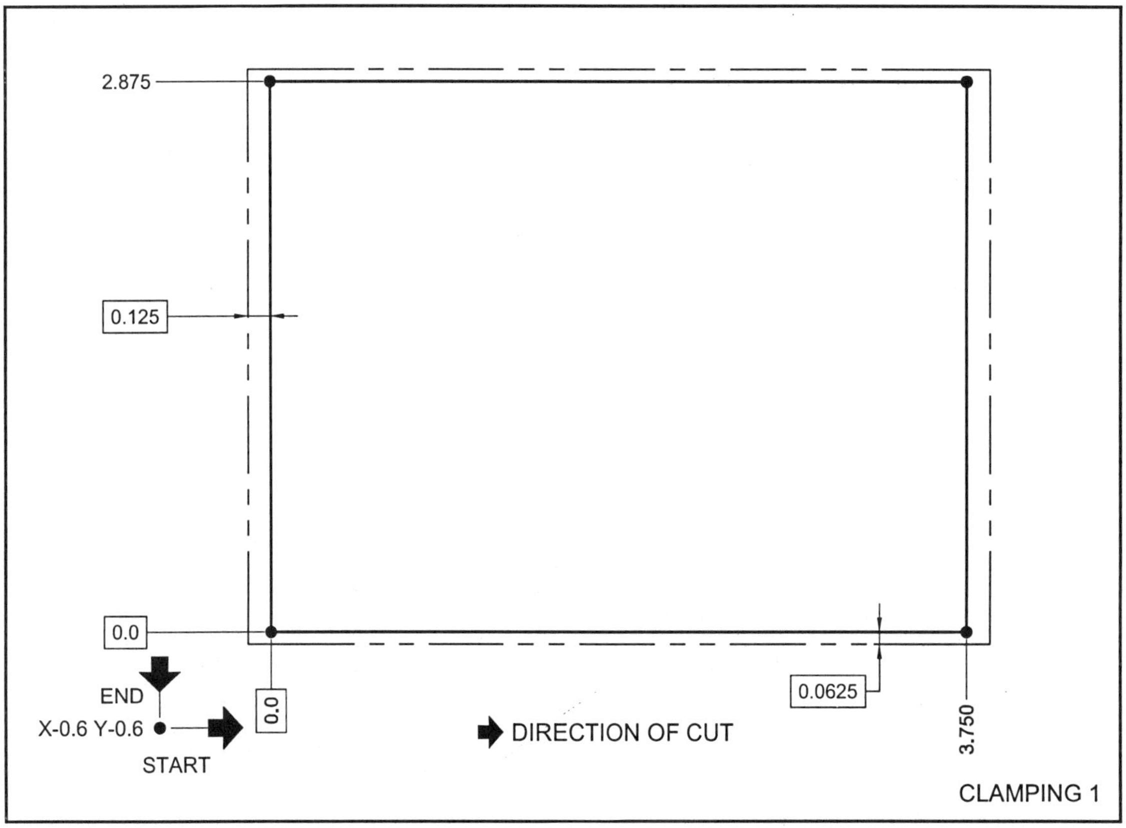

The tool starts at the location X-0.6 Y-0.6, which is suitable enough to accommodate the cutter diameter and provide sufficient clearance - particularly away from the initial stock.

Several common points should be considered:

1. **Part zero is on the future finished corner at the lower left of the part**
2. **Cutting will be done with right hand tools in climb milling mode**
3. **Offset values - when listed - are only suggested values or ideal values**
4. **Tool numbers and offsets are the same unless change is necessary**

The second illustration shows the initial setup and toolpath for the *second* clamping of the part. This is the result of the part being flipped over (upside down) within the same setup (after M00 in the program). General point from the first part applies here as well.

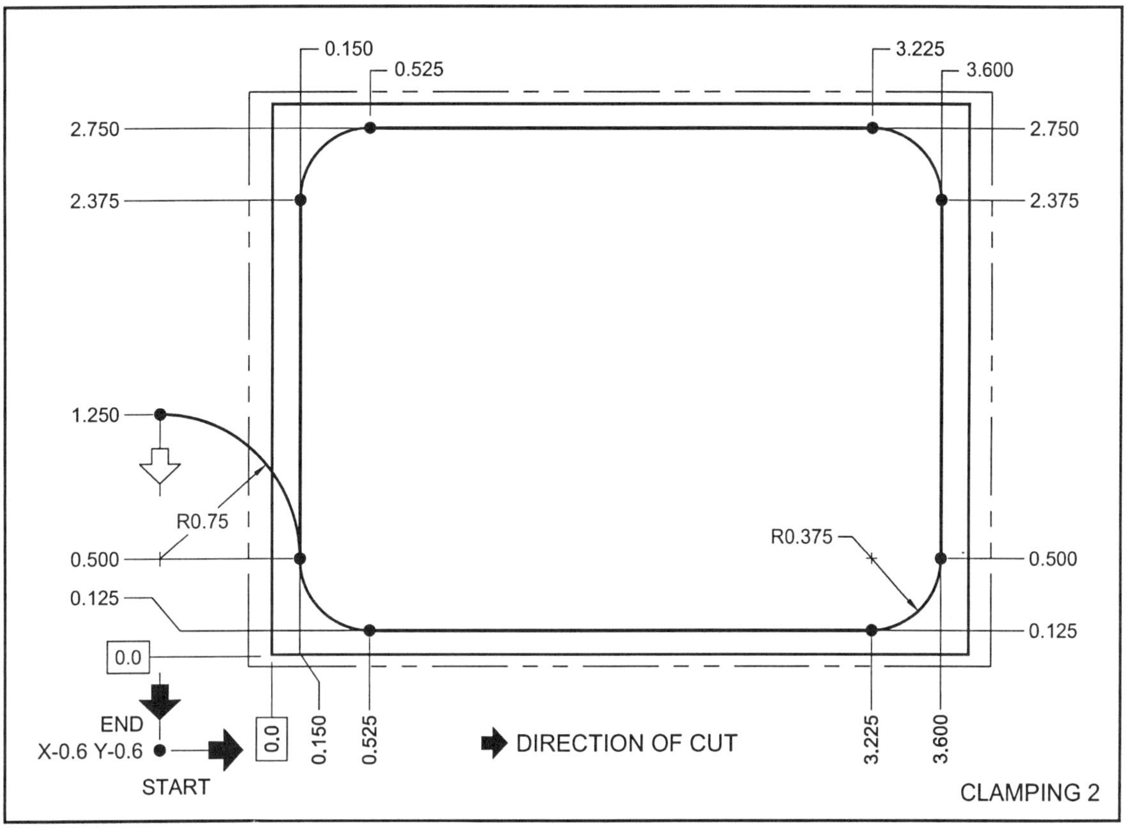

Program Listing - WORK OFFSETS G54-G55

```
(PART REVERSAL - USING WORK OFFSETS G54-G59)

(G54 = THE LOWER LEFT CORNER OF THE VISE - USED FOR CLAMPING 2 ONLY)
(ALL TOOL LENGTHS SET TO THE TOP FACE OF THE STOCK)

(===>>> CLAMPING 1 <<<===)
(G55 X.. Y.. SETTING ...)
(... TO BE EXACTLY 0.125 SHORTER IN X AND 0.0625 SHORTER IN Y THEN G54)
(G55 Z.. SET TO Z-0.025)

(T01 - 2.5 DIA FACE MILL - MINIMUM FACE CUT - 0.025)
N1 G20
N2 G17 G40 G80 T01
N3 M06
N4 G90 G55 G00 X5.5 Y1.0 S2500 M03 T02
N5 G43 Z1.0 H01 M08
N6 G01 Z0 F50.0
N7 X-1.5 F10.0
N8 G00 Y2.0
N9 G01 X5.5
N10 G00 Z1.0 M09
N11 G28 Z1.0 M05
N12 M01                          (CHECK THICKNESS - SHOULD BE 0.475)
```

```
(T02 - 0.5 DIA END MILL - CONTOUR BASE RECTANGLE TO SIZE)
N13 T02
N14 M06
N15 G90 G55 G00 X-0.6 Y-0.6 S2600 M03 T03
N16 G43 Z1.0 H02 M08
N17 Z0.1
N18 G01 Z-0.275 F50.0
N19 G41 X0 D52                  (SUGGESTED D52 = 0.2500)
N20 Y2.875 F10.0
N21 X3.75
N22 Y0
N23 X-0.6
N24 G40 G00 Y-0.6
N25 Z1.0 M09
N26 G28 Z1.0 M05
N27 M01                         (CHECK - LENGTH=3.750 WIDTH=2.875)

(T03 - 0.375 DIA CHAMFER MILL - 0.010-0.015 CHAMFER)
N28 T03
N29 M06
N30 G90 G55 G00 X-0.6 Y-0.6 S3200 M03 T01
N31 G43 Z1.0 H03 M08
N32 Z0.1
N33 G01 Z-0.1 F50.0
N34 G41 X0 D53                  (SUGGESTED D53 = 0.1)
N35 Y2.875 F11.0
N36 X3.75
N37 Y0
N38 X-0.6
N39 G40 G00 Y-0.6
N40 Z1.0 M09
N41 G28 Z1.0 M05
N42 X-2.0 Y9.0                  (PART REVERSAL POSITION)
N43 M00                         (CHECK CHAMFER - FLIP PART UPSIDE DOWN)

(===>>> CLAMPING 2 <<<===)
(G54 X.. Y.. SET TO THE LOWER LEFT CORNER OF THE VISE)
(G54 Z.. SET TO Z-0.1)
(T01 - 2.5 DIA FACE MILL - TWO FACE CUTS - 0.045 AND 0.03)
N44 T01
N45 M06
N46 G90 G54 G00 X5.5 Y1.0 S2500 M03 T02
N47 G43 Z1.0 H01 M08
N48 G01 Z0.03 F50.0             (REMOVES 0.045 OFF THE TOP FACE)
N49 X-0.5 F10.0                 (CURRENT PART THICKNESS = 0.430)
N50 Y2.0
N51 X5.5
N52 G00 Y1.0
N53 G01 Z0                      (REMOVES FINAL 0.030 OFF THE TOP FACE)
N54 X-1.5                       (FINAL PART THICKNESS = 0.400)
N55 G00 Y2.0
N56 G01 X5.5
N57 G00 Z1.0 M09
N58 G28 Z1.0 M05
N59 M01                         (CHECK PART THICKNESS - SHOULD BE 0.400)

(T02 - 0.5 DIA END MILL - CONTOUR ROUNDED RECTANGLE TO SIZE)
N60 T02
N61 M06
N62 G90 G54 G00 X-0.6 Y-0.6 S2600 M03 T03
N63 G43 Z1.0 H02 M08
N64 Z0.1
N65 G01 Z-0.15 F50.0
N66 G41 X0.15 D52               (SUGGESTED D52 = 0.2500)
```

```
N67 Y2.375 F8.0
N68 G02 X0.525 Y2.75 I0.375 J0
N69 G01 X3.225
N70 G02 X3.6 Y2.375 I0 J-0.375
N71 G01 Y0.5
N72 G02 X3.225 Y0.125 I-0.375 J0
N73 G01 X0.525
N74 G02 X0.15 Y0.5 I0 J0.375
N75 G03 X-0.6 Y1.25 I-0.75 J0
N76 G40 G00 Y-0.6
N77 Z1.0 M09
N78 G28 Z1.0 M05
N79 M01                          (ROUNDED RECTANGLE LGTH=3.450 WIDTH=2.625)

(T03 - 0.375 DIA CHAMFER MILL - 0.010-0.015 CHAMFER)
N80 T03
N81 M06
N82 G90 G54 G00 X-0.6 Y-0.6 S3200 M03 T01
N83 G43 Z1.0 H03 M08
N84 Z0.1
N85 G01 Z-0.1 F50.0
N86 G41 X0.15 D53                (SUGGESTED D53 = 0.1)
N87 Y2.375 F11.0
N88 G02 X0.525 Y2.75 I0.375 J0
N89 G01 X3.225
N90 G02 X3.6 Y2.375 I0 J-0.375
N91 G01 Y0.5
N92 G02 X3.225 Y0.125 I-0.375 J0
N93 G01 X0.525
N94 G02 X0.15 Y0.5 I0 J0.375
N95 G03 X-0.6 Y1.25 I-0.75 J0
N96 G40 G00 Y-0.6
N97 Z1.0 M09
N98 G28 Z1.0 M05
N99 X-2.0 Y9.0                   (PART REMOVAL POSITION)
N100 M30
%
```

Program Listing - WORK OFFSETS G54-G55 - with Subprograms

Subprograms offer additional flexibility and significantly shorten the program length. The following program is identical to the previous one, but it uses subprograms for repeated machining operations:

```
(PART REVERSAL - USING WORK OFFSETS G54-G59 WITH SUBPROGRAMS)

(G54 = THE LOWER LEFT CORNER OF THE VISE - USED FOR CLAMPING 2 ONLY)
(ALL TOOL LENGTHS SET TO THE TOP FACE OF THE STOCK)

(===>>> CLAMPING 1 <<<===)
(G55 X.. Y.. SETTING ...)
(... TO BE EXACTLY 0.125 SHORTER IN X AND 0.0625 SHORTER IN Y THAN G54)
(G55 Z.. SET TO Z-0.025)

(T01 - 2.5 DIA FACE MILL - MINIMUM FACE CUT - 0.025)
N1 G20
N2 G17 G40 G80 T01
N3 M06
N4 G90 G55 G00 X5.5 Y1.0 S2500 M03 T02
N5 G43 Z1.0 H01 M08
N6 G01 Z0 F50.0
N7 X-1.5 F10.0
```

```
N8 G00 Y2.0
N9 G01 X5.5
N10 G00 Z1.0 M09
N11 G28 Z1.0 M05
N12 M01                        (CHECK THICKNESS - SHOULD BE 0.475)

(T02 - 0.5 DIA END MILL - CONTOUR BASE RECTANGLE TO SIZE)
N13 T02
N14 M06
N15 G90 G55 G00 X-0.6 Y-0.6 S2600 M03 T03
N16 G43 Z1.0 H02 M08
N17 Z0.1
N18 G01 Z-0.275 F50.0
N19 M98 P7101 D52 F10.0        (SUGGESTED D52 = 0.2500)
N20 G00 Z1.0 M09
N21 G28 Z1.0 M05
N22 M01                        (CHECK - LENGTH=3.750 WIDTH=2.875)

(T03 - 0.375 DIA CHAMFER MILL - 0.010-0.015 CHAMFER)
N23 T03
N24 M06
N25 G90 G55 G00 X-0.6 Y-0.6 S3200 M03 T01
N26 G43 Z1.0 H03 M08
N27 Z0.1
N28 G01 Z-0.1 F50.0
N29 M98 P7101 D53 F11.0        (SUGGESTED D53 = 0.1)
N30 G00 Z1.0 M09
N31 G28 Z1.0 M05
N32 X-2.0 Y9.0                 (PART REVERSAL POSITION)
N33 M00                        (CHECK CHAMFER - FLIP PART UPSIDE DOWN)

(===>>> CLAMPING 2 <<<===)
(G54 X.. Y.. SET TO THE LOWER LEFT CORNER OF THE VISE)
(G54 Z.. SET TO Z-0.1)

(T01 - 2.5 DIA FACE MILL - TWO FACE CUTS - 0.045 AND 0.03)
N34 T01
N35 M06
N36 G90 G54 G00 X5.5 Y1.0 S2500 M03 T02
N37 G43 Z1.0 H01 M08
N38 G01 Z0.03 F50.0            (REMOVES 0.045 OFF THE TOP FACE)
N39 X-0.5 F10.0                (CURRENT PART THICKNESS = 0.430)
N40 Y2.0
N41 X5.5
N42 G00 Y1.0
N43 G01 Z0                     (REMOVES FINAL 0.030 OFF THE TOP FACE)
N44 X-1.5                      (FINAL PART THICKNESS = 0.400)
N45 G00 Y2.0
N46 G01 X5.5
N47 G00 Z1.0 M09
N48 G28 Z1.0 M05
N49 M01                        (CHECK PART THICKNESS - SHOULD BE 0.400)

(T02 - 0.5 DIA END MILL - CONTOUR ROUNDED RECTANGLE TO SIZE)
N50 T02
N51 M06
N52 G90 G54 G00 X-0.6 Y-0.6 S2600 M03 T03
N53 G43 Z1.0 H02 M08
N54 Z0.1
N55 G01 Z-0.15 F50.0
N56 M98 P7102 D52 F8.0         (SUGGESTED D52 = 0.2500)
N57 G00 Z1.0 M09
N58 G28 Z1.0 M05
N59 M01                        (ROUNDED RECTANGLE LGTH=3.450 WIDTH=2.625)
```

```
(T03 - 0.375 DIA CHAMFER MILL - 0.010-0.015 CHAMFER)
N60 T03
N61 M06
N62 G90 G54 G00 X-0.6 Y-0.6 S3200 M03 T01
N63 G43 Z1.0 H03 M08
N64 Z0.1
N65 G01 Z-0.1 F50.0
N66 M98 P7102 D53 F11.0          (SUGGESTED D53 = 0.1)
N67 Z1.0 M09
N68 G28 Z1.0 M05
N69 X-2.0 Y9.0                   (PART REMOVAL POSITION)
N70 M30
%

O7101 (SUBPROGRAM FOR CLAMPING 1)
N101 G41 G01 X0
N102 Y2.875
N103 X3.75
N104 Y0
N105 X-0.6
N106 G40 G00 Y-0.6
N107 M99
%

O7102 (SUBPROGRAM FOR CLAMPING 2)
N201 G41 G01 X0.15
N202 Y2.375
N203 G02 X0.525 Y2.75 I0.375 J0
N204 G01 X3.225
N205 G02 X3.6 Y2.375 I0 J-0.375
N206 G01 Y0.5
N207 G02 X3.225 Y0.125 I-0.375 J0
N208 G01 X0.525
N209 G02 X0.15 Y0.5 I0 J0.375
N210 G03 X-0.6 Y1.25 I-0.75 J0
N211 G40 G00 Y-0.6
N212 M99
%
```

Note - if each subprogram is entered separately into the control system via DNC, don't forget to place the % sign (end of file transfer) after each M99 and M30 function.

Using the LOCAL COORDINATES Method - G52

Providing the control unit supports *Local Coordinate System* feature (called by G52 command), this is the best programming method to use for any part reversal application (in milling or turning). It is equally simple to use and understand for both the CNC programmer and the CNC machine operator. G52 command does nothing more than *internally* shifts the part zero *within the current work offset* by a specified amount of one or more of the three axes.

What are the benefits of programming a local coordinate system? There are several. First, there is no need to set and change the work offset at the machine. In this special project, only the G54 work offset will be used - for *both* clampings. Second, there is no need to adjust the Z-setting for the work offset - it can remain at Z0.0000, even if the tools were touched-off at the top of stock. Finally, all tool length offsets are set to their actual measured values. The greatest benefit to the CNC operator is that a program containing G52 does not require any special effort during machine setup. *In a nutshell, the program takes control of all offset changes internally and automatically!*

Here are the proper offset settings, using previously measured dimensions:

G54		LENGTH OFFSET	
X	-13.6000	H01	-11.0000
Y	-11.4000	H02	-12.0000
Z	0.0000	H03	-11.5000

Keep in mind that G52 command is described as a *'local'* coordinate system for a very good reason - *it cannot exist on its own* - it needs an active work offset of the G54-G59 series. In this case, the active - or current - work offset will be G54 - for *both* clampings. G52 is always programmed on a temporary basis only, after the main work offset was activated - it remains in effect until cancelled by a block containing **G52 X0 Y0**. By cancelling local coordinate system, the control returns to normal operation, using the work offset exclusively.

There are four special arrows in the following illustrations. They show how the internal shift will be applied in this project for both the first and second clampings:

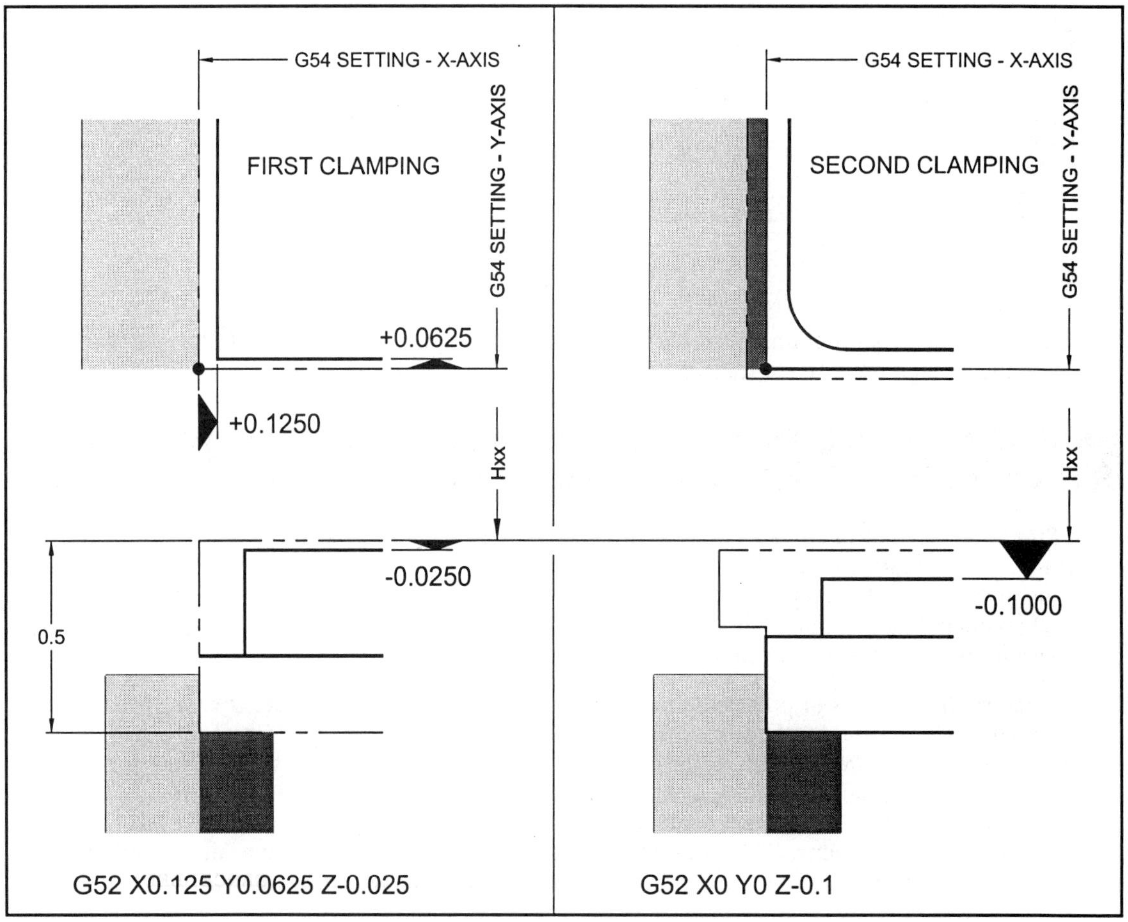

Program Listing - LOCAL COORDINATE SYSTEM G52

```
(PART REVERSAL - USING LOCAL COORDINATE SYSTEM G52)

(G54 = THE LOWER LEFT CORNER OF THE VISE - USED FOR CLAMPING 1 AND 2)
(ALL TOOL LENGTHS SET TO THE TOP FACE OF THE STOCK)

(===>>> CLAMPING 1 <<<===)
(G52 X0.125 Y0.0625 Z-0.025 SHIFTS THE ORIGIN FOR CLAMPING 1 ONLY)

(T01 - 2.5 DIA FACE MILL - MINIMUM FACE CUT - 0.025)
N1 G20
N2 G17 G40 G80 T01
N3 M06
N4 G52 X0 Y0 Z0                (TEMPORARY WORK SHIFT CANCELLED IF ACTIVE)
N5 G90 G54 G00 X5.5 Y1.0 S2500 M03 T02
N6 G43 Z1.0 H01 M08
N7 G52 X0.125 Y0.0625 Z-0.025  (TEMPORARY SHIFT OF G54 ORIGIN)
N8 G01 Z0 F50.0
N9 X-1.5 F10.0
N10 G00 Y2.0
N11 G01 X5.5
N12 G52 X0 Y0 Z0               (TEMPORARY ORIGIN SHIFT CANCELLED)
N13 G00 Z1.0 M09
N14 G28 Z1.0 M05
N15 M01                        (CHECK THICKNESS - SHOULD BE 0.475)

(T02 - 0.5 DIA END MILL - CONTOUR BASE RECTANGLE TO SIZE)
N16 T02
N17 M06
N18 G90 G54 G00 X-0.6 Y-0.6 S2600 M03 T03
N19 G43 Z1.0 H02 M08
N20 G52 X0.125 Y0.0625 Z-0.025 (TEMPORARY SHIFT OF G54 ORIGIN)
N21 Z0.1
N22 G01 Z-0.275 F50.0
N23 G41 X0 D52                 (SUGGESTED D52 = 0.2500)
N24 Y2.875 F10.0
N25 X3.75
N26 Y0
N27 X-0.6
N28 G40 G00 Y-0.6
N29 G52 X0 Y0 Z0               (TEMPORARY ORIGIN SHIFT CANCELLED)
N30 Z1.0 M09
N31 G28 Z1.0 M05
N32 M01                        (CHECK - LENGTH=3.750 WIDTH=2.875)

(T03 - 0.375 DIA CHAMFER MILL - 0.010-0.015 CHAMFER)
N33 T03
N34 M06
N35 G90 G54 G00 X-0.6 Y-0.6 S3200 M03 T01
N36 G43 Z1.0 H03 M08
N37 G52 X0.125 Y0.0625 Z-0.025 (TEMPORARY SHIFT OF G54 ORIGIN)
N38 Z0.1
N39 G01 Z-0.1 F50.0
N40 G41 X0 D53                 (SUGGESTED D53 = 0.1)
N41 Y2.875 F11.0
N42 X3.75
N43 Y0
N44 X-0.6
N45 G40 G00 Y-0.6
N46 G52 X0 Y0 Z0               (TEMPORARY ORIGIN SHIFT CANCELLED)
N47 Z1.0 M09
N48 G28 Z1.0 M05
```

```
N49 X-2.0 Y9.0                    (PART REVERSAL POSITION)
N50 M00                           (CHECK CHAMFER - FLIP PART UPSIDE DOWN)

(===>>> CLAMPING 2 <<<===)
(G52 X0 Y0 Z-0.1 SHIFTS THE ORIGIN FOR CLAMPING 2 ONLY)
(T01 - 2.5 DIA FACE MILL - TWO FACE CUTS - 0.045 AND 0.03)
N51 T01
N52 M06
N53 G52 X0 Y0 Z0                  (TEMPORARY WORK SHIFT CANCELLED IF ACTIVE)
N54 G90 G54 G00 X5.5 Y1.0 S2500 M03 T02
N55 G43 Z1.0 H01 M08
N56 G52 Z-0.1                     (TEMPORARY SHIFT OF G54 ORIGIN)
N57 G01 Z0.03 F50.0               (REMOVES 0.045 OFF THE TOP FACE)
N58 X-0.5 F10.0                   (CURRENT PART THICKNESS = 0.430)
N59 Y2.0
N60 X5.5
N61 G00 Y1.0
N62 G01 Z0                        (REMOVES FINAL 0.030 OFF THE TOP FACE)
N63 X-1.5                         (FINAL PART THICKNESS = 0.400)
N64 G00 Y2.0
N65 G01 X5.5
N66 G52 Z0                        (TEMPORARY ORIGIN SHIFT CANCELLED)
N67 G00 Z1.0 M09
N68 G28 Z1.0 M05
N69 M01                           (CHECK PART THICKNESS - SHOULD BE 0.400)

(T02 - 0.5 DIA END MILL - CONTOUR ROUNDED RECTANGLE TO SIZE)
N70 T02
N71 M06
N72 G90 G54 G00 X-0.6 Y-0.6 S2600 M03 T03
N73 G43 Z1.0 H02 M08
N74 G52 Z-0.1                     (TEMPORARY SHIFT OF G54 ORIGIN)
N75 Z0.1
N76 G01 Z-0.15 F50.0
N77 G41 X0.15 D52                 (SUGGESTED D52 = 0.2500)
N78 Y2.375 F8.0
N79 G02 X0.525 Y2.75 I0.375 J0
N80 G01 X3.225
N81 G02 X3.6 Y2.375 I0 J-0.375
N82 G01 Y0.5
N83 G02 X3.225 Y0.125 I-0.375 J0
N84 G01 X0.525
N85 G02 X0.15 Y0.5 I0 J0.375
N86 G03 X-0.6 Y1.25 I-0.75 J0
N87 G40 G00 Y-0.6
N88 G52 Z0                        (TEMPORARY ORIGIN SHIFT CANCELLED)
N89 Z1.0 M09
N90 G28 Z1.0 M05
N91 M01                           (ROUNDED RECTANGLE LGTH=3.450 WIDTH=2.625)

(T03 - 0.375 DIA CHAMFER MILL - 0.010-0.015 CHAMFER)
N92 T03
N93 M06
N94 G90 G54 G00 X-0.6 Y-0.6 S3200 M03 T01
N95 G43 Z1.0 H03 M08
N96 G52 Z-0.1                     (TEMPORARY SHIFT OF G54 ORIGIN)
N97 Z0.1
N98 G01 Z-0.1 F50.0
N99 G41 X0.15 D53                 (SUGGESTED D53 = 0.1)
N100 Y2.375 F11.0
N101 G02 X0.525 Y2.75 I0.375 J0
N102 G01 X3.225
N103 G02 X3.6 Y2.375 I0 J-0.375
N104 G01 Y0.5
N105 G02 X3.225 Y0.125 I-0.375 J0
```

```
N106 G01 X0.525
N107 G02 X0.15 Y0.5 I0 J0.375
N108 G03 X-0.6 Y1.25 I-0.75 J0
N109 G40 G00 Y-0.6
N110 G52 Z0                    (TEMPORARY ORIGIN SHIFT CANCELLED)
N111 Z1.0 M09
N112 G28 Z1.0 M05
N113 X-2.0 Y9.0                (PART REMOVAL POSITION)
N114 M30
%
```

These two past examples illustrate the advantages and disadvantages of some common methods of programming a part reversal. When compared, the work offset method is the most widely available. If available, G52 - the local coordinate method - places all control of the program flow into the hands of the programmer. CNC operator works with normal work offsets only. You can check whether your control supports G52 by trying an example in the MDI (Manual Data Input) mode.

There is a third method of programming a part reversal, which also uses an optional control feature - that method uses so called *datum shift* and its associated command G10.

Using the DATUM SHIFT Method - G10

Another method to change offsets through the CNC program is somewhat less common, but equally powerful. It uses a preparatory command G10, commonly described as the *Datum Shift* command. G10 can be used to actually *set* or *change* an existing contents of any offset available in the control - work offset, length offset, and radius offset.

The main requirement of using the G10 method is that any particular offset value has to be known to the CNC programmer at the time of program development. Incorporating this value into the program, a new offset can be set (absolute value), or any previously set offset can be changed by a specified value (incremental amount). This incremental method will be used in the program.

The datum shift command G10 has several flavors - it has to distinguish between different types of offsets, between different types of control memory, even between the types of control system itself. For this reason, it is very important to study the machine and control manuals for a particular application. For this project, the G10 is used to *update* the currently set work offset only. Keep in mind that the programmer does not have to know the actual work offset setting.

The following general format is the same for all program entries:

G91	Incremental input - a current offset will be updated by the specified XYZ data
G10	Datum shift command
L2	Arbitrarily selects work offset to be set or updated (ignores tool length or radius offsets)
P1	Select the first of the six standard work offsets - G54=P1, G55=P2, G56=P3, etc.
X... Y... Z...	Axis data to change - values are either absolute (replacements) or incremental (updates)

These two settings are used for the *first clamping* in the program listing below:

```
G91 G10 L2 P1 X0.125 Y0.0625 Z-0.025  (TEMPORARY CHANGE OF G54 SETTINGS)
G91 G10 L2 P1 X-0.125 Y-0.0625 Z0.025 (RESTORE ORIGINAL G54 SETTINGS)
```

There are two settings used *after* the part reversal - applied to the *second clamping.*

☞ **The following block *shifts* the G54 settings for CLAMPING 2 only:**

```
G91 G10 L2 P1 Z-0.1
```

☞ **The following block *restores* the G54 settings for CLAMPING 2 only:**

```
G91 G10 L2 P1 Z0.1
```

Study the completed program carefully - it offers a wealth of information that can be used for many other programming jobs.

Program Listing - DATUM SHIFT G10

```
(PART REVERSAL - USING DATUM SHIFT G10)

(G54 = THE LOWER LEFT CORNER OF THE VISE - USED FOR CLAMPING 1 AND 2)
(ALL TOOL LENGTHS SET TO THE TOP FACE OF THE STOCK)

(===>>> CLAMPING 1 <<<===)
(G91 G10 L2 P1 X0.125 Y0.0625 Z-0.025 SHIFTS THE G54 SETTINGS FOR CLAMPING
1 ONLY)
(G91 G10 L2 P1 X-0.125 Y-0.0625 Z0.025 RESTORES THE G54 SETTINGS FOR
CLAMPING 1 ONLY)

(T01 - 2.5 DIA FACE MILL - MINIMUM FACE CUT - 0.025)
N1 G20
N2 G17 G40 G80 T01
N3 M06
N4 G91 G10 L2 P1 X0.125 Y0.0625 Z-0.025     (TEMPORARY CHANGE OF G54)
N5 G90 G54 G00 X5.5 Y1.0 S2500 M03 T02
N6 G43 Z1.0 H01 M08
N7 G01 Z0 F50.0
N8 X-1.5 F10.0
N9 G00 Y2.0
N10 G01 X5.5
N11 G91 G10 L2 P1 X-0.125 Y-0.0625 Z0.025   (RESTORE ORIGINAL G54 SETTINGS)
N12 G90 G00 Z1.0 M09
N13 G28 Z1.0 M05
N14 M01                                     (CHECK THICKNESS OF 0.475)

(T02 - 0.5 DIA END MILL - CONTOUR BASE RECTANGLE TO SIZE)
N15 T02
N16 M06
N17 G91 G10 L2 P1 X0.125 Y0.0625 Z-0.025    (TEMPORARY CHANGE OF G54)
N18 G90 G54 G00 X-0.6 Y-0.6 S2600 M03 T03
N19 G43 Z1.0 H02 M08
N20 Z0.1
N21 G01 Z-0.275 F50.0
N22 G41 X0 D52                              (SUGGESTED D52 = 0.2500)
N23 Y2.875 F10.0
N24 X3.75
N25 Y0
N26 X-0.6
N27 G40 G00 Y-0.6
N28 G91 G10 L2 P1 X-0.125 Y-0.0625 Z0.025   (RESTORE ORIGINAL G54 SETTINGS)
N29 G90 Z1.0 M09
N30 G28 Z1.0 M05
N31 M01                                     (CHECK LGTH=3.750 WIDTH=2.875)
```

```
(T03 - 0.375 DIA CHAMFER MILL - 0.010-0.015 CHAMFER)
N32 T03
N33 M06
N34 G91 G10 L2 P1 X0.125 Y0.0625 Z-0.025  (TEMPORARY CHANGE OF G54)
N35 G90 G54 G00 X-0.6 Y-0.6 S3200 M03 T01
N36 G43 Z1.0 H03 M08
N37 Z0.1
N38 G01 Z-0.1 F50.0
N39 G41 X0 D53                            (SUGGESTED D53 = 0.1)
N40 Y2.875 F11.0
N41 X3.75
N42 Y0
N43 X-0.6
N44 G40 G00 Y-0.6
N45 G91 G10 L2 P1 X-0.125 Y-0.0625 Z0.025 (RESTORE ORIGINAL G54 SETTINGS)
N46 G90 Z1.0 M09
N47 G28 Z1.0 M05
N48 X-2.0 Y9.0                            (PART REVERSAL POSITION)
N49 M00                                   (CHECK CHFR - FLIP PART)

(===>>> CLAMPING 2 <<<===)
(G91 G10 L2 P1 Z-0.1 SHIFTS THE G54 SETTINGS FOR CLAMPING 2 ONLY)
(G91 G10 L2 P1 Z0.1 RESTORES THE G54 SETTINGS FOR CLAMPING 2 ONLY)

(T01 - 2.5 DIA FACE MILL - TWO FACE CUTS - 0.045 AND 0.03)
N50 T01
N51 M06
N52 G91 G10 L2 P1 Z-0.1                   (TEMP CHANGE OF G54 SETTING IN Z)
N53 G90 G54 G00 X5.5 Y1.0 S2500 M03 T02
N54 G43 Z1.0 H01 M08
N55 G01 Z0.03 F50.0                       (REMOVES 0.045 OFF THE TOP FACE)
N56 X-0.5 F10.0                           (CURRENT PART THICKNESS = 0.430)
N57 Y2.0
N58 X5.5
N59 G00 Y1.0
N60 G01 Z0                                (REMOVES FINAL 0.03 OFF TOP FACE)
N61 X-1.5                                 (FINAL PART THICKNESS = 0.400)
N62 G00 Y2.0
N63 G01 X5.5
N64 G91 G10 L2 P1 Z0.1                    (RESTORE ORIGINAL G54 SETTINGS)
N65 G90 G00 Z1.0 M09
N66 G28 Z1.0 M05
N67 M01                                   (CHECK PART THICKNESS FOR 0.400)

(T02 - 0.5 DIA END MILL - CONTOUR ROUNDED RECTANGLE TO SIZE)
N68 T02
N69 M06
N70 G91 G10 L2 P1 Z-0.1                   (TEMP CHANGE OF G54 SETTING IN Z)
N71 G90 G54 G00 X-0.6 Y-0.6 S2600 M03 T03
N72 G43 Z1.0 H02 M08
N73 Z0.1
N74 G01 Z-0.15 F50.0
N75 G41 X0.15 D52                         (SUGGESTED D52 = 0.2500)
N76 Y2.375 F8.0
N77 G02 X0.525 Y2.75 I0.375 J0
N78 G01 X3.225
N79 G02 X3.6 Y2.375 I0 J-0.375
N80 G01 Y0.5
N81 G02 X3.225 Y0.125 I-0.375 J0
N82 G01 X0.525
N83 G02 X0.15 Y0.5 I0 J0.375
N84 G03 X-0.6 Y1.25 I-0.75 J0
N85 G40 G00 Y-0.6
N86 G91 G10 L2 P1 Z0.1                    (RESTORE ORIGINAL G54 SETTINGS)
N87 G90 Z1.0 M09
```

```
N88 G28 Z1.0 M05
N89 M01                                    (ROUNDED RECT LG=3.450 WDT=2.625)

(T03 - 0.375 DIA CHAMFER MILL - 0.010-0.015 CHAMFER)
N90 T03
N91 M06
N92 G91 G10 L2 P1 Z-0.1                    (TEMP CHANGE OF G54 SETTING IN Z)
N93 G90 G54 G00 X-0.6 Y-0.6 S3200 M03 T01
N94 G43 Z1.0 H03 M08
N95 Z0.1
N96 G01 Z-0.1 F50.0
N97 G41 X0.15 D53                          (SUGGESTED D53 = 0.1)
N98 Y2.375 F11.0
N99 G02 X0.525 Y2.75 I0.375 J0
N100 G01 X3.225
N101 G02 X3.6 Y2.375 I0 J-0.375
N102 G01 Y0.5
N103 G02 X3.225 Y0.125 I-0.375 J0
N104 G01 X0.525
N105 G02 X0.15 Y0.5 I0 J0.375
N106 G03 X-0.6 Y1.25 I-0.75 J0
N107 G40 G00 Y-0.6
N108 G91 G10 L2 P1 Z0.1                    (RESTORE ORIGINAL G54 SETTINGS)
N109 G90 Z1.0 M09
N110 G28 Z1.0 M05
N111 X-2.0 Y9.0                            (PART REMOVAL POSITION)
N112 M30
%
```

Summary

With the correct program, properly input control settings and accurate physical location of the part, this could be a very rewarding exercise to study. Keep in mind that on some lower level controls, the G52 command and the G10 command may be optional features. Also keep in mind that both G52 and G10 can only be used only if there is enough information available. For example, neither command can be used if there are several parts located randomly on the machine table. Often, the original part reference location is determined during setup, then either G52 or G10 can be used *within* the program to change the reference point.

Although this project involves a relatively simple machining operation, it requires careful planning and execution. The main reason to use the G52 or G10 method rather than the G54-G59 method is to remove unnecessary activities from the CNC operator and speed up the setup procedure. For parts that are often repeated, this approach can be a real time saver.

6 USING TAPERED END MILLS

In CNC machining, taper end mills are used in many industries for a large number of applications, such as walls with draft or clearance angle, tool and die work, mold work, even for 'reaming' holes to make them conical. This section covers the applications of both flat and ball types of tapered end mills in standard, *two dimensional* machining (usually in the XY plane). Tapered walls can be created as a single side milling operation or as a continuous contour. In mold making and other 3D machining, tapered end mills are used differently - only the small portion of the end mill tip is used for roughing or finishing surfaces.

Types of Tapered End Mills

There are two types of taper end mills, manufactured from high-speed steel (HSS) or solid carbide:

- **... tapered end mills with a flat tip** *also called tapered end mills with* ***square end***
- **... tapered end mills with a ball tip** *also called tapered end mills with* ***ball nose end***

In two-dimensional work (2D), their usage typically involves a selected depth of cut, followed by the required external or internal contouring. From the illustration below, it is apparent that the contour will be different at the top than at the bottom (at the tool tip), because of the taper. The part drawing determines the size and shape of the contour, which could have dimensions defined either at the bottom of the cut or at the top of the cut, with a given slope angle. The programming job is usually easier if the contour dimensions are defined at the tool tip (given depth).

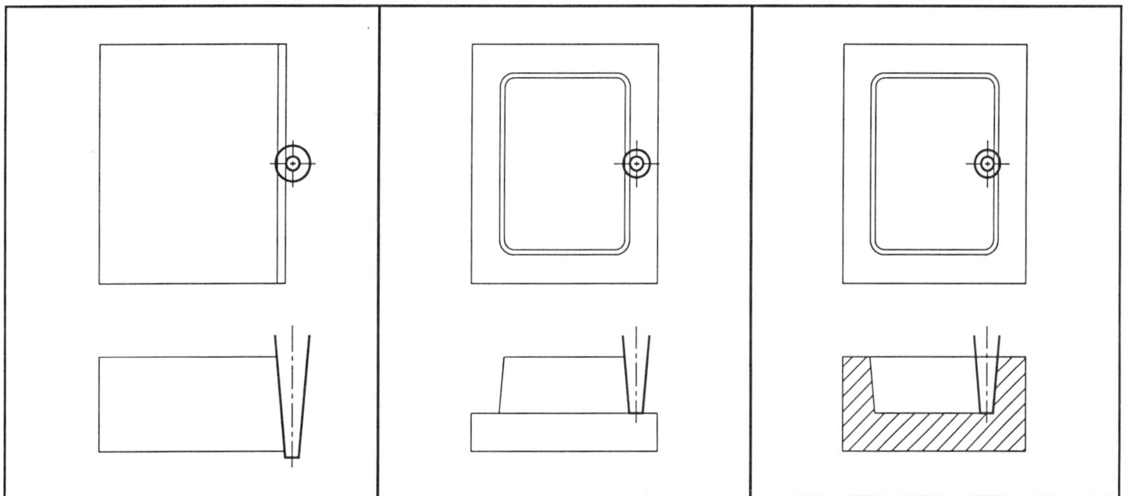

SIDE CUTTING	EXTERNAL CONTOUR	INTERNAL CONTOUR

One rule that applies to all tools - *tapered end mills included* - is the relationship between the tool length and the tool diameter. Combination of longer tool length and smaller diameter contributes to weaker machining. The main culprits are the tool deflection and possible chatter. Shorter tool length and larger diameter will increase the cutting tool efficiency. If the tool diameter must be small to do a particular job, select a series of 'light' cuts rather than a single 'heavy' cut.

Chatter is not an uncommon occurrence with tapered end mills, but it is just as undesirable as with more common tools. In order to eliminate chatter, try variations of speeds and feeds, but also seriously consider the tapered end mill geometry. Selecting a tool with suitable flute design will go a long way to elimination of chatter.

Tool Material

Many small tools (drills, end mills, reamers, etc.) are commonly available in two material selections - *high steel speed* (HSS) and *solid carbide.* When deciding on the proper selection of tapered end mills, there are no significant differences from most other types of tools. High steel speed tools are cheaper, offer somewhat greater flexibility than carbide cutters, but lack the durability and efficiency of the carbide.

Range of Taper Angles

In some industries, taper angles of more than 10 degrees are considered 'large'. In other industries, taper angles much larger are considered 'normal'. To satisfy the need of all industrial needs, tooling manufacturers offer a great range of taper angles.

For smaller tools, you will find some taper angles starting with a minor ½ degree taper per side to larger tools - and larger taper angles - designed with greater increments. Typically, the generally available tapered end mills range from ½ degree taper per side (TPS) to full 45 degrees taper per side. Some industries use a particular taper angle more often than they use other tapers.

Flat Tip Tapered End Mills

Flat tip tapered end mills are defined in various tooling catalogues by their main characteristics:

- **Tool material (HSS or carbide)**
- **Overall length**
- **Number of flutes**
- **Flute geometry (left hand or right hand, straight or spiral)**
- **Center cutting** *or* **Non-center cutting type**
- **Shank diameter (large diameter)**
- **Flat tip diameter (small diameter)**
- **Maximum flute length (usually measured *parallel* with the tool centerline)**
- **Taper per side (often designated as an abbreviation TPS) = draft angle**

Although all characteristics are important, from the perspective of a CNC programmer, the taper angle, the flat tip diameter and the maximum flute length are the most important features.

The difference between the large and small diameters, combined with the flute length, generally DO NOT provide the accurate taper per side. Always use the flute length as a *guide* only - the flute length usually implies the *maximum* flute length only

Ball Tip Tapered End Mills

The tooling catalogue designation of ball tip tapered end mills is similar to those with the flat tip. Virtually all characteristics of the flat end apply here as well, including the comment relating to the flute length. The main difference is the definition of the tool tip radius. Watch for a direct radius specification versus a radius based on the diameter of a flat tip tapered end mill - *they are different.*

Effective Diameter Calculation

Whether a particular job requires the flat tip or the ball tip type of tapered end mill, the programmed cutting depth must always be *smaller than the maximum flute length.* With tapered cutters, the programmed cutting depth is in direct relationship with the effective tool diameter, located somewhere within the flute length. The obvious conclusion is that the deeper the cutting depth, the larger the effective diameter will be. Some jobs do not require this calculation, but many jobs do.

With standard end mills (constant diameter), the end mill diameter is equal to its effective diameter, so there is no need for any special calculations. That is not the case for end mills with a built-in taper.

Effective cutting diameter *D* is defined as the actual diameter cutting at the largest portion of the taper

For tapered end mills with relatively high taper angle, the effective cutting diameter may help to determine optimal speeds and feedrates.

In this chapter, a number of abbreviations will be used in formulas, as they related to flat and ball tapered end mills.

- **Known dimensions - flat tip:**

W T A

- **Known dimensions - ball tip:**

W R A

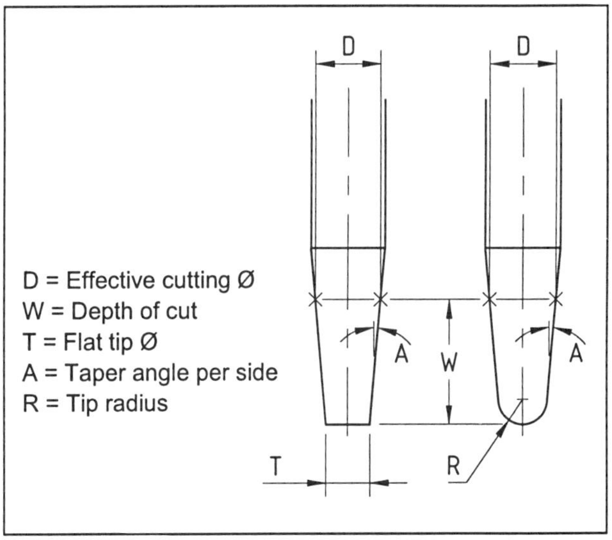

The illustration shows the meaning of each abbreviation

Flat Tip

Calculations related to the flat end mill with a taper are the easiest. From the tooling catalogue, all data are important for the program development, but not all data are equally important for some specific calculations. The most common need in programming tapered end mills is to calculate the effective cutting diameter *D*.

In order to find the effective cutting diameter of a flat tip tapered end mill, the following information has to be known for programming purposes (as shown earlier):

- **Desired depth of cut (W)** ... *measured from the tool tip*
- **End mill taper per side (A)** ... *as provided in the tooling catalogue*
- **Flat tip diameter (T)** ... *as provided in the tooling catalogue or measured*

The following table can be used for all specified taper angles (taper per side):

Taper per side (TPS)	Constant *K* ...		... based on formula
0.5°	0.017454	0.0175	2 × tan0.5
1°	0.034910	0.0349	2 × tan1
1.5°	0.052372	0.0524	2 × tan1.5
2°	0.069842	0.0698	2 × tan2
3°	0.104816	0.1048	2 × tan3
4°	0.139854	0.1399	2 × tan4
5°	0.174977	0.1750	2 × tan5
6°	0.210208	0.2102	2 × tan6
7°	0.245569	0.2456	2 × tan7
10°	0.352654	0.3527	2 × tan10
14°	0.498656	0.4987	2 × tan14
15°	0.535898	0.5359	2 × tan15
20°	0.727940	0.7279	2 × tan20
25°	0.932615	0.9326	2 × tan25
30°	1.154701	1.1547	2 × tan30
45°	2.000000	2.0000	2 × tan45

The constant *K* is the multiplying factor that represents the double tangent amount of the specified angle per side; for example, calculations for a 5 degree taper per side will use the following constant, based on this general formula:

$$K = 2 \times \tan A$$

☞ ***for example ...***

```
2 × tan5 = 0.174977 = 0.1750
```

In order to calculate the large diameter *D* (effective cutting diameter) of such a tool for a given cutting depth *W*, just multiply the constant *K* by the desired depth of cut *W*, then add the flat tip diameter *T* of the tool:

$$D = K \times W + T$$

Using a 5 degree tapered end mill as an example (A), with the depth of cut of 18 mm (W) and the flat tip diameter of 6 mm (T), the large diameter (D) will be (based on the constant K):

```
D = 0.174977 × 18 + 6
  = 9.149592
  = 9.150 mm
```

The formula for any angle is:

$$\mathbf{D = 2 \times tanA \times W + T}$$

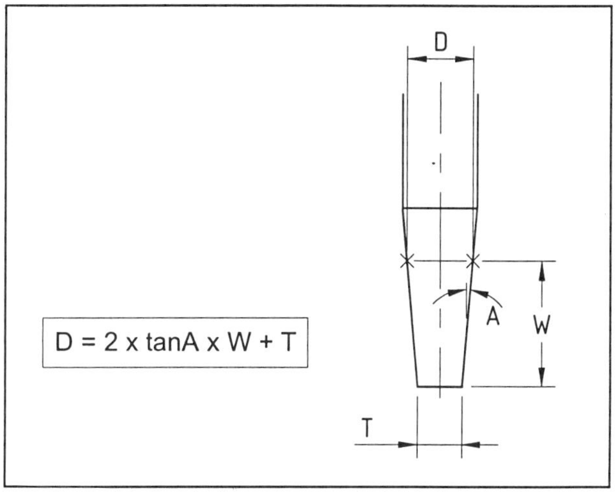

The illustration shows the formula for any taper angle A, without the previously established constant K

Calculations for the flat tip tapered mills are quite easy - the similar calculations for ball tip tapered end mills are a little more complex.

Stock Removal

A straight end mill removes the same amount of material over the full depth of cut. With tapered end mills that is not the case - more material will be cut at the top (at the effective cutting diameter), than at the bottom. Some stock will be left over from roughing or semi-finishing operation, it will also have to be considered. Of course, the greater the taper an of the tool, the more material will have to be removed. If the material is too heavy to remove in a single cut, multiple cutter radius offsets may be used with two or more repeating cuts.

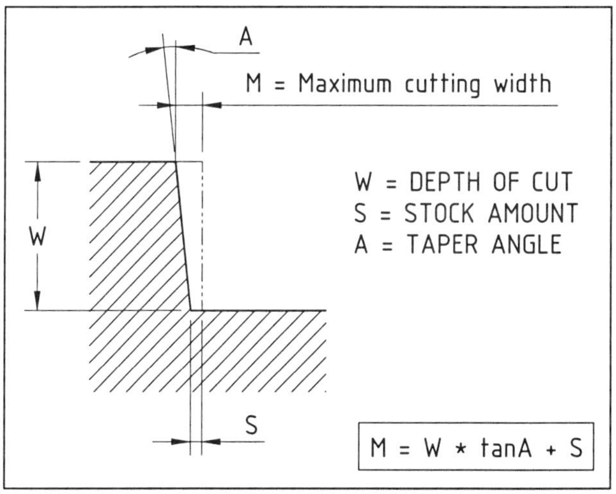

The illustration shows the calculations necessary to establish maximum width of cut.

Ball Tip with Specified Radius

Some tapered end mills, such as those for die sinking and mold cavities, are often defined with a specific tip radius. In order to find the effective cutting diameter *D* of such a tool, the following information has to be known for programming purposes:

- **Desired depth of cut (W)** *... measured from the tool tip*
- **End mill taper per side (A)** *... as provided in the tooling catalogue*
- **Tip radius ®)** *... as provided in the tooling catalogue*

Unlike the flat tip tapered end mill, which is quite easy to calculate, there is no formula attached to the illustration at right. That is because a single formula would be too long and not very practical. A much better method is to separate the rather complex calculations into individual steps, then follow the steps the arrive at the purpose of the calculation, establishing the effective cutting diameter *D*.

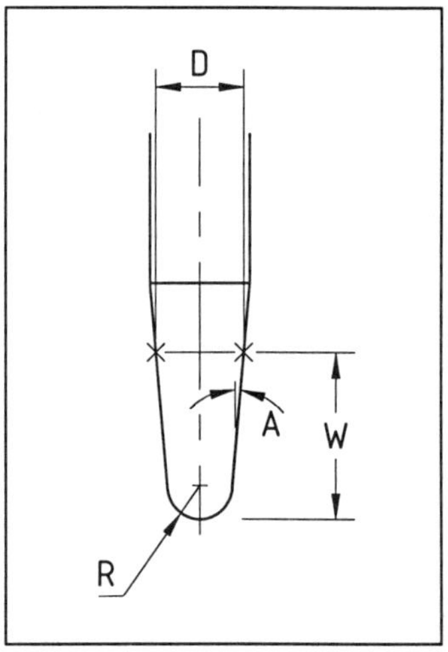

The process described in this section works only if the ball radius R is known

A detail section of the tapered tool end shown at right, lists individual calculations (in form of right triangles) with the following mathematical process to establish the effective cutting diameter *D* when *W, A, R* are known:

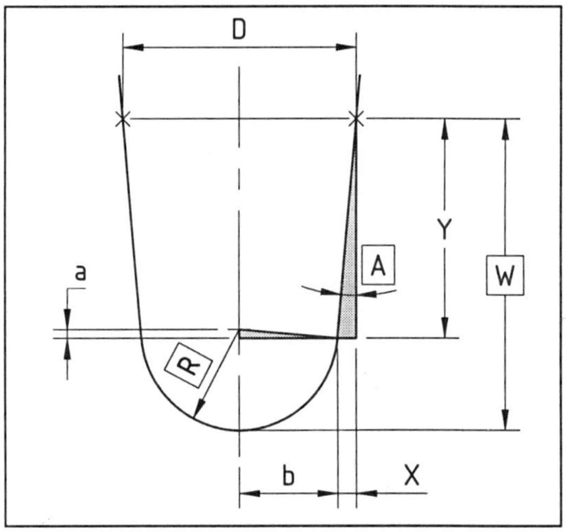


```
a = R × sinA
b = R × cosA
Y = W - R + a
X = Y × tanA

W = 2 × (X + b)
```

Example:

W = 15 mm, A = 6°, R = 4 mm

in this case, the calculated values will be:

```
a = 4 × sin6 = 0.41811                b = 4 × cos6 = 3.97809
Y = 15 - 4 + 0.41811 = 11.41811       X = 11.41811 × tan6 = 1.20009
```

`W = 2 × (1.20009 + 3.97809) = 10.35636 = 10.356` mm effective cutting diameter

Flat Tip with Added Blend Radius

When the end mill tip radius is based on a given flat tip diameter, the actual radius amount will not be known or specified directly. This radius remains unknown and may have to be calculated, if necessary. Such type of a tapered end mill is often the result of a flat tip tapered end mill that has been modified (reground) at the end, from a flat design to a blend radius design.

In order to find the effective cutting diameter of such a tool, the calculation is the same as for the flat tip tapered end mill, as described earlier - the added radius has no bearing on the effective cutting diameter, unless the flat diameter has changed. The following information will normally be known for programming purposes (note that it is the same as for the flat tip tool):

- **Desired depth of cut (W)** ... *measured from the tool tip (not required for this type of tool)*
- **End mill taper per side (A)** ... *as provided in the tooling catalogue*
- **Flat tip diameter (T)** ... *as provided in the tooling catalogue or measured*

What puts a little twist on this type of tool is not the calculation of the effective cutting diameter, but of the *blend radius* that has been formed from the flat end by grinding (blend radius is always tangent to all surrounding entities). If the radius is not important, then there is nothing to do and the tapered end mill can be used the same way as a flat tip tapered end mill. On the other hand, for work setup, for fine finishing, particularly if cutter radius offset is in effect, or even when using this type of tool in CAD/CAM program development, the radius suddenly may become very important.

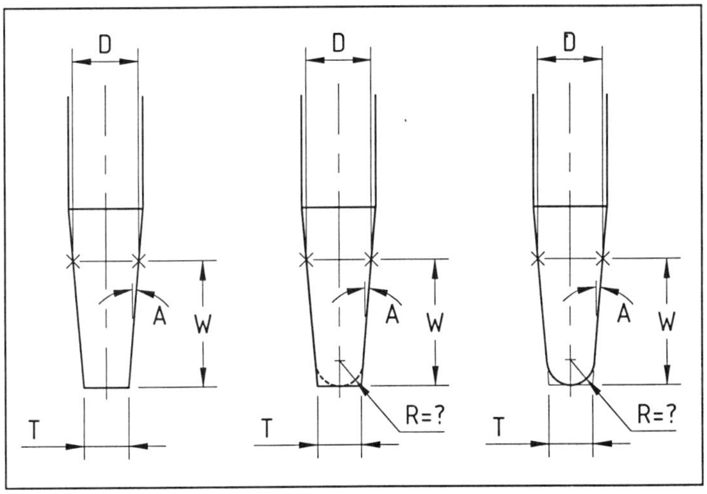

The illustration shows development progress of a tapered end mill with known flat tip and an in-ground radius (unknown)

In order to find the tool tip radius, the calculation will be only slightly more involved - see the formula shown at right

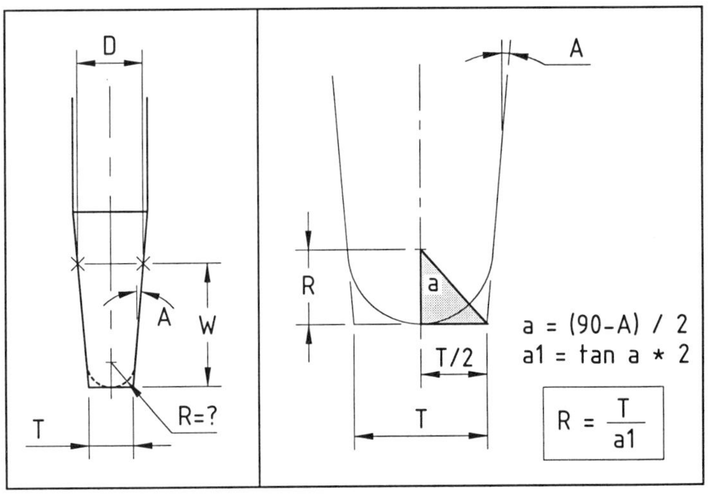

Example:

```
T = 8 mm
A = 5°

a = (90 - 5) / 2 = 42.5
a1 = tan42.5 x 2 =
1.83266

R = 8 / 1.83266
R = 4.36523

R = 4.365
```

Tapered Holes

A tapered end mill can also be used to enlarge a straight hole (predrilled) and make it tapered, based on the tool selection. Both *through holes* and *blind holes* can be successfully tapered using this method, with the proper tool selection. Typically, through holes with a diameter of 6 mm or greater will benefit using a tapered end mill with a left hand spiral flute design. Because the hole is through, the chips will flow down by gravity, not clogging the working part of the tool. For blind holes, the opposite is true - a right hand spiral design is normally a better choice, to cause the chips to flow up and prevent a certain degree of pulling. The number of flutes may also be important in certain cases, and experimenting with even and odd number of flutes may bring the desired results for best quality of tapered holes.

7 SPECIAL PURPOSE G-CODES

Every CNC programmer and most of CNC machine operators have a simple chart of all common G-commands (G-codes) and M-functions (M-codes), usually tucked away somewhere under the lid of their tool box or they have them posted on any convenient machine side or cork board. This chapter covers most of those G-codes that are either uncommon, seldom used, special, or outright mysterious. Keep in mind that machine manufacturers often add G-codes and M-codes of their own. These special codes or functions cannot be covered in a general publication, such as this handbook.

Miscellaneous functions (M-functions) are not covered here at all, as they are often very much dependent on the machine tool manufacturer - for that reason, they are not part of this chapter. The situation is much different with various G-codes, some standard, some optional - they are covered here.

These special and less frequently used G-codes are as important as those used on a daily basis, even if only as accepting them for possible future use. Programmers often forget that there are many preparatory commands available that are not used very frequently. In this chapter, the focus will be on those G-codes that may sometimes become the key to solving a particular problem or achieving a particular programming goal. Some of these preparatory G-codes have a direct relationship with each other, in which case, all related commands will be considered together and explained together.

Not all controls support the preparatory commands described - check your control documentation

Divided into seven groups, seventeen preparatory commands covered in this chapter are:

G60	G09/G61/G62/G63/G64	G22/G23	G25/G26	G27/G28/G29/G30	G92/G50	G31

Single Direction Positioning - G60

Seldom used command G60 can be very practical if the machine tool suffers from a problem called *backlash*. Backlash is the result of wearing out the machine slides over a period of time. For that reason, the backlash problem is associated with older machines or machines that were exposed to some heavy duty machining, misuse and even some abuse. Virtually every control system offers a feature called *backlash compensation* - this is *not* a programmable feature, rather it is aimed at service technicians as it relates to the system parameter settings, sometimes combined with physical adjustments. Whether the backlash is compensated by software or by some physical means, it may not be enough to provide the optimum machining conditions for older machines. For that purpose, the control offers one special command - G60 - called the *uni-directional approach* or *single direction positioning*.

The command G60 belongs to the G-code *Group 00*, which means it is a one-shot command (used only in the block that includes it).

The purpose of G60 is to always approach the target position from the *same* direction

G60 does not eliminate backlash! During single direction positioning, two settings are important - direction and overrun stroke. Both are handled by a system parameter. Direction is actually less important, because the fact that the tool always approached from the *same* direction is sufficient enough.

How to use G60? As its purpose is *tool positioning* - not cutting - it replaces the G00 rapid motion command. Its most common uses are applied to fixed cycles where the distance between hole positions is extremely critical. In fixed cycles, G60 has no effect on the Z-axis motions. Also, in fixed cycles that contain shift from the center (G76 and G87), the single direction positioning does *not* apply to the shift. Absolute or incremental mode can be used with G60 the same way as with the G00. If mirror image is used, the positioning direction does not change. For the example shown, system parameter setting for the direction is X+Y+, and the amount of overrun as 1 mm (X+1.0 and Y+1.0).

Example - Spot drilling of 4 holes to the depth of 2 mm:

```
N1 G21
N2 G17 G40 G80 T01
N3 M06
N4 G90 G54 G60 X12.0 Y15.0
   S1200 M03 T02
N5 G43 Z2.5 H01 M08
N6 G99 G82 R2.5 Z-2.0 P200
   F150.0
N7 G60 Y36.0
N8 G60 X44.0
N9 G60 Y15.0
N10 G80 Z2.5 M09
N11 G28 Z2.5 M05
N12 M01
...
```

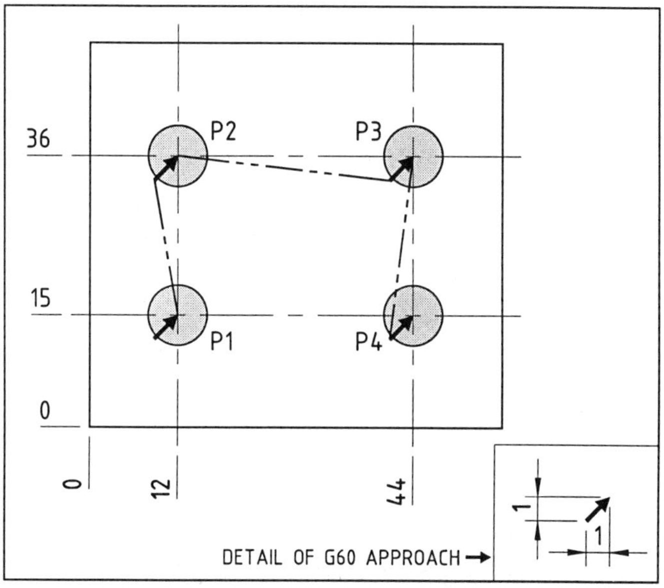

Note:
Because G60 is a non-modal command, it does *not* cancel the active fixed cycle

Special Cutting Modes

Fanuc controls provide several features that can be considered together under the general description *special cutting modes*. The term 'cutting mode' refers to the status of the control selected for the cutting toolpath. Combined to a table, there are five cutting modes available:

G-code	Description	Active for - One Block or Modal Mode
G09	Exact stop check	One block only
G61	Exact stop check mode	Modal - canceled by G62, G63, G64
G62	Automatic corner override mode	Modal - canceled by G61, G63, G64
G63	Tapping mode	Modal - canceled by G61, G62, G64
G64	Normal cutting mode	Cancels G61, G62, and G63 (G64 is the default)

All commands in the G61 to G64 series belong to the same modal G-code group, so one mode cancels another mode. Only one special cutting mode can be in effect at any moment, and all modal commands are cancelled by each other - or brought to normal cutting mode by programming G64.

Exact Stop Check G09 - G61

Commands G09 and G61 are used for the same purpose - to control the positioning accuracy at the end of cutting motions (at sharp corners). The difference between the two commands is modality:

- **G09 - Exact stop check** ... **non-modal - effective for one block only**
- **G61 - Exact stop check mode** ... **modal - remains effective until canceled**

Either command can be used to control how the cutting tool approaches the programmed end point. When used in the program, the control system provides a slight deceleration before reaching the contour end point. Both G09 and G61 will cause the control system to check if the cutting tool is positioned exactly where it should be. When the position is confirmed, the next program block will be processed. The illustration shows *(greatly exaggerated)* the effect of normal cutting and cutting with exact check programmed. *Use only when necessary!*

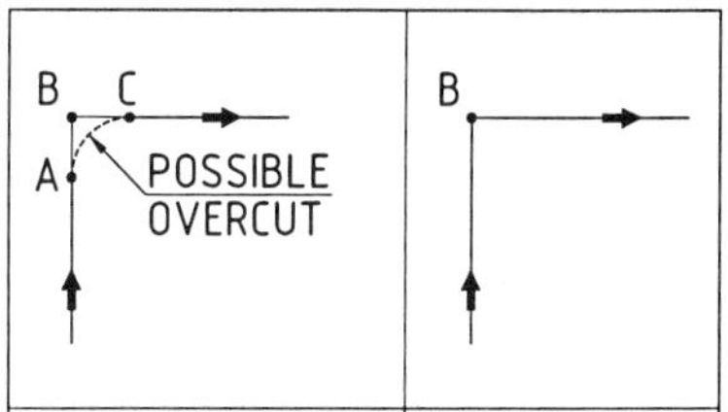

MOTION A-C WITHOUT G09/G61	MOTION B WITH G09/G61

Once the exact position has been reached, the control system will automatically accelerate at the beginning of the next block. Exact check (single or modal) is normally programmed for linear motion (G01) as well as circular motions (G02 and G03). It is most effective at sharp corners (intersections), not at tangencies (blend radius), particularly when fast cutting feedrate is used.

G09 - G61 Example

A simple 75 × 50 mm rectangle has four corners, but only three may require the use of exact check command - there is no corner cutting at the beginning and at the end:

Using G09 compared	.... with using G61
N1 G21	N1 G21
N2 G17 G40 G80	N2 G17 G40 G80
N3 G90 G54 G00 X-10.0 Y-10.0 S900 M03	N3 G90 G54 G00 X-10.0 Y-10.0 S900 M03
N4 G43 Z-5.0 H01 M08	N4 G43 Z-5.0 H01 M08
N5 G41 X0 D51	N5 G61
N6 G09 G01 Y50.0 F500.0	N6 G41 X0 D51
N7 G09 X75.0	N7 G01 Y50.0 F500.0
N8 G09 Y0	N8 X75.0
N9 X-10.0	N9 Y0
N10 G00 G40 Y-10.0 M09	N10 X-10.0
N11 G28 Z-5.0	N11 G64 G00 G40 Y-10.0 M09
N12 M30	N12 G28 Z-5.0
%	N13 M30
	%

Expect a very short delay at each corner, while the G61 command remains in effect. The modal mode G61 is canceled by G62, G63 or G64 commands and no other values are required in the block.

Automatic Corner Override - G62

G62 is another not very common G-code that may come useful under the right circumstances. It is defined as *automatic corner override* command. The main purpose of G62 is to control the feedrate for inside corners *(sharp inner corners)*, when the cutter radius offset is in effect. G62 will force an adjustment to the feedrate, resulting in an improved quality of the surface finish and improves the tool life at the same time. Once activated in the program, the override of inner corners is automatic, until another cutting mode is used (G61, G63, or G64). There are no other values programmed with G62. In terms of contour change, see illustration for the possibilities. There are four types of inner corners that are supported by the G62 command - listed in the direction of cutting motion:

- **Inner corner between a line and another line** ... ***Line-Line transition***
- **Inner corner between a line and an arc** ... ***Line-Arc transition***
- **Inner corner between an arc and a line** ... ***Arc-Line transition***
- **Inner corner between an arc and another arc** ... ***Arc-Arc transition***

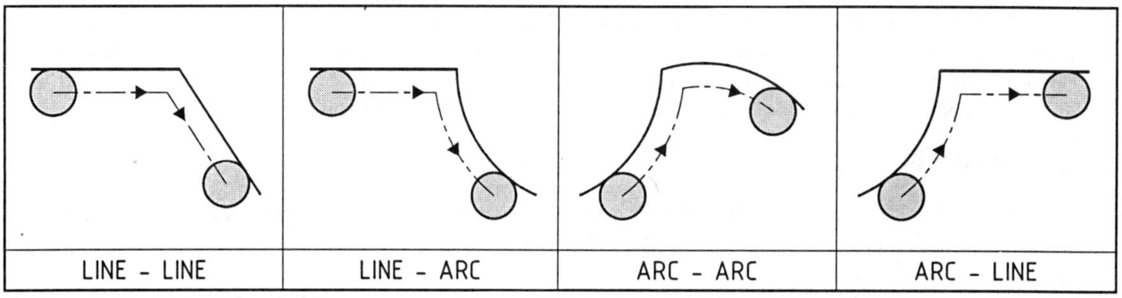

Tapping Mode - G63

Feedrate override switch and the feedhold button are not active in the tapping mode G63. In tapping mode, the tool does not decelerate at the contour change point - the next block is processed immediately. There are no other values programmed with G63.

In practical usage, this command can be used when a tapping sequence needs to be programmed the *long way* - without the benefit of cycles (G74 or G84). Such a sequence will include standard G00 and G01 commands, which can be manipulated during machining operation. G63 command prevents such manipulation, eliminating source of a possible problem. A typical programming example when this mode will be very useful is when the tapping requires different feedrate on the way in than on the way out. This technique is often employed for very small and fine threads, usually less than 0.5 mm pitch. In the example, 0.35 mm pitch will be used, at 700 r/min. The requirement is 80% of the feedrate in, 120% of the normal feedrate out, so G84 cycle cannot be used. Illustration is not necessary.

```
G90 G54 G00 X50.0 Y60.0 S800 M03   (HOLE LOCATION)
G43 Z4.0 H03 M08                   (START POINT AT X50.0 Y60.0 Z4.0)
G63                                (TAPPING MODE BECOMES EFFECTIVE)
G01 Z-12.0 F224.0 M05              (FEED-IN AT 80 PERCENT OF FEEDRATE)
Z5.0 F336.0 M04                    (FEED-OUT AT 120 PERCENT OF FEEDRATE)
G64                                (NORMAL CUTTING MODE - G63 MODE CANCELED)
M05 ...
```

Standard tapping feedrate is *800 r/min × 0.35 pitch = 280 mm/min*. To feed-in at 80%, the feedrate will be 224 mm/min, and to feed-out at 120%, the feedrate will be 336 mm/min.

Normal Cutting Mode - G64

This is the mode that is used in programming the most often. If you never use other modes listed in this chapter, you may not even need G64. When the power is supplied to the control system, the control starts in the state of *'normal cutting mode'*. Normal cutting mode means there are no special conditions required to accommodate single direction positioning, stop check, corner override, or tapping modes. In this case, there is no need to include G64 in the CNC program, unless one of the special cutting mode is used or the programmer chooses to include it. In that case, the G64 should always be included at the program start, along with other initialization commands:

```
G21
G17 G40 G64 G80                (CUTTING MODE AT THE START)
...
G61                            (EXACT STOP CHECK MODE IN EFFECT)
...
   <machining uses the exact stop check mode>
...
G64                            (CUTTING MODE CANCELS G61 - NORMAL MODE)
...
```

Including the cancellation at the beginning (G64), the program is guaranteed to start in the normal cutting mode. G64 also cancels all other modal commands (G61, G62, and G63).

Stored Stroke Limits Definitions - G22 - G23

Another practical, but rarely used and optional feature of the CNC system is the definition of special zones that the tool can enter *(allowed)* or those that the tool cannot enter *(disallowed)*:

G-code	Description	Usage
G22	Stored stroke check ON	Start checking for special zones (allowed or disallowed)
G23	Stored stroke check OFF	Stop checking for special zones

The definition of custom limits for a three-axis machining center is specified by diagonally opposed coordinate points, consisting of three axes, and resulting in a box - rectangular areas for CNC lathes are based on the same principle. For any machine, the specially defined zone establishes the required boundary. Control system parameter determines whether the selected zone will be protected internally (inside the boundary) or externally (outside the boundary). For example, parameter *1300*, bit *#0* is used for this purpose on Fanuc 16-18-21 controls (additional parameter settings may also apply).

The programming format for the G22 command is:

```
G22 X.. Y.. Z.. I.. J.. K..
```

☞ where ...

G22	=	Stored stroke limits activated	... defined zone
XYZ	=	Start point of the limited area	... measured from machine zero
IJK	=	End point of the limited area	... measured from machine zero

The active units of measurement are used as input values. G23 cancels the stored stroke check and is programmed by itself in a block. No more checking will be performed after the G23 command. There are two important conditions to be satisfied when specifying the G22 corner values:

X > I	Y > J	Z > K
X - I > 2 mm	Y - J > 2 mm	Z - K > 2 mm

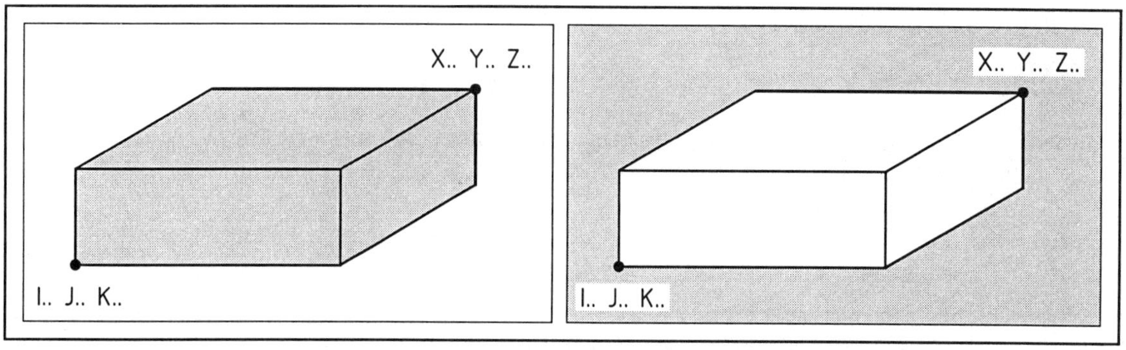

The shaded area in the illustration above identifies the *prohibited* zone.

Example:

```
G21
G17 G23 G40 G80
G22 X-150.0 Y-75.0 K-100.0 I-280.0 J-175.0 K-125.0   (STROKE CHECK ON)
G90 G54 G00 X.. Y.. S.. M03
...
   <machining with stored stroke check active>
...
G23                                                    (STROKE CHECK OFF)
...
```

Can you tell how large zone has been selected for protection in the program above? It should be easy: 130 mm along X-axis, 100 mm along the Y-axis, and 25 mm along the Z-axis. Just take the difference between X-I, Y-J, and Z-K amounts in the G22 statement. Before any programmed tool motion attempts to enter the forbidden zone, the control system issues an alarm and stops further program processing *before* the tool enters the prohibited zone. The most common use of this feature is to program a protective zone around fixtures, during multi-part setups and other undesirable items that may be in the way of toolpath (even part of the current work may be defined as prohibited).

There is a rather good reason why this apparently powerful control feature has not been commonly used. The main reason is that it's just *not powerful enough*; mainly when working with tool length and radius offsets. Neither offset is evaluated during the stroke check, because all the command does is following the XYZ motions. Also, in practice, it is not always easy to actually establish the zone that is to be protected by the program, particularly when programming away from the machine.

G23 cancellation command stands on its own - it does not accept any arguments and must be programmed in a separate block. One final word of caution:

The G22 command is active when the machine is powered up

Spindle Fluctuation G25 - G26

Before studying this section, please take note:

This is one G-code that may be programmed after consultation with a service technician

Spindle speed is programmed in revolutions per minute (r/min or rpm). For example, the program may contain S1000 M03, selecting 1000 r/min at normal rotation. Ideally, the spindle should run at the constant 1000 r/min, however, that is not always the case. The actual spindle speed will fluctuate slightly, which is normal. Just watch the display of the spindle speed at the control. In most cases, the speed fluctuation (also called spindle speed variation) can be ignored, as it presents no problem. For those cases, when the fluctuation is high, the control system provides two preparatory commands:

G-code	Description	Usage
G25	Spindle speed fluctuation detection OFF	Stop checking for spindle fluctuation
G26	Spindle speed fluctuation detection ON	Start checking for spindle fluctuation

Note that the G26 command does not actually remove the fluctuation, it only *detects it*. It only monitors it, and issues and alarm if the spindle is *overheated* - the main reason for spindle fluctuation is heat. The format for G26 uses three entries (Fanuc 16-18 format shown:

```
G26 P.. Q.. R..
```

☞ *where ...*

P = **Time in milliseconds to start the check**
Q = **Percentage of allowed tolerance**
R = **Percentage of spindle speed fluctuation**

The above descriptions are rather brief, and do not indicate what exactly happens. Using an example for spindle speed of 1000 r/min, the meaning of individual entries should becomes clearer.

```
G26 P2000 Q5 R10
```

The *P* value is the time in milliseconds (ms) that is counted from the spindle rotation function *S* to the start of check for possible overheating. The checking will start only if the programmed spindle speed is actually established during the time specified by *P*. If *P* is *P2000* it is selected as 2000 ms; the system will check if the programmed spindle speed is reached within 2 seconds.

The *Q* value is specified as a percentage of allowed tolerance from the programmed spindle speed. For example, *Q5* is 5% of S1000, which is $1000/100 \times 5 = 50$. Subtracting 50 from the programmed speed results in 950 r/min for the spindle speed to reach. If the programmed speed is within 950 and 1000 r/min, the check of actual spindle speed begins.

The *R* value is the percentage of spindle speed fluctuation where the spindle speed is too fast and overheating is possible. *R10* is 10% of the programmed spindle speed, and $1000/100 \times 10 = 100$. Subtracting 100 from the programmed speed results in 900 r/min as the speed when alarm will be generated. Alarm will be issued if overheating is likely to occur.

Often the critical speeds are known, in which case the individual values of *Q* and *R* have to be calculated. There are two formulas to use for this purpose:

$$Q = (1 - A / S) \times 100$$

$$R = (1 - H / S) \times 100$$

☞ where ...

A = Actual spindle speed
H = High speed at which overheating is possible
S = Programmed spindle speed

For example, if *A = 950, H = 900* and *S = 1000*, then:

$Q = (1 - 950 / 1000) \times 100 = 5$
$R = (1 - 900 / 1000) \times 100 = 10$

Machine Zero Commands - G27- G28 - G29 - G30

Although included in the title, the G28 command is definitely not special - it is used on a daily basis, virtually in every CNC program. Still, it is described here, along with its related commands, to provide a single reference to all commands related to machine zero return. First, some definitions:

G27	Machine zero return position check
G28	Primary machine zero return command
G29	Return from machine zero
G30	Secondary machine zero return command

Primary Machine Zero Return - G28

The most common G-code of the four commands listed above is definitely the G28. To a large extent, G30 shares the same characteristics as G28, except it relates to a secondary machine zero, if available. For the record, machine zero is the position of all axes that always remains the same. Machine zero is the master origin of the machine - origin that is specified by the machine manufacturer. Machine zero always has the coordinates of X0Y0Z0. Normally, this position is never changed and is used as a reference point for other required measurements and adjustments, such as work offsets and tool length offset, for example. In order to measure work offset or tool length offset accurately, the measuring process must start from a common point - from a zero - *machine zero, that is.*

As a command, G28 is different than other G-codes, in the sense that it functions as *two commands rolled into one*. Once you understand this very basic concept, using G28 will be problem free. The following definition should shed some light on the *'two commands in one'* statement:

G28 command will cause the tool to return to machine zero along one or more axes,
THROUGH AN INTERMEDIATE POINT

The meaning of the definition should be clear - the tool motion will indeed terminate at machine zero, but it will take a break along the way - through another location - *through an intermediate point.* Consider the two G28 versions - one in absolute mode, the other in incremental mode (return along the Z-axis only is shown, where Z0 is the top of part):

`G90 G28 Z20.0` ... *the tool moves to the position 20 mm above part, then continues to machine zero*

`G91 G28 Z20.0` ... *the tool moves by the distance of 20 mm, then continues to machine zero*

This type of program input is also one of the most common. After all, in order to make an automatic tool change, the tool must be located at machine zero for the Z-axis (vertical machining centers) or for the Y-axis (horizontal machining centers). What makes this command sometimes hard to understand is that it does not seem to be necessary, it does not seem to have any purpose. In the two common examples above, that is true. Actually, if the Z-axis alone is used with G28, the motion to the intermediate point is practically useless. In case of two or three axes used simultaneously with G28, the situation is much different. The following example and illustration show the *intended* use of G28:

```
...
G90 G00 X14.4 Y9.0        (CURRENT POINT - XY)
Z2.0                      (CURRENT POINT - Z)
G01 Z-13.3 F150.0
G28 X14.4 Y9.0 Z2.0       (INTERMEDIATE POINT)
...
```

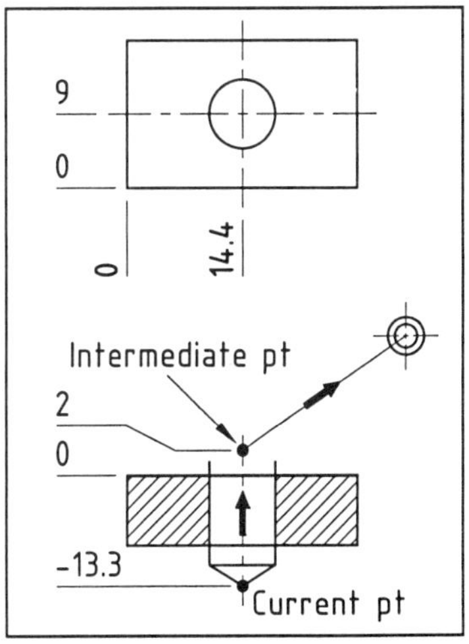

Many programmers have wondered at the actual benefits of this command, as it saves only a single block of the program. There is a simple idea behind the G28 command - to *bypass obstacles* on the way to machine zero when moving two or more axes simultaneously. In a single axis machine zero return, there are no obstacles to consider, so the intermediate point is unnecessary, but it still must be programmed. The solution? *The intermediate tool position will be the same as the current tool position.*

The previous example will be used with a retract move added:

```
...
G90 G00 X14.4 Y9.0
Z2.0
G01 Z-13.3 F150.0
G00 Z2.0                  (SAFE RETRACT)
G28 X14.4 Y9.0 Z2.0       (INTERMEDIATE POINT)
...
```

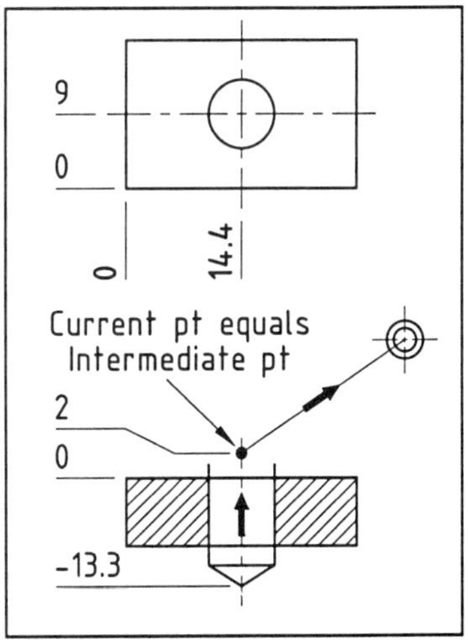

Some programmers prefer to use incremental zero return, with exactly the same effect:

```
...
G90 G00 X14.4 Y9.0
Z2.0
G01 Z-13.3 F150.0
G00 Z2.0                  (SAFE RETRACT)
G91 G28 X0 Y0 Z0          (INTERMEDIATE POINT)
G90 ...                   (REINSTATE ABSOLUTE MODE)
...
```

The two versions with the safe retract are preferred methods - the G90 method is favored over the G91 version, as there is no need to change the mode. Use G91 only if absolute position is unknown.

Keep in mind that *X0Y0Z0* in absolute mode refer to part zero location, *X0Y0Z0* in incremental mode refer to *no motion*, that is a motion that has a zero length. Zero motion during machine zero return applies only to the intermediate point - G91 G28 Z0 simply means that there will be no motion to the intermediate point and the tool will move to machine zero directly. Make sure to include G91 in the block. Both G90 and G91 examples will have the same effect - *the Z-axis retract above part is very strongly recommended.* The best way to see the effect of this often mysterious command is in the single block mode, where both parts of the zero return motion will be apparent.

Return from Machine Zero - G29

G29 is the virtual opposite of G28 - it's almost that simple. Practical usage of this command is very questionable. While G28 can be described as the return to machine zero from the current tool position through an intermediate point, the G29 is the return *from* the machine zero to the specified tool position *through the last intermediate point.* Of course the last intermediate point must be defined first, often for the previous tool. For example, here are all critical blocks for two tools:

```
...                           (TOOL 01 WORKING)
G90 G00 X50.0 Y30.0
Z2.0
G01 Z-15.0 F150.0             (TOOL IS BELOW THE PART)
G28 X50.0 Y30.0 Z2.0          (XY IS THE CURRENT POINT THROUGH Z2.0 FIRST)
...
M06                           (TOOL 02 WORKING)
...
G90 G29 X75.0 Y40.0 Z2.0      (COORDINATES SPECIFY THE NEW TARGET LOCATION)
...
```

What will be actually happening at the machine? For the first tool T01, machining will take place at X50.0Y30.0, to the depth of Z-15.0. The tool retracts to Z2.0, and from this location, all three axes return to machine zero (for whatever reason). Now, the next tool T02 is working and G29 specifies target location of X75.0Y40.0 and Z2.0. How the tool will move there is the very purpose of G29. The control will use the last intermediate point (X50.0Y30.0Z2.0), moves the tool to this position first, *then* moves to the actual target of X75.0Y40.0Z2.0. Why? Honestly, beats me.

The difficulty is not much in the understanding of this command; the difficulty is mainly related to its application. It is much easier to understand a command that is used all the time (G28) when compared with a command that is practically never used (G29). Here is the simple conclusion:

Strongly recommended => Avoid using the G29 command completely !

Machine Zero Return Position Check - G27

Although listed first in the table, this G-code can only be truly understood if several other G-codes are understood first, particularly G90, G91, and G92 (or G50 on lathes).

Since the G27 command is listed in this chapter, it belongs to one of those commands seldom - if ever - used in CNC programs. First, its definition in the section title identifies this command as machine zero return *position check.* This command has no practical purpose in modern programming, but it can be useful to those programmers who still have to program G92 or G50 position registers for some very old control systems (those that do *not* support work offsets). Details about G92/G50 are described in the next section. For now, just accept the fact that G92 registers the specified XYZ coordinates as the *current tool position from part zero - not from machine zero.* Here is an example of a typical G27 use in incremental mode:

```
...
G90                          (ABSOLUTE MODE)
G92 X300.0 Y200.0 Z100.0     (CURRENT TOOL LOCATION AT MACHINE ZERO)
G00 X20.0 Y15.0 Z2.0         (START LOCATION FOR CUTTING)
G01 Z-5.0 F120.0             (MACHINE TO ABSOLUTE DEPTH OF 5 MM)
G91 G01 X40.0 F150.0         (MACHINE PART IN INCREMENTAL MODE - CUT ¼)
Y30.0                        (MACHINE PART IN INCREMENTAL MODE - CUT 2/4)
X-40.0                       (MACHINE PART IN INCREMENTAL MODE - CUT ¾)
Y-30.0                       (MACHINE PART IN INCREMENTAL MODE - CUT 4/4)
G00 Z7.0                     (RETRACT BY 7 MM)
G27 X280.0 Y185.0 Z98.0      (CHECK IF THE MOTION TERMINATES AT MACHINE ZERO)
...
```

In the above program, look at the G27 block - it is of special interest. Once the machining has taken place, all axes will move to the position specified in the G27 block, which is the incremental motion to the machine zero, from the current tool position. The big question is - did all axes actually reached the machine zero? Well, upon the block completion, G27 will perform the check, and if the machine zero position has been reached, axes confirmation lights will turn on; if the machine zero position has not been reached, the control will issue an alarm, and no further program processing will be allowed.

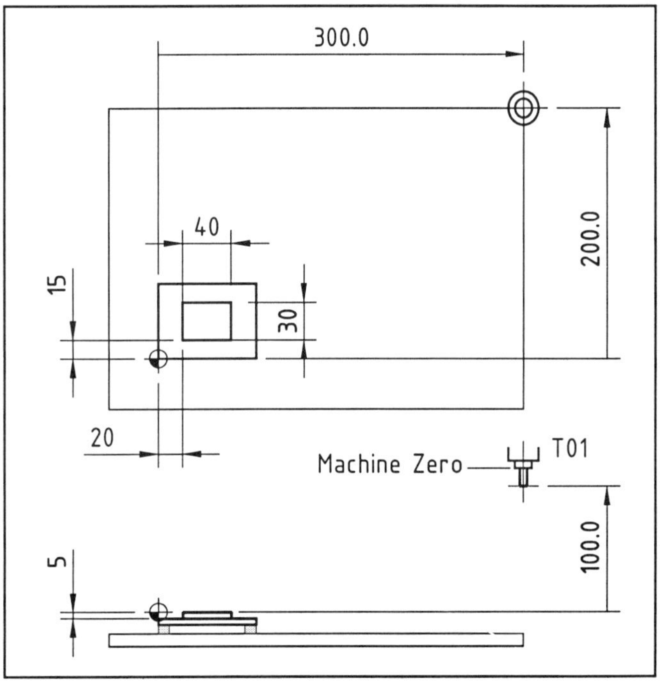

Now, let's look at the above example more carefully - *will the machine zero lights come on or will there be an alarm?*

The best approach to understand G27 is to convert the whole program to absolute mode:

```
...
G90                          (ABSOLUTE MODE)
G92 X300.0 Y200.0 Z100.0     (CURRENT TOOL LOCATION AT MACHINE ZERO)
G00 X20.0 Y15.0 Z2.0         (START LOCATION FOR CUTTING)
G01 Z-5.0 F120.0             (MACHINE TO ABSOLUTE DEPTH OF 5 MM)
G01 X60.0 F150.0             (MACHINE PART IN ABSOLUTE MODE - CUT ¼)
Y45.0                        (MACHINE PART IN ABSOLUTE MODE - CUT 2/4)
X20.0                        (MACHINE PART IN ABSOLUTE MODE - CUT ¾)
Y15.0                        (MACHINE PART IN ABSOLUTE MODE - CUT 4/4)
G00 Z2.0                     (RETRACT 2 MM ABOVE PART)
G27 X280.0 Y185.0 Z98.0      (CHECK IF THE MOTION TERMINATES AT MACHINE ZERO)
...
```

Write down the last positions of XY and Z - they are usually in two blocks, just before the G27 command - they are X20.0 Y15.0 Z2.0. Next, subtract each position from the respective axis in the G92:

```
X200.0 - 20.0 = X280.0     Y100.0 - 15.0 = Y185.0     Z100.0 - 2.0 = Z198.0
```

This example is correct, G27 will confirm that machine zero has been reached in all three axes.

Using G27

G27 should always be used in the G40 mode (cutter radius offset canceled), otherwise the tool will not be confirmed at machine zero. Also watch for cases when the machine lock switch is turned on, for example, during program verification. In this case, there will be no machine zero check performed. If you find a practical application for this command, you can use it for both mills and lathes with same logical approach. Programs written in incremental mode (G91) benefit the most from this G-code. Under normal situation, use G28 in order to return to machine zero - the position checking is already built in and it is a much more convenient way of programming.

Secondary Machine Zero Return - G30

Most vertical machines have only one machine zero. One machine zero is always the primary machine zero, as there is no other choice. Some horizontal machining centers - particularly those used with pallet changers - often need a secondary machine zero location, in order to align two independent pallets, for example. The applications for both G28 and G30 are exactly the same - in order to return to the primary machine zero, use G28; in order to return to the secondary machine zero, use G30. All other conditions listed under the G28 heading apply equally to G30. Third and fourth machine zero commands are also available on some controls, but rather rare.

Position Register - G92/G50

In the earlier section describing the G27 command, a small example of the G92 command was presented. On the lathes, the most common equivalent command is G50. Both these commands are considered obsolete by any modern standards - they have been replaced by the much more flexible work offsets G54-G59. The reason there is at least a miniature version of these commands described here, came from the many requests to cover this feature for the users who still work with older controls.

All machines that support work offsets also support G92, mainly for the compatibility with older programs. This statement - while correct - also requires a word of caution first:

Never use G92/G50 together with work offsets G54-G59 in the same program

Yes, it is possible to mix work offsets with position register commands in the same program, but it can cause more headaches than it's worth. The most common definition for G92/G50 is *position register command*. What does it really mean? Keep in mind these commands are dated *before* any work offsets existed, although the G43 command for tool length offset was already available. While work offset is always measured from machine zero to part zero along an axis (usually as a negative dimension), position register is measured from part zero to machine zero (usually as a positive dimension). That is an important difference, but not really the major difference between the two methods of coordinate setting. The major difference is in the respective applications of both command types:

- **Position register G92/G50 is set in the program for each tool**
  ... actual dimensions for each axis must be known during program development
- **Work offsets G54-G59 are set at the machine and only called by the program**
  ... actual dimensions are not known during program development
- **Tool length offset G43 is set in the program for both G92 and G54-G56 (mill controls only)**
  ... actual dimensions are not known during programming

In summary, there is only one - *and only one* - reason for either G92 or G50 command in the program:

Position register always defines the CURRENT tool position from part zero location

Example:

```
G92 X195.0 Y88.0 Z20.0     (POSITION REGISTER COMMAND - THREE AXES SELECTED)
```

In this block, the programmed data *'tell'* the control that the tool current position is located 195 mm from the part zero along the X-axis, 88 mm along the Y-axis, and 20 mm along the Z-axis.

☞ **It is extremely important to understand that position register G92/G50 does *not* result in any actual tool motion. It is the programmer's responsibility to assure the provided data correspond to the actual - *that is physical* - tool position**

The major problem with position register commands is the fact that for most setups, the programmer has no way of knowing where exactly the tool is located. In order to make this feature work, several rather elaborate methods have been devised, to shorten the setup time. Programmers soon found out that it is easier to leave the G92 block in the program empty and let the operator to fill-in the data, based on actual setup. However, this method had only been possible when software based CNC systems replaced the hardwired NC systems. Of course, work offsets have solved this problem for all but those users who still have no choice but to program in the very old fashioned way.

G92 Position Register for Milling

Some older CNC lathes also accept G92 as the position register command, but generally G92 is used for milling, while G50 is used for turning. The best way to illustrate the G92 command is to use it in a program with two tools and follow its settings from one tool to the next. The illustration below shows top and front setup of a simple part. *Note the arrow directions and the dimensions required by setup.*

A suitable spot drill and a ∅5 mm drill will be programmed to machine the two holes shown. The first tool - *spot drill* - starts from machine zero. This position is a normal start for any first tool, because the machine has to be zeroed for every setup at startup. When the tool is located at machine zero, the CNC operator must set the part zero *exactly* 300 mm along the X-axis and 200 mm along the Y-axis. These are the main conditions of G92 command - *this is how G92 works!*

Based on extensive experience of many users, this setup task is virtually impossible to do at the machine, with any degree of precision - at least in a reasonable time frame. However, that was - *and still is, if necessary* - the main drawback of G92.

The following programming example is based on the provided setup, including both tools, with appropriate comments:

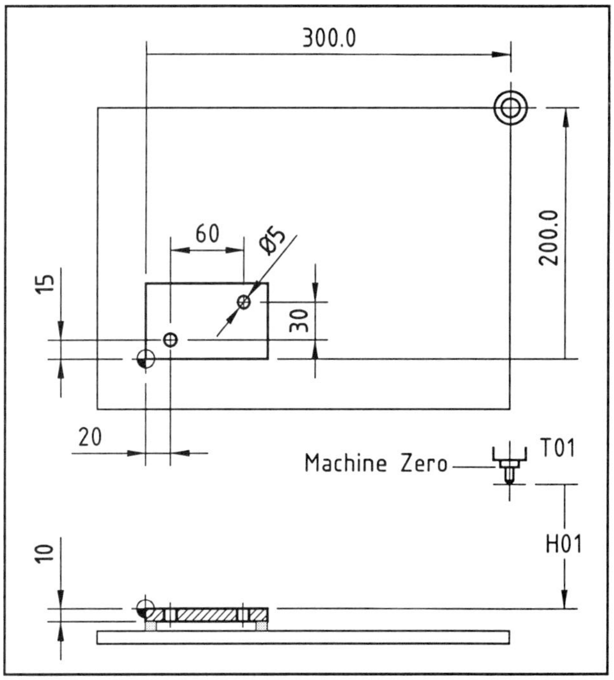

```
N1 G21
N2 G17 G40 G80 G90 T01
N3 M06 (SPOT DRILL)
N4 G92 X300.0 Y200.0 Z0 S1200 M03 T02    (CURRENT TOOL AT MACHINE ZERO)
N5 G00 X20.0 Y15.0                       (FIRST HOLE LOCATION)
N6 G43 Z10.0 H01 M08                     (TOOL 10 MM ABOVE WORK)
N7 G92 Z10.0                             (CURRENT TOOL POSITION REGISTERED)
N8 G99 G82 R2.0 Z-2.75 P200 F120.0       (SPOT DRILL AT CURRENT LOCATION)
N9 X80.0 Y45.0                           (SECOND HOLE LOCATION)
N10 G80 Z10.0 M09                        (CYCLE COMPLETED)
N11 G28 Z10.0 M05                        (ONLY Z-AXIS TO MACHINE ZERO)
N12 M01
```

After the first tool in the program (T01 is the spot drill in the example) completes the machining operation, the location of th T01 is critical within G92 environment and must always be known. Based on the program developed so far, the tool T01 is located at X80.0 Y45.0, when block N12 is processed. The tool T01 is also located at machine zero in the Z-axis.

The illustration at right shows the *current* tool position of T02, which replaced tool T01 during automatic tool change.

Following the normal rules, G92 command data must contain the current tool location - as it relates to part zero. In this programming example, the second tool T02 is located exactly at the same XY position as the previous tool T01, so no special programming effort is necessary. If G92 for the X and Y coordinates is used in the program, its values will be:

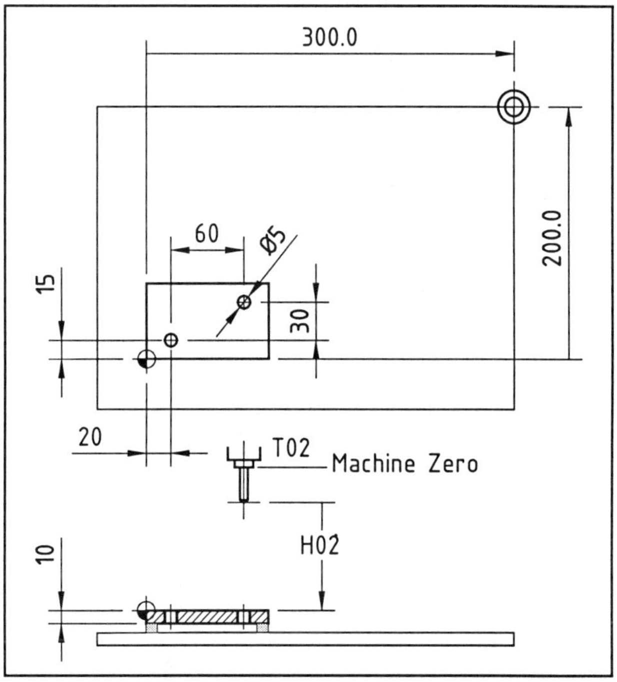


```
G92 X80.0 Y45.0 Z0                       (CORRECT BUT UNNECESSARY SETTING)
```

The remainder of the program can now be written, considering the settings described so far:

```
N13 T02
N14 M06 (5 MM DRILL)
N15 G90 G92 Z0                           (CANCEL Z-AXIS SETTING)
N16 G00 G43 X80.0 Y45.0 Z10.0 H02 M08    (MOVE TO SECOND HOLE 10 MM ABOVE)
N17 G92 Z10.0 S1350 M03                  (CURRENT TOOL POSITION REGISTERED)
N18 G99 G81 R2.0 Z-12.5 F150.0           (DRILL AT CURRENT LOCATION)
N19 X20.0 Y15.0                          (FIRST HOLE LOCATION)
N20 G80 Z10.0 M09                        (CYCLE COMPLETED)
N21 G28 Z10.0 M05                        (ONLY Z-AXIS TO MACHINE ZERO)
N22 G28 X20.0 Y15.0                      (=> WHAT IS THE G92 AT THIS BLOCK?)
N23 M30
%
```

Of course, the comment in block N22 is unnecessary, but included here anyway - *what is the current tool position?* Since the current tool T02 is located at machine zero, it means it is located 300 mm and 200 mm respectively, along the X and Y axes - *from part zero.* The Z-axis is the original setting as determined by the tool length.

G50 Position Register for Turning

The G50 command shares most programming features with its G92 cousin - its definition, its purpose, and its method of programming. The only visible difference in the program is the axes designation as X and Z rather than X, Y and Z. Although an obsolete command, many older programs do contain this type of tool setting and still use it. The part programmer should convert the program to the modern method, which is very easy - just eliminate the G50 X.. Z.. reference. For those who still need to work with the G50 position register, here is an example that should make its use clearer.

To illustrate the G50 concept, a simple drawing shown at right will be used for the example. For actual machining, three typical and unique tools will be selected, to show the various G50 relationships:

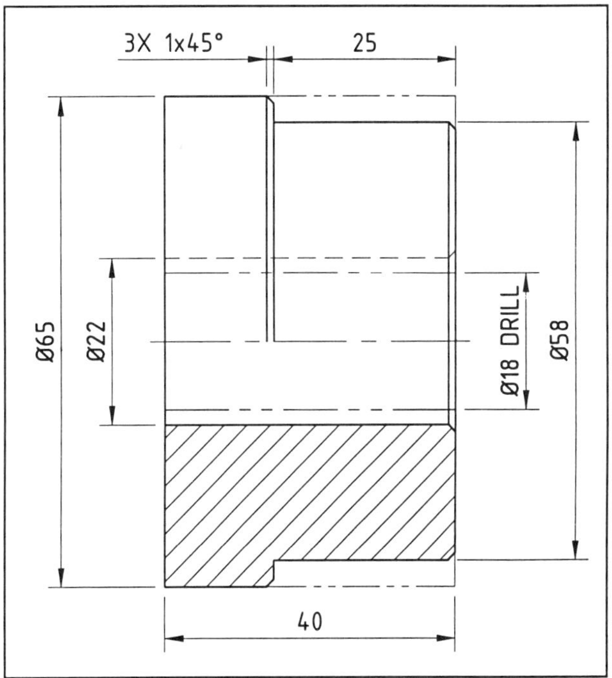

- **T01 = 80° Turning tool**
- **T02 = ∅18 mm Drill**
- **T04 = ∅16 mm Boring bar**

For simplicity (and better clarity), only the necessary toolpath will be covered in the program. Facing and center-drilling operations have been skipped. Spindle speeds and cutting feedrates are only reasonable and *not* important at this time.

Tool Change Position

During tool change, it is quite common on a CNC lathe to move the current tool to a safe position away from the part, but not necessarily to machine zero. A variation on the same topic will be a tool change at machine zero in the X-axis only - but close to the part in the Z-axis. The main reason is that the X-axis travel is much shorter than the Z-axis travel on most machines. Also, this method adds to the safety aspects of the automatic tool change process by adding additional clearance. This is also the programming method used in the examples provided in this chapter.

***** SAFETY ADVISORY *****

When selecting a tool change position, the minimum Z-clearance must always be established for the *longest* tool used by the program

The above safety message applies to any type of lathe programming, not just when the G50 is used.

Initial Data

Unlike the modern programming method, using G50 means using several dimensions known to the part programmer. These dimensions relate to the CNC lathe itself, as well as to all tools used in the program. The schematic illustrations on the next page show dimensional data for the CNC lathe used for this example, with attached chuck, jaws and turret *without and with* tools. The dimensions used are realistic but ***not to scale***, and are always known from the data supplied by the lathe manufacturer.

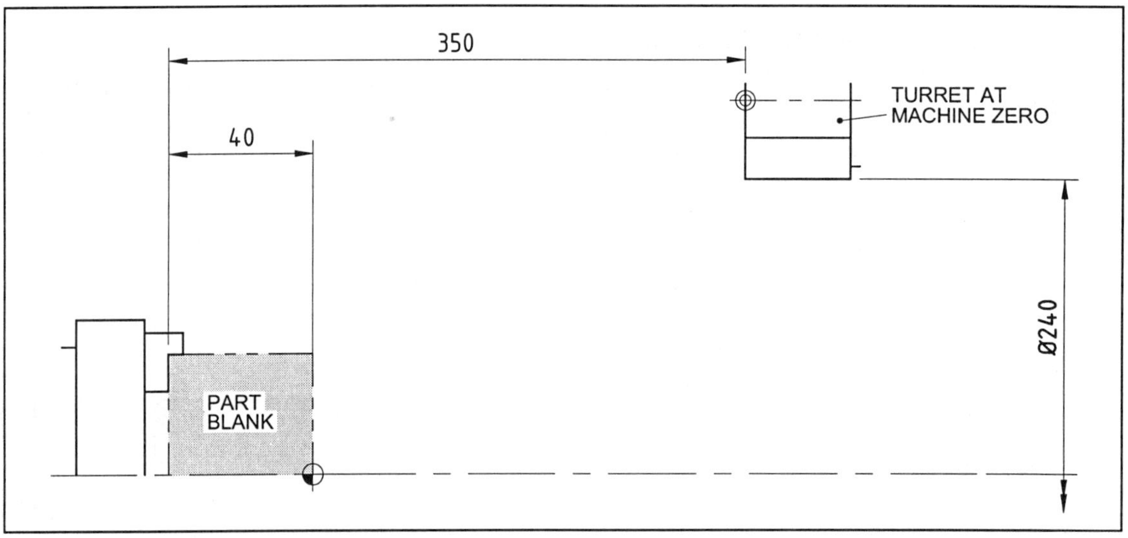

The first piece of the 'puzzle' defines the fixed machine dimensions provided by the manufacturer, when the empty tool turret is located at machine zero. If the Z-dimension is given to the chuck face, just add the width of the standard jaws for all related calculations. The width of the part is included in the illustration as part of the current setup, *not* part of the machine specifications.

When the cutting tools are mounted in the turret, their ***dimensions must also be known*** to the CNC programmer. This is the program section that makes the G50 programming method very inconvenient and perhaps even *hard* in some opinions (by today's standards).

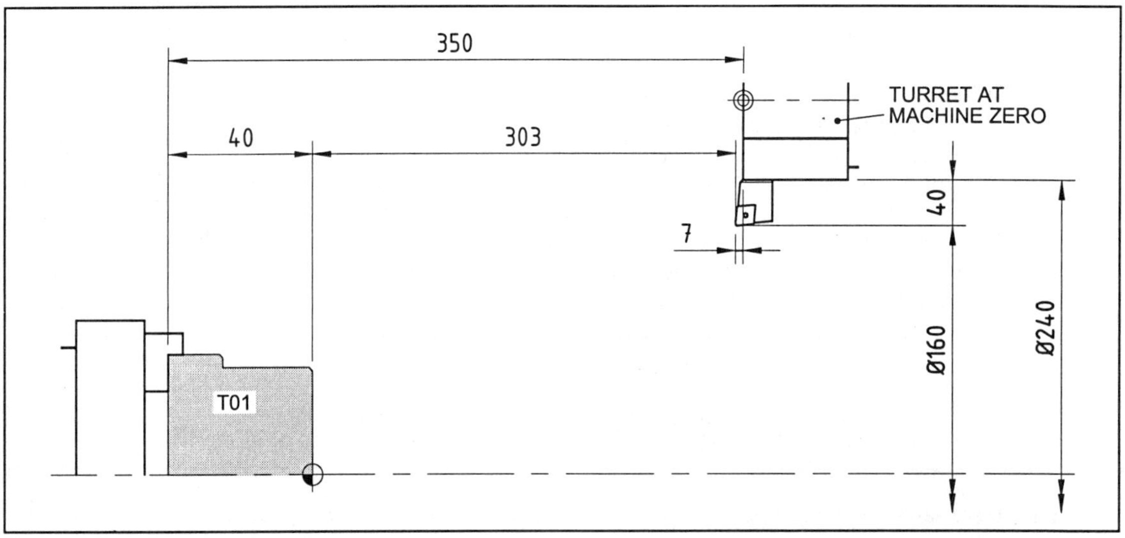

As there are three tools used by the program, the critical dimensions of each tool must be known *before* the actual program can be written. The illustration above shows the critical dimensions for tool T01 (turning tool) that are important to the overall setup and related program development.

The remaining two illustrations that follow show the equivalent *critical* dimensions - *also known when the program is written* - for the other two tools used by the program. Note that the data for all three tools listed are dimensioned from machine zero, and their X/Z dimensions are always known.

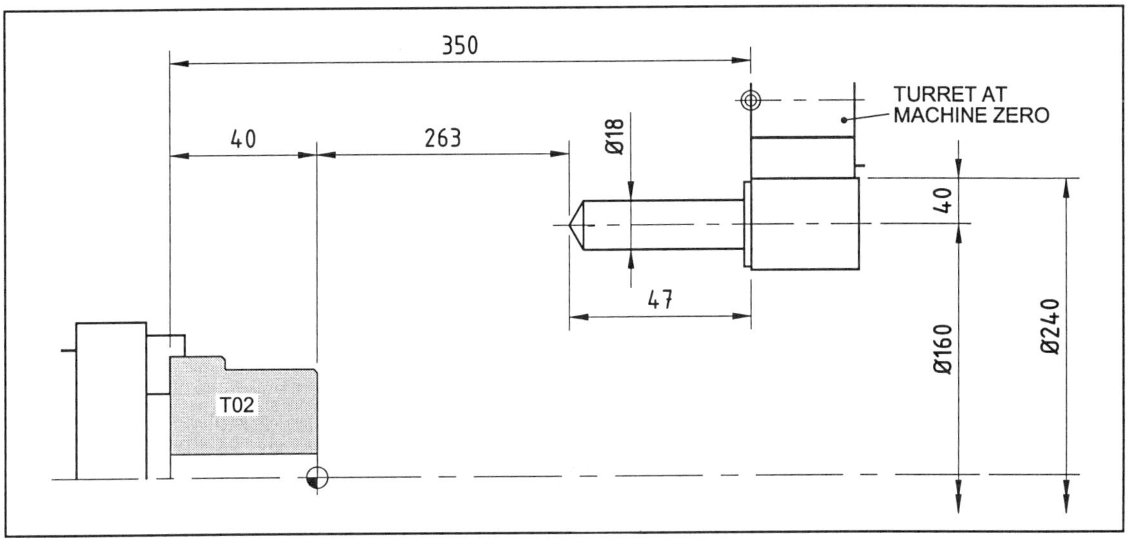

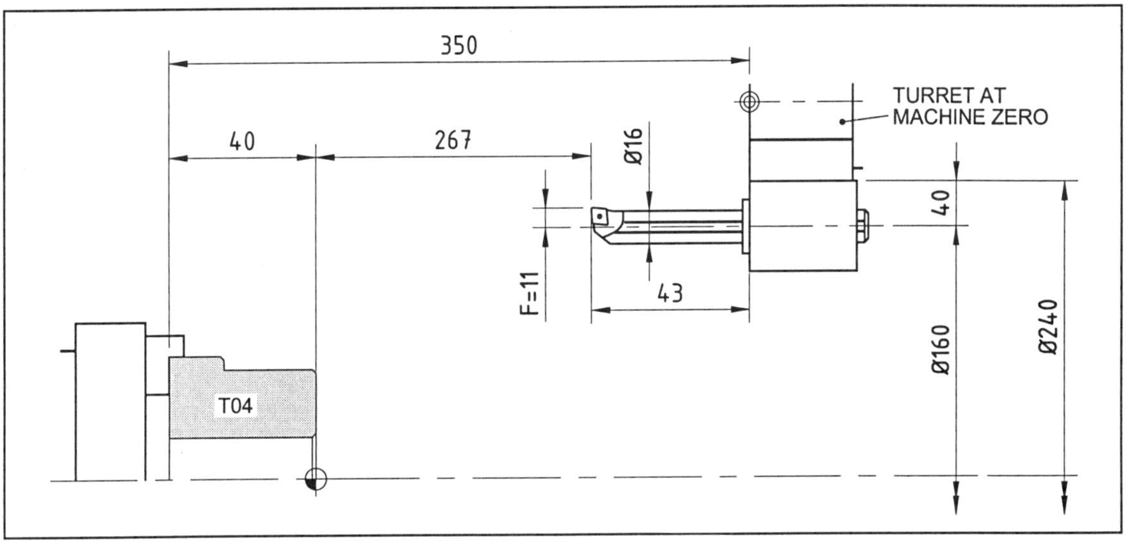

From the CNC programmer's viewpoint, once the individual tool dimensions are known relative to the provided machine and part dimensions, the program using the position register command G50 can be written and used for tool change at any reasonable location.

Tool Change Position

It is not necessary to return to machine zero after each tool has completed its work. The travel can be unnecessarily long, particularly along the Z-axis. That means the programmer has to select a tool change position for each tool. The basic rule (as defined earlier) is to make sure that the minimum clearance is applied to the longest tool. If the longest tool has enough clearance, every other tool will have even more clearance.

For the purposes of comparison, this example will be presented in two stages - in the first stage, each tool change will be at machine zero (easier but less efficient), the second stage will show how the tool change can be done close to the part in Z-axis (the shorter X-axis will still be done at machine zero position).

Tool Change at Machine Zero

In this case, the longest tool length is not important, as machine zero position is a fair distance from the part. The following program is listed in the order of described tools - T01, T02, T04:

```
N1 G21                              (METRIC UNITS)
N2 T0100                            (TOOL 01 LOCATED AT MACHINE ZERO)
N3 G50 X160.0 Z303.0                (O.D. TOOL TIP POSITION FROM PART ZERO)
N4 G96 S120 M03                     (SPINDLE SPEED AND ROTATION)
N5 G00 G42 X52.0 Z3.0 T0101 M08     (CHAMFER START POINT)
N6 G01 X58.0 Z-1.0 F0.125           (FRONT CHAMFER)
N7 Z-25.0 F0.3                      (OUTSIDE DIAMETER CUTTING)
N8 X63.0                            (FACE CUTTING)
N9 X67.0 Z-27.0 F0.125              (REAR CHAMFER)
N10 U5.0 F0.5                       (CLEARANCE RETRACT)
N11 G00 G40 X160.0 Z303.0 T0100     (RETURN TO THE MACHINE ZERO POSITION)
N12 G27 X160.0 Z303.0               (CHECK IF THE TOOL IS AT MACHINE ZERO)
N13 M01                             (OPTIONAL STOP)

N14 T0200                           (TOOL 02 LOCATED AT MACHINE ZERO)
N15 G50 X160.0 Z263.0               (DRILL TIP POSITION FROM PART ZERO)
N16 G97 S900 M03                    (SPINDLE SPEED AND ROTATION)
N17 G00 X0 Z5.0 T0202 M08           (START LOCATION FOR DRILLING)
N18 G01 Z-47.0 F0.375               (END LOCATION FOR DRILLING)
N19 G00 Z5.0                        (RETURN BACK TO START LOCATION)
N20 X160.0 Z263.0 T0200             (RETURN TO THE MACHINE ZERO POSITION)
N21 G27 X160.0 Z303.0               (CHECK IF THE TOOL IS AT MACHINE ZERO)
N22 M01                             (OPTIONAL STOP)

N23 T0400                           (TOOL 04 LOCATED AT MACHINE ZERO)
N24 G50 X182.0 Z267.0               (BORING BAR TIP POSITION FROM PART ZERO)
N25 G96 S110 M03                    (SPINDLE SPEED AND ROTATION)
N26 G00 G41 X30.0 Z3.0 T0404 M08    (CHAMFER START POINT)
N27 G01 X22.0 Z-1.0 F0.125          (FRONT CHAMFER)
N28 Z-42.0 F0.25                    (INSIDE DIAMETER CUTTING)
N29 U-5.0                           (CLEARANCE RETRACT)
N30 G00 Z3.0                        (RETRACT FROM THE HOLE)
N31 G40 X182.0 Z267.0 T0400         (RETURN TO THE MACHINE ZERO POSITION)
N32 G27 X182.0 Z267.0               (CHECK IF THE TOOL IS AT MACHINE ZERO)
N33 M30                             (END OF PROGRAM)
%
```

Note the inclusion of G27 command - it checks if the current tool position is actually at machine zero. In a way, the command in this program is redundant, but at least it brings some piece of mind - for example, if there is an error in the data input (data entry error). The G27 command has been described earlier in this chapter.

Tool Change Close to the Part

When the tool change takes place close to the part, the longest tool length *is* important. In the example used, the longest tool is the drill, with 47 mm overhang. The programmer selects any suitable clearance - for the example, 50 mm Z-clearance will used for the longest tool (T02). That means a very special care must be taken to program the G50 coordinates for all other tools. When the G50 method was the only method available to establish the current tool position, this particular reason was the weakest point of the whole concept. CNC programmers often resorted to the method of returning all tools to machine zero when each operation was completed. Convenience won over efficiency.

As the tool does not return back to the machine zero position, there is no need for the G27 command in this version. Although only the Z-clearance is shown, X-clearance will work the same way.

The last program will not change in *any way* that relates to the toolpath and cutting conditions. Only the G50 will change, based on the current tool position. Watch the underlined settings carefully - even when combined with the illustrations, this subject may take a while to fully understand. The most difficult part of this programming method is to determine the G50 position of the next tool, relative to the position of the current tool. For safety reasons, tool change position of any tool must be at least 50 mm away from the front face of part, that is from Z0. The result is a guarantee that no tool will be located behind the chuck during tool change, causing a possible collision.

In a summary, each tool will make a tool change at the exact location of the longest tool. However, because the length of each tool is different, the G50 Z-setting will also be different to maintain this minimum clearance position.

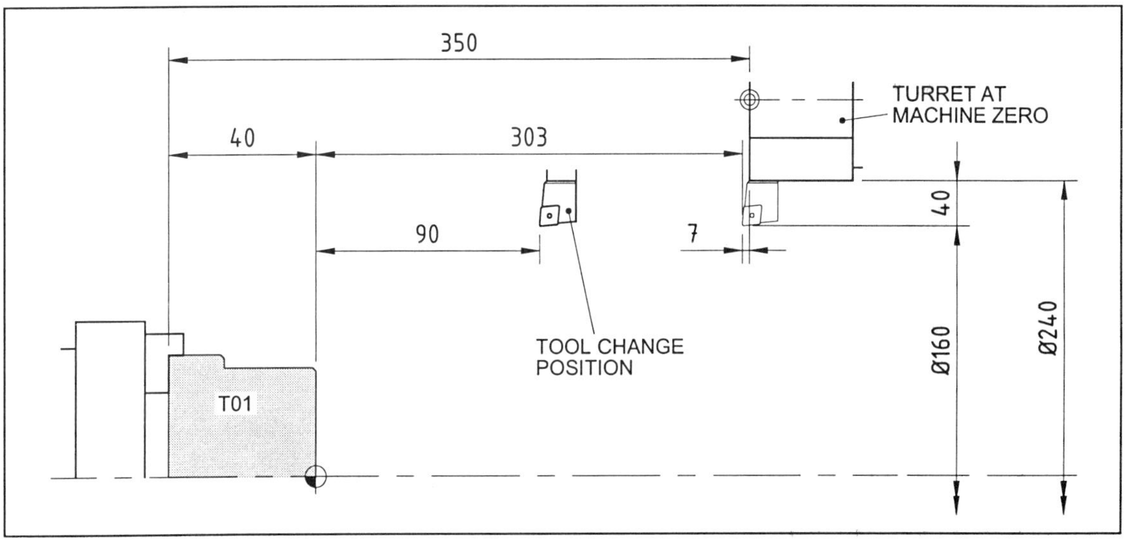

The calculation of the G50 at the tool change position of any tool considers the known settings:

G50 = Total distance of the current tool − Total distance of the longest tool + Clearance from Z0

For the first tool, the formula can be used with exact dimensions:

```
G50 = 303 - 263 + 50 = 90 = G50 X.. Z90.0
```

The longest tool (T02) total distance is 263 mm, and the clearance from Z0 for that tool is 50 mm. The program for the first tool still originates at machine zero, but terminates at a Z90.0 tool change position:

```
N1 G21                              (METRIC UNITS)
N2 T0100                            (TOOL 01 LOCATED AT MACHINE ZERO)
N3 G50 X160.0 Z303.0                (O.D. TOOL TIP POSITION FROM PART ZERO)
N4 G96 S120 M03                     (SPINDLE SPEED AND ROTATION)
N5 G00 G42 X52.0 Z3.0 T0101 M08     (CHAMFER START POINT)
N6 G01 X58.0 Z-1.0 F0.125           (FRONT CHAMFER)
N7 Z-25.0 F0.3                      (OUTSIDE DIAMETER CUTTING)
N8 X63.0                            (FACE CUTTING)
N9 X67.0 Z-27.0 F0.125              (REAR CHAMFER)
N10 U5.0 F0.5                       (CLEARANCE RETRACT)
N11 G00 G40 X160.0 Z90.0 T0100      (RETURN TO THE TOOL CHANGE POSITION)
N12 M01                             (OPTIONAL STOP)
```

The second tool - T02 - is also the longest tool, therefore its G50 Z.. must be G50 Z50.0. The formula listed above applies equally for this tool as well:

```
G50 = 263 - 263 + 50 = 50 = G50 X.. Z50.0
```

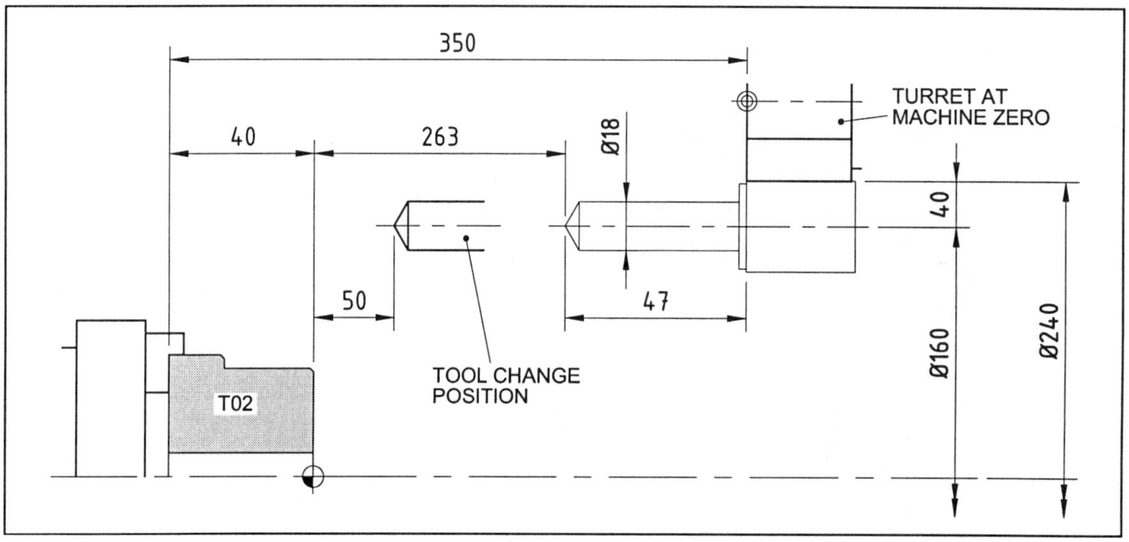

The program listing for the second tool shows the setting at the program beginning (block N14) and the return motion block N19, using the same coordinates.

```
N13 T0200                        (TOOL 02 LOCATED AT TOOL CHANGE POSITION)
N14 G50 X160.0 Z50.0             (DRILL TIP POSITION FROM PART ZERO)
N15 G97 S900 M03                 (SPINDLE SPEED AND ROTATION)
N16 G00 X0 Z5.0 T0202 M08        (START LOCATION FOR DRILLING)
N17 G01 Z-47.0 F0.375            (END LOCATION FOR DRILLING)
N18 G00 Z5.0                     (RETURN BACK TO START LOCATION)
N19 X160.0 Z50.0 T0200           (RETURN TO THE TOOL CHANGE POSITION)
N20 M01                          (OPTIONAL STOP)
```

When the original drawing was presented and three tools had been selected, there was a good reason for such a selection. In terms of the program structure, every first tool of the program represents program formatting at beginning (of the first tool), every second tool represents the formatting for all tools between the first one and the last one (the second tool), and the third tool represents the last tool format.

When this method is applied to the G50 X.. Z.. position register command, the original intent remains the same - the G50 for any middle tool (all tools after the first and before the last) will start and end at the same position - see above illustration.

The last tool (see program and illustration on the next page) is always treated somewhat differently, and programmer's preferences may not work here very well. When the last tool has completed all machining, the programmer may choose to return the tool to the tool change position, as any previous tool - *this is a wrong decision*. Returning the tool to a position other than machine zero means inconsistency between the position of the last tool for one part, and the position of the first tool for the next part. Since the program is written in such a way that the first tool starts from machine zero, the last tool *must* return to the same position - that means to machine zero. The suggestion is simple:

Start and finish each program at machine zero position

Based on the explanation for programming the first and last tool, the boring bar (T04) - the last tool from the example - can be programmed.

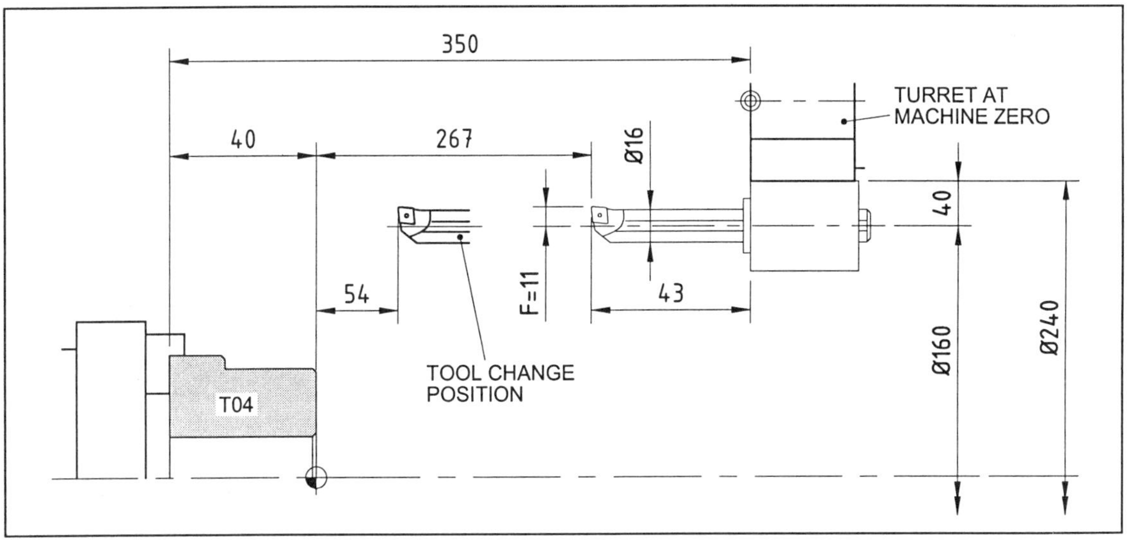


```
N21 T0400                          (TOOL 04 LOCATED AT TOOL CHANGE POSITION)
N22 G50 X182.0 Z54.0               (BORING BAR TIP POSITION FROM PART ZERO)
N23 G96 S110 M03                   (SPINDLE SPEED AND ROTATION)
N24 G00 G41 X30.0 Z3.0 T0404 M08   (CHAMFER START POINT)
N25 G01 X22.0 Z-1.0 F0.125         (FRONT CHAMFER)
N26 Z-42.0 F0.25                   (INSIDE DIAMETER CUTTING)
N27 U-5.0                          (CLEARANCE RETRACT)
N28 G00 Z3.0                       (RETRACT FROM THE HOLE)
N29 G40 X182.0 Z267.0 T0400        (RETURN TO THE MACHINE ZERO POSITION)
N30 G27 X182.0 Z267.0              (CHECK IF THE TOOL IS AT MACHINE ZERO)
N31 M30                            (END OF PROGRAM)
%
```

Only the last tool returns to the machine zero position, so the G27 command is justified, if required.

Conversion of G50 to Geometry Offset

Although the word 'conversion' in the title may seem a bit overstated (there is no actual conversion), how do we change a program listing that contains G50 to the modern method of programming, using *Geometry* and *Wear* offsets? Simple - take the G50 and its related blocks out of the program altogether. Here is the same program as listed above, but using the modern method - *no G50 at all.* Study it carefully, particularly the tool position at the end of each tool - it offers a reasonable position to provide a clearance for the tool change, considering all tools.

```
N1 G21                             (METRIC UNITS)
N2 T0100                           (TOOL 01 LOCATED AT MACHINE ZERO)
N3 G96 S120 M03                    (SPINDLE SPEED AND ROTATION)
N4 G00 G42 X52.0 Z3.0 T0101 M08    (CHAMFER START POINT)
N5 G01 X58.0 Z-1.0 F0.125          (FRONT CHAMFER)
N6 Z-25.0 F0.3                     (OUTSIDE DIAMETER CUTTING)
N7 X63.0                           (FACE CUTTING)
N8 X67.0 Z-27.0 F0.125             (REAR CHAMFER)
N9 U5.0 F0.5                       (CLEARANCE RETRACT)
N10 G00 G40 X150.0 Z100.0 T0100    (RETURN TO A SAFE TOOL CHANGE POSITION)
N11 M01                            (OPTIONAL STOP)
```

```
N12 T0200                          (TOOL 02 LOCATED AT TOOL CHANGE POSITION)
N13 G97 S900 M03                   (SPINDLE SPEED AND ROTATION)
N14 G00 X0 Z5.0 T0202 M08          (START LOCATION FOR DRILLING)
N15 G01 Z-47.0 F0.375              (END LOCATION FOR DRILLING)
N16 G00 Z5.0                       (RETURN BACK TO START LOCATION)
N17 X150.0 Z50.0 T0200             (RETURN TO A SAFE TOOL CHANGE POSITION)
N18 M01                            (OPTIONAL STOP)

N19 T0400                          (TOOL 04 LOCATED AT MACHINE ZERO)
N20 G96 S110 M03                   (SPINDLE SPEED AND ROTATION)
N21 G00 G41 X30.0 Z3.0 T0404 M08   (CHAMFER START POINT)
N22 G01 X22.0 Z-1.0 F0.125         (FRONT CHAMFER)
N23 Z-42.0 F0.25                   (INSIDE DIAMETER CUTTING)
N24 U-5.0                          (CLEARANCE RETRACT)
N25 G00 Z3.0                       (RETRACT FROM THE HOLE)
N26 G40 G28 U0 W0 T0400            (RETURN TO THE MACHINE ZERO POSITION)
N27 M30                            (END OF PROGRAM)
%
```

Summary

In a summary, the G50 command is very impractical and inefficient but it reflect the progress of its time - yes, it was a great progress, in spite of the appearances. Its main disadvantage is quite obvious - many dimensions must be known to the part programmer at the time of program development, which creates a situation not always practical or even possible. Repeating a tool is often very difficult and the whole concept is prone to many errors. Yet, millions of parts have been machined with the G50 used in the program.

Skip Command - G31

Do not confuse the terms *Skip Command* with the term *Block Skip Command* - they have nothing to do with each other. While block skip command is identified with a slash (usually as the first block character), the G31 is a programmable motion command, mainly used with probing devices on CNC machines. Other applications also include automatic tool length measurement and part alignment, where a similar sensor is part of the configuration. G31 in a command is very similar to G01, linear interpolation - it cannot be used for arc motions, and it cannot be used in regular machining. Also, it must always be used in the G40 mode (cutter radius cancel).

In order to program probes (or other devices), you will need a really good knowledge of custom macros as well as the methods how touch probing devices actually function. Without going into details - and without a complete programming example - here are some comments to offer some idea about this command:

In probing, the main purpose is to find a dimensional feature, for example width, depth, length, etc. The touch probe is an electronic device that is interfaced with the control system; it can send a signal to the control when a certain probing (measurement) is done. The signal is sent when the probe touches the part in a particular location, sending - for example - the XY location of the point. In order to do that, the programmer has to provide a certain travel motion for the probe, that covers the point to be measured. This location is known as a general position, but not precise position - that is the job of the probe. In other words, the precise location must be within the travel of the probe. When the probe moves from the start point to the end point of the travel, it encounters the measured point (feature). At that precise moment, the probe will convey that position to the control system, to be used for further processing However, there is some travel left - travel that is no longer necessary. This leftover travel must be - yes, skipped. That is when the G31 command does its job.

Other Seldom Used G-codes

Of course, the words *'rarely'* and *'seldom'* are very subjective and rather relative. A preparatory command listed here as 'seldom used' may be used daily by some programmers. The opposite is true as well - commands not listed in this chapter are considered common - or at least *more* common - yet, some programmers may not use them at all.

In this conclusion of the 'special' G-codes subject, just a few brief notes relating to those G-codes that could be listed here but were not:

Tool Length Offset Negative - G44

It is virtually impossible to consider G44 without considering the related command G43. Let's look at their standard definitions first:

G43 - Tool length offset positive

G44 - Tool length offset negative

The G44 command is (almost) never used in daily CNC programming. Think of the G44 command as the ***exact*** opposite of G43 command, and at the same time think of some reasonable and practical applications for this command - most likely, you will find none. In order to use G44 over G43, there must be a benefit that is not found in the G43 application - then the G44 may be justified..

Tool Length Offset Cancel - G49

This is one command that asks a simple question - *to use it or not to use it*? It has its firm adherents and its equally determined opponents - both camps claim that their approach is better. Without prejudice, let me state this - you programmers in both camps are right - up to a point, at least. In that respect, G49 becomes a rather strange command, perhaps even a bit foggy. Its single purpose is also a simple purpose - to *cancel the tool length offset*, initialized by the G43 (or G44) command, usually programmed at the beginning of each tool. So what is all the fuss about?

A lot depends on the part program itself - how it is written, particularly in terms of its structure. Program structure determines the *order* and *placement* of various commands and functions within the program body at the most convenient locations, particularly before and after the many machining sequences. In my *CNC Programming Handbook*, I have promoted a program structure that has been proven to work for the majority of programming situations. Here is its generic format:

```
G21
G17 G40 G80 T01
M06
G90 G54 G00 X.. Y.. S.. M03 T02
G43 Z2.0 H01 M08
G01 Z-... F...
   <... machining commands ...>
G00 Z2.0 M09
G28 Z2.0 M05
M01
```

No G49 at all. Why? Because the program structure is designed to *allow* its omission. What is a seldom known fact is that the G28 (machine zero return) command *automatically cancels* the tool length offset. This statement should satisfy the opponents of programming G49. I plead guilty here.

Like a coin with two sides, there are also two sides of this argument, and fair is fair. If you wish to include the G49 in the part program, why not? Go for it. The above structure can be modified in two places, just by adding the G49 command:

```
N1 G21
N2 G17 G40 G49 G80 T01                          (G49 USED)
N3 M06
N4 G90 G54 G00 X.. Y.. S.. M03 T02
N5 G43 Z2.0 H01 M08
N6 G01 Z-5.0 F...
   <... machining commands ...>
N27 00 Z2.0 M09
N28 G49 G28 Z2.0 M05                            (G49 USED)
N29 M01
```

If that is not *much ado about nothing*, then what is? While both approaches are very acceptable, and the program structure is followed - where is the catch? There is none - *providing the structure is followed!* The most important part of the program structure is the call of G43, before the machining sequences take over. Now consider this - what will happen *if* the G43 block is forgotten, simply missing for the next tool. It should not happen, but what if it *does* happen? There are two possibilities:

- **Possibility #1** ... **G49 is used in the program**
- **Possibility #2** ... **G49 is *not* used in the program**

Going from one tool to another within a single program always means repeating many commands, even if they are modal. This is a very worthwhile practice and saves many headaches later. So, here is a program structure for two tools, where G49 is used, but G43 is *'forgotten'* for the second tool:

```
(=== G49 IS USED IN THE PROGRAM ===)
N1 G21
N2 G17 G40 G49 G80 T01                          (G49 USED)
N3 M06
N4 G90 G54 G00 X.. Y.. S.. M03 T02
N5 G43 Z2.0 H01 M08
N6 G01 Z-5.0 F...
   <... machining commands for T01 ...>
N27 G00 Z2.0 M09
N28 G49 G28 Z2.0 M05                            (G49 USED)
N29 M01

N30 T02
N31 M06
N32 G90 G54 G00 X.. Y.. S.. M03 T03
   <... missing G43 block ... >                 (N33 G43 Z2.0 H02 M08 missing)
N34 G01 Z-5.0 F...
   <... machining commands for T02 ...>
N52 G00 Z2.0 M09
N53 G49 G28 Z2.0 M05                            (G49 USED)
N54 M01
```

Apart from a relatively minor problem of losing the coolant function, the consequences of the G43 omission for the second tool are far more severe. This is what will happen (correct setup is assumed) for *Possibility #1* (G49 is used in the program):

G49 IS Programmed

The first tool will machine the part just fine. Then, the next tool - T02 - is brought into the spindle and moves to the first XY location. So far, all goes as planned. The block after N32 should be G43 block, etc., - but that block is *missing!* There will be no alarm, no message - no notification whatsoever - in fact, there will be *no warning*. The control system will simply interpret the Z-axis motion in block N34 as originating from machine zero - there is no tool to be considered, there is no tool length to be considered. Because of that, the point at the spindle gage line (spindle reference point at machine zero) will act as the tool tip and will try to move to the Z-5.0 location. That may not present a problem, *if the spindle were empty*. Of course, that is not the case at all - the spindle is *not* empty - there is a real tool - set and designated as T02 - and this tool has a length that reaches far below the spindle gage line. Situations such as this one provide good reasons for real worries, because without an alert reaction of the CNC operator, a disaster is ready to strike.

☞ ***The conclusion is that unless the CNC operator is very alert, collision is practically unavoidable***

G49 IS NOT Programmed

What about the other alternative - the one where G49 is *not* included in the program? Is it a better alternative? Does it solve the problem? Should you prefer it?

Evaluate the following program - it is the same one as above, but no G49 is programmed.

```
(=== G49 IS NOT USED IN THE PROGRAM ===)
N1 G21
N2 G17 G40 G80 T01
N3 M06
N4 G90 G54 G00 X.. Y.. S.. M03 T02
N5 G43 Z2.0 H01 M08
N6 G01 Z-5.0 F...
   <... machining commands for T01 ...>
N27 G00 Z2.0 M09
N28 G28 Z2.0 M05
N29 M01

N30 T02
N31 M06
N32 G90 G54 G00 X.. Y.. S.. M03 T03
   <... missing G43 block ... >            (N33 G43 Z2.0 H02 M08 missing)
N34 G01 Z-5.0 F...
   <... machining commands for T02 ...>
N52 G00 Z2.0 M09
N53 G28 Z2.0 M05
N54 M01
```

To answer the earlier questions - yes, *this is* a better alternative to the previous version, and no, it *does not* solve the problem. Whether you should prefer this version largely depends on the major consideration, described next. Don't make too many assumptions here and don't think that all is clear - the situation is still potentially dangerous, but *a lot less* dangerous than in the previous example. How much is less? Actually 50% less - and here is why.

Again, the first tool will perform flawlessly, as before. This time, there is no G49 tool length cancellation transferred to the next tool (T02), so the current tool length will still be in effect. In this case, the tool length offset for the tool T02 will be exactly the same as the tool length offset for tool T01. Not exactly an ideal situation, but a situation that has a chance to succeed.

The benefit of ***not*** programming the G49 is based on the rather dubious chance of 50/50 in case of a program error. If the *actual* physical length of T02 is shorter the *actual* physical length of T01, no collision will take place - and no machining will take place either. Of course, this is not the expected activity, but at least it is an activity that does no harm. Corrections to the program still have to made, but under friendly overall conditions.

If the program does include G49 and the situation is equal to the one described, the potential for a collision is much greater. Please, do not make the mistake of thinking that 50/50 chance is better - it is not. Both examples create problems and are potentially dangerous situations. The focus of this section was to come to a single conclusion:

Excluding G49 from the program may prove to be more beneficial than including it

As you see from the demonstration, even an apparently insignificant missing of a single G-code often makes a great difference in the final results. So what is the moral of this section ... ?

Always program G43 H.. for each tool in the program - G49 then becomes irrelevant

Conclusion

Other G-codes can be added to this chapter, depending on the exact interpretation of the term *'seldom used G-codes'*. In my *CNC Programming Handbook*, subtitled *A Comprehensive Guide to Practical CNC Programming*, most G-codes were explained in great detail, such as coordinate rotation G68 and G69, data input G10, scaling commands G50 and G51. Also included were some G-codes that are not used very often or not used any more, such as position compensation codes G45 through G48. Fanuc Macro B related G-codes G65 through G67 and the whole subject of macros are covered in a separate book, *Fanuc CNC Custom Macros*, with the subtitle *Programming Resources for Fanuc Custom B Users*. Both books have been published by Industrial Press, Inc., New York, NY, USA.

In this handbook, a special chapter has been set aside for the plane selection commands G17 through G19, containing many details, techniques and special applications.

8 TOOL LENGTH OFFSET CHANGE

During typical everyday program development, the number of tool length offsets in a program will correspond to the number of different tools used in the same program. Normally, each tool requires - *and uses* - only a single length offset. It is also typical, that the tool number and the offset number are the same, for a logical association during setup: T01 will be associated with H01, T16 with H16, etc. In some cases, the CNC programmer will benefit from programming two (or even more) tool length offsets for a single tool. Also, there are other methods available that achieve the same goal, so consider the methods and techniques described in this chapter as several different options.

Tool Length Offset

The concept of programming tool length should be familiar to all programmers and operators. In its simplest form, tool length offset is the measured distance from a tool reference point (command point) to part reference point (Z0). In reality, it is not a measurement of an actual tool, but its *distance* only. This method is the most popular, particularly in small shops and job jobs, as it does not require any special setup equipment. Another method, commonly used in high production environment, uses a special tool presetter that allows tool length (and other tool data) to be set off-machine. For the purposes of this chapter, no setup equipment is needed, and the tool setup will use the common 'touch-off' method, measured from the bottom of the tool cutting edge to Z0 of the part.

The drawing shows a part with a slot that is wider than the tool width:

Slot width:

14 - 6 = 8 mm

Cutter width:

5 mm

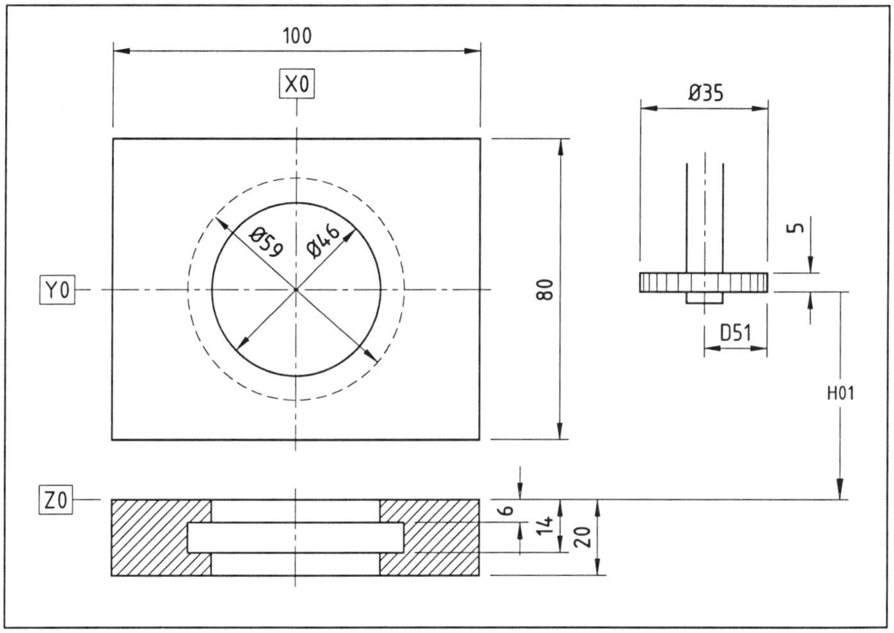

The selected tool in the program will use G43 H01 for the tool length offset and G41 D51 for the cutter radius offset. A subprogram that would shorten the program could be developed for the slot machining, but it will not be used for this project.

Offset Adjustment

Practical Application

The drawing shows a 59 mm diameter slot groove in a 46 mm diameter bored hole. The programming objective is to use a circular slot mill in a circular interpolation mode, in order to machine the slot. The goal is to illustrate programming techniques to complete the circular slot, using the common touch-off tool length offset method.

For the example, the selected slot mill has the diameter of 35 mm and the width of 5 mm. It is necessary to select a cutter that has diameter *smaller* than the bored hole. The slot mill width of 5 mm is not enough to machine the full groove width of 8 mm. The 8 mm width is not shown on the drawing directly, but can be calculated from the given absolute values. It can be calculated as the difference between slot top and the slot bottom, which is 14 - 6 = 8 mm.

There are several methods to approach this project. Three different examples will be shown in this chapter for comparison. All examples will share the same cutter and machining technique - the bottom edge of the cutter will finish the 14 mm absolute depth, the top edge of the tool will finish the 6 mm absolute depth. All programming methods will use the cutter already mounted in the spindle (no tool change required in the program), and the center of the bore and top face of the part as program zero (X0 Y0 Z0). The ∅46 mm bored hole is assumed to exist in the part. For testing purposes, the G54 setting will have Z-offset set to Z0.0000, and the H01 offset will be set as -200.000. XY settings are important for the actual setup but not important for the example.

All three methods will use the following machining process (steps are matched in each program) :

STEP 01	**From machine zero, rapid to the center of the bore (XY only)**
STEP 02	**Apply tool length offset and rapid to a suitable clearance above the part**
STEP 03	**Feed-in to the depth of the first circular cut (for the bottom edge of the slot)**
STEP 04	**Apply cutter radius offset, and lead in towards the start point of the slot**
STEP 05	**Cut the bottom part of the slot in climb milling mode - full circle cutting**
STEP 06	**Lead out from the slot - cancel cutter radius offset during motion to the center of the bore**
STEP 07	**Feed-in to the depth of the second circular cut (for the top edge of the slot)**
STEP 08	**Apply cutter radius offset, and lead in towards the start point of the slot**
STEP 09	**Cut the top part of the slot in climb milling mode - full circle cutting**
STEP 10	**Lead out from the slot - cancel cutter radius offset during motion to the center of the bore**
STEP 11	**Retract above the part**
STEP 12	**Return the Z-axis to the machine zero**
STEP 13	**Return the XY axes to the machine zero**
STEP 14	**End the program**

For more precision, three cuts could have been used, as well as roughing and finishing operations, even using two tools. However, for the purpose of this project, the described procedure will be used as the basis for other types of machining, depending on the work required.

Programming Method 1 - No Offset Adjustment

This is the basic machining method that works very well, providing the slot is machined for general purposes, rather than a precision slot. This method *does not* use two tool length offsets, it uses only the *standard* single offset (H01). It is presented here as the most common method, and for comparison with the other two methods. The steps established prior to the program development are matched:

```
(CIRCULAR SLOT - SINGLE TOOL LENGTH OFFSET METHOD)
N1 G21                                (METRIC MODE)
N2 G17 G40 G80                        (INITIAL SETTINGS)
N3 G90 G00 G54 X0 Y0 S1200 M03        (STEP 01 - CENTER OF BORE)
N4 G43 Z5.0 H01 M08                   (STEP 02 - H01 ACTIVATED)
N5 G01 Z-14.0 F600.0                  (STEP 03 - H01 IN EFFECT - SLOT BOTTOM)
N6 G41 X4.5 Y-25.0 D51 F300.0         (STEP 04 - LEAD-IN LINE)
N7 G03 X29.5 Y0 R25.0                 (STEP 04 - LEAD-IN ARC)
N8 I-29.5                             (STEP 05 - FULL CIRCLE CUTTING)
N9 X4.5 Y25.0 R25.0                   (STEP 06 - LEAD-OUT ARC)
N10 G40 G01 X0 Y0 F600.0              (STEP 06 - LEAD-OUT LINE)
N11 Z-11.0                            (STEP 07 - CALCULATED ADJUSTMENT - TOP)
N12 G41 X4.5 Y-25.0 D51 F300.0        (STEP 08 - LEAD-IN LINE)
N13 G03 X29.5 Y0 R25.0                (STEP 08 - LEAD-IN ARC)
N14 I-29.5                            (STEP 09 - FULL CIRCLE CUTTING)
N15 X4.5 Y25.0 R25.0                  (STEP 10 - LEAD-OUT ARC)
N16 G40 G01 X0 Y0 F600.0              (STEP 10 - LEAD-OUT LINE)
N17 G00 Z25.0 M09                     (STEP 11 - RETRACT ABOVE PART)
N18 G28 Z25.0 M05                     (STEP 12 - Z AXIS TO MACHINE ZERO)
N19 G28 X0 Y0                         (STEP 13 - XY AXES TO MACHINE ZERO)
N20 M30                               (STEP 14 - PROGRAM END)
%
```

In this example, block N11 is the key block - it is the block that fully controls the cutter position for the *second* slot cut. Since the second slot cut is located at absolute depth of Z-6.0 - and the cutter width is 5 mm - the bottom edge of the slot mill *must be* at Z-11.0 location:

6 mm top as dimensioned + 5 mm cutter width = 11 mm

In machining terms, there is nothing wrong with this method. Providing there is no significant tool wear expected and the slot width tolerances are not excessive, this programming method will do just fine. Unfortunately, tool wear and drawing tolerances are part of programmer's everyday life - and they are not reflected in this method. In short, as good as the program is overall, it lacks one important feature - *flexibility*. Since the width of the cutter is built into the programmed dimensions, even a slight change of the cutter width will require a change within the program itself.

The next programming method eliminates any change withing the program altogether. Using the second method, any change(-s) required can be performed at the control system, by a skilled CNC operator. *This method introduces a **secondary** tool length offset.*

Programming Method 2 - With Offset Adjustment

Since *two* tool length offsets are used in the *same* program - a significant departure from common programming - it is important to understand the purpose they will serve. As in the previous example, the H01 offset (set to -200.0) will control the *bottom edge* of the slot. The second offset - H91 - will control the *top edge* of the slot, and *must be* set 5 mm higher on the cutter (Z+ direction). For the actual setting process itself, it means finding the tool length for the bottom edge (stored in H01 offset), then storing the secondary tool length offset in H91 registry. This setting will still have a negative amount, but 5 mm *further away from Z0*. In the example presented in this chapter, if the tool length offset H01 is set to -200.0, then the offset H91 must be set to -205.0 !

One key element of this approach is important to understand before using this particular programming technique. The tool length offset is activated by G43 command, which is a *modal* command - it means it remains in effect until canceled (usually with G28 Z.. or G49).

The illustration shows the setup measurements of both offsets. The secondary offset H91 is measured 5 mm higher than offset H01. Both offsets are stored as negative.

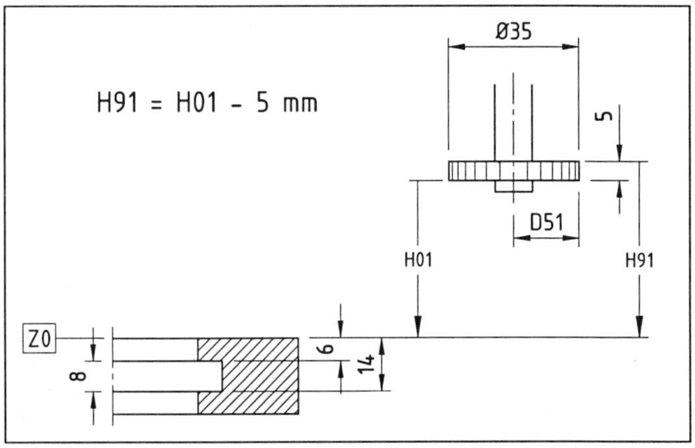

The initial part program will be changed in the block N11 only:

```
(CIRCULAR SLOT - DOUBLE TOOL LENGTH OFFSET METHOD)
N1 G21                              (METRIC MODE)
N2 G17 G40 G80                      (INITIAL SETTINGS)
N3 G90 G00 G54 X0 Y0 S1200 M03      (STEP 01 - CENTER OF BORE)
N4 G43 Z5.0 H01 M08                 (STEP 02 - H01 ACTIVATED)
N5 G01 Z-14.0 F600.0                (STEP 03 - H01 IN EFFECT - SLOT BOTTOM)
N6 G41 X4.5 Y-25.0 D51 F300.0       (STEP 04 - LEAD-IN LINE)
N7 G03 X29.5 Y0 R25.0               (STEP 04 - LEAD-IN ARC)
N8 I-29.5                           (STEP 05 - FULL CIRCLE CUTTING)
N9 X4.5 Y25.0 R25.0                 (STEP 06 - LEAD-OUT ARC)
N10 G40 G01 X0 Y0 F600.0            (STEP 06 - LEAD-OUT LINE)
N11 Z-6.0 H91                       (STEP 07 - H91 IN EFFECT - SLOT TOP)
N12 G41 X4.5 Y-25.0 D51 F300.0      (STEP 08 - LEAD-IN LINE)
N13 G03 X29.5 Y0 R25.0              (STEP 08 - LEAD-IN ARC)
N14 I-29.5                          (STEP 09 - FULL CIRCLE CUTTING)
N15 X4.5 Y25.0 R25.0                (STEP 10 - LEAD-OUT ARC)
N16 G40 G01 X0 Y0 F600.0            (STEP 10 - LEAD-OUT LINE)
N17 G00 Z25.0 M09                   (STEP 11 - RETRACT ABOVE PART)
N18 G28 Z25.0 M05                   (STEP 12 - Z AXIS TO MACHINE ZERO)
N19 G28 X0 Y0                       (STEP 13 - XY AXES TO MACHINE ZERO)
N20 M30                             (STEP 14 - PROGRAM END)
%
```

In this version, the part program contents will never change, regardless of the cutter used. A slight disadvantage of this method is the additional work the CNC operator must perform (setting two length offsets rather than one), resulting in possible human errors during frequently repeated machine setups.

As the cutter width (5 mm) has to be considered for setting of the H91 offset, errors can happen. Solution to this problem is in the third programming method, one that uses *macro features* of the control system, providing they are available (*Custom Macro B* is an optional feature on most Fanuc controls). The macro will calculate the second offset value automatically - all input needed is the cutter width.

Programming Method 3 - Advanced Macro Method

The key to this less common method exists in two advanced CNC programming techniques - one technique uses the concept of variables, the other uses the data input command G10. There are several variations on this method available as well, because it offers great flexibility. This method was tested on Fanuc 16MB control system, with tool offset memory *Type C* (separate *Geometry* and *Wear* offsets for tool length offset and cutter radius offset).

The program is not a true macro similar to a subprogram, but it does use macro features:

```
(CIRCULAR SLOT - ADVANCED MACRO METHOD)
(-------------------------------------------------------------------)
#100 = 5.0                         (STORE THE CUTTER WIDTH)
(-------------------------------------------------------------------)
N1 G21                             (METRIC MODE)
N2 G17 G40 G80                     (INITIAL SETTINGS)
N3 G90 G10 L10 P91 R0              (CLEAR LENGTH OFFSET H91 SETTING TO 0)
N4 #101 = #2201                    (STORE SETTING OF LENGTH OFFSET H01)
N5 #101 = #101-#100                (CALCULATE NEW SETTING OF OFFSET H91)
N6 G10 L10 P91 R#101               (STORE NEW SETTING AS OFFSET H91)
N7 G00 G54 X0 Y0 S1200 M03         (STEP 01 - CENTER OF BORE)
N8 G43 Z5.0 H01 M08                (STEP 02 - H01 ACTIVATED)
N9 G01 Z-14.0 F600.0               (STEP 03 - H01 IN EFFECT - SLOT BOTTOM)
N10 G41 X4.5 Y-25.0 D51 F300.0     (STEP 04 - LEAD-IN LINE)
N11 G03 X29.5 Y0 R25.0             (STEP 04 - LEAD-IN ARC)
N12 I-29.5                         (STEP 05 - FULL CIRCLE CUTTING)
N13 X4.5 Y25.0 R25.0               (STEP 06 - LEAD-OUT ARC)
N14 G40 G01 X0 Y0 F600.0           (STEP 06 - LEAD-OUT LINE)
N15 Z-6.0 H91                      (STEP 07 - H91 IN EFFECT - SLOT TOP)
N16 G41 X4.5 Y-25.0 D51 F300.0     (STEP 08 - LEAD-IN LINE)
N17 G03 X29.5 Y0 R25.0             (STEP 08 - LEAD-IN ARC)
N18 I-29.5                         (STEP 09 - FULL CIRCLE CUTTING)
N19 X4.5 Y25.0 R25.0               (STEP 10 - LEAD-OUT ARC)
N20 G40 G01 X0 Y0 F600.0           (STEP 10 - LEAD-OUT LINE)
N21 G00 Z25.0 M09                  (STEP 11 - RETRACT ABOVE PART)
N22 G28 Z25.0 M05                  (STEP 12 - Z AXIS TO MACHINE ZERO)
N23 G28 X0 Y0                      (STEP 13 - XY AXES TO MACHINE ZERO)
N24 M30                            (STEP 14 - PROGRAM END)
%
```

In this method, the CNC operator sets the slot mill exactly the same way as any other tool, by measuring the distance from the tool tip (tool bottom edge) to the top of the part (Z0) from machine zero position. If the measured distance is 200.0 mm, the H01 will be set to -200.0. The value of H91 is totally irrelevant during setup - ***it can contain literally any value***. Once the program is processed, the block N3 will always clear any amount stored in the geometry setting of offset H91. H91 offset setting will now be 0.000. The following block, N4, will store the current value of offset H01 (automatically registered in the *system* variable #2201) into variable #101. The next block, N5, will redefine the amount of #101 by subtracting the cutter width, stored as a positive value at the top of the program, in variable #100. If #100=5.0, and #101 was -200.0, the block N5 formula will calculate #101-#100 as -200.0-5.0, with the result of -205.0. This is the value of the *second* tool length offset H91. In order to store this amount in the control system, G10 data input command is used in block N6. The rest of the program is unchanged, except for the block numbers.

Although this method also requires a program change (#100 = <varies>), it is done in a much safer way than the previous method. With the use of true Fanuc macros, this technique can be expanded upon and many error trapping routines can be added and built-in, for example, if the tool width is greater than the groove width. Tool wear offset can be added, as well as other improvements. *Chapter 24* explains the basic macro concepts, but to fully understand macros, a special purpose publication will be required, such as *Fanuc CNC Custom Macros*, written by yours truly.

Unless you are familiar with macro programming, keep away from the last example. For the beginner, the concept of system variables (such a #2201) may be a difficult one. The subject itself is quite simple, but the difficulty is in the inconsistency of the system variables used by different Fanuc control systems. In simple terms, the system variables for the same purpose vary with different controls.

Offset Adjustment - Setup for Two Parts

In any CNC program, various heights of the part are associated with the tool length offset, relative to the part Z0 location. The previous section of this chapter has shown that it is possible in CNC programming to change the tool length offset within a single program. The focus of the previous example was on programming a single part. In this section, the focus will be on two (or more) parts that use only one cutting tool and apply its motions to parts of *different heights - Part A* and *Part B*.

There are three common methods to achieve the stated goal:

- **Method 1 - Use only one work offset and one tool length offset (differences changed by program)**
- **Method 2 - Use two work offsets and one tool length offset (differences changed by work offsets)**
- **Method 3 - Use two work offsets and two tool length offsets (differences changed by both offsets)**

Regardless of which method the programmer may consider most practical for a given part, the *Method 1* is the weakest and is not recommended, although some may consider it to be the simplest. Toolpath that includes fixed adjustment is not flexible for editing, and is also subject to errors of omission. In addition, the exact XY distance between the two parts would have to be known, which is generally not the case. On the other hand, *Method 2* is the most commonly used, and highly recommended. *Method 3* is also correct but less convenient, and is included here mainly to illustrate the concept. In the two last cases, two work offsets are required (G54 and G55, for example).

Four holes will be machined through each part (A and B), using only a ∅12 mm drill for the example. In both cases, the tool will start at 2mm above the top face (R2.0 in the G81 cycle), will have breakthrough of 1 mm below the part and additional 3.6 mm (12 x 0.3) for drill point length, using Z-20.6 depth for *Part A* and Z-28.6 for *Part B*.

Both parts will be setup at the same time. In *Method 1* (shown below), the difference between XYZ origin of *Part A* and XYZ origin of *Part B* will be known.

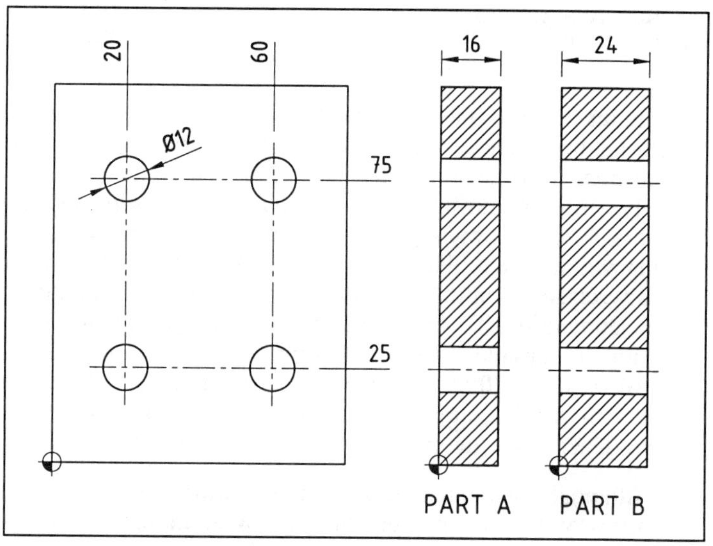

Method 1 - One Work Offset + One Length Offset

This method is not flexible and cannot be used for all applications. Because only one work offset is programmed, the X, Y and Z distances between the origins of the two parts must be known and incorporated into the program. That is not the case with majority of multiple parts setup. Unless using a specially designed fixture, it is very unlikely that the distances (particularly XY distance) will be known at the time of programming. There is, however, a situation when these distances will be known, but that will only apply to a single part that contains both patterns at different heights. If - and when - the conditions are favorable, the program for the four holes can be developed.

In the illustration on the next page, complete setup is shown.

Note that only one part zero is provided. The underlined entries are most important for the actual machining.

```
N1 G21
N2 G17 G40 G80
N3 G90 G54 G00 X20.0 Y75.0 S800 M03
N4 G43 Z25.0 H01 M08
N5 G99 G81 R2.0 Z-20.6 F180.0          ( A )
N6 Y25.0                               ( B )
N7 X60.0                               ( C )
N8 G98 Y75.0                           ( D )
N9 G99 X133.0 R10.0                    ( E )
N10 Y25.0                              ( F )
N11 X173.0                             ( G )
N12 Y75.0                              ( H )
N13 G80 Z25.0 M09
N14 G28 Z25.0 M05
N15 M30
%
```

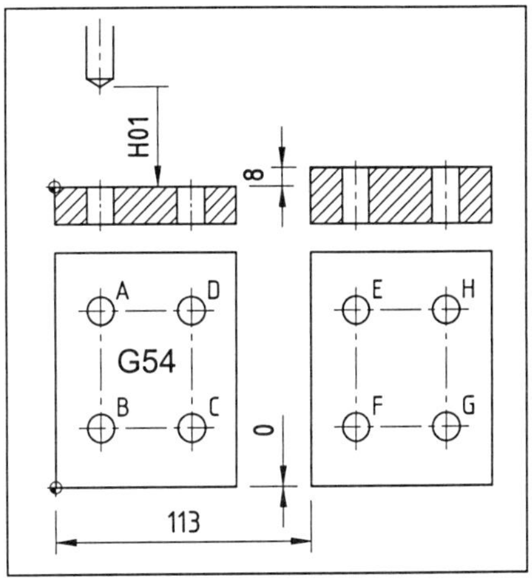

The order of drilling is the same for both parts. Note that the Z-20.6 depth has not changed, only the starting position *R* in the G81 cycle. Also, 133 mm have been added to all X-coordinates for *Part B*.

At least two other programming techniques can be used for this particular setup. One is the use of local coordinate offset G52, the other is to use a subprogram. Various combinations are possible.

Method 2 - Two Work Offsets + One Length Offset

Programming two work offsets makes the setup marginally longer, but has certain advantages. The most important one is that the four hole locations are the same for both parts.

```
N1 G21
N2 G17
N3 G90 G54 G00 X20.0 Y75.0 S800 M03
N4 G43 Z25.0 H01 M08
N5 G99 G81 R2.0 Z-20.6 F180.0          ( A )
N6 Y25.0                               ( B )
N7 X60.0                               ( C )
N8 G98 Y75.0                           ( D )
N9 G55 G99 X20.0 Y75.0 R2.0 Z-28.6     ( E )
N10 Y25.0                              ( F )
N11 X60.0                              ( G )
N12 Y75.0                              ( H )
N13 G80 Z25.0 M09
N14 G28 Z25.0 M05
N15 M30
%
```

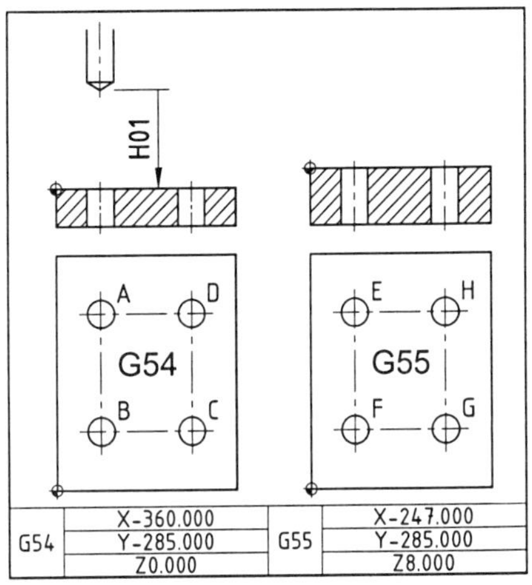

Both G54 and G55 work offset settings are only examples

When setting up the job, the part zero for each part will be established by normal means, for example, using an edge finder. In this example, the tool length offset H01 was measured to the Z0 of the part on the left, so the Z-setting in G54 is equal to zero. Once the difference in parts heights is known (it may not always be dimensioned in the drawing), this difference is entered into the Z-setting of work offset G55. If the *Part B* is higher than *Part A*, the Z-setting will be positive; if the *Part B* is lower than *Part A*, the Z-setting will be negative.

It is also possible to use two work offsets and two tool length offsets, as the next example shows.

Method 3 - Two Work Offsets + Two Length Offsets

There is only a very small difference in the program when compared with the *Method 2*. The secondary tool length offset H91 has been added to block N9. In this block, the tool moves from *Part A* to *Part B*. Also, the G98 return to initial level guarantees sufficient clearance above *Part B*.

```
N1 G21
N2 G17
N3 G90 G54 G00 X20.0 Y75.0 S800 M03
N4 G43 Z25.0 H01 M08
N5 G99 G81 R2.0 Z-20.6 F180.0           ( A )
N6 Y25.0                                ( B )
N7 X60.0                                ( C )
N8 G98 Y75.0                            ( D )
N9 G55 G99 X20.0 Y75.0 R2.0 Z-28.6 H91  ( E )
N10 Y25.0                               ( F )
N11 X60.0                               ( G )
N12 Y75.0                               ( H )
N13 G80 Z25.0 M09
N14 G28 Z25.0 M05
N15 M30
%
```

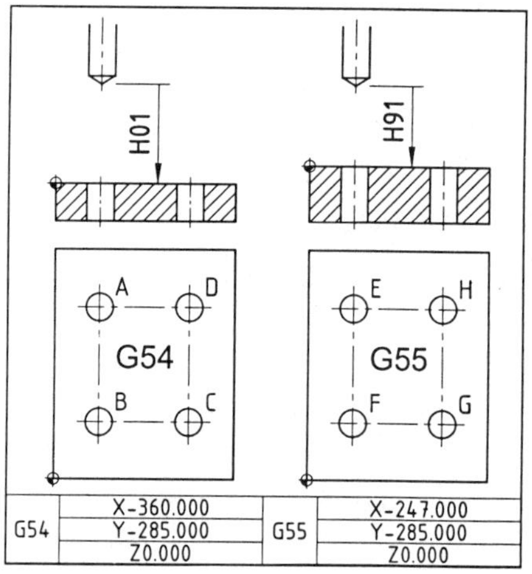

Both G54 and G55 work offset settings are only example

Note that there is no need to change the Z-settings of either work offset - the tool length offset settings take care of the proper tool position.

In this chapter, several methods have been presented to demonstrate various techniques of changing the tool length setting for parts of different heights. Other techniques are also available, generally as variations of those presented in this chapter.

9 BLOCK SKIP APPLICATIONS

Since the very early days of NC and CNC in the years 1970-1980's, many control systems manufacturers provided a number of special features to their products, some were standard, some were optional. Many features were developed as innovative improvements, others became new features - and several others may also have been included for the reasons of *'keeping up with the competition'*. Regardless of the addition of various features throughout the years, one special feature stands out to this day and presents many constructive possibilities to a imaginative CNC programmer. This feature has been available to all controls virtually from the beginning - it is called the *block skip function*. Another common name for the same function is *block delete function*. This latter definition presents a rather unfortunate choice of words for a function description that does not delete anything - keep in mind that there are *no* program blocks to be actually deleted from the CNC program, although they may be skipped on demand, that is optionally - yes, this is a big difference.

Functions that are simple to program are often pushed away, forgotten, or at least given low priority by programmers who have to rush a job to the production line. Block skip qualifies for this category, because it is generally quite simple to program. All that is needed in the CNC program is a slash symbol (yes, a symbol represented the forward / - slash)!

General Applications

There are several programmable features of the control system that can only be used in conjunction with the *Operation Panel* of the machine tool itself. The most common example is the use of miscellaneous function *M01 - Optional Stop*. In order to activate the optional program stop, two conditions have to be satisfied - *M01* function must be included in the program, and the *Optional Stop* switch that is located on the *Operation Panel* must be turned *ON*. The *Block Skip* function also requires two conditions to be satisfied:

- ***Block Skip* function (the slash symbol '/') must be included in the part program**
- ***Block Skip* toggle switch on the *Operation Panel* must be set to provide the desired result**

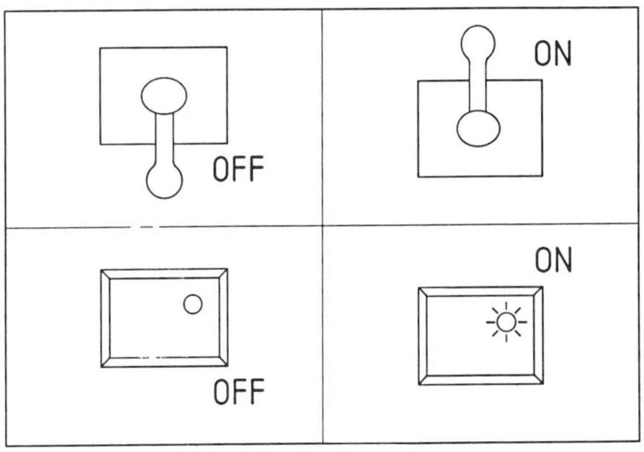

The *Block Skip* switch is normally located on the *Operation Panel* of the CNC machine, usually designed as a *toggle* switch (*up* and *down* setting), or an illuminated *push button* switch (*in* or *out* setting) - *see example*.

Regardless of the design, when the switch is set to the *ON* position, all program data following the slash symbol ***will be ignored*** by the control system during program execution.

When the switch is set to the *OFF* position, all program data will be processed, including all those following the slash symbol.

During actual machining, one of the main responsibilities of CNC operator to select such a switch mode setting that corresponds to the current work and its related requirements.

When using a program containing the block skip command, always keep in mind one main rule:

Always apply block skip switch *before* the slash symbol is encountered during program processing

In general, block skip applications may vary quite a bit, but in most cases, they are used for selecting whether a particular tool or operation will be used or not, partially or completely. The rest of this chapter shows at least some of the most common possibilities the CNC programmer can employ. A simple slash symbol in the part program can provide some serious and powerful results.

Similar Parts Applications

It is not unusual in CNC programming to discover that some toolpaths are *similar* to other toolpaths, for example, in terms of their coordinate location or orientation. In certain manufacturing sectors, a drawing for one part may not be much different from a drawing for another part. All such drawings should be made available to the CNC programmer, so the similarity between the drawings can be used as an advantage to develop more than one program with utmost efficiency.

Once the CNC programmer has an access to all these similar drawing, it is up to him or her to decide how the similarity will be handled. Using the block skip method is only one of the options available - the others include the use of subprograms or macros.

The two drawings below shop *Part A* with nine holes and *Part B* with six holes of a simple hole pattern. This is an example where only one part program can be written rather than two, and the block skip function used very effectively.

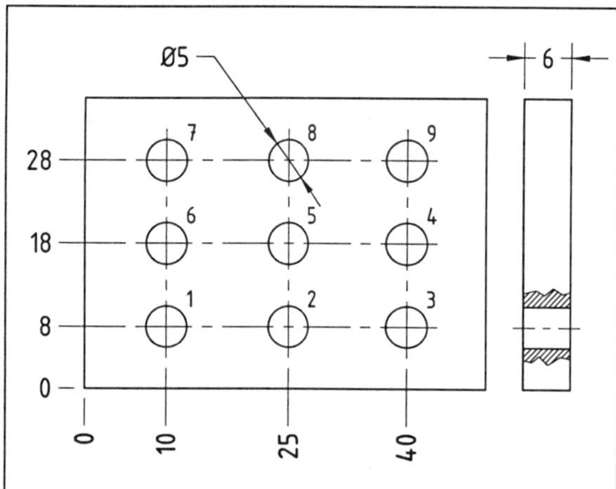

Part A - 9 holes

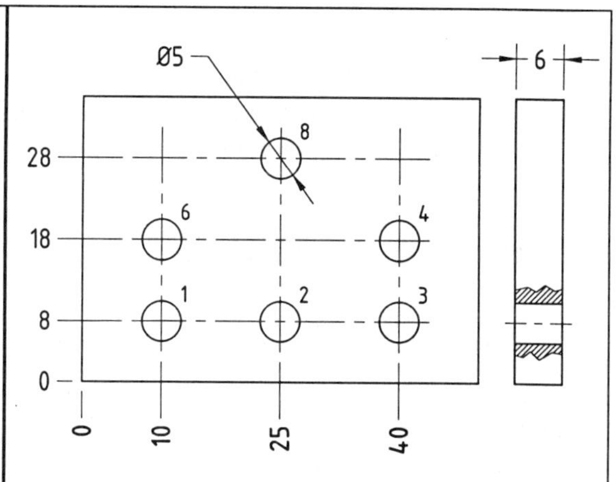

Part B - 6 holes

The best approach to developing a part program with a block skip is - *to forget the block skip*, at least initially - and always program the complete toolpath. For the drilling example above, it means programming all nine holes *(Part A)* first. Once this initial program is developed, block skip symbol (/) can be added to the blocks representing the holes that are not available in *Part B*. On a simple level of this example, that will be the extent of the changes. In only slightly more complex applications, the blocks identified by the slash symbol, may contain certain modal data, such as motion commands, coolant functions, spindle speed and feedrate functions, etc. These functions must be repeated in subsequent blocks, in order to prevent any program data being incorrect. In the example, all holes have numbered for convenience; these number would not be a part of the actual drawing.

The program with and without block skip is listed in the table below - only the spot drilling tool is used to illustrate the concept. The drilling program that follows will have the same structure.

<pre>N1 G21   (PART A ONLY)
N2 G17 G40 G80 T01
N3 M06
N4 G90 G54 G00 X10.0 Y8.0 S900 M03
N5 G43 Z25.0 H01 M08 T02
N6 G99 G82 R2.0 Z-2.8 P200 F120.0
N7 X25.0   (HOLE 2)
N8 X40.0   (HOLE 3)
N9 Y18.0   (HOLE 4)
N10 X25.0   (HOLE 5)
N11 X10.0   (HOLE 6)
N12 Y28.0   (HOLE 7)
N13 X25.0   (HOLE 8)
N14 X40.0   (HOLE 9)
N15 G80 Z25.0 M09
N16 G28 Z25.0 M05
N17 M01
...</pre>
	<pre>N1 G21 (PARTS A AND B)
	N2 G17 G40 G80 T01
	N3 M06
	N4 G90 G54 G00 X10.0 Y8.0 S900 M03
	N5 G43 Z25.0 H01 M08 T02
	N6 G99 G82 R2.0 Z-2.8 P200 F120.0
	N7 X25.0 (HOLE 2)
	N8 X40.0 (HOLE 3)
	N9 Y18.0 (HOLE 4)
	/ N10 X25.0 (HOLE 5)
	N11 X10.0 (HOLE 6)
	/ N12 Y28.0 (HOLE 7)
	N13 X25.0 (HOLE 8)
	/ N14 X40.0 (HOLE 9)
	N15 G80 Z25.0 M09
	N16 G28 Z25.0 M05
	N17 M01
	...</pre>

All nine holes have been machined for *Part A*, but three holes (5, 7, and 9) have been skipped in the program for *Part B*. Assuming the CNC operator runs both versions during the same setup, here is a very useful tip:

Always run the program with BLOCK SKIP switch set to ON first !

When the block skip switch is set to ON, *Part B* will be machined - that is the part with *fewer* holes. If the operator forgets to turn the block skip switch ON, *all holes* will be machined, which means *Part A*. Providing the oversight is caught in time, there will still be enough material to complete *Part B* as ordered by the customer. Machining *Part A* as first of the two does not allow this flexibility.

In this example, adding the slash symbol was enough and no other changes have been necessary, as no modal data were affected by the block skip function. That is not the same in the second example.

Although the same basic drawing has been used for this example, there is no relationship with the previous variations. The objective is to machine *Part C* and *Part D*, using the block skip function, but this time with some obstacles between holes. Obstacles can have several forms - for example - welds, spigots, fixtures, clamps, or any other protruding objects. Changing between modal values of G98 and G99 is expected for the most efficient machining but it should be done carefully.

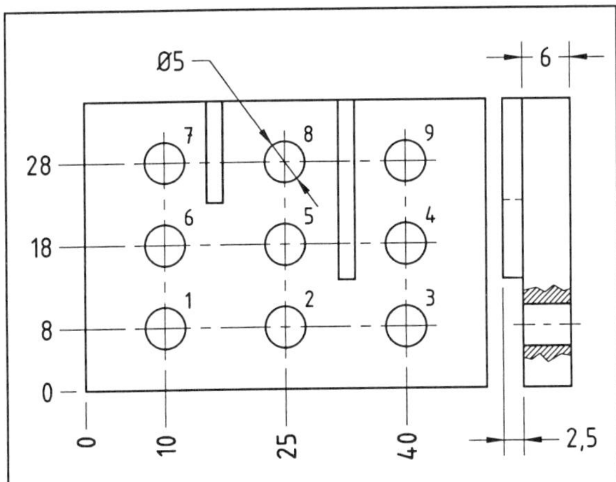

Part C - 9 holes

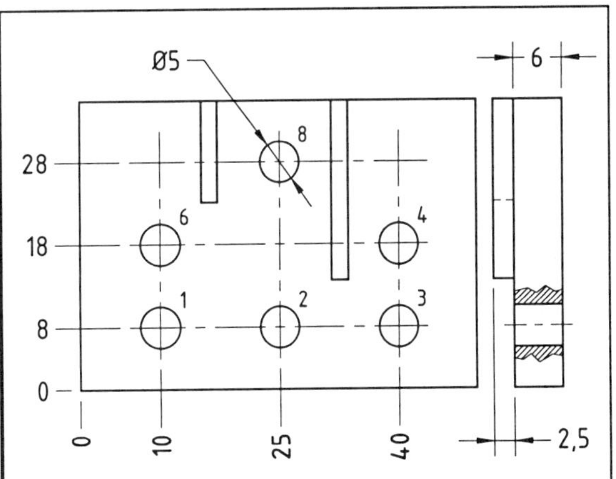

Part D - 6 holes

The exact locations of obstacles are not important for the example. What is more important is their actual height and their location between holes. Since their height is greater than the R-amount in the fixed cycle, G98 and G99 commands will be used in the program for *Part C* and *Part D*:

```
N1 G21                    (PART C ONLY)
N2 G17 G40 G80 T01
N3 M06
N4 G90 G54 G00 X10.0 Y8.0 S900 M03
N5 G43 Z25.0 H01 M08 T02
N6 G99 G82 R2.0 Z-2.8 P200 F120.0
N7 X25.0                      (HOLE 2)
N8 X40.0                      (HOLE 3)
N9 G98 Y18.0                  (HOLE 4)
N10 G99 X25.0                 (HOLE 5)
N11 X10.0                     (HOLE 6)
N12 G98 Y28.0                 (HOLE 7)
N13 X25.0                     (HOLE 8)
N14 G99 X40.0                 (HOLE 9)
N15 G80 Z25.0 M09
N16 G28 Z25.0 M05
N17 M01
...
```

```
N1 G21                    (PARTS C AND D)
N2 G17 G40 G80 T01
N3 M06
N4 G90 G54 G00 X10.0 Y8.0 S900 M03
N5 G43 Z25.0 H01 M08 T02
N6 G99 G82 R2.0 Z-2.8 P200 F120.0
N7 X25.0                      (HOLE 2)
N8 X40.0                      (HOLE 3)
N9 G98 Y18.0                  (HOLE 4)
/ N10 G99 X25.0               (HOLE 5)
N11 G98 X10.0                 (HOLE 6)
/ N12 G98 Y28.0               (HOLE 7)
N13 X25.0                     (HOLE 8)
/ N14 G99 X40.0               (HOLE 9)
N15 G80 Z25.0 M09
N16 G28 Z25.0 M05
N17 M01
...
```

In block N14 (for both versions), the G99 command is not really necessary - this block represents the *last* hole of the pattern, so the motion that follows will always be along the Z-axis - to machine zero reference position, from *any* current Z-position of the tool. Note the G98 command in block N11 (underlined). Its inclusion is necessary, if both parts are to be machined from a single program.

Programming for CNC lathes uses exactly the same logic as for drilling or milling applications. Typical examples are similar contours, pockets, and grooves.

Programming a Trial Cut

When the CNC operator completes the setup for a given part, there is no guarantee that the first part will be perfect (as per drawing specifications). The reason is not necessarily the lack of CNC operator's skills, but several external possibilities that are unknown or unpredictable at the time of setup. Such possibilities include, for example, the actual cutter radius amount as being different than the cutter radius stored in the offset register. Other unknown reasons that can make a difference between a part machined within specifications and a part machined outside the specifications (*i.e,* scrap) are tool deflection, influence of heat, tool wear, effect of the coolant, and several other factors. Programming a trial cut simply means making a test cut, measuring it, and fine tuning offsets based on the results. As the offsets are set precisely *before* the finishing operation takes place, trial cut is an excellent way to help eliminate scraps.

Trial Cut for Milling

The one cutting method that benefits the most from a trial cut for CNC machining center applications is the *face cut*. The main reason for using a trial cutting in such situations in the first place is to control the tolerances of the *part height*. Later in this chapter, there is a section describing the use of block skip function for parts of variable stock size. The trial cut method described here can be easily adapted for such parts.

In the drawing illustrated at right, a simple face cut has to be programmed, maintaining the tolerance on the height. There are approximately 2 mm of stock to be removed from the top. The trial cut will take place above the part zero (*i.e.,* finished top face) and above the 2 mm stock.

The cutter selected for this job will be ∅100 mm multi-insert face mill, and will start off the part center in the clear area at right. A single face cut will be programmed at Z0 to the clear area at left of the part. Z0 location is the finished face top.

Both steps of the part have been premachined and are not included in the program. The trial cut included in the program has one objective - to guarantee the 0.1 mm tolerance on the 40 mm step height. It will take place between Z0 and Z2.0 - with the stock to be eventually removed.

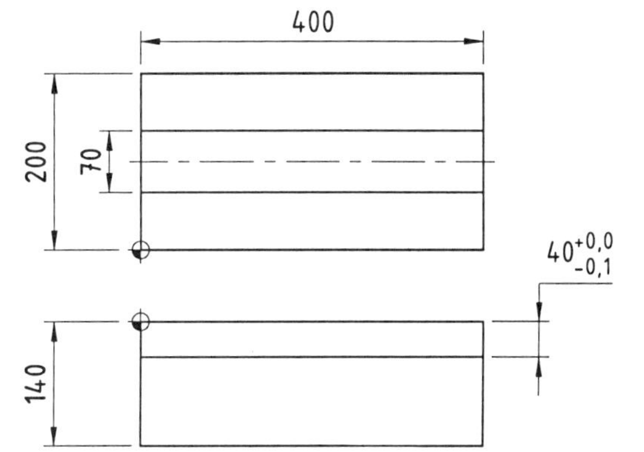

STOCK: 400x200x142 mm - with steps

First, the program *without* a trial cut - just one cut across and a simple program:

```
N1 G21
N2 G17 G40 G80
N3 G90 G54 G00 X455.0 Y110.0 S1200 M03
N4 G43 Z25.0 H01 M08
N5 Z5.0
N6 G01 Z0 F350.0                    (FINISHED FACE = TOP OF PART)
N7 X-55.0 F250.0 M09
N8 G00 G28 Z0 M05
N9 M01
```

The second version of the program will include the trial cut and will be a little more complex:

```
N1 G21
N2 G17 G40 G80                      (SWITCH OFF=TRIAL CUT,ON=PRODUCTION)
N3 G90 G54 G00 X455.0 Y110.0 S1200 M03
N4 G43 Z25.0 H01 M08
N5 Z5.0
/ N6 G01 Z1.0 F350.0                (TRIAL CUT AT Z1.0)
/ N7 X-55.0 F250.0 M09
/ N8 G00 G28 Z1.0 M05
/ N9 M00                            (CHECK DEPTH = BETWEEN 41.0 AND 40.9)
/ N10 G00 X455.0 Y110.0 S1200 M03
/ N11 G43 Z5.0 H01 M08
N12 G01 Z0 F350.0
N13 X-55.0 F250.0 M09
N14 G00 G28 Z0 M05
N15 M01
```

Note all program entries that are repeated within the blocks N6 through N11. This section of the program applies specifically to the trial cut. M00 (Program Stop) will cancel both the coolant and the spindle rotation, even if M09 and M05 were not programmed. After the M00, the spindle speed has to be reinstated - in fact, only M03 was necessary, as the control still keeps the spindle rotation of 1200 r/min in memory. One more important note relates to the repetition of the G43 block. Is it necessary of G00 Z5.0 M08 would be sufficient? Yes, the G43 block repetition is necessary, because G28 motion to machine zero in the Z-axis *cancels the tool length offset*. No G49 is necessary.

The whole length of the face has been machined during the trial cut, so the depth can be measured at both ends of the part. An adjustment in setup will be necessary if one end is deeper than the other end.

Trial Cut for Turning

In the previous milling example, the programming approach towards the trial cut toolpath was to make a measurable face cut slightly above the part zero - a cut that does not interfere with any other cut and still provides the required information (size) during a dimension check. The same technique will be used for a trial cut for finish turning, using the supplied drawing.

In a typical turning or boring application, the roughing cycle (usually G71) will leave a certain amount of stock on the diameters and faces of the part (U and W addresses in the cycle). When the finishing tool uses the G70 finishing cycle, this stock is removed and the part is completed. In order to make a trial cut on diameter, a small cut must be made within the area of the stock left on the diameter. The purpose is the same as for milling - take a test cut, measure the diameter and adjust offsets as necessary. All this can be done *before* the final contouring takes place, using block skip function.

Only the initial roughing tool and subsequent finishing tool are explained for this example - the part-off tool is not relevant to the subject.

During the roughing cuts, the front face will be completed and the part roughed out (using G71 cycle), leaving 1 mm per side on diameters. The Z-stock will be set to 0.25 mm. The 1 mm stock per side provides enough material to make the trial cut (test cut) before completing the part with the finishing cycle G70. The complete program is listed below:

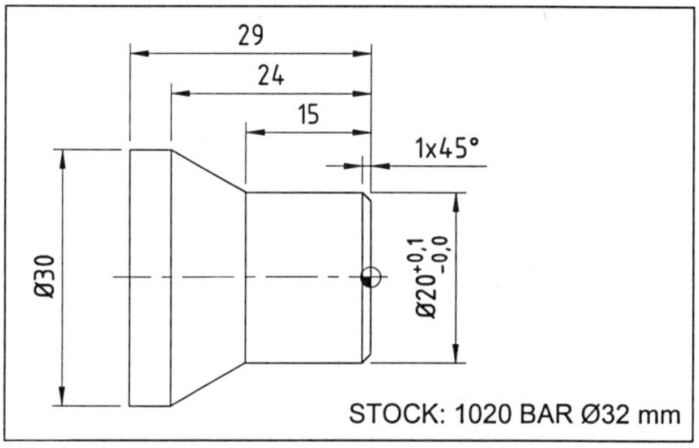


```
N1 G21 T0100                            (ROUGHING TOOL - TOOL 1)
N2 G96 S160 M03                         (160 M/MIN SURFACE SPEED)
N3 G00 X35.0 Z0 T0101 M08               (FACE START + OFFSET + M08)
N4 G01 X-1.8 F0.25                      (FACE BELOW CENTERLINE)
N5 G00 Z2.5                             (CLEAR OFF-FACE - Z-START FOR G71)
N6 G42 X31.0                            (RADIUS OFFSET + X-START FOR G71)
N7 G71 U2.5 R1.0                        (G71 - BLOCK 1 OF 2)
N8 G71 P9 Q14 U2.0 W0.25 F0.3           (G71 - BLOCK 2 OF 2)
N9 G00 X13.0                            (P-BLOCK - START OF CHAMFER)
N10 G01 X20.0 Z-1.0 F0.125              (CHAMFER CUTTING)
N11 Z-15.0 F0.2                         (FIRST DIAMETER CUTTING)
N12 X30.0 Z-24.0                        (TAPER CUTTING)
N13 Z-33.0                              (SECOND DIAMETER CUTTING)
N14 U5.0 F0.3                           (Q-BLOCK - END OF CONTOUR)
N15 G40 G00 X100.0 Z50.0 T0100          (TOOL CHANGE POSITION + NO OFFSET)
N16 M01                                 (OPTIONAL STOP)

N17 T0300                               (FINISHING TOOL - TOOL 3)
N18 G96 S200 M03                        (200 M/MIN SURFACE SPEED)
/ N19 G42 G00 X21.0 Z2.5 T0303 M08      (TRIAL CUT START + OFFSET + M08)
/ N20 G01 Z-7.0 F0.15                   (TRIAL CUT DIAMETER IS 21 MM)
/ N21 U5.0 F0.2                         (END OF TRIAL CUT)
/ N22 G40 G00 X100.0 Z100.0 T0300       (CLEAR POSITION FOR MEASURING PART)
/ N23 M00                               (MEASURE DIAMETER - MUST BE 21 MM)
/ N24 G96 S200 M03                      (REINSTATE SPINDLE SPEED)
N25 G42 G00 X31.0 Z2.5 T0303 M08        (RADIUS OFFSET + XZ-START FOR G70)
N26 G70 P9 Q14                          (FINISHING CYCLE - PART COMPLETION)
N27 G40 G00 X100.0 Z50.0 T0300          (TOOL CHANGE POSITION + NO OFFSET)
N28 M01
```

At this point in the program, the part-off tool will be called and completes the part. The program containing the block skip is not difficult at all - study the example above. The most important aspect of developing such a program is to make sure all program entries cancelled in the block skip mode are reinstated in the standard program mode.

One note relating to the block N19 of the above example. The X-address contains the diameter to be measured. In the example, the diameter is 21 mm. Because of the stock left for finishing (1 mm per side), the programmer has a choice of programming any diameter larger than 20 mm (part diameter) and 22 mm (stock diameter). Choosing the middle value is rather arbitrary and may not always be the best choice, mainly for two reasons:

- ***Reason 1*** **- The actual depth of trial cut is too different from the finishing depth of cut**
- ***Reason 2*** **- The middle value does not imply the importance of the measurement**

As for *Reason 1*, the actual depth of the trial cut will *always* be different from the final contour depth. That is the nature of any trial cut - *it is not meant* to machine the final size (diameter). Even with this consideration, the trial cut depth can be increased, for example, from X21.0 to X20.25, bringing it much closer to the final contour depth (when trial cut is not used).

The *Reason 2* is a bit of a psychological exercise. If you program the trial cut diameter as X21.0, the one decimal place may not imply the urgency of a rather tight tolerance. On the other hand, if the CNC operator is instructed to check a diameter - say - of 20.65 (X20.65), the psychological impact of the two decimal places may be significant.

Irregular Stock Removal

When the rough stock for a particular job is delivered to the machine shop, its condition should be fairly uniform in shape, size, and condition. In practice, that is not always the case, and for many reasons. Regardless of the reasons, the CNC programmer is faced with a question - how to handle the variations in stock sizes? There is always the simple answer - separate the stock into groups of sizes that can be machined together and make individual programs for each group. While this option should always be available in extreme situations, there is also a good chance that all stock variations can be handled by a single program. The method is, of course, using the *block skip function*. Although there may be only a single program to machine all the variable stock, it will still be necessary to separate the provided stock into groups on the floor. Block skip function can be used very successfully if there are only two groups for machining. This is also the subject of this section.

Variable Stock in Milling

A rough material (stock) delivered to any CNC machine should be uniform, with only minimum deviations in size. So what happens if a typical rectangular (cubical) stock, for example, is delivered with rather large variations in height? Using a face mill, the excessive stock can easily be brought to final size, as required by the drawing. However, the CNC programmer has to know how much stock is to be removed. Based on the maximum cutting depth for the selected face mill, the programmer must decide how many face cut will be necessary. Ideally, only one face should be required, otherwise too much material will have to be removed, adding unnecessary cost to the machining process.

Let's have a look at a situation, where the available stock must be divided into two groups for machining. One group will require a single face cut, the other group will require two face cuts. When developing the program for machining, the programmer has to determine the maximum safe depth for each face milling pass. The programmer must also know the *maximum* and *minimum* size of the stock. This is often done the old fashioned way, by actually *measuring* the available stock.

Once the maximum and minimum stock sizes are known for the batch, the CNC programmer will determine not only the actual *number of facing cuts*, but also *how many groups* of the stock allowance have to be used for the toolpath development.

The drawing example presented at right is a modified version of the earlier drawing, but there is no direct connection between the two.

The note identifying the stock variations will not be found in the drawing - the stock size has to be assured by performing a physical check.

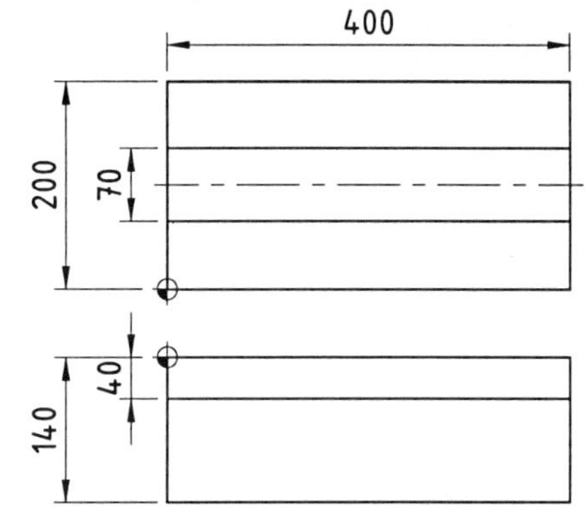

STOCK: 400x200 mm
Note: Stock height varies between 141 and 145 mm

The available stock is divided into two groups, based on the maximum depth of each face mill:

- **Maximum facing depth** ... **3 mm for the selected face mill**
- **Group 1 - Stock height is less than 3 mm or 3 mm** ... **Block skip function ON will *ignore* the two cuts**
- **Group 2 - Stock height is between 3 and 5 mm** ... **Block skip function OFF will *force* the two cuts**

Since no stock height exceeds 6 mm (2x the maximum cutting depth), it will be possible to machine the *Group 1* parts with a single face cut, and the *Group 2* parts with two face cuts.

The programming technique in this case - and similar cases - is to always program *all cuts required*, without any block skip function included. After such a program has been written, the block skip symbols may be added and program modified for consistency (always check the program integrity). Such a program can be developed for the part shown in the illustration above:

```
N1 G21
N2 G17 G40 G80
N3 G90 G54 G00 X455.0 Y110.0 S1200 M03
N4 G43 Z10.0 H01 M08
/ N5 G01 Z3.0 F500.0              (3 MM ABOVE TOP FACE - TWO FACE CUTS)
/ N6 X-55.0 F250.0 M09
/ N7 G00 Z5.0
/ N8 X455.0
N9 G01 Z0 F500.0
N10 X-55.0 F250.0                 (FINAL CUT AT Z0)
N11 G00 Z5.0 M09
N12 G28 Z5.0 M05
N13 M01
```

Programming examples of this type are quite common and also quite simple. Yet, even in a simple application of the block skip function, it is easy to overlook a small item. In particular, watch for modal settings and how they carry over between a run with block skip switch set to ON, and block skip switch set to OFF. The most important observation for the above example is one that relates to the *direction* of cut. If two facing cuts are part of the toolpath all the time, for all parts, it makes sense to cut from one end to the other (right to left) for the first pass, change the Z-depth, and cut from the opposite direction (left to right) for the second pass. When the block skip function is programmed, this method is not possible - both facing cuts must originate at the same XY location, for example, at the right side of the part.

Variable Stock in Turning

In turning, the most common stock used for machining is round in shape. As a round stock is fairly consistent from one supplier to another for a given size, the most common difficulty a CNC programmer will face is the *stock length*. Actually, it is the *billet* length. The word *billet* is used in machine shop environment to identify a round stock that has been cut into suitable smaller lengths for machining in a chuck, and clamped in the jaws. Such stock is often prepared on a saw, and the sawing operation may result in some irregularities in the length actually cut. While small irregularities are expected for billets - after all, a saw is not exactly a precision equipment - larger irregularities will present problems, generally during a facing operation.

In the illustration at right, only the two critical dimensions are shown. They both relate to the stock size and face cut(-s) toolpath. If the 5.5 mm amount of stock can be machined in a single cut, the part program will reflect that and no other considerations will be required. On the other hand, if the maximum face cut width is only 3 mm, for example, two cuts will have to be made. With irregular stock length, it means some parts will require only a single face cut, while others will require two face cuts. The following program shows how to use the block skip function for such situations:

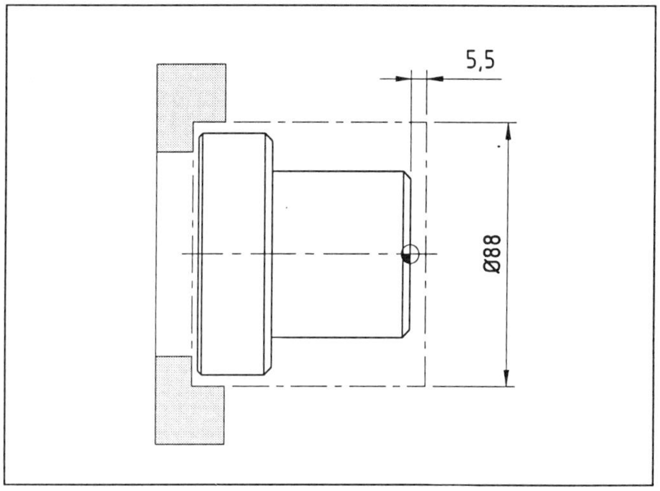


```
N1 G21 T0100                      (TOOL 1 SELECTED FOR FACING)
N2 G96 S200 M03                   (200 M/MIN SURFACE SPEED)
N3 G00 X92.0 Z3.0 T0101 M08       (READY TO FACE 3 MM OFF FRONT FACE Z0)
/ N4 G01 Z-1.8 F0.1               (FACE TO BELOW CENTER LINE)
/ N5 G00 W3.0                     (MOVE OFF THE FACE BY 3 MM)
/ N6 X92.0                        (RAPID BACK TO X-START POSITION)
N7 Z0                             (READY TO FACE THE FINISHED FRONT FACE)
N8 G01 Z-1.8                      (FACE TO BELOW CENTER LINE)
N9 G00 W3.0                       (MOVE OFF THE FACE BY 3 MM)
N10 G42 X..                       (MOVE TO NEXT LOCATION - FOR TURNING)
...
```

There is nothing special to consider in the program listed. Depending on the exact situation at the time of programming, considering the current modal settings in the program may be the only necessary addition. It is always a good idea to check the batch of unknown material for variations in sizes - other dimensional deviations may be found at the same time and influence the program development.

Summary of Rules

Whenever a block skip function is used to control irregular stock - or any other procedure - the best approach is to make a program for all tool motions, disregarding the blocks to be skipped. When such a program is developed, all blocks to be skipped should be marked. From this point on, make sure you always check both versions of the program - once with the block skip function ON, and again, with the block skip function OFF. If you find even a minor error, correct it, and do the double check again - it will save you a few headaches later on.

Block Skip Within A Block

Normally, in the majority of block skip applications, the block skip symbol (/) is programmed as the first character of a block, even *before* the block number *Nxx* (for example, as **/N23 G28 Z2.0**). Some higher level controls add an additional flexibility to this very useful feature, by allowing its use *within a block*. If the control system allows this method of programming, the block skip function can become even more powerful and will offer extended flexibility in part program development.

As the purpose of the block skip function is to optionally skip all program data in the block *following* the slash symbol, it also means that if the slash is placed at the beginning of the block, the *whole* block will be skipped or the *whole* block will be processed, depending on the block skip switch setting. If the control allows programming of a block skip function *anywhere within* the block, only the data located *after the slash* can be optionally skipped. This situation presents a number of opportunities, but there is one item the programmer must be well aware of - *how to handle conflicting words in a block.*

Conflicting Words

With the exception of G-codes (and M-functions on some controls), there is no reason to program more than one address in a block. After all, what would be the purpose of having a block containing two spindle speeds? For example:

```
N46 G90 G54 G00 X45.0 Y90.0 S600 S950 M03
```

Writing this block intentionally does not make any sense. Most likely, such a block is the result of a programming error, an oversight. It is also a classic example of two words in a single block that are in *conflict* with each other. This brings up an expected question - *how does the control system handle such a conflict?* If you answered that an alarm will be generated, you were wrong. The control will process the block and generates no alarm. As it is impossible to have two different spindle speeds for one tool at the same time, *the **latter** of the two conflicting words will be accepted as active*. In the example, the active spindle speed will be 950 r/min and the 600 r/min will be ignored. Why the latter and not the former? The situation is similar to *"changing one's mind"*. Initially, you may want a speed of 600, than change your mind to a speed of 950 as a better choice. Since the speed of 950 r/min supercedes the 600 r/min, it becomes *'more important - a better choice'*. Fanuc takes the same approach - the latter of the two conflicting words in a block will become active and remains such until changed or cancelled. The fact that the control system will accept two conflicting commands in the same block can be used creatively - using the block skip function *within* a block, rather than as the first function in the block.

One Program - Two Materials

It may not be every day that a part program will be required for two different materials in a production environment. However, there is a very common application when such a situation is quite common - *during testing of expensive or complex parts*. Take, for example, a part to be machined from a tool steel. Tool steels can be quite expensive and require much slower speeds and feedrates than softer materials. Making a test part - or a prototype - often involves using a cheaper material, not only in terms of cost, but also in terms of the added cost of the testing itself. Two programs can always be an option - one for the hard material (production), another for the soft material (testing). What will be different between the two programs? Since the purpose of such a program is to be used for the actual production, consideration regarding depth of cuts, width of cuts, etc., cannot be changed. That leaves only two major items that will not significantly alter the intent of the program - *spindle speeds and feedrates*. The objective is to write a production program that can also be used for testing. The following example demonstrates the programming approach.

As only the spindle speeds and cutting feedrates are being different, selecting these two items makes a logical beginning:

Application	Material	Spindle speed (S)	Feedrate (F)
Testing	Sotfter	950	250.0
Production	Harder	600	100.0

For clarity reasons, compare the two programs that follow - one drills a hole in a soft material for testing purposes (program at left), the other drills the same hole in a hard material for production (program at right) - differences between the two programs are underlined:

N44 T03	N44 T03
N45 M06	N45 M06
N46 G90 G54 G00 X45.0 Y90.0 S950 M03	N46 G90 G54 G00 X45.0 Y90.0 S600 M03
N47 G43 Z25.0 H03 M08	N47 G43 Z25.0 H03 M08
N48 G99 G81 R2.0 Z-15.4 F250.0	N48 G99 G81 R2.0 Z-15.4 F100.0
N49 G80 Z25.0 M09	N49 G80 Z25.0 M09
N50 G28 Z25.0 M05	N50 G28 Z25.0 M05
N51 M01	N51 M01

If the two programs are combined into a single program, *both* spindle speeds and *both* feedrates will have to be incorporated:

```
N44 T03
N45 M06
N46 G90 G54 G00 X45.0 Y90.0 S950 S600 M03
N47 G43 Z25.0 H03 M08
N48 G99 G81 R2.0 Z-15.4 F250.0 F100.0
N49 G80 Z25.0 M09
N50 G28 Z25.0 M05
N51 M01
```

Of course, the program in this form will work well for production but the speeds and feeds will be too slow for testing. The slash symbol within the block will come into effect:

```
N44 T03
N45 M06
N46 G90 G54 G00 X45.0 Y90.0 S950 / S600 M03
N47 G43 Z25.0 H03 M08
N48 G99 G81 R2.0 Z-15.4 F250.0 / F100.0
N49 G80 Z25.0 M09
N50 G28 Z25.0 M05
N51 M01
```

Note the added slash functions - the block skip switch on the operation panel can now be used:

- **Block skip switch set to ON** ... ***TEST* or *TRIAL* machining will take place**
- **Block skip switch set to OFF** ... ***PRODUCTION machining will take place***

There is one significant problem - there will be *no spindle rotation* for the production version. The reason is that the M03 function is located *after* the block skip function! In order to fix this error, program the spindle rotation *before* the slash symbol, as it is common to both types of machining.

The same logical approach has to be taken for any other words in the program. Also note that it is important to have all items from one group on the same side of the slash symbol - note the correct use in the example - *order is important*:

... `S950 / S600` ... *and* ... `F250.0 / F100.0` ...

Spindle speed of 950 r/min is tied up with the feedrate of 250.0 mm/min, whereby the spindle speed of 600 r/min is tied up with the feedrate of 100.0 mm/min.

Here is the complete program so far, including the repositioning of the M03 function:

```
N44 T03
N45 M06
N46 G90 G54 G00 X45.0 Y90.0 S950 M03 / S600
N47 G43 Z25.0 H03 M08
N48 G99 G81 R2.0 Z-15.4 F250.0 / F100.0
N49 G80 Z25.0 M09
N50 G28 Z25.0 M05
N51 M01
```

If the test material is wood or similar material that does not require coolant, you may also include the coolant function after the slash - it will only be available for production machining, not the test machining:

```
N44 T03
N45 M06
N46 G90 G54 G00 X45.0 Y90.0 S950 M03 / S600
N47 G43 Z25.0 H03 / M08
N48 G99 G81 R2.0 Z-15.4 F250.0 / F100.0
N49 G80 Z25.0 M09
N50 G28 Z25.0 M05
N51 M01
```

As the final example shows, any data to the right of the slash symbol relate to *production* machining, while conflicting data in the same block will be ignored.

Having the option to program a block skip within a block offers many opportunities not available when the block skip function is restricted to the block beginning.

Numbered Block Skip Functions

This is a special block skip related technique in CNC programming that is hardly ever used. Partly, because it is not generally available on the control system (optional feature), but mainly because it needs constant attention of the CNC operator. In short, this control feature allows programming of block *groups* that have to be skipped. For example, the CNC operator may skip only group number 2, and process groups number 1 or 3, etc. This feature is called the *Optional Block Skip*, and uses a number (1 to 9) following the slash, for example, `/2N46 X50.0` or `/1/3N46 G00 X50.0`. The numbered block must always be *before* all other data in the block. This way, the operator can set which group or groups should be processed - or not processed - when the block skip switch is set on. The settings for each group are normally done through the control software, on the display screen.

10 STANDARD AND RIGID TAPPING

Tapping operation on conventional mills, lathes, and CNC machines has been one of the basic operations to machine holes. To this day, with all its disadvantages, one of the most common methods of tapping uses special *tapping heads*, known as the *push-pull* type in the shop, or *tension-compression* type more formally. Another term for this type of tool is a *floating tap holder*.

Standard Tapping Method

In a typical tension-compression tap holder, the tap is held firmly so it cannot fall out, but loose at the same time to provide axial flexibility upon entering a drilled hole. Thread in the hole is formed by the size and shape of the tap. Although still very popular, floating tap holders present one major problem - they make it difficult to maintain precise thread depth, particularly in blind holes.

Basic Principles

The main principle of standard tapping is the synchronization of spindle speed and tapping feedrate. That is achieved by multiplying the spindle speed (r/min) by the tap lead (pitch) resulting in a feedrate in inches or millimeters per minute. At the beginning of the cut, the tap has to accelerate from zero feedrate to a rather heavy feedrate. The time required to reach the programmed feedrate is called *acceleration*. In order to reach the programmed feedrate *before* any contact with the material, the tap must start further away from the hole - this is the R-level in fixed cycles. As the tap enters the hole, it encounters a resistence, and the tap is pushed slightly into the holder (compression). When the tap reaches the hole bottom and the spindle stops, the *deceleration* of the spindle forces the tap to be slightly pulled out of the tool holder.

Standard tapping is almost always programmed using one of two tapping cycles - G84 (right hand tapping) and G74 (left hand tapping). The typical format of either cycle is:

```
G98 G84 X.. Y.. R.. Z.. F..        (RIGHT HAND - RETRACTS TO INITIAL LEVEL)
G99 G84 X.. Y.. R.. Z.. F..        (RIGHT HAND - RETRACTS TO R-LEVEL)

G98 G74 X.. Y.. R.. Z.. F..        (LEFT HAND - RETRACTS TO INITIAL LEVEL)
G99 G74 X.. Y.. R.. Z.. F..        (LEFT HAND - RETRACTS TO R-LEVEL)
```

☞ where ...

G98	-	Retracts to initial level (last programmed Z-axis position before cycle call)
G99	-	Retracts to level specified by R-address in the fixed cycle
X.. Y..	-	Hole location
R	-	Feedrate plane (programmed feedrate starts from this position)
Z	-	Final depth of tap (thread)
F	-	Tapping feedrate (synchronized with spindle r/min)

As you will find shortly, many control systems support the same format for rigid tapping either as provided or slightly modified. The major difference is not much in the programming technique as it is in machine design and suitable tooling.

Why Underfeed?

Most experienced CNC programmers have faced various tapping problems in their careers. The floating tap holder is an imperfect tool to say the least. It is a mechanical device and, as such, subject to some variables, including spindle fluctuations. If the tapping operation is not successful the first time (assuming the program is correct in principle), the programmer often employs a simple technique for a quick fix - *tapping feedrate reduction*. This reduction is called *underfeeding*, because normal tapping feedrate is always *r/min* x *pitch*. Here are the reasons why this technique often works.

Most spindles may fluctuate up to plus/minus 10 r/min, particularly for high speeds. Because the tapping feed is always a direct relationship of the spindle speed and the tap pitch, any spindle fluctuation throws this relationship out of balance. When the tap in a floating holder enters the hole, it is pushed a bit into the holder, before the tap has a chance to 'bite' into the material. This part of tapping is called *compression*. When a tap reaches the Z-depth, the spindle stops, but not right away - it has to 'wind down' first. In order not to break the tap, the floating holder allows the tap to be slightly pulled out of the holder by its own force, after the Z-axis motion stopped. This is called *tension*. Compression means *pushing*, tension means *pulling* (may be another way to explain the subject). Now back to the spindle fluctuation.

Take a floating tap holder and try to push, then pull, the part that holds the tap. It is much easier to pull it out than push it in. In technical terms, it means tension presents fewer, if any, problems. In order to eliminate thread damage upon tap entry or exit, the feedrate can be reduced by 2% to 5%, anywhere between 98% and 95% of the nominal tapping feedrate. In this case, the tap is pulling from the tap holder, and compensates for the spindle speed differences.

Keep in mind that a slight underfeeding when using floating tap holders is not always a solution to better tapping, but it may be the quickest solution in certain cases.

Floating tap holders are on the way out. Although they have served the industry for many years, a new technology has been ushered in. Since the introduction of *rigid tapping* in the early 1990's, the tapping methods and techniques have changed dramatically.

Feed-In Slower - Feed-Out Faster

When applying the underfeeding technique, feedrate is the same for the tap motion *into* the material as it is for the motion *out*. What about programming a different feedrate for the in-motion than for the out-motion. Of course, this is not possible with standard fixed cycles, but it is possible to write a special program (for a few parts) or develop a custom macro (for many parts). This technique is very useful, as this actual case demonstrates.

A customer purchased a new CNC vertical machining center for the sole purpose of tapping very fine threads (56 to 64 threads per inch - pitch 0.45 to 0.4 mm) in aluminum. All holes were through holes in material averaging about 0.1 inches (2.5 mm) in thickness. The problem was a frequent thread stripping. After many unsuccessful methods, attempts and various tests, including machine integrity tests, the cutting tool representative suggested a rather unusual method: *tap-in at a **slower** feedrate than normal, then tap-out at a feedrate **faster** than normal*. Using a standard floating tap holder (no rigid tapping then), the job required a special program.

There are two example provided here; both will use 90% feed-in and 110% feed-out. These are not rates guaranteed to work for every problem, but the examples can be used as a basis for a specific tapping program.

One program will illustrate a standard method of programming, the other will use a special macro, developed just for such a purpose. Macro feature must be available at the control system.

Standard Tapping

In its most simple form, to tap a hole at the current XY location requires only a few blocks - feed-in to depth, stop and reverse the spindle, feed-out and stop spindle again, so it can be resumed in normal mode. In this form, the CNC operator will have to be very careful and run the tapping tool at 100% overrides settings. Even then, pressing the feedhold button in the middle of tapping presents a problem. Fortunately, Fanuc offers a special G-code - G63 - *tapping mode.*

When the tapping mode G63 is used in the program, the control system will ignore any setting of the *feedrate override switch*, except the 100% setting. The feedhold key, located on the operation panel, will also be disabled. When the tapping mode is no longer necessary, the G64 command will cancel it - G64 sets the normal cutting mode. Nominal feedrate is 750 speed x 1/56 tpi = 13.4 in/min.

```
(SPECIAL TAPPING - 0.1 PART THICKNESS - 56 TPI THREAD)
(USE AT THE CURRENT XY LOCATION)
G63                          (TAPPING MODE SELECTION)
S750 M03                     (TAPPING SPINDLE SPEED AND NORMAL ROTATION)
G00 Z0.1                     (START OF TAPPING)
G01 Z-0.15 F12.1 M05         (F13.4 IS NOMINAL - F12.1 IS 90 PERCENT)
Z0.1 F14.7 M04               (F13.4 IS NOMINAL - F14.7 IS 110 PERCENT)
M05                          (SPINDLE STOP ABOVE HOLE)
M03                          (RESUME NORMAL ROTATION)
G64                          (CUTTING MODE SELECTION)
```

Note the inclusion of M05 and M04 functions in the motion block, not in separate blocks. This is very important. M05 will stop the spindle at the motion completion, M04 will start the spindle simultaneously with the motion. Equally important is to start the cutting with G63 and end it with G64 commands, both in separate blocks.

Macro Tapping

When a macro is used for this tapping method, any pitch can be defined in the main body of the program, within the G65 statement. No changes to macro will take place.

This section is aimed at those programmers who have good knowledge of *Fanuc Custom Macros*
(see separate chapter for general introduction to this subject)

The purpose of this macro is to simulate the effect of the G84 fixed cycle, where the cutting feedrate on the way in will be 95% of the given feedrate and on the way out 105% of the given feedrate. The tapping macro will be called at the current XY tool location. Macro call is for imperial units:

```
G65 P8000 R0.35 Z1.15 S750 T56
```

☞ where ...

R	=	(#18)	R-level clearance equivalent
Z	=	(#26)	Z-depth
S	=	(#19)	Spindle speed (r/min)
T	=	(#20)	Number of threads per inch (TPI)

Other arguments could be added, for example, calculation of the tapping depth, automatic calculation of the top clearance (known as the R-level in G84 cycle), etc. For the purposes of the subject presented, there is no benefit in complicating the macro. Z0 of the part is the top face - note the positive value of the Z-depth (forced as negative value in the macro). The special tapping macro listing:

```
O8000 (SPECIAL TAPPING MACRO)
#3003 = 1                              Disable single block setting
G00 Z[ABS[#18]] S#19 M03               Rapid to positive R-level + spindle rotation
#3004 = 7                              Disable feedhold, feedrate override, exact stop check
G01 Z-[ABS[#26]] F[#19/#20*0.9] M05    Feed to depth at 90% of feedrate - stop spindle
Z#18 F[#19/#20*1.1] M04                Feed back to R-level at 110% feedrate - reverse spindle
#3004 = 0                              Enable feedhold, feedrate override, and exact stop check
M05                                    Stop spindle
M03                                    Restore normal spindle rotation
#3003 = 0                              Enable single block setting
M99                                    End of macro
%
```

Even if macros are not a feature you use at all, the example should be clear enough, particularly with the attached comments. The purpose of the 3003 and 3004 system variables is to disable (or enable) single block, feedhold, feedrate override, and exact stop check during operation.

Rigid Tapping Method

Rigid tapping is a common description of a technology known as *synchronized* tapping. Most CNC manufacturers offer rigid tapping method as a standard feature on their equipment. It is important to realize that the machine tool has to be designed for such purpose, including the control system.

Rigid tapping must be supported by the machine design and related software

Basic Principles

The most significant difference between standard and rigid tapping is the tool holder. While standard tapping requires a spring loaded floating tap holder, rigid tapping can use any solid tool holder used for end mills, including collets (watch for possible tap spinning when tapping tough materials). The reason why a floating tap holder is not required in rigid tapping is that the spindle itself does the required synchronization via a built-in encoder. Tapping using the synchronized method also uses fixed cycles of the control system, often the same ones as for standard tapping. It is the computer software in the control system, not the cutting tool, that synchronizes the programmed spindle speed (r/min) with the programmed tapping feedrate (mm/min or in/min), to match the thread lead (pitch). Machine design has to support these features.

Benefits

If there is only one benefit of rigid tapping over standard tapping, it would be the *depth control*. Unlike tension-compression holders, the machining power for rigid tapping is provided by the spindle, rather than torque of the cutter. As any tool wear requires higher torque, the tapping depth suffers as a result. This is not an issue with rigid tapping. In fact, the horsepower (or kilowatt) rating of the machine tool is the only limiting factor.

There are other benefits as well. Rigid tapping method is more suitable for making threads in tough and hardened materials, as well as in deep holes. Some space age and other exotic materials may even be tapped using the 'pecking' method, common to drilling deep holes. In this case, a special part program will have to be developed, but that is not a difficult task. Re-tapping is also possible, as long as the starting point and depth never change. This is a benefit for holes that require a certain tolerance on the tapped depth. Overall, a significant reduction of cost per hole can be achieved with rigid tapping over tapping with floating holders.

Setup

Unlike floating tap holders, rigid taps are held in solid holders. It is very important to make sure that the deviation of the tool from the spindle centerline is an absolute minimum, and does not exceed 0.005 mm (5 microns or about 0.0002 inches) of *total indicator reading* (TIR). Equally important is to use high quality taps, which are less likely to run-out - they may cost more than standard taps but are more economical in overall. Large run-out will cause an oversized thread and may even break the tap, particularly for small sizes.

Possible Problems

Although rigid tapping is supported by the majority of modern machining centers, there are significant differences in their design. It would be a mistake to assume that synchronization of every machine spindle is perfect and that a solid tap holder solves all tapping problems. Various physical factors contribute greatly to the quality of the thread tool life, particularly in volume production. No significant difficulties are generally encountered in small batches, but tap wear over many holes has to be considered as a serious potential problem, as the quality of tapped holes will gradually deteriorate. The reason for reduced tool life are mechanical forces applied mainly during the Z-axis motion. There is also the issue of efficiency when using rigid tapping for a large number of holes.

Programming Approach

As in every effort to develop part programs for specific machine tool features, rigid tapping also requires a thorough knowledge of machine specifications and related control features, including programming structure and format. The first consideration is *spindle speed*, the second is *feedrate*.

Spindle speed selection is always a matter of selecting the most suitable spindle speed within a certain range. For any machining operation, the material being machined is a significant factor influencing the spindle speed. In rigid tapping, there is another factor to consider - as the spindle speed is directly related to the tapping feed, a constant torque range of the spindle motor is required. General recommendations vary somewhat between machine manufacturers, but the general consensus is that rigid tapping should not exceed 2000 r/min (higher is also possible), depending on material.

Feedrate selection is the same as for the older tapping method:

F = r/min × P	*Imperial or metric units*
F = r/min / TPI	*Imperial units only*

☞ **where ...**

F	=	Tapping feedrate	(mm/min or in/min)
r/min	=	Selected spindle speed	(revolutions per minute)
P	=	Pitch of the thread	(distance between threads)
TPI	=	Threads per inch	(number of threads per one inch length)

In the earlier section, the subject of under-feeding was described. Keep in mind that the under-feed is used in certain cases for floating type tap holders - *never for rigid tapping*.

Rigid tapping feedrate must always be the exact result of *r/min × pitch*

Special Functions

If a standard G84 or G74 cycle can be used for both floating and rigid devices, the question is how does the control recognize which one is the programmer's intent? The answer depends on the manufacturer. In fact, some manufacturers do not allow G84/G74 for rigid tapping at all, and provide a special G-code instead. Other allow G84/G74, but also require a special command or function to be used. For example, one manufacturer requires M29 function to be programmed before the tapping fixed cycle is used:

```
T04                                   (T04 - M6x0.75 METRIC - RIGID TAP)
M06                                   (TOOL CHANGE)
G90 G54 G00 X200.0 Y100.0 S1400 M03   (1400 R/MIN SPINDLE SPEED)
G43 Z25.0 H04 M08                     (INITIAL CLEARANCE)
M29                                   (RIGID MODE ACTIVE)
G84 R5.0 Z-12.0 F1050.0               (F1050.0 = 1400 R/MIN X 0.75 PITCH)
X.. Y..                               (NEXT HOLE)
G80 Z25.0 M09                         (CANCELS CYCLE AND RIGID TAPPING MODE)
G28 Z25.0 M05                         (MACHINE ZERO RETURN IN Z-AXIS)
M01                                   (OPTIONAL STOP)
```

This is just a small example - a very simple one. Consult the machine manuals for exact program format. Rigid tapping is mostly a machine function and the additional programming requirements are minimal.

11 POLAR COORDINATES

In the CNC world, point coordinates are the key to develop a program using any type of available motion. *The single purpose of coordinates is to define a point*, then use various motion commands to connect two or more points and develop a unique toolpath. While the word 'coordinate' is heard almost daily, it still does not hurt to look at its definition, which is quite simple:

A coordinate is any set of two or more numbers that determine the position of a point with respect to established references

A reference position is a fixed position, such as machine zero (machine origin) or a part zero (part origin). Mathematically, there are four types of coordinates:

- **Rectangular Coordinates**
- **Polar Coordinates**
- **Cylindrical coordinates**
- **Spherical Coordinates**

All four types are used in CNC applications, for different machines and different kinds of work. Without a doubt, rectangular coordinates are the 'bread and butter' of CNC programming for machining centers, lathes, wire EDM, laser cutters, and a dozen of other machine tools. Polar coordinates are very common on non-Fanuc controls, mainly because they are standard feature. Fanuc offers polar coordinates only as a special control option, which is unfortunate, considering their many applications advantages that all users could benefit from. The remaining two coordinate systems - cylindrical and spherical - are not supported by the majority of controls and those controls that do support these coordinate methods offer them as optional features. In this case, the option offer is justified, because neither cylindrical or spherical coordinate system is used very often for general applications. This chapter focuses on polar coordinates and the method of using them in a CNC program.

Definition and G-codes

Mathematical definitions of polar coordinates are often much more involved, but a simpler definition should be sufficient for the purposes of CNC programming:

A point in polar coordinates is defined as the endpoint of a line that starts from another point with a given length and angle

The line is often called the *radius vector*, and the point where the line starts is called the *pole* and the distance between the pole and the defined point is called the *radial distance*. Polar angle is measured from the polar axis, as the positive angle to the line that passes the defined point. Radius vector is always a distance with direction.

In CNC, the polar axis is the X-axis (in G17 plane), and the positive angle is measured from zero degrees of this axis, located to the right of X-axis origin. This position is typically called the *3 o'clock* position or the *East* position. The illustration on the page shows these relationships graphically.

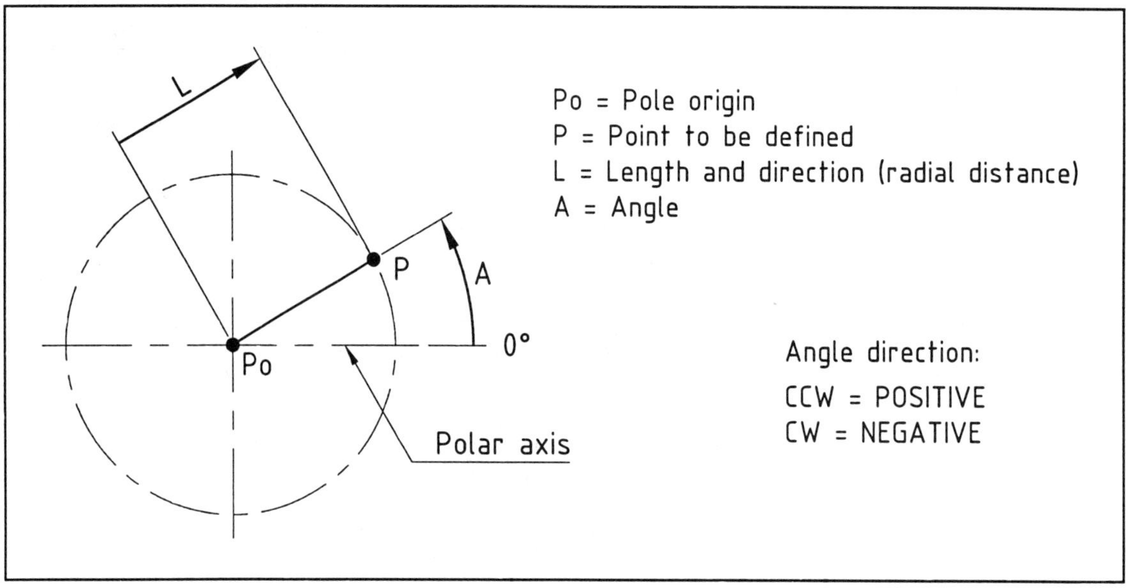

The main benefit of using polar coordinates in CNC programming is that they save many calculations for objects located on an arc or circle. The example used will be the application of polar coordinates on a bolt hole circle pattern. This is also the most suitable example in point-to-point machining.

Polar Coordinates and Planes

The selected working plane has a great effect on the way angle is measured in polar coordinates. See a separate chapter dedicated to the subject of planes. The above illustration shows the polar axis and the zero degrees direction on that axis - it does not show the name of the machine axis. The next illustration shows the plane in which the angle will be active:

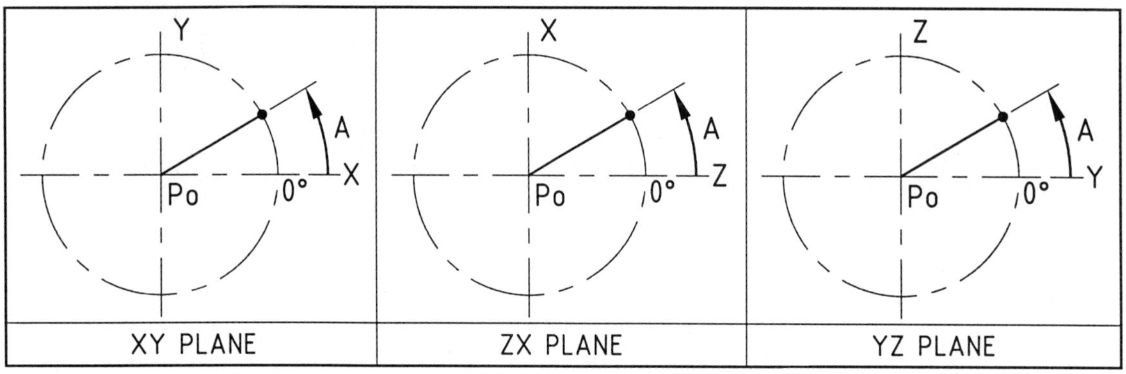

Note:

Although the XY plane is programmed as G17, and the YZ plane as G19, the plane on the machine that is programmed as G18 is not ZX plane, but XZ plane. This is due to the fact that the mathematical plane does not correspond to the machine plane. Angle definition in G18 plane will be affected.

Within the selected plane G17-G19, there are two preparatory commands - G-codes - available for programming polar coordinates, providing the control system supports this option.

G15 - G16 Polar Coordinates

By definition, polar coordinates are either active or inactive (ON or OFF):

G15	Polar coordinate system cancel	OFF = inactive
G16	Polar coordinate system	ON = active

The programming format requires either G-code to be entered as the only command available in the block. Between the two blocks G16 and G15, the machining takes place with polar coordinates active. From the previous section dealing with active plane selection, it is important the format of polar coordinates. An example of seven equally spaced holes will be used. If rectangular coordinates are used, calculation of odd number of holes on a bolt circle is always more time consuming than calculation of even number of holes. Using polar coordinates, there is no difference - only the angle between holes has to be known:

360 / 7 = 51.42857143 = 51.429 degrees

The rounding of such a number to many decimal places does make a difference in very high precision applications, although there is a very small error in this case - 0.003 degrees maximum:

Hole Position	Angles rounding based on 8 decimal places	Angles rounding based on 3 decimal places
1	0.000	0.000
2	51.429	51.429
3	102.857	102.858
4	154.286	154.287
5	205.714	205.716
6	257.143	257.145
7	308.571	308.574

Programming Format

Fixed cycles are normally used to machine holes. This programming method does not change, regardless of whether the program uses rectangular coordinates or polar coordinates. The practical difference is in the *meaning of the axes* (within the specified work plane):

In the *rectangular coordinate system*, the XY program entries are normally defined as absolute positions, as measured from origin (part zero), along each axis.

In the *polar coordinate system*, the XY program entries are normally defined as the arc radius (X) and the hole position angle (Y).

The illustration on the next page will be used as an example. An odd number of holes in a bolt circle is less common than an even number, but it better represents the benefits of polar coordinates.

The illustration at right shows a typical bolt hole pattern - 50 mm bolt circle diameter with seven equally spaced holes, distributed along the bolt circle. The first hole starts at zero degrees, as per standard angle orientation.

The center of the bolt circle diameter is the part zero (X0Y). Using the rectangular coordinate system, the absolute location of each hole will have to be calculated from the bolt circle center, along each axis. This goal represents fourteen individual trigonometric calculations. These calculations are not difficult mathematically, but they certainly do require some precious time. With more calculations, the possibility of errors increases. Using the polar method eliminates most problems.

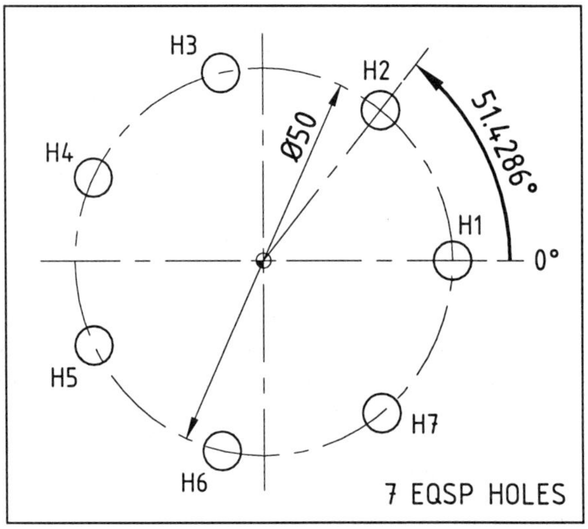

The programming format of polar coordinates can be best described in the following example. In order to make a complete program for one tool, spot drilling operation will applied to each hole, for the depth of 2.8 mm:

```
N1 G21
N2 G15 G17 G40 G80 T01                      (NOTE G15 AND G17 IN THIS BLOCK)
N3 M06
N4 G90 G54 G00 X0 Y0 S1200 M03 T02          (MOTION TO THE POLE - XY CENTER)
N5 G43 Z2.0 H01 M08
N6 G16                                      (POLAR COORDINATES ON)
N7 G99 G82 X25.0 Y0 R2.0 Z-2.8 P200 F125.0  (HOLE H1 OF 7)
N8 X25.0 Y51.429                            (HOLE H2 OF 7)
N9 X25.0 Y102.857                           (HOLE H3 OF 7)
N10 X25.0 Y154.286                          (HOLE H4 OF 7)
N11 X25.0 Y205.714                          (HOLE H5 OF 7)
N12 X25.0 Y257.143                          (HOLE H6 OF 7)
N13 X25.0 Y308.571                          (HOLE H7 OF 7)
N14 G15                                     (POLAR COORDINATES OFF)
N15 G80 M09
N16 G91 G28 Z0 M05
N17 M01
...
```

The programming pattern is quite obvious from the example. While the X address does not change, the Y address increases by the amount of angle between holes. No additional calculations are necessary and the program is not only easy to understand, it is also easy to change, if necessary. Upon a closer look, there are five important parts in the program that require some explanation:

1. **G15 has been programmed at the beginning, in order to cancel polar coordinate input. This is a precautionary measure, mostly not needed, but it may play a role if the fixed cycle has been interrupted before its completion**

2. **The first major tool motion is always to the pole - the bolt hole circle center**

3. **For a bolt hole pattern, the X-radius does not change, only the Y-angle increment**

4. **Either absolute or incremental mode can be used while polar coordinates are in effect**

5. **Programming the angle with 0.001° accuracy makes a better program (three decimal place for angle)**

Toolpath Direction

In most cases, the toolpath direction is not critical when machining the holes, using polar coordinates. At the same time, it is important to understand how the machining direction affects the overall cycle time, particularly when large volume of parts are involved.

It is a common programming practice to change the toolpath direction for two tools; for example, a spot drill may be programmed in the order of H1 to H7. The drill that follows starts at the last hole (H7), so it makes sense to reverse the toolpath in the order of H7 to H1. Polar coordinates support either direction by a simple change of the angle sign:

Angular increment from the first axis in the counter-clockwise direction is measured as positive

Angular increment from the first axis in the clockwise direction is measured as negative

The first part of the part program was listed earlier - in it, the cutting direction was H1 to H7, counter-clockwise (blocks N1 to N17). The second part shows the opposite direction programming the seven holes, from H7 to H1 (blocks N18 to N33):

```
N1 G21
N2 G15 G17 G40 G80 T01                    (NOTE G15 AND G17 IN THIS BLOCK)
N3 M06
N4 G90 G54 G00 X0 Y0 S1200 M03 T02        (MOTION TO THE POLE - XY CENTER)
N5 G43 Z2.0 H01 M08
N6 G16                                        (POLAR COORDINATES ON)
N7 G99 G82 X25.0 Y0 R2.0 Z-2.8 P200 F125.0    (HOLE H1 OF 7)
N8 X25.0 Y51.429                              (HOLE H2 OF 7)
N9 X25.0 Y102.857                             (HOLE H3 OF 7)
N10 X25.0 Y154.286                            (HOLE H4 OF 7)
N11 X25.0 Y205.714                            (HOLE H5 OF 7)
N12 X25.0 Y257.143                            (HOLE H6 OF 7)
N13 X25.0 Y308.571                            (HOLE H7 OF 7)
N14 G15                                       (POLAR COORDINATES OFF)
N15 G80 M09
N16 G91 G28 Z0 M05
N17 M01

N18 T02
N19 M06
N20 G90 G54 G00 X0 Y0 S900 M03 T03        (MOTION TO THE POLE - XY CENTER)
N21 G43 Z2.0 H02 M08
N22 G16                                       (POLAR COORDINATES ON)
N23 G99 G81 X25.0 Y308.571 R2.0 Z-12.5 F150.0 (HOLE H7 OF 7)
N24 X25.0 Y-257.143                           (HOLE H6 OF 7)
N25 X25.0 Y-205.714                           (HOLE H5 OF 7)
N26 X25.0 Y-154.286                           (HOLE H4 OF 7)
N27 X25.0 Y-102.857                           (HOLE H3 OF 7)
N28 X25.0 Y-51.429                            (HOLE H2 OF 7)
N29 X25.0 Y0                                  (HOLE H1 OF 7)
N30 G15                                       (POLAR COORDINATES OFF)
N31 G80 M09
N32 G91 G28 Z0 M05
N33 M01
...
```

Note that only the angular direction changes, not the bolt circle radius. For programming convenience, the angle between holes can be programmed either as an absolute angle (shown in the example), using the G90 mode or as an incremental angle, using the G91 mode.

Applications in Planes

Polar coordinates can be used in any one of the three work planes available (G17, G18, G19). Although the majority of jobs using polar coordinates are done using the G17 (XY) plane, the CNC programmer should be aware of the differences when using the other two planes. Some vertical machining centers are designed to accept a 90° head spindle adapter, where the drilling axis is shifted by 90°. This feature does affect the plane selection for machining holes. Over all, keep in mind the basic - and most important - principles relating to polar coordinates:

- **Plane selection does matter when polar coordinates are used**
- **The first axis of the selected plane is represented by the same axis in the program**
- **The second axis of the selected plane is represented by the hole position angle**

The summary of these principles can be provided in a table:

G-code programmed	Plane selected	First axis = radius	Second axis = angle
G17	XY	X	Y
G18	ZX	Z	X
G19	YZ	Y	Z

The seven equally spaced holes presented earlier and programmed in all three available planes will share most of the program data, but the plane selection - and the corresponding axes within that plane - will be different. The following example shows the comparison of programming the *second* hole of the pattern used in this chapter, in all three planes:

```
G17                               (G17 XY PLANE SELECTED)
...
N6 G16                            (POLAR COORDINATES ON)
...
N8 X25.0 Y51.429                  (HOLE H2 OF 7)
...
N14 G15                           (POLAR COORDINATES OFF)
...

G18                               (G18 ZX PLANE SELECTED)
...
N6 G16                            (POLAR COORDINATES ON)
...
N8 Z25.0 X51.429                  (HOLE H2 OF 7)
...
N14 G15                           (POLAR COORDINATES OFF)
...

G19                               (G19 YZ PLANE SELECTED)
...
N6 G16                            (POLAR COORDINATES ON)
...
N8 Y25.0 Z51.429                  (HOLE H2 OF 7)
...
N14 G15                           (POLAR COORDINATES OFF)
...
```

The programming format does not change - just the axis selection.

12 SUBPROGRAM DEVELOPMENT

Although the subject of subprograms has been used in applications described in other chapters, it is a subject too important to leave it at that level only. In this chapter, specially designed for the subject of subprogram development, all logical steps will be explained on actual examples.

Definition and Usage

In the simplest terms, *a subprogram is a program segment that runs under another program.* While quite true, the simple definition falls short of explaining the purpose - *and need* - of subprograms in CNC programming. Subprograms are used for a number of reasons - two most important reasons are:

- **Overall length of the part program is smaller**
- **Changes to toolpath related data are much easier to make**

There are other benefits as well. Subprograms are very powerful tool for CNC programming. They can be used in absolute or incremental mode, called by any program, and they can even be nested, which means one subprogram calls another subprogram, up to four levels deep on most controls. The best way to understand subprogram development is to work with an example:

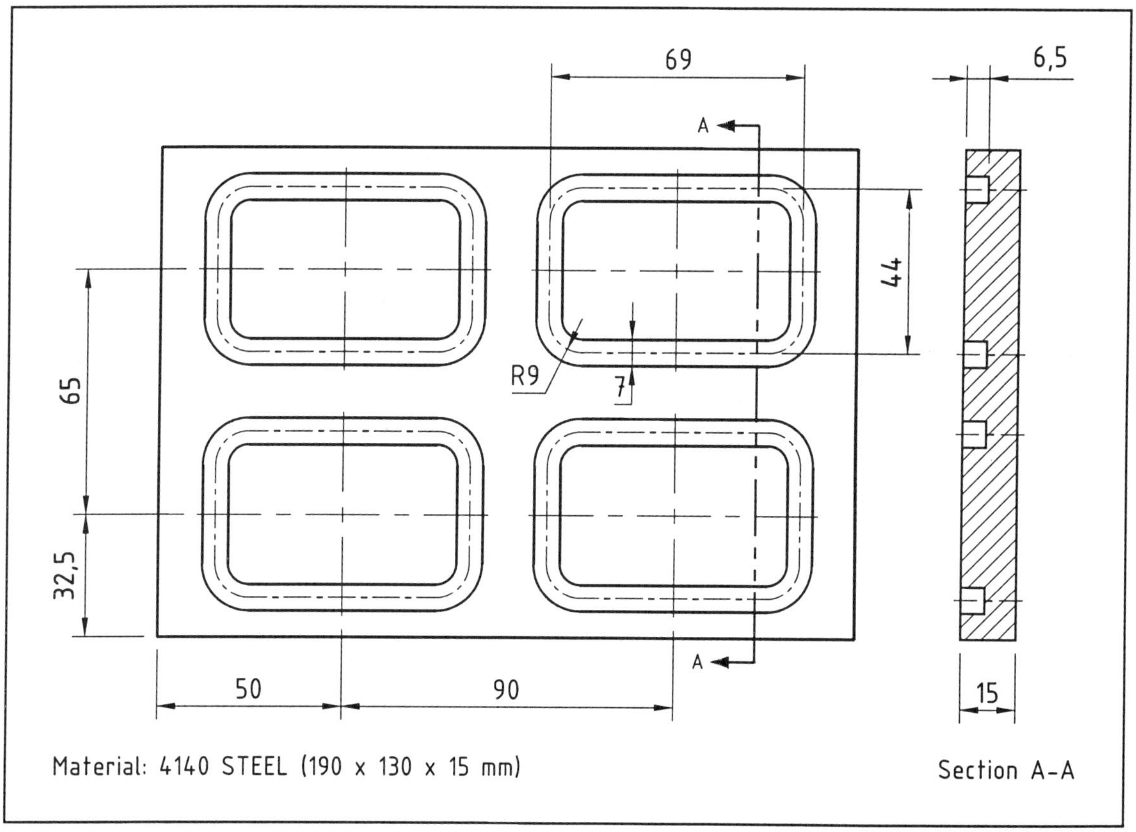

Drawing Evaluation

In the drawing, there are four identical grooves, at four different locations. That means there are *four identical patterns* that have to be translated into a suitable toolpath used with one or more cutting tools. Each groove is defined by its centerline length (69 mm), its centerline width (44 mm) and its total depth (6.5 mm). When studying this or a similar drawing, it is also important to identify the method of *how* each pattern is dimensioned. In the example, the distances between each pattern are known - 90 mm along the X-axis, 65 mm along the Y-axis, from one middle point another. Finally, the location of the lower left pattern is also defined, as X50.0Y32.5.

Each groove is also defined, as a simple rectangle with four round corners and the width of 7 mm. The last items to consider is the material specification, which is an important factor used in deciding on the best machining method. As machining method influences the subprogram development, the material specification is important for the same reason. Material specification of 4140 steel indicates a fairly tough material, but a material with quite good machinability characteristics.

A careful drawing evaluation is important for any CNC part program development, but takes on additional considerations when a subprogram is planned. Having four identical patterns in the drawing is definitely a case for planning a main CNC program using at least one subprogram. The next step covers some organizational items and the determination of ***how the part is to be machined.***

Subprogram Planning

There are four patterns of the same groove provided by the drawing. By evaluating the drawing, there are no reasons to machine the four grooves in any particular order, so the actual order of machining used will be arbitrarily selected, and any order would work equally well.

The illustration shows the order of machining by identifying each groove by a letter. *Groove A* will be machined first, *Groove B* will be next, etc.

It also shows the part zero and the XY location of each start point *(middle of the bottom groove)*. The start point is the location at which the tool enters the groove along the Z-axis. The direction of cut from the start point is also shown, at any given depth. In subprograms, the conditions for each pattern must be identical, as they will be stored in the subprogram body. *In summary:*

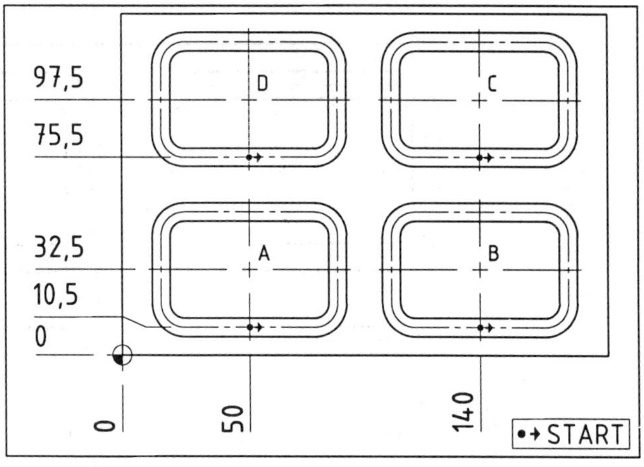

Groove	Start X-location	Start Y-location
A	X50.0	Y10.5
B	X140.0	Y10.5
C	X140.0	Y75.5
D	X50.0	Y75.5

Depth Control

With the same care as the XY locations have been established, so must be the grooving toolpath. The groove is only 7 mm wide, and almost that deep (6.5 mm). In a tough material, even a carbide tool will have to be used carefully, and plunging into the full depth is not an option (assuming the end mill selected is center-cutting). The programmer has to select each depth that is reasonable for the tool used. The first thing to consider is dividing the actual groove depth into equal number of segments. For example, 2 segment depths of 3.25 mm, 4 segment depths of 1.625 mm, 5 segment depths of 1.3 mm, etc. There are several methods how to plunge into a material, particularly one that is fairly tough to machine:

- **Using a center-cutting end mill and plunge into the solid material in small increments**
- **Using any suitable end mill and ramp from a clear position at 2° to 4° ramping angle**
- **Pre-machine and entry hole, typically with a drill or a drill and end mill**

Not all depth segments have to be equal - for example, the final depth may be smaller than all preceding depths. Depending on the exact segmentation, the Z-axis start position is very critical to the overall quality of the groove.

Z-axis Start Position

Subprograms for CNC are usually developed in the incremental mode G91, using actual distances or depths. This example is no different - once the depth of each cut and the number of cuts have been established, the starting position of the Z-axis must also be established, in order to reach the depth as per drawing. Also important is to make sure there are no cuts in the air, above the part, cutting nothing.

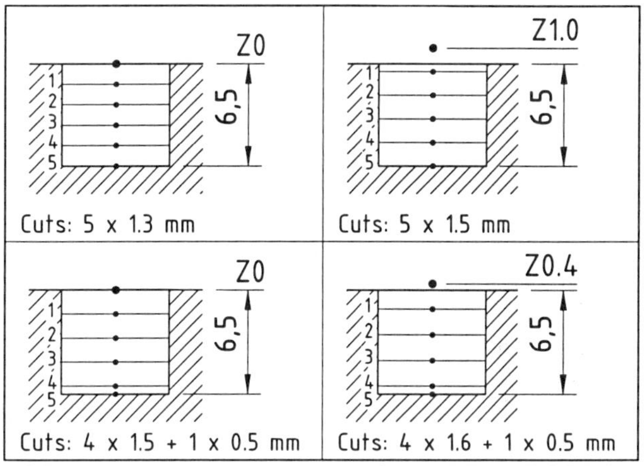

In the illustration, there are four typical possibilities, each with five depth cuts but different distribution of individual depth cuts.

For the example, five equal depths of 1.3 mm will be used, as shown in the upper left corner.

Width of Cut Control

Another important part of planning a subprogram is to determine the *width of cut*. If a large area has to be removed by roughing, the width of cut is also called *stepover*. In case of the 7 mm wide groove in the example, selecting a ∅7 mm center-cutting end mill will cut the groove but its size and quality will be rather poor. A better selection is a smaller tool - one that will rough-out the groove in the middle and than finish each groove wall separately. Two tools for this purpose would be even better.

Cutting Tool Selection

For the subprogram development in the example, the common selection of using a center-cutting end mill with small increments will be also the most practical. ∅6 mm tool will be the choice.

Developing the Subprogram

Once the planning stage has been completed, the subprogram can be developed, applying all the decisions made so far. When learning a new subject, it is never the best idea to learn on a complex level, but start with a simple application. Although the groove cutting job presented cannot be considered complex, three approach methods will be presented, including the one with the full width. Each version will bring a new programming method that could be used on other subprograms.

Method 1 - Full Width and Full Depth

First, a short note. This is not a method either suitable or recommended to cut the four grooves. The benefit of its inclusion is in its simplicity. Consider the final goal - *to cut four identical grooves at different locations, with a **single** plunge at each location*. Now, transfer the verbose goal and adapt it to programming steps for a single groove:

- **STEP 1 - Rapid to XY+Z location of Groove A** ... **X50.0 Y10.5 + Z2.5 (Z is an arbitrary clearance)**
- **STEP 2 - Plunge to the full grove depth`** ... **Z-6.5**
- **STEP 3 - Machine the groove** ... **XY motions in linear and arc mode**
- **STEP 4 - Rapid above part to clear Z-location** ... **Z2.5 (an arbitrary clearance)**

Repeat the four steps for the remaining three grooves - what will change? ***Only the XY location will change - nothing else!*** This is the key to understanding subprograms - the key is to identify what machining is common to all parts (or grooves in this case). In the above procedure, steps 2-4 are common to all four grooves. They will be programmed only once and they will become the main contents of the subprogram. The subprogram will then be called at different locations, as per *Step 1*. What about the *Step 1*? Individual motions will be programmed in the *main program*, which will call the new subprogram four times at four places. Here is the complete program listing for the *Method 1*:

```
N1 G21 (7 MM TOOL IN THE SPINDLE)
N2 G17 G40 G80
N3 G90 G54 G00 X50.0 Y10.5 S600 M03
N4 G43 Z25.0 H01 M08
N5 G01 Z2.5 F250.0
N6 M98 P7301                (GROOVE A)
N7 G90 G00 X140.0
N8 M98 P7301                (GROOVE B)
N9 G90 G00 Y75.5
N10 M98 P7301               (GROOVE C)
N11 G90 G00 X50.0
N12 M98 P7301               (GROOVE D)
N13 G90 G00 Z25.0 M09
N14 G28 Z25.0
N15 G28 X50.0 Y75.5 M05
N16 M30
%
```

```
O7301 (SUBPROGRAM FOR ONE GROOVE)

(START FROM Z2.5 MIDDLE OF BOTTOM)
(ABSOLUTE DEPTH WILL BE Z-6.25)

N101 G90 G01 Z-6.5 F80.0
N102 G91 X25.5 F100.0
N103 G03 X9.0 Y9.0 I0 J9.0
N104 G01 Y26.0
N105 G03 X-9.0 Y9.0 I-9.0 J0
N106 G01 X-51.0
N107 G03 X-9.0 Y-9.0 I0 J-9.0
N108 G01 Y-26.0
N109 G03 X9.0 Y-9.0 I9.0 J0
N110 G01 X25.5
N111 G90 G00 Z2.5
N112 M99
%
```

Both, the main program and the subprogram contain a few special items. In the main program (left), both G90 and G00 commands are repeated. Rather than counting on the subprogram content, it is safer to program important G-codes twice, even if it is not necessary, as in this case. The same applies to feedrates and other important features. It also makes it easier to interpret the main program or subprogram and edit them, if necessary.

In the subprogram (right), the first and last motions are absolute (G90), while the motions between are all incremental (G91). **Always be careful when changing the programming mode!**

Method 2 - Full Width and Divided Depth

This is almost the same approach as in the first method - there is a definite improvement in cutting the depth, but the quality of the groove on its walls will still be poor. The main difference is that the depth will be divided into five equal cuts of 1.3 mm each, making it easier to plunge. Again, consider the final goal - *to cut four identical grooves at different locations, with a* ***multiple*** *plunge at each location*. The completed program uses the letter *L* (try *K* for some controls) to repeat the subprogram five times (M98 P7401 L5). Each subprogram must start from Z0 (see blocks N6, N10, N14 and N18).

Another very important observation is that the subprogram O7401 *does not* - and ***must not*** - include the return Z-motion. The reason is obvious - only the *last depth* requires the retract. Obvious or not, this is a common and expensive error in subprogramming - even experienced programmers make it.

The programming steps for the program follow these steps (to be repeated for all grooves):

- **STEP 1 - Rapid to XY+Z location of Groove A** ... **X50.0 Y10.5 + Z25.0 (Z is an arbitrary clearance)**
- **STEP 2 - Rapid to Z-axis location close to the top** ... **Z1.0 is an arbitrary clearance**
- **STEP 3 - Feed-in to Z0** ... **Z0 is a mandatory start point**
- **STEP 4 - Plunge *BY* the 1.3 mm grove depth** ... **Z-1.3 is an incremental depth in the subprogram**
- **STEP 5 - Machine the groove at the current depth** ... **XY motions in linear and arc mode (one depth only)**
- **STEP 6 - Repeat steps 4 and 5 four more times** ... **Total number of repetitions will be five**
- **STEP 7 - Rapid above part to clear Z-location** ... **Z25.0 (an arbitrary clearance) - in the MAIN program**

The complete listing of the main program and its related subprogram is listed below:

```
N1 G21
N2 G17 G40 G80
N3 G90 G54 G00 X50.0 Y10.5 S600 M03
N4 G43 Z25.0 H01 M08
N5 Z1.0
N6 G01 Z0 F100.0
N7 M98 P7401 L5    (GROOVE A 5 CUTS)
N8 G90 G00 Z1.0
N9 X140.0
N10 G01 Z0 F100.0
N11 M98 P7401 L5   (GROOVE B 5 CUTS)
N12 G90 G00 Z1.0
N13 Y75.5
N14 G01 Z0 F100.0
N15 M98 P7401 L5   (GROOVE C 5 CUTS)
N16 G90 G00 Z1.0
N17 X50.0
N18 G01 Z0 F100.0
N19 M98 P7401 L5   (GROOVE D 5 CUTS)
N20 G90 G00 Z25.0 M09
N21 G28 Z25.0
N22 G28 X50.0 Y75.5 M05
N23 M30
%
```

```
O7401 (SUBPROGRAM FOR ONE GROOVE)

(START FROM Z0)
(EACH DEPTH IS 1.3 MM 5X REPEATED)
(ABSOLUTE DEPTH WILL BE Z-6.5)

N101 G91 G01 Z-1.3 F80.0
N102 X25.5 F100.0
N103 G03 X9.0 Y9.0 I0 J9.0
N104 G01 Y26.0
N105 G03 X-9.0 Y9.0 I-9.0 J0
N106 G01 X-51.0
N107 G03 X-9.0 Y-9.0 I0 J-9.0
N108 G01 Y-26.0
N109 G03 X9.0 Y-9.0 I9.0 J0
N110 G01 X25.5
N111 M99
%
```

If the start point at Z0 is not your preferred way of programming, use 1.5 mm depth and start the subprogram at Z1.0. Earlier in the chapter, this method was presented in an illustration (upper right).

Any change in either program requires careful verification

Method 3 - Smaller Width and Full Depth

This third and final method used in this example provides the best quality groove. It will be presented using two tools rather than one. Using one tool for both roughing and finishing is a not a very good choice - the groove quality and tool life both suffer. Adapting the program from one tool to two should be easy, as no major change in the overall approach is needed and some steps have already been described before.

What programming approach will make the groove better? The answer is not only the selection of two tools already mentioned, but also using a smaller tool or tools. For a 7 mm wide groove, selecting the roughing tool as ∅6 mm center-cutting end mill makes sense. For finishing the contour of walls, another ∅6 mm end mill can be used, or a even a smaller tool of ∅5 mm, as in this example.

The previous step-by-step method must be modified as two tools and operations are used.

First tool - ∅6 mm center-cutting end mill used for roughing (repeat for all grooves):

- **STEP 1 - Rapid to XY+Z location of Groove A** ... **X50.0 Y10.5 + Z25.0 (Z is an arbitrary clearance)**
- **STEP 2 - Rapid to Z-axis location close to the top** ... **Z1.0 is an arbitrary clearance**
- **STEP 3 - Feed-in to Z0.25** ... **Z0.25 is a mandatory start point for *roughing***
- **STEP 4 - Plunge *BY* the 1.3 mm grove depth** ... **Z-1.3 is an incremental depth in the subprogram**
- **STEP 5 - Machine the groove at the current depth** ... **XY motions in linear and arc mode (one depth only)**
- **STEP 6 - Repeat steps 4 and 5 four more times** ... **Total number of repetitions will be five**
- **STEP 7 - Rapid above part to clear Z-location** ... **Z25.0 (an arbitrary clearance) - in the MAIN program**

All four grooves will be 6 mm wide, leaving 0.5 mm on each wall for finishing. They will also have 0.25 mm extra material at the bottom, also to be removed by the second tool, during finishing.

```
N1 G21
N2 G17 G40 G80 T01
N3 M06          (T01 = 6 MM ROUGHING)
N4 G90 G54 G00 X50.0 Y10.5 S600 M03
N5 G43 Z25.0 H01 M08 T02
N6 G01 Z1.0 F250.0
N7 Z0.25 F100.0
N8 M98 P7501 L5    (GROOVE A 5 CUTS)
N9 G90 G00 Z1.0
N10 X140.0
N11 G01 Z0.25 F100.0
N12 M98 P7501 L5   (GROOVE B 5 CUTS)
N13 G90 G00 Z1.0
N14 Y75.5
N15 G01 Z0.25 F100.0
N16 M98 P7501 L5   (GROOVE C 5 CUTS)
N17 G90 G00 Z1.0
N18 X50.0
N19 G01 Z0.25 F100.0
N20 M98 P7501 L5   (GROOVE D 5 CUTS)
N21 G90 G00 Z25.0 M09
N22 G28 Z25.0 M05
N23 M01
```

```
O7501 (SUBPROGRAM FOR ONE GROOVE)

(START FROM Z0.25)
(EACH DEPTH IS 1.3 MM 5X REPEATED)
(ABSOLUTE DEPTH WILL BE Z-6.25)

N101 G91 G01 Z-1.3 F80.0
N102 X25.5 F100.0
N103 G03 X9.0 Y9.0 I0 J9.0
N104 G01 Y26.0
N105 G03 X-9.0 Y9.0 I-9.0 J0
N106 G01 X-51.0
N107 G03 X-9.0 Y-9.0 I0 J-9.0
N108 G01 Y-26.0
N109 G03 X9.0 Y-9.0 I9.0 J0
N110 G01 X25.5
N111 M99
%
```

Note that the two programs are almost identical to the programs used in the example shown as *Method 2*. The main difference is the starting position in Z-axis, that leaves 0.25 mm extra material at the bottom of each groove for finishing.

Second tool - ∅5 mm center-cutting end mill used for finishing (repeat for all grooves):

Adding more steps to the procedure is always required when more machining motions are needed. The second tool will require two separate toolpaths, one for finishing the inner wall, the other for finishing the outer wall of each groove. The first groove will be *Groove D*, the current tool location:

- **STEP 1 - Rapid to XY+Z location of Groove D** ... **X50.0 Y75.5 + Z25.0 (Z is an arbitrary clearance)**
- **STEP 1 - Feed-in to the final depth** ... **Final depth is 0.25 mm deeper than current depth**
- **STEP 2 - In climb milling mode machine inner wall** ... **Use G41 D52 for fine tuning during production**
- **STEP 3 - In climb milling mode machine outer wall** ... **Use G41 D62 for fine tuning during production**
- **STEP 4 - Rapid above part to clear Z-location** ... **Z25.0 (an arbitrary clearance) - in the MAIN program**

The method of controlling the depth cuts will be exactly the same as in the *Method 2* - 1.3 mm for each depth, starting at Z0.

```
N24 T02
N25 M06          (T01 = 5 MM FINISHING)
N26 G90 G54 G00 X50.0 Y75.5 S850 M03
N27 G43 Z25.0 H02 M08 T01
N28 G01 Z1.0 F250.0
N29 M98 P7502                  (GROOVE D)
N30 X140.0 M98 P7502           (GROOVE C)
N31 Y10.0 M98 P7502            (GROOVE B)
N32 X50.0 M98 P7502            (GROOVE A)
N33 G28 Z25.0 M09
N34 G28 X50.0 Y10.5 M05
N35 M30
%
```

```
O7501 (SUBPROGRAM FOR ONE GROOVE)

(FINISH INNER CONTOUR - OFFSET D52)
N201 G90 G01 Z-6.0 F200.0
N202 Z-6.5 F50.0
N203 G91 G41 X3.5 D52 F75.0
N204 G03 X-3.5 Y3.5 I-3.5 J0
N205 G01 X-25.5
N206 G02 X-5.5 Y5.5 I0 J5.5
N207 G01 Y26.0
N208 G02 X5.5 Y5.5 I5.5 J0
N209 G01 X51.0
N210 G02 X5.5 Y-5.5 I0 J-5.5
N211 G01 Y-26.0
N212 G02 X-5.5 Y-5.5 I-5.5 J0
N213 G01 X-25.5
N214 G03 X-3.5 Y-3.5 I0 J-3.5
N215 G40 G01 X3.5
(FINISH OUTER CONTOUR - OFFSET D62)
N216 G41 X-3.5 D62
N217 G03 X3.5 Y-3.5 I3.5 J0
N218 G01 X25.5
N219 G03 X12.5 Y12.5 I0 J12.5
N220 G01 Y26.0
N221 G03 X-12.5 Y12.5 I-12.5 J0
N222 G01 X-51.0
N223 G03 X-12.5 Y-12.5 I0 J-12.5
N224 G01 Y-26.0
N225 G03 X12.5 Y-12.5 I12.5 J0
N226 G01 X25.5
N227 G03 X3.5 Y3.5 I0 J3.5
N228 G40 G01 X-3.5
N229 G90 G00 Z1.0
N230 M99
%
```

Lead-in/lead-out toolpaths - finish cuts - are shown at right:

D52/D62 setting should be the same for both contours

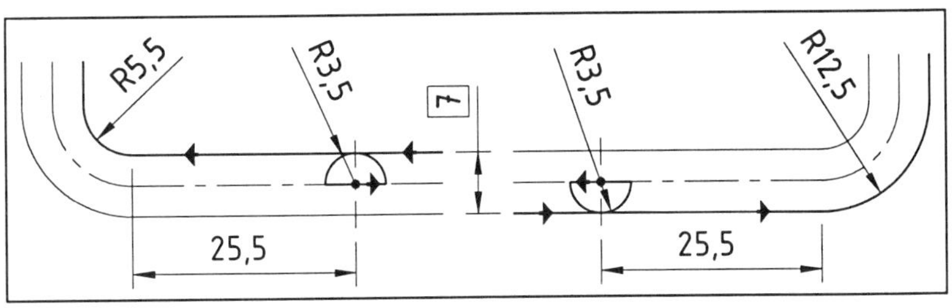

Round Pocket Subprogram

The previous program could not be considered very difficult, but still had some interesting features. The next example will be a little simpler (it will be offered in two versions). In both version, the subprogram will be developed for the ∅1.0 round pocket, adhering to the following machining requirements (programmer's decision):

- **Version 1** ***Single depth*** **round pocket with stepovers**
- **Version 2** ***Multi depth*** **round pocket with stepovers**

The sample drawing at right is dimensioned in inches and shows a single round pocket, ∅1.0 and 0.3 deep. Center of the pocket and its top face will be part zero, both the most logical locations.

Both examples will also use the same tool - specified as *T01* in the program:

T01 = ∅0.25 center cutting end mill - slot drill

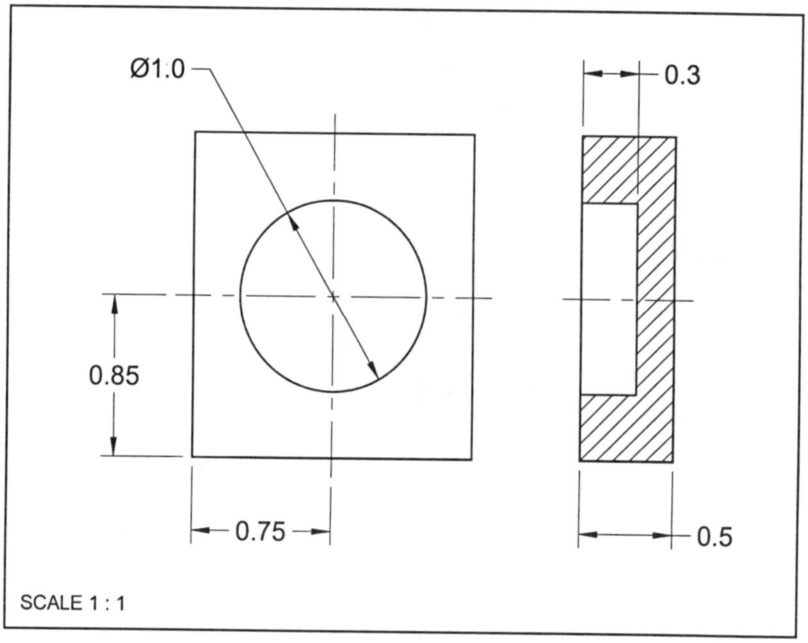

Single Depth Pocket with Stepovers

The first version of the example illustrates complete programming approach to a subprogram development, when the tool has to cut a round pocket in steps at the same plane as the depth. These steps are commonly known as *stepovers* or - the *width of cut*. Using stepovers may solve a typical situation when a large area and a small tool are necessary to remove excessive material. Using a round pocket as an example is a simple way to present the programming approach, which can be adapted to more complex toolpath.

For the subprogram development, the most important decision is the *stepover amount* - the actual *width of cut*. If possible, try to select the stepovers of equal width for each pass. In some cases, you may want to control the width of the *last pass*, perhaps to achieve a better surface finish. The following calculations shows the process:

- **Radius of the pocket (P)** = **1.0 / 2 = 0.5**
- **Radius of the tool (T)** = **0.25 / 2 = 0.125**
- **Total amount of stepover (S)** = **0.5 - 0.125 = 0.375**

The stepover amount can be easily calculated by selecting the number of required stepovers (passes) and calculating the actual stepover amount. Some experimentation may be necessary for best results.

Example 1 - Calculating each stepover as equal width:

Each stepover = S / number of stepovers

Each stepover = 0.375 / 3 = 0.125, 0.375 / 5 = 0.075, 0.375 / 6 = 0.0625 ...

Note the absence of 0.375 / 4. The result is 0.09375, which would have to be rounded. Rounded amount is usually applied to the first or the last stepover. If the amount of the last pass should be smaller than each stepover for machining reasons, the calculation is similar but must include the selected finishing amount. Again, some experimentation with the numbers may be expected.

Example 2a - Calculating the total stepover amount using 0.015 for the finishing pass *(F)*:

Stepover amount S = P - T - F = 0.5 - 0.125 - 0.015 = 0.360

Example 2b - Calculating each stepover based on 0.015 for the finishing pass *(F)*:

Each stepover = S / number of stepovers

Each stepover = 360 / 3 = 0.12, 360 / 4 = 0.09, 0.36 / 5 = 0.072 ...

Keep in mind that in this particular case, there will always be one more pass added. For example, if the stepover width is 0.09, that means four passes of 0.09 ***plus*** one pass of 0.015 for finishing:

4 x 0.09 + 0.015 = 0.375

Round Pocket Example - Decisions

The pocket shown on the previous page will use the total of six steps, into X+ direction, while each step will have the same width:

0.375 / 6 = 0.0625

Once the preliminary decisions and calculations have been completed, both the main program as well as the subprogram can be developed.

A single tool ∅0.25 will be used at the full pocket depth of 0.300. In the example that follows, not only the six stepovers will be considered, but also the depth control.

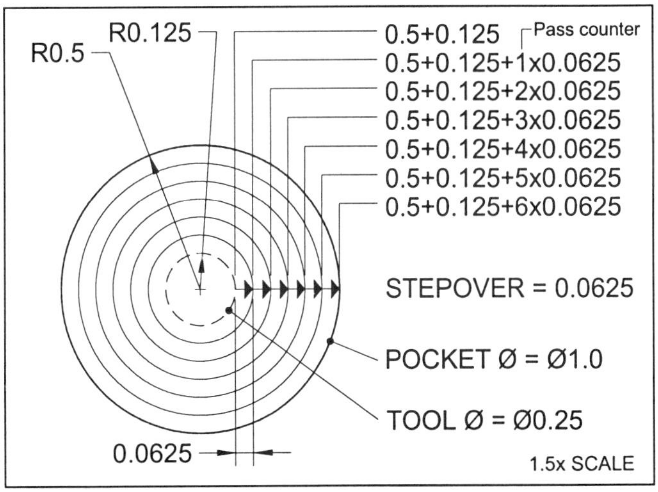

Round Pocket Example - Programs

```
(MAIN PROGRAM FOR ROUND POCKET)
(*** SINGLE DEPTH + STEPOVERS ***)

(T01 - 0.25 CENTER CUTTING E-MILL)
N1 G20
N2 G17 G40 G80
N3 G90 G54 G00 X0.75 Y0.85 S1800 M03
N4 G43 Z0.1 H01 M08
N5 G01 Z-0.05 F7.0
N6 M98 P7601
N7 G90 Z0.1 M09
N8 G28 Z0.1 M05
N9 G28 X0.75 Y2.15
N10 M30
%
```

```
O7601 (SUBPROGRAM SINGLE DEPTH)

N101 G91 X0.0625      (STEP 1/6)
N102 G03 I-0.0625     (ROUND CUT 1/6)
N103 G01 X0.0625      (STEP 2/6)
N104 G03 I-0.125      (ROUND CUT 2/6)
N105 G01 X0.0625      (STEP 3/6)
N106 G03 I-0.1875     (ROUND CUT 3/6)
N107 G01 X0.0625      (STEP 4/6)
N108 G03 I-0.25       (ROUND CUT 4/6)
N109 G01 X0.0625      (STEP 5/6)
N110 G03 I-0.3125     (ROUND CUT 5/6)
N111 G01 X0.0625      (STEP 6/6)
N112 G03 I-0.375      (ROUND CUT 6/6)
N113 G01 X-0.375      (BACK TO CENTER)
N114 M99
%
```

As shown above, this version of subprogram development does not really offer any obvious benefit from a version that does *not* use a subprogram. That is true of the above example, *as shown*. The real and tangible benefit of this method will only be apparent if the *same pocket* is machined at *several locations*. That is the area where subprograms are of real benefit.

Multidepth Pocket with Stepovers

The second version will apply the basic stepovers developed previously, but also adds several cutting depths. In this sense, it will be similar to the first subprogram example in this chapter. Even one pocket at a single location will benefit from a subprogram, when its total depth must be separated into smaller cuts. If such a combined toolpath takes place at several part locations, the subprogramming method shines even more.

Selecting the individual depth of each cut follows the same basic principles as selecting the stepover (width of cut), described earlier. The total depth may be segmented into a number of equal depths, or it can be distributed in such a way that the last depth will be smaller, perhaps to provide better machining conditions.

In this example, the total depth of the round pocket is 0.3 inches. Such a depth can be distributed in several ways, so the selection of six depth cuts of 0.05 each is reasonable - and also selected for this example.

Illustration at right shows the concept

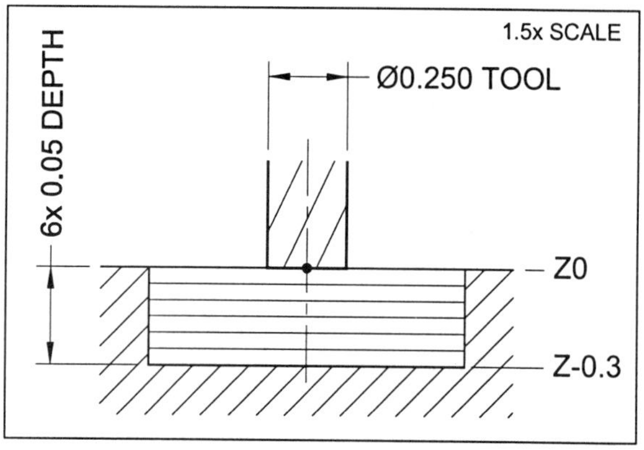

The starting Z-axis position of tool is also very critical, because the depth cuts will be programmed in *incremental* mode (G91). As shown, the position of the end mill before the subprogram is called *must start* at Z0, in this particular example.

On the next page is complete listing of the program that uses a subprogram to cut a round pocket with stepovers *and* multiple depth cuts. Although the example does not require it, the program also shows how such a pocket would be used at *different locations*. It is really not the pocket itself that is important here - it only serves as an example for practical subprogram development.

```
(MAIN PROGRAM FOR ROUND POCKET)
(*** MULTI DEPTH + STEPOVERS ***)

(T01 - 0.25 CENTER CUTTING E-MILL)
N1 G20
N2 G17 G40 G80
N3 G90 G54 G00 X0.75 Y0.85 S1800 M03
N4 G43 Z0.1 H01 M08
N5 G01 Z0 F7.0
N6 M98 P7602 L6       (POCKET 1)

  N7 G90 Z0.05
  N8 X2.0             (NEXT LOCATION)
  N9 G01 Z0
  N10 M98 P7602 L6    (POCKET 2)
  N11 G90 Z0.05
  N12 X3.25           (NEXT LOCATION)
  N13 G01 Z0
  N14 M98 P7602 L6    (POCKET 3)
  N15 G90 Z0.05
  N16 Y2.15           (NEXT LOCATION)
  N17 G01 Z0
  N18 M98 P7602 L6    (POCKET 4)
  N19 G90 Z0.05
  N20 X2.0            (NEXT LOCATION)
  N21 G01 Z0
  N22 M98 P7602 L6    (POCKET 5)
  N23 G90 Z0.05
  N24 X0.75           (NEXT LOCATION)
  N25 G01 Z0
  N26 M98 P7602 L6    (POCKET 6)

N7 (N27) G90 Z0.1 M09
N8 (N28) G28 Z0.1 M05
N9 (N29) G91 G28 X0 Y0
N10 (N30) M30
%
```

```
O7602 (SUBPROGRAM FOR MULTI DEPTH)

N101 G91 Z-0.05     (0.05 DEPTH)
N102 X0.0625        (STEP 1/6)
N103 G03 I-0.0625   (ROUND CUT 1/6)
N104 G01 X0.0625    (STEP 2/6)
N105 G03 I-0.125    (ROUND CUT 2/6)
N106 G01 X0.0625    (STEP 3/6)
N107 G03 I-0.1875   (ROUND CUT 3/6)
N108 G01 X0.0625    (STEP 4/6)
N109 G03 I-0.25     (ROUND CUT 4/6)
N110 G01 X0.0625    (STEP 5/6)
N111 G03 I-0.3125   (ROUND CUT 5/6)
N112 G01 X0.0625    (STEP 6/6)
N113 G03 I-0.375    (ROUND CUT 6/6)
N114 G01 X-0.375    (BACK TO CENTER)
N115 M99
%
```

Blocks N7 through N26 are not necessary to machine only the one pocket defined in the original drawing. They have been included to show how the program would be structured, if the pocket was machined at six different locations.

When comparing with the previous subprogram, note that the 0.05 depth of each pocket was moved from the main program to the subprogram itself - in ***incremental*** mode - while the tool motions at any given depth have not changed at all. The main difference is that the subprogram is repeated six times (M98 P7602 L6). Without the subprogram, the repetitions would have to be written individually, at each of the six depths. Needless to say, the complete program would be long and hard to manage.

Rough and Finish Cuts with a Subprogram

In spite of being somewhat different, the two subprogramming methods shown so far shared several common characteristics. The next programming technique using subprograms will be much different, although the general principles of subprogram development will be maintained.

One Toolpath for Two Cuts

In CNC programming of contour toolpaths, it is normal to program all XYZ coordinates as reflecting the finished dimensions of the part. When such contour toolpaths have to be roughed first, and then finished, programming two different toolpaths is impractical and inefficient. *Solution?*

The answer is - using a *subprogram*, of course. Even a simple contour, as the one shown below, will benefit from a subprogram that contains the *complete* toolpath used for finishing - that is, using *drawing dimensions*. Subsequently, this program can be used as a subprogram and called repeatedly for such operations as roughing, finishing, chamfering etc. The key to success in these cases is the use of *cutter radius offset* feature of the control system.

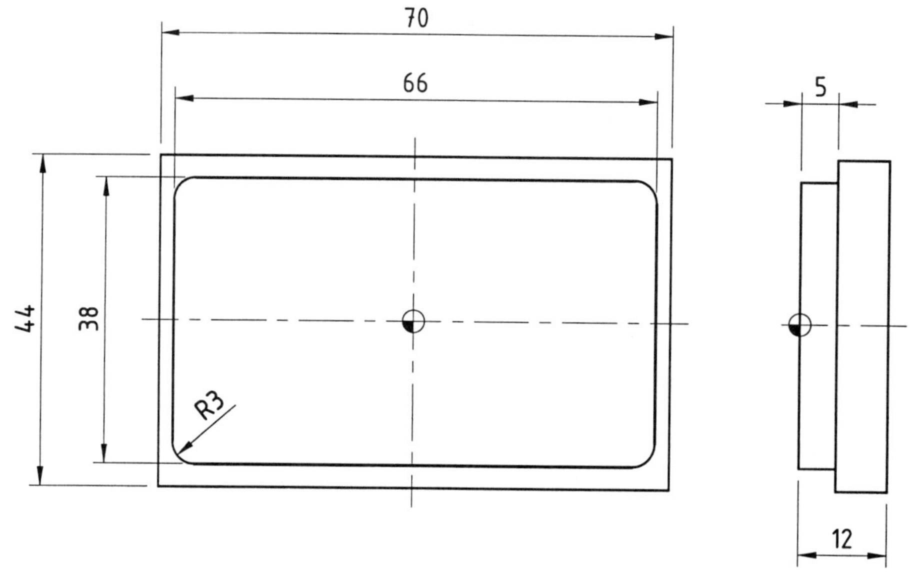

Material: 1020 STEEL (70 x 44 x 12 mm)

Cutter radius offset for standard tools in climb milling mode is G41 D.., where the address D identifies the offset number, where the actual offset value is stored. In this example, there will be two tools used, one for roughing, the other for finishing:

- **T01 = ∅15 mm end mill for roughing the contour ... will use D51 offset**
- **T02 = ∅12 mm end mill for finishing the contour ... will use D52 offset**

As roughing requires that some stock is left over on the contour for the finishing operation, the programmer selects a suggested amount for the XY-stock, and another for the Z-stock. In this example, the aim will be to leave 0.5 mm stock on the contour walls (XY), and 0.25 mm on the contour bottom (Z). The reals benefit of this programming technique is that the stock is only *suggested* to the CNC operator, but *not* programmed directly. Programmer only enters suggested amounts of stock and the offset numbers where these amounts will be stored, once the program reaches the CNC machine:

T01 will use D51 as its offset number for roughing, with the suggested amount to be stored equal to the *cutter radius minus the stock amount* in X and Y axes. T02 will use D52 as its offset number for finishing, with the suggested amount to be stored *equal to the stock amount* in the X and Y axes:

D51 cutter radius offset = ∅15 / 2 = 7.5	**D51 suggested setting = 7.5 - 0.5 = 7.000**
D52 cutter radius offset = ∅12 / 2 = 6.0	**D52 suggested setting = 6.000**

What about the stock left over for finishing at the bottom of the contour? The 0.25 mm stock left will be built directly into the main program. The next decision - *a very critical decision* - is the determination of the lead-in and lead-out for the programmed finishing contour.

Lead-In and Lead-Out

It is a poor machining practice to enter the selected tool into depth while touching the contour. Using a lead-in (and lead-out) motions allow the tool to enter (and exit) the contour smoothly, without damaging its surface finish.

For the example described in this section, both leads consist of a linear and arc tool motions. Programming any lead using a cutter radius offset, make sure the lead radius is *greater than the largest tool radius used.* In this case, the lead radius is 10 mm, while the radius of the roughing tool is 7.5 mm (only 6 mm for the finishing tool).

The leads in the part program can located at any suitable location of the contour. The illustration shows the leads at the *middle* location of the *bottom* contour. SP is the start point of the contour (also its last point).

The next step is to develop the contour toolpath, based on the drawing dimensions.

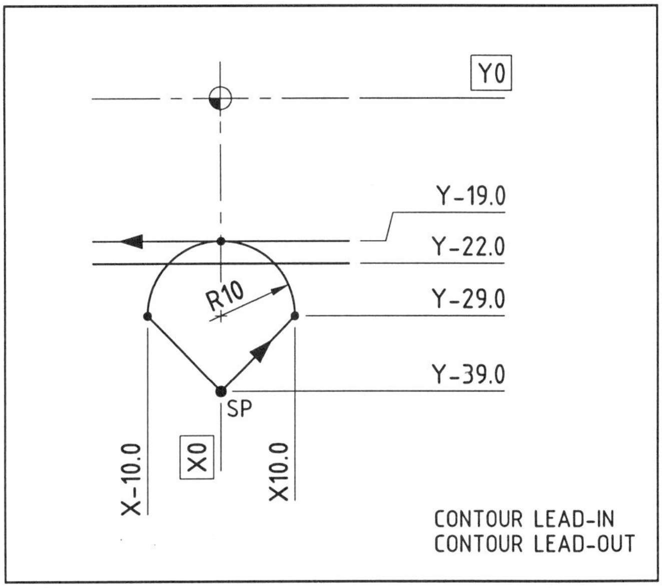

Common Contour Toolpath

The common contour toolpath is quite simple - it will start and end at the start point SP. The result is the subprogram, used for *both* roughing and finishing. *Non-changing* feedrate can also be included.

Note the absence of spindle speed (S), depth (Z), feedrate (F), and cutter radius offset number (D)

```
O7603 (CONTOUR SUBPROGRAM)

N101 G00 G41 X10.0 Y-29.0
N102 G03 X0 Y-19.0 I-10.0 J0
N103 G01 X-30.0
N104 G02 X-33.0 Y-16.0 I0 J3.0
N105 G01 Y16.0
N106 G02 X-30.0 Y19.0 I3.0 J0
N107 G01 X30.0
N108 G02 X33.0 Y16.0 I0 J-3.0
N109 G01 Y-19.0
N110 G02 X30.0 Y-19.0 I-3.0 J0
N111 G01 X0
N112 G03 X-10.0 Y-29.0 I0 J-10.0
N113 G40 G00 X0 Y-39.0
N114 M99
%
```

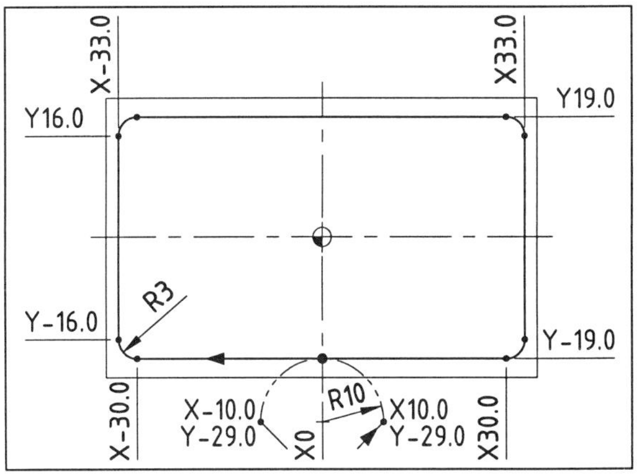

Main Program

Both previously selected cutting tools will now be used in a standard format main program. When it comes to the actual cutting, the four features missing in the subprogram must be defined in the main program. Spindle speed and feedrate will depend on the actual setup and cutting conditions, and may be changed as required. The table shows the data that will be used:

	Roughing with T01	Finishing with T02
Spindle	S950	S1200
Z-depth	Z-4.75	Z-5.0
Feedrate	F150.0	F200.0
D-offset	D51	D52

When merging a subprogram with the main program, the start position must be selected properly. In the example, either tool has to move to the start point SP, reach the desired depth and apply the D-offset *before* the subprogram is called. Selected spindle speed and feedrate have to be also in effect:

```
(MAIN PROGRAM)
(T01 - 15 MM END MILL - ROUGHING - USES D51)
(T02 - 12 MM END MILL - FINISHING - USES D52)
N1 G21
N2 G17 G40 G80 T01
N3 M06                                   (ROUGHING TOOL SELECTED)
N4 G90 G54 G00 X0 Y-39.0 S950 M03 T02    (RAPID TO START POINT SP)
N5 G43 Z2.0 H01 M08                      (CLEARANCE ABOVE PART)
N6 G01 Z-4.75 F300.0                     (FEED TO DEPTH - 0.25 STOCK LEFT)
N7 M98 P7603 D51 F150.0                  (*** ROUGH WITH D51 AND F150.0)
N8 G90 G00 Z2.0 M09                      (CLEARANCE ABOVE PART)
N9 G28 Z2.0 M05                          (TOOL CHANGE POSITION)
N10 M01                                  (OPTIONAL STOP)

N11 T02
N12 M06                                  (FINISHING TOOL SELECTED)
N13 G90 G54 G00 X0 Y-39.0 S1200 M03 T01 (RAPID TO START POINT SP)
N14 G43 Z2.0 H02 M08                     (CLEARANCE ABOVE PART)
N15 G01 Z-5.0 F300.0                     (FEED TO FULL DEPTH)
N16 M98 P7603 D52 F200.0                 (*** FINISH WITH D52 AND F200.0)
N17 G90 G00 Z2.0 M09                     (CLEARANCE ABOVE PART)
N18 G28 Z2.0 M05                         (TOOL CHANGE POSITION)
N19 G28 X0 Y-39.0                        (RETURN TO MACHINE ZERO)
N20 M30                                  (END OF MAIN PROGRAM)
%
```

For emphasis, the critical word have been highlighted. This method of programming is very flexible, for example, if a chamfer is required at the top of the contour, the same subprogram can be used, and only the data in the main program changed.

13 TURNING AND BORING IN DEPTH

External and internal machining, commonly known as turning and boring, are the most common operations on CNC lathes. Either operation involves roughing and finishing toolpaths and can range from very simple to very complex. Roughing means removal of a significant portion of the material and forming the shape that will be finalized during finishing. Multiple repetitive cycles G70, G71, G72 and - a to a much smaller extent - G73, are very useful for this type of work. Depending on the job setup and complexity, the CNC programmer has to deal with many issues that relate to roughing and finishing, with or without using multiple repetitive cycles. This chapter describes various techniques that can be used in many turning and boring operations on CNC lathes.

Program Zero Selection

Selecting program zero (part zero) for lathe parts is always related to the Z-axis. Without an exception, the program zero for the X-axis *must be* the spindle centerline. When it comes to the Z-axis zero selection, there are several methods that may at least be evaluated and either rejected or accepted.

The illustration at right shows five reasonable options, identified by a reference letter. The Z-zero position can be anywhere, but there is some credibility in the five options.

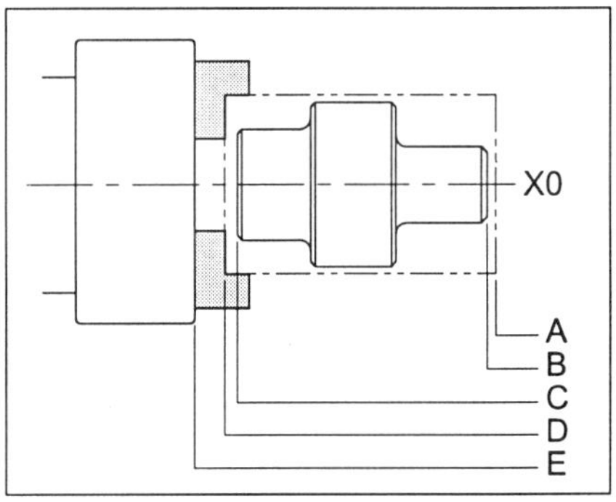

- **Position A - Front face of the stock**

Although this face is easy to reach during tool setup, selecting the stock front face has several disadvantages. First one is that the stock will be removed by facing in the early stages of machining, but the amount removed will have to be included in all absolute Z-axis motions. The second disadvantage is that the stock amount is not always consistent.

- **Position B - Front face of the part**

This is the most popular overall selection, and for a good reason. Many, if not all, dimensions on the drawing originate front the front face and can be used directly. Another advantage is that all Z-axis dimensions in the drawing will be negative - forgetting the minus sign in the program will force the cutting tool *away* from the part. Setup using the touch-off method is not difficult.

- **Position C - Back face of the part**

This is probably the least favorable location from the five possibilities listed. It is hard to measure and provides virtually no benefits, even if some drawing dimensions originated from this face. Keep in mind that when the part is turned around, this will be the preferred face - see *Position B*.

- **Position D - Jaws face**

In some cases, this is face may be the rough stock, in other it may be the finished back face. Face of jaws is easy to measure and may benefit some castings or forgings where the stock length varies. On the negative side, the remaining rough stock must be always accounted for, if present.

- **Position E - Chuck face**

Unless the chuck is replaced, the chuck face remains the same for any job. No other benefits.

Corner Radius and Back Angle Selection

CNC programmers do not directly include the tool nose radius in the program, as the toolpath uses drawing dimensions and the actual radius is handled by cutter radius offset. That does not mean the programmer should totally ignore the corner radius or even the back angle of the selected tool.

The most commonly used corner radiuses for turning and boring are 0.4 mm, 0.8 mm, and 1.2 mm or their Imperial equivalents of 1/64, 1/32, and 3/64. The middle size (0.8 mm or 1/32 inch) is the most common radius used, but that also depends on the machine size and the type of work. Typically, larger radiuses (up to 1.6 mm or 1/16 inch) are common for large CNC lathes with higher power rating. Large radiuses are also selected for heavy work on castings and forgings, to minimize tool wear. In addition, it is not just the nose radius itself that has to be considered, but the insert size as well. The larger the inscribed circle of the insert (as per tooling catalog), the heavier cut may be taken, providing the machine has enough power to handle it (power is measured in kW or HP).

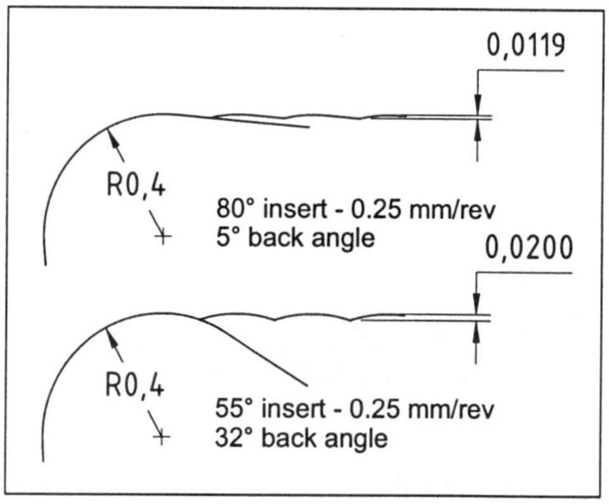

The above illustration shows an example of a 80° insert with 0.4 mm radius (0.0156 inch) and a 5° back angle, as well as a 55° insert, also with a 0.4 mm radius but 32° back angle. At the common feedrate of F0.25 mm/rev (0.0984 in/rev), note the depth of the scallops caused by each tool in millimeters. The following metric and Imperial tables show the differences for all three common radiuses:

Insert	Back Angle	R0.4 mm		R0.8 mm		R1.2 mm	
		Radius only	Radius +back angle	Radius only	Radius +back angle	Radius only	Radius +back angle
80°	5°	0.0200	0.0119	0.0098	0.0086	0.0065	0.0064
55°	32°	0.0200	0.0200	0.0098	0.0098	0.0065	0.0065

Insert	Back Angle	R0.0156 inches		R0.0313 inches		R0.0469 inches	
		Radius only	Radius +back angle	Radius only	Radius +back angle	Radius only	Radius +back angle
80°	5°	0.00079	0.00047	0.00039	0.00034	0.00026	0.00025
55°	32°	0.00079	0.00079	0.00039	0.00039	0.00026	0.00026

The example shows quite clearly that selecting the turning or boring tool based on radius alone is not always sufficient, particularly if a certain surface finish is to be achieved. There is a clear relationship between the tool nose radius, the back angle and cutting feedrate.

Every machinist knows that a larger corner radius provides a better surface finish as compared to a smaller corner radius under identical conditions. The reason is that a smaller corner radius will develop a deeper ridge than a larger corner radius - again, under identical cutting conditions, as demonstrated in the tables.

The purpose of the nose radius is to minimize impact at the weakest part of the insert

Selecting the proper corner radius for a particular turning or boring operation should take a few realities for consideration:

- **Smaller radius has much smaller inclination to cause chatter than a large corner radius**
- **Chip formed by a smaller corner radius is more uniform than a chip formed by a large radius**
- **Rigid setup, including a strong tool holder, increase in importance for larger corner radiuses**
- **Material support is important, particularly for a large stock extension to stock diameter ratio**
- **Internal corner radius is not the only reason for small corner radius - small radius is better for thin material**
- **Smaller corner radius is also beneficial for longer stock extension (such as 1:4 or 1:5 diameter to length ratio)**
- **Tool life increases for inserts with greater included angle**

These are just some of the observations determined by the tooling companies and actual users. Sometime it may take a bit of an experimentation to select the best toll - and corner radius - but it also shows that a corner radius selection should be considered as carefully as making other decisions.

Cutter Radius Offset

Cutter radius offset is also known as the *tool nose radius offset* on CNC lathes. Its purpose and overall functionality is the same as for milling applications and has been described in great detail earlier, in a separate chapter ***'Using Cutter Radius Offset'***. This section will only cover some additional items important for lathe applications, mainly the effect of *not* programming cutter radius offset.

Surprisingly, a significant number of machine shops do not use cutter radius offset on CNC lathes at all or using it poorly. Regardless of the reasons (some are mere excuses), the message is simple - in manual programming you *do* need to compensate for the tool nose radius. If you don't, the finished part just will not have the right dimensions - *it can't have.*

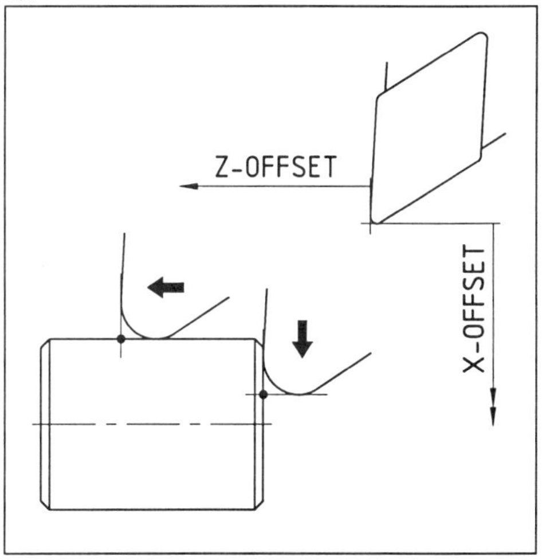

The main reason is the method of setting up the tools for turning and boring. In majority of setups, the mounted tool is touched-off the Z0 face to get the Z-axis geometry offset. The same approach is applied to the X-axis as well, but since it is impossible to touch the center line, the measuring process takes some extra steps, but the final result - the X-axis geometry offset - is the same as if the center line could had been touched. Both measurements are made between the extremities of the tool tip and the X0 or Z0. In practice, it means there is a significant difference between setting the radius offset for milling applications and doing the same for turning and boring applications. Rather than the center of cutter radius (as in milling), the lathe setup (geometry offset) is measured to a *corner* point of the tool nose, often called the *imaginary tool point* or the *virtual tool point.*

The illustration shows the offset measurement for both axes, as well as their effect on a diameter cutting (Z-only direction) and the face or shoulder cutting (X-only direction). When it comes to cutting radiuses, chamfers and tapers, the control system requires a different type of internal calculations, therefore it also requires a few adjustments in the programming (and setup) method.

Imaginary Tool Point

In the illustration at right, three points are shown as examples of imaginary points of typical lathe tools. These three points have one thing in common - they both are the only setup points the controls system *'knows about'*. These points are often called the *command points*, as they are commanded by the program. Each XZ coordinate in the program determines the actual position of the command point on the contour.

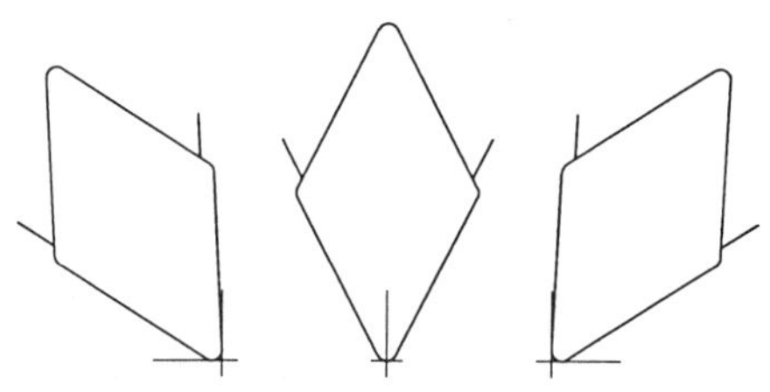

Typical orientation of lathe tools

Position of Command Points on a Contour

A typical contour is not just a series of diameters and faces, it also includes radiuses, chamfers and tapers. Without cutter radius offset, the command point will always be positioned at the *endpoint* of the contour element. As the illustration on previous page shows, there is no problem with diameter or face cutting. The illustration below shows the position of the command point at the end point of a radius, chamfer or a taper, in different configurations. Only an external contour is shown, but it applies equally to an internal contour as well.

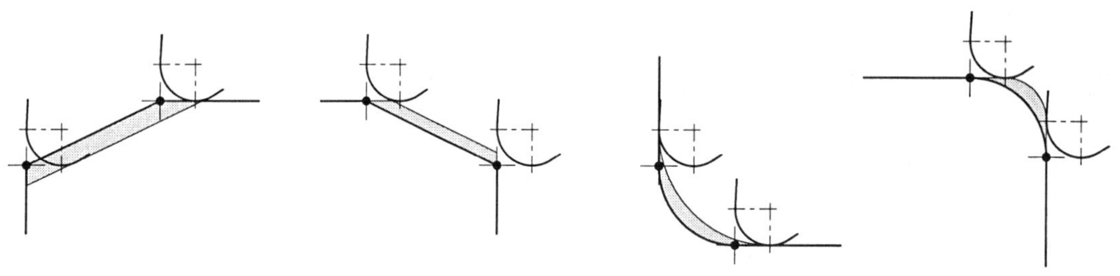

Excessive cutting Insufficient cutting Insufficient cutting Insufficient cutting

Without a cutter radius offset, either *excessive* cutting or *insufficient* cutting will take place, resulting in dimensions that do not conform to the drawing.

Stock Allowance

In a machining environment, *stock allowance* is a term that defines the amount of material left for finishing after rough cutting had been completed. Although it seems a subject simple enough to ignore, there are some considerations that will influence the programming techniques applied to roughing, particularly the last tool passes, where the stock will applied.

The major considerations of stock selection and control are the nature of the cut, particularly the *shape* of finished contour, and the *type* of tool used. Leaving stock on the roughed contour is generally designed for the finishing cut, but there could be other reasons, for example, as a grinding allowance. When using the G71/G72 lathe cycles, stock allowance can be defined in the cycle. If a cycle is not used, the required stock allowance must be programmed directly, even if it means a few additional calculations.

Contour Shape

Most contour toolpaths will be a combination of lines and arcs, where the lines represent diameters, faces, shoulders, tapers and chamfers, while arcs represent fillet radiuses, partial radiuses, undercuts, etc. Stock left on an external contour will increase its nominal size, whereby stock left on internal contour will decrease its nominal size.

Cutting Tool Used

A typical cutting tool (external or internal) used on a CNC lathe may have several tip orientations, depending of the tool design and its direction when mounted in the turret. The standard turning and boring tools (with 80°, 55°, 35° inserts) mounted for typical working conditions are the ones to watch for. The main problem is with their front clearance angle (lead angle), particularly when a face or a shoulder is machined. Unlike in conventional machining, where tool cutting direction can be changed frequently, on a CNC lathe the situation is different, particularly when using lathe cycles. Tool cutting direction can change as per programmer's decision, but not with cycles. Both G71 and G72 (and G70) maintain a uniform toolpath and changing direction is rather limited. It is possible to program part of a contour using G71 cycle, and another part using the G72 cycle.

Stock Allowance in X and Z axes

Having a uniform stock over the whole contour is not only impractical, it can even be dangerous. This is no overstatement and the reason is the tool geometry and its use in the part program. Typical lead angle for 80° inserts is 5°, while the lead angle for 55° and 35° inserts is 3°. When facing up, the edge of the tool is almost parallel with the face, leaving virtually no clearance. When the stock on the face is too heavy, the insert can literally be 'buried' in it and cause severe problems. The solution is to leave the Z-stock very small - see below.

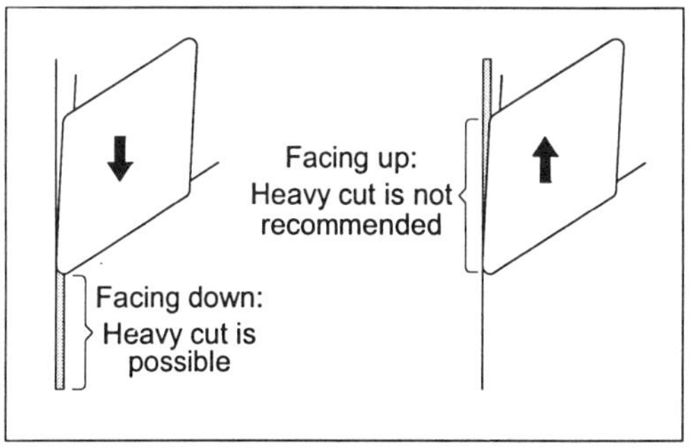

Both the G71 and G72 cycles allow stock allowance specification to be selected individually for the X-axis and the Z-axis. Depending on the type of G71/G72 cycle, the format is either a single block or a double block:

- **Single block format:**

```
G71 P.. Q.. I.. K.. U.. W.. D.. F.. S..
```
U and W are stock allowances

```
G72 P.. Q.. I.. K.. U.. W.. D.. F.. S..
```
U and W are stock allowances

- **Double block format:**

```
G71 U.. R..
G71 P.. Q.. U.. W.. F.. S..
```
U and W are stock allowances

```
G72 W.. R..
G72 P.. Q.. U.. W.. F.. S..
```
U and W are stock allowances

Do not confuse the U or W from the first block with the U and W in the second block!

By definition, the U-address is the stock left on the diameter in the positive or negative direction, while the W-address is the stock left on the faces or shoulders in the positive or negative direction. On a rear type lathe, typical U-address will be *positive* for external cutting and *negative* for internal cutting. For the majority of jobs, the W-address will be *positive* for both external and internal toolpaths.

The U-address for stock allowance is always measured on the diameter !

The actual amount of stock left on diameters will depend to a large extent on the type of job, material, tool nose radius and general setup. For thin or tubular stock, a smaller amount of stock may prevent chatter during finishing, particularly when combined with a smaller tool nose radius. If cutting conditions are favorable, stock allowance per side is commonly programmed close to the tool nose radius. For example, U1.6 will leave 0.08 mm of stock for finishing, per side. In Imperial units, similar stock allowance would be programmed as U0.06, leaving 0.03' per side for finishing.

As the illustration on previous page shows, when cutting with G71 cycle, the facing will be in the *up* direction, and the W stock allowance must be very small, typically between 0.08 to 0.15 mm or 0.003 to 0.006 inches. With such small amounts, the tool lead-in angle does not produce heavy depth of cut. There is way to calculate the actual depth of cut on a face, using the following formula:

$$D = \tan\frac{A}{2} \times R + \frac{W}{\tan A} + R$$

where ...

D = Depth of cut on the face
A = Insert lead angle
R = Corner radius of insert
W = Stock left on face

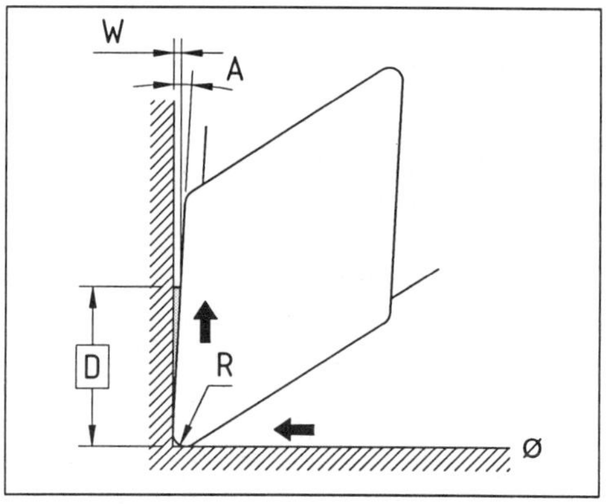

Use care and common sense when selecting W stock.

Compound Stock

When the stock allowance on diameters is combined with the stock allowance on faces, the actual stock allowance is equivalent to neither the U nor the W amount of the G71/G72 cycle - it becomes a *compound stock*. In simpler terms, compound stock is a combination of the U and W amounts applied to angles and radiuses. The illustration shows a typical compound stock ©). Although the programmer should be aware of its existence, there is no reason to pay special attention to compound stock in daily programming.

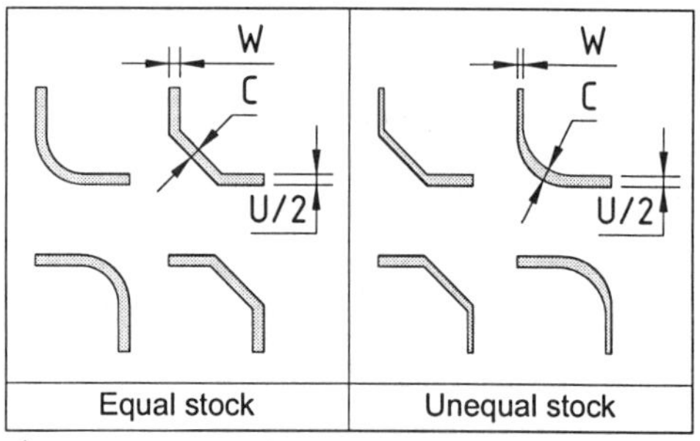

Grinding Allowance

Leaving stock for subsequent turning or boring operations is the most common application of stock allowance. In some cases, no stock is left on the part for turning or boring, but for *grinding* - this is a somewhat different type of stock allowance, called *grinding allowance*. Neither G71 nor G72 cycle supports grinding allowance as such, although it can be used for such purpose in certain cases.

A true grinding allowance is a suitable amount of stock left for the sole purposes of grinding on selected diameters and/or faces and shoulders. The key word here is *selected* entities, as compared to *all* entities for turning or boring. Take - as an example - the drawing shown at right (sizes are not to scale and exaggerated for clarity) - the diameter will be ground later, after the CNC lathe machining is completed. The taper that precedes the diameter is turned to size - it is *not* ground. For a smooth transition between the finished taper and the unfinished diameter, the taper should be extended. Programmer selects the amount of grinding allowance, in this example as 0.05 mm (~0.002").

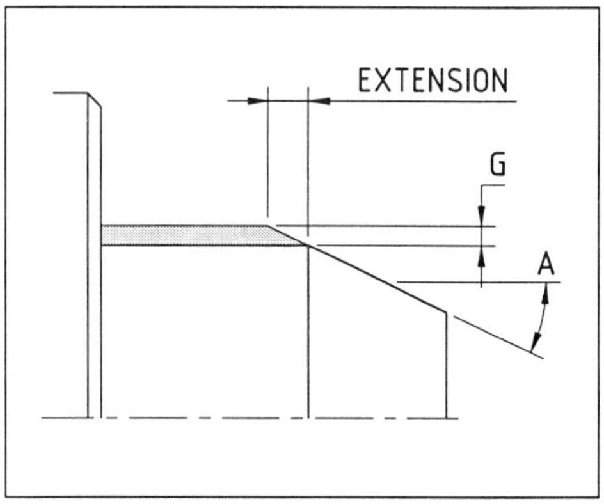

The taper (or a chamfer) extension can be easily calculated, based on the grinding allowance *G* and the taper (chamfer) angle *A*, using simple trigonometric function:

EXTENSION = G / tanA

Example - Grinding allowance is 0.05 mm (0.002"), Taper angle is 25° - then, the extension length is:

EXTENSION = 0.05 / tan25 = 0.10723 = 0.107 mm

This extension will be added to a known Z-axis location, as defined in the drawing between the taper and the *finished* diameter (not shown). At the same time, the specified drawing diameter will be increased by the double grinding allowance *G* (or decreased for boring). The effect on the part face (shoulder) was not considered in this example.

In many cases, there will be some radius required in the corner between the part diameter and the face or shoulder. In this case, a smooth transition is required, so a new radius has to be programmed - a temporary one for the grinding allowance. Note that the grinding allowance is applied to the diameter only, not the face.

The center of the temporary radius *r* has to be shifted from the center of the defined radius *R* by the amount of grinding allowance *G*. Also, the temporary radius *r* must be smaller than the defined radius *R*, by the amount of grinding allowance *G*.

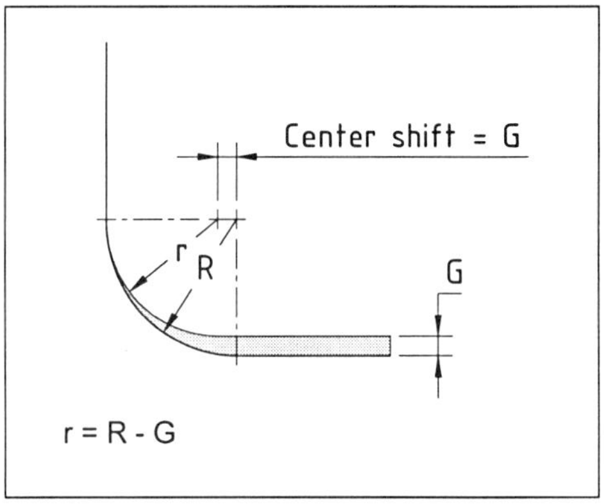

Both of the presented examples are quite easy to calculate. The main focus is to correctly evaluate the engineering or production demands in terms of grinding allowance requirements.

Example: R = 5 mm, G = 0.750 mm

The center shift between drawing radius *R* and temporary radius *r* will be 0.75 mm and the radius *r* will be 5 - 0.75 = 4.25 mm

Although some grinding operations may be eliminated by the *hard turning* method (see a special section at the end of this chapter), grinding is an important part of many post-CNC operations.

Tool Approach Techniques - Lead-In

Series of tool motions programmed *before the first element* of the contour is machined is called tool approach or *lead-in* motion. When the tool approaches the contour, it can approach a face, diameter, chamfer or a taper, or a radius. A series of similar drawings shows the main features of a tool approach towards a part (not to scale for clarity). Note that the approach direction is only an example, and can originate from another point.

Approaching the Front Face

In most turning jobs, there will be a need to program at least one facing cut, to remove the front face stock. The tool should start well above the stock diameter by a clearance *C* of 2.5 mm (or 0.1 inches) per side - or more, if necessary. Keep in mind that not all parts are perfectly round and the rotation increases the effective stock diameter. Using cutter radius offset for the face alone is nor necessary, but will do no harm if programmed correctly.

Example:

```
N1 G21
N2 T0300
N3 G96 S250 M03
N4 G41 G00 X35.0 Z0 T0303 M08
N5 G01 X..
...
```

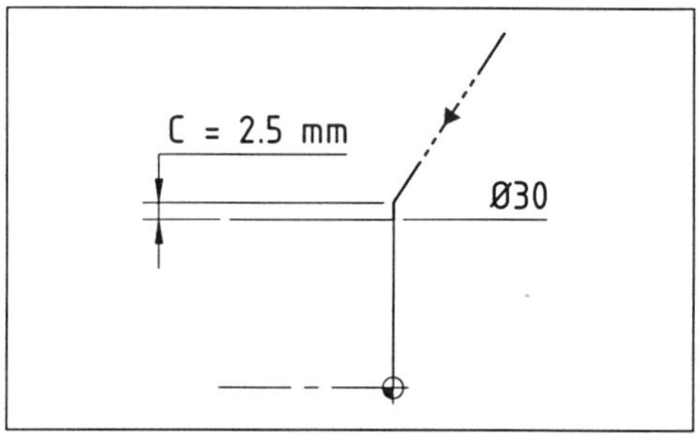

In this case, it is not necessary to include the G41 command, but there is no harm in doing so.

Approaching a Diameter

Any tool motion towards a diameter is the easiest to program and requires no special considerations. It is similar to the previous example, except for the cutting direction. For both external and internal cuts, the tool should have a suitable clearance from the part, along the Z-axis. Typical clearance *C* of 2.5 mm or (0.1 inches) is generally sufficient. Cutter radius offset should always be in effect, even for the simplest of contours. The approach motion is also the motion that activates cutter radius offset for turning or boring.

Example:

```
N1 G21
N2 T0300
N3 G96 S250 M03
N4 G42 G00 X30.0 Z2.5 T0303 M08
N5 G01 Z-..
...
```

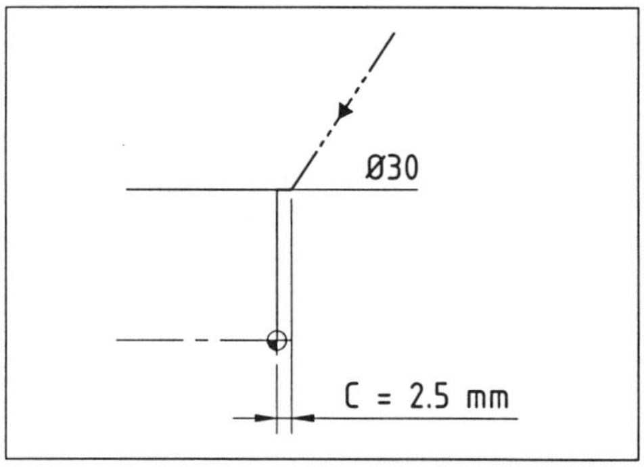

There is no need to program G42 for the ∅30 diameter alone, but as it is more than likely that such a diameter will not be the only contour element, it is strongly recommended to program the cutter radius offset.

Actual clearance of 2.5 mm (0.1") is recommended as the minimum clearance

Approaching a Chamfer

Many contours start with a chamfer or a taper. Approach towards a taper is a little more involved and has been defined in detail in the chapter ***Programming Tapers***. Chamfers are much more frequent as the first entity of a contour. In order to maintain the angle of the chamfer, the cutting tool has to approach the part at a *selected Z-axis* clearance and a *calculated X-axis* diameter. Cutter radius offset should always be in effect for precision machining.

Example:

```
N1 G21
N2 T0300
N3 G96 S250 M03
N4 G42 G00 X22.5 Z2.5 T0303 M08
N5 G01 X30.0 Z-1.25 F0.25
N6 Z-..
...
```

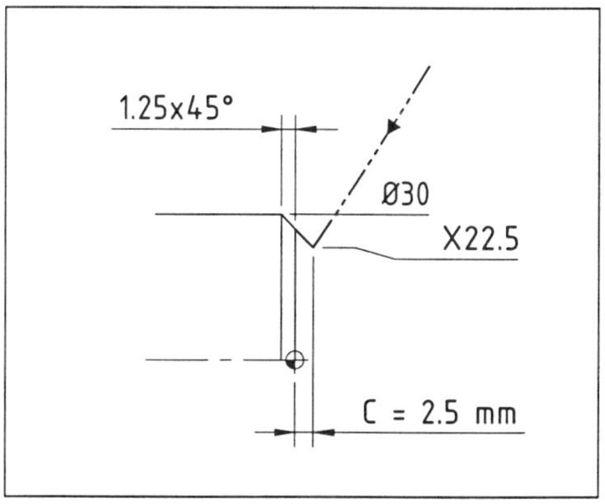

For an approach towards a chamfer, the front clearance has to be selected (2.5 mm), and the start of the chamfer calculated. The calculation takes into consideration the chamfer size and the clearance:

X = 30 - 2 × 1.25 - 2 × 2.5 = X22.5

Note that to calculate a diameter, both the chamfer amount and the clearance must be ***doubled***.

Approaching a Radius

If the first element of the machined contour is a radius, the tool approach is unique in the sense that it must include *two* motions during the lead-in. The reason is that cutter radius offset (which should become effective during the approach) *cannot* be started on an arc. That means a lead-in linear motion must be programmed first, and during this motion the radius offset will be activated.

There are three typical approach methods to an arc, all shown here with examples. All methods are similar, whether tool moves towards a tangential arc or an intersecting (partial) arc. This section explains typical programming techniques of approaching an arc as the first entity of a finished contour.

Example 01 - Tangential corner radius:

```
N1 G21
N2 T0300
N3 G96 S250 M03
N4 G42 G00 X20.5 Z4.25 T0303 M08
N5 G01 Z1.75 F0.25
N6 G02 X24.0 Z0 R1.75
N7 G03 X30.0 Z-3.0 R3.0
N8 Z-..
...
```

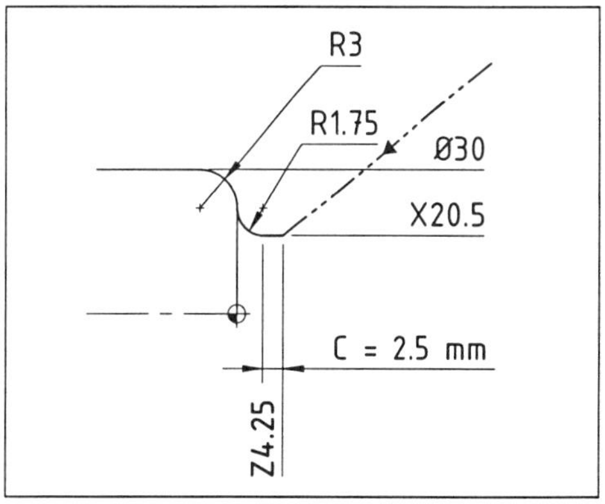

The starting diameter X20.5 was calculated in a similar fashion as in the previous example - from the main diameter of ∅30 and the two tangent radiuses:

X = 30 - 2 × 3 - 2 × 1.75 = X20.5

The *lead-in* radius must be larger than the tool nose radius

For the next example, the initial tool approach will also be towards a tangential radius, but located not at a corner, but at the centerline. This type of radius is called *spherical radius.*

Example 02 - Tangential spherical radius:

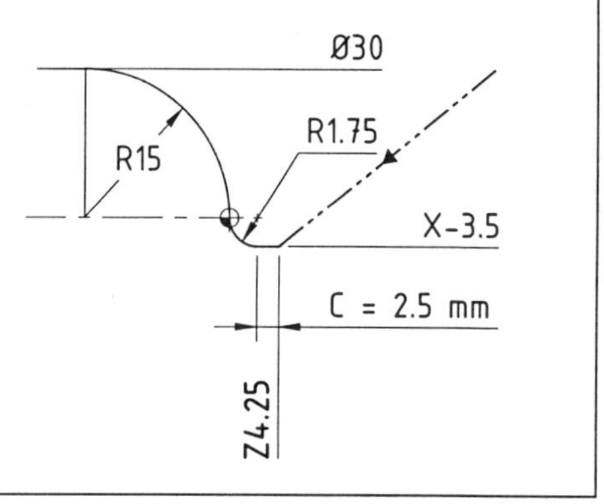


```
N1 G21
N2 T0300
N3 G96 S250 M03
N4 G42 G00 X-3.5 Z4.25 T0303 M08
N5 G01 Z1.75 F0.25
N6 G02 X0 Z0 R1.75
N7 G03 X30.0 Z-15.0 R15.0
N8 Z-..
...
```

Almost identical calculations as in the previous example. The starting X-diameter is below the centerline, at the double amount of the lead-in arc.

Another radius that can be the first element of a finishing contour is an *intersecting arc*, sometimes called a *partial arc.*

Example 03 - Intersecting radius:

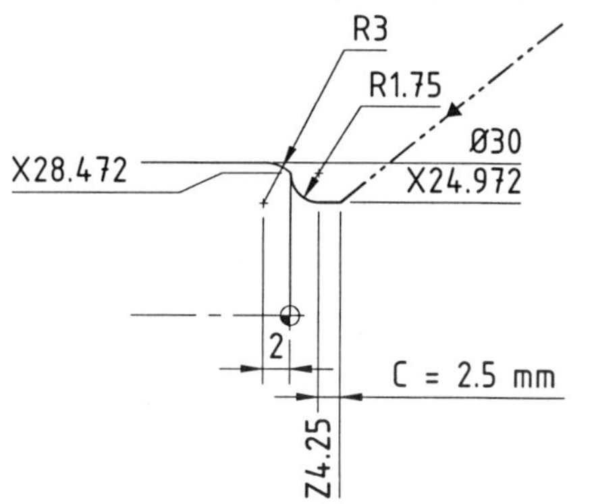


```
N1 G21
N2 T0300
N3 G96 S250 M03
N4 G42 G00 X24.972 Z4.25 T0303 M08
N5 G01 Z1.75 F0.25
N6 G02 X28.472 Z0 R1.75
N7 G03 X30.0 Z-2.0 R3.0
N8 Z-..
...
```

Because of the intersecting point between the front face and the arc, some additional calculations will be necessary in order to calculate the intersecting diameter as well as the starting diameter.

The first calculation must be the diameter at the intersection. Once this diameter is known, the starting diameter can be calculated very easily.

Pythagorean Theorem method will be used to calculate the dimension *W*. Once the *W* dimension is known, the unknown diameter ∅X can be calculated by subtracting the double amount of the lead-in arc from the new diameter.

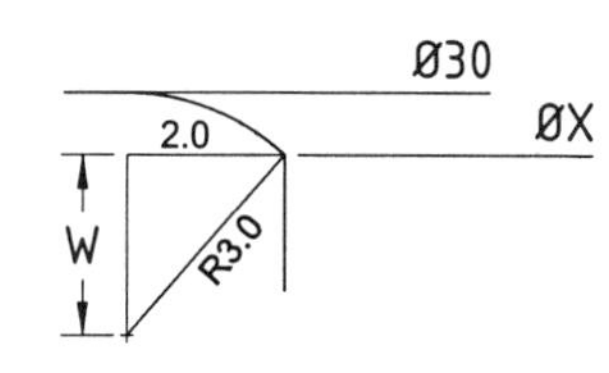

$$W = \sqrt{3^2 - 2^2} = 2.236$$

$$ØX = 30 - 2 \times (3 - W) = 28.472$$

From the examples shown, the various methods of programming a tool approach towards a part, generally for finishing operations, should be clear. It is impossible to cover all possibilities, but these representative examples show the most common situations. One common denominator should be apparent from all the previous examples:

Always approach the contour from a clearance location

Approaches to Avoid

The following illustration shows what tool approaches to avoid. In all cases, the G42 command (or G41 for boring) is activated during the approach towards the part.

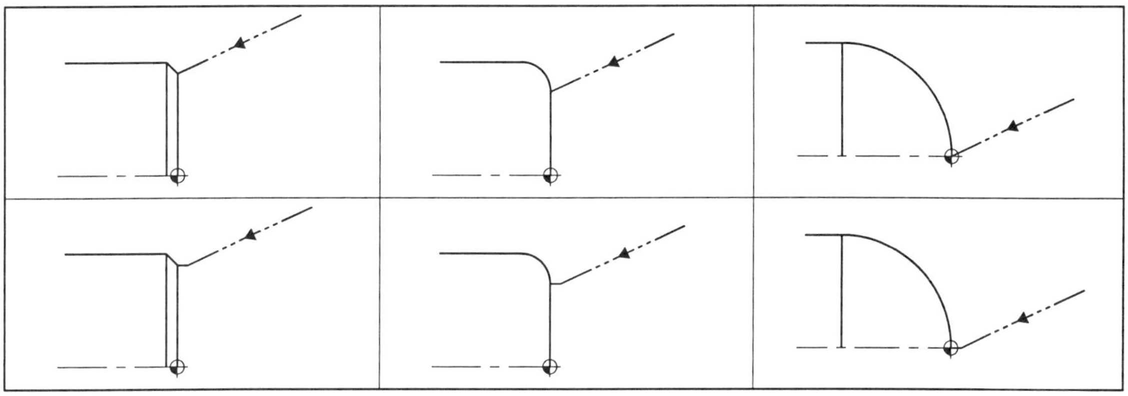

One main reason for the avoidance is to prevent improper application of the cutter radius offset. The bottom left image will be used as an example - see detailed illustration.

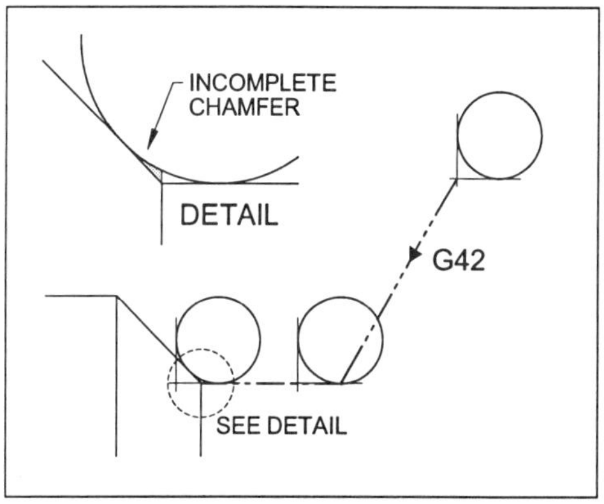

Although the cutter radius is applied correctly, and the Z-clearance is also correct, the tool tip approaches the chamfer at the same diameter as the lower chamfer point, resulting in an incomplete chamfer. The error will be greater when a large tool radius is used and will also be influenced by the chamfer (taper) angle.

Compare with the correct approach methods shown earlier in this section.

Tool Retract Techniques - Lead-Out

Series of tool motions programmed *after the last element* of the contour had been machined is called tool retract or *lead-out* motion. In many ways, the methods and suggestions for the lead-out motions are similar to the lead-in methods. The major considerations are - as before - the cutter radius offset and sufficient clearances.

Retract from a Face

If the last element of the contour is a face (also called a *shoulder,* if it is in the middle of a contour), the retract is very simple, and there are no special considerations required. All that is needed is to *continue* the face cutting 'into the air' - into a X-axis clear position (clearance diameter).

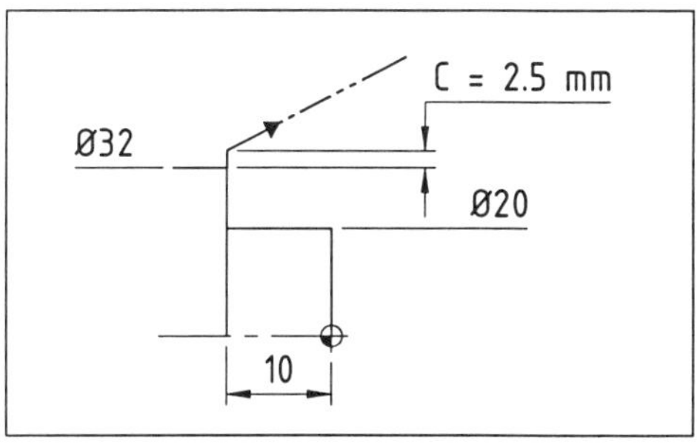

The following program illustrates the method.

Example:

```
...
N11 G01 Z-10.0 F0.25
N12 X37.0
N13 G00 G40 X.. Z.. T0300
...
```

Note the retract diameter - although the part diameter is only 32 mm, another 2.5 mm per side has been added as a standard procedure.

Retract from a Diameter

When the last element of a contour is a diameter (specified by the X-axis only), program any tool retraction along the X-axis first, *before* moving away from the part. This X-motion should *always* be minimum of 2.5 mm (or 0.1 inches) per side (5 mm or 0.2" on diameter). As mentioned several times before, this clearance is necessary to accommodate the three standard tool nose radius sizes - 0.4, 0.8, and 1.2 mm (0.0156, 0.0313, and 0.0469 inches).

Example:

```
...
N11 G01 Z-10.0 F0.25
N12 X35.0
N13 Z-18.0
N14 U5.0                        (SAME AS X40.0)
N15 G00 G40 X.. Z.. T0300
...
```

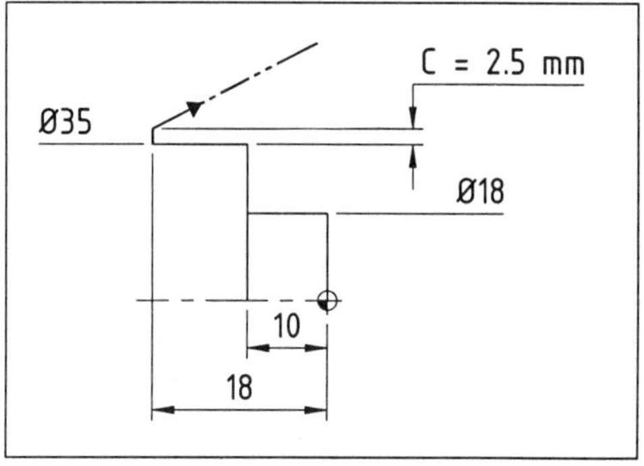

Note the block N14 - it uses an *incremental* amount of motion, rather than absolute dimension. Either one can be used, but U5.0 does not require calculations.

Retract from a Chamfer

A chamfer (or a taper) should always be extended, if it is the last element of the contour. This extension is necessary to guarantee completion of the contour. Typically, the part programmer selects the clearance above the diameter and calculates the final X-axis position, based on the clearance amount and the chamfer (or taper) angle. There are two common methods of programming a tool retract from a chamfer, both correct, if used properly. In both examples, the chamfer is extended *above* the final diameter by 0.75 mm (0.03 in inches).

Example 01 - End of chamfer only:

```
...
N11 G01 Z-12.0 F0.25
N12 X30.0                       (CHFR START)
N13 X32.5 Z-13.25               (CHFR END)
N14 G00 G40 X.. Z.. T0300
...
```

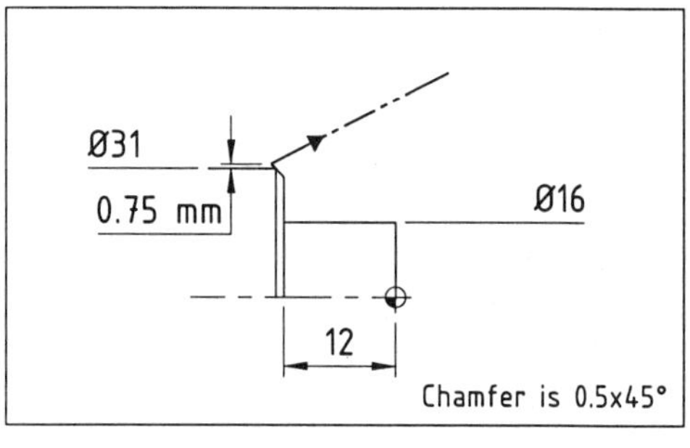

The start of chamfer is calculated by *subtracting* the double chamfer size from the top diameter (31 - 2 × 0.5 = 30.0), while the diameter at the chamfer end required a similar calculation, *adding* the double clearance to the top diameter(31 + 2 × 0.75 = 32.5).

Example 02 - End of chamfer with retract:

```
...
N11 G01 Z-12.0 F0.25
N12 X30.0                   (CHFR START)
N13 X32.5 Z-13.25           (CHFR END)
N14 G00 U5.0                (SAME AS X37.5)
N15 G40 X.. Z.. T0300
...
```

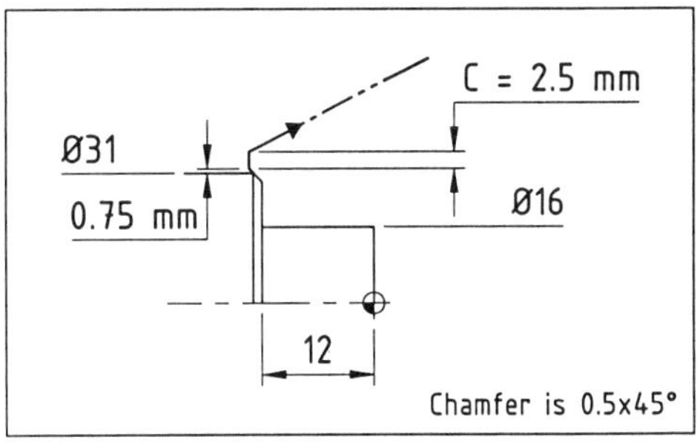

Chamfer calculations are the same as the last method, only the additional *rapid* retract has been added.

Retract from a Radius

Retracting from a radius is similar to the lead-in motions, there are two common methods. If the last element of the contour is a radius, it should be followed by a lead-out radius *and* a linear or rapid motion, where the cutter radius offset will be canceled. This is the most common method and also one that will always work, providing the lead-out radius is greater than the tool nose radius. Programming this radius as 1.5 mm or even 2 mm will satisfy that condition.

Example 01 - End of radius only:

```
...
N11 G01 Z-12.0 F0.25
N12 X26.6                   (RADIUS START)
N13 G03 X31.0 Z-14.2 R2.2   (RADIUS END)
N14 G02 U3.0 W-1.5 R1.5     (LEAD-OUT)
N15 G00 G40 X.. Z.. T0300
...
```

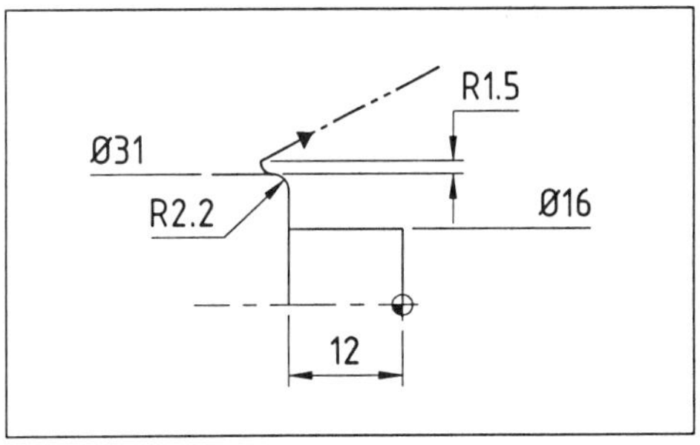

The rapid motion from the end of lead-out radius is the necessary retract motion.

Example 02 - End of radius with extra retract:

```
...
N11 G01 Z-12.0 F0.25
N12 X26.6                   (RADIUS START)
N13 G03 X31.0 Z-14.2 R2.2   (RADIUS END)
N14 G02 U3.0 W-1.5 R1.5     (LEAD-OUT)
N15 G00 U5.0
N16 G40 X.. Z.. T0300
...
```

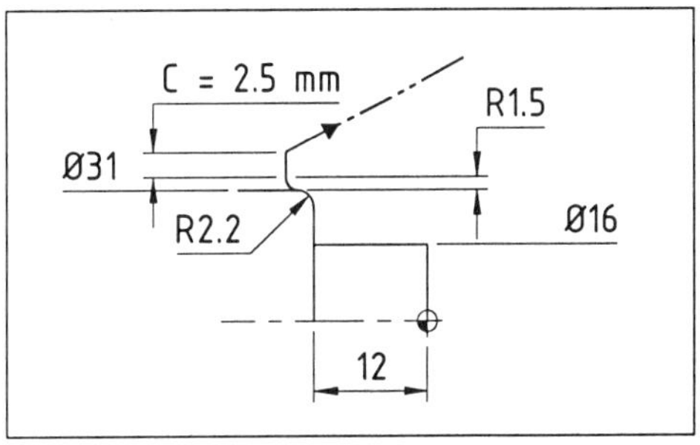

Only the G00 U5.0 motion has been added.

Retracts to Avoid

Certain retract motions from the last chamfer or a radius are not recommended and, when used in the program, they can cause problems. The illustration shows some of the most common lead-out methods to avoid.

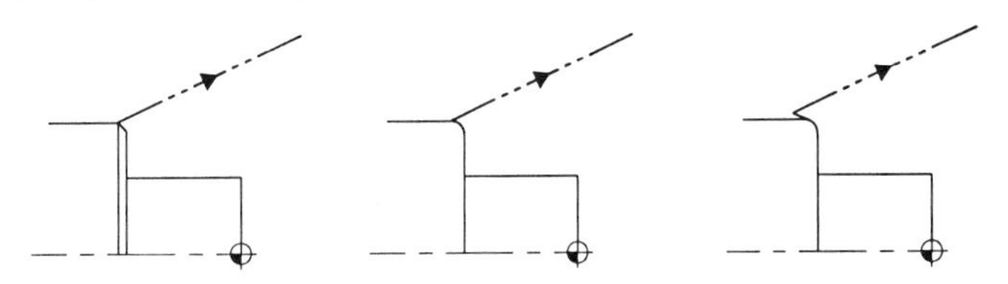

CHAMFER OR RADIUS WILL NOT BE COMPLETED

One Job - Two Operations

In a typical CNC lathe work, it is not unusual - in fact, it is quite common - to fully complete machining of the required part by clamping it more than once. CNC machine shops are used to a certain terminology when it comes to this type of machining. The main objective is to focus on the difference between reversing the part in the *same setup* and having *two separate setups*. For the purposes of this chapter, two terms will be used, with the following meaning:

- **First chucking / Second chucking** ... **one setup - part is reversed in the *middle* of program**
- **First operation / Second operation** ... **two setups - only a *portion* of the program is completed**

These are by no means firmly established terms, but they are used in CNC machine shops quite frequently, even if the local meaning may be a bit unclear. Using these terms in this chapter with a specific definition should make it easier to distinguish between the two different types of machining.

About Jaws

On the basic level, there are two types of jaws for CNC lathes - *hard jaws* and *soft jaws*. As the material they are made of indicates, hard jaws are generally used for roughing and semifinishing toolpaths, whereby soft jaws are used for semifinishing and finishing toolpaths. Because of their hard material, hard jaws will *'bite'* into the material and, in many cases, leave a physical mark on the part diameter, as they are mainly used with maximum chucking pressure. That is the main reason why hard jaws are not suitable for clamping of finished surfaces. Another reason is that hard jaws do no guarantee concentricity of the part as well as soft jaws do.

Soft jaws are made from mild steel, usually 1018 or 1020, with only a small percentage of carbon difference between them. The main benefit of soft jaws is that they can be bored to the exact diameter required with excellent concentricity of the clamped part. Since boring soft jaws has to be done *between* the stroke limits of the jaws travel, either an external or an internal ring should be used to provide the required clamping diameter.

Boring soft jaws is one of the most important skills a CNC lathe operator should have. Jaws can be bored manually, or with the help of a simple subprogram or a macro. An undercut between the jaws diameter and the locating face may be a necessary addition.

The illustration at right shows the correct bore of soft jaws (top), where the bored diameter *matches* the part diameter. The bottom figures show how the part will be clamped if the jaws bore is *smaller* than the part (left) or *greater* than the part (right). In both cases, the physical contact between the jaws and the part will be only at one or two points. Concentricity will also be hard to maintain.

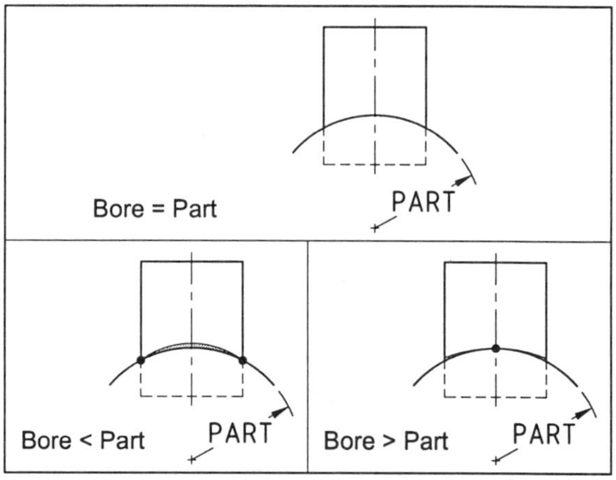

Another very important information relating to soft jaws is the *minimum depth of grip*. Apart from time consuming study of the program, the CNC operator has no way of knowing how far along the Z-axis the tool is programmed. Many a toolpath is programmed very close to the front face of the jaws, and the programmer should provide all information relating to safe setup.

Single Setup - Two Chuckings

Two *chuckings* usually mean machining one end of the part, implementing a program stop M00 function in a suitable place within the program, reverse the part and continue with the machining using the *same* program. In this type of setup, each part will be *completely* finished when it leaves the CNC lathe.

There is no need for a complete example - the program structure should illustrate the concept quite clearly. Note the *comment* following the M00 function. Anytime the *Program Stop* (M00) is used in the program, the CNC operator should be aware of its purpose. The best place to convey this information is the comment section, programmed in the same block as the M00 function.

The example shown will show a program structure for three tools - after the first tool, there will be a part reversal, followed by two other tools:

```
N1 G21
N2 T0100
N3 G96 S.. M03
N4 G00 X.. Z.. T0101 M08
N5 ...
N6 <... actual machining - tool 01 ...>
...
N16 G00 X.. Z.. T0100   (CLEAR POSITION FOR TOOL CHANGE AND PART REVERSAL)
N17 M00 (*** REVERSE PART FOR SECOND CHUCKING ***)

(================================================)

N18 T0200
N19 G96 S.. M03
N20 G00 X.. Z.. T0202 M08
N21 ...
N22 <... actual machining - tool 02 ...>
...
N33 G00 X.. Z.. T0200   (CLEAR POSITION FOR TOOL CHANGE)
N34 M01                 (OPTIONAL STOP)

N35 T0300
N36 G96 S.. M03
N37 G00 X.. Z.. T0303 M08
N38 ...
N39 <... actual machining - tool 03 ...>
...
N50 G00 X.. Z.. T0300   (CLEAR POSITION FOR TOOL CHANGE AND PART REMOVAL)
N51 M30                 (PROGRAM END)
%
```

Make sure to understand what exactly happens when the *Program Stop* function is activated:

- **All axis motions will stop**
- **Spindle will stop**
- **Coolant will be turned off**

These three activities are equally vital to successful program processing. Do not forget to reinstate them in the tool that follows the part reversal. Although the last feedrate, spindle speed itself, motion mode, and several other functions will still be active (in the computer memory), it is always advisable to program them anyway. Consider the likely possibility that the program will start ***after*** the reversal, for example, when the machining was interrupted or at the beginning of another shift. It is always the best approach *not* to count on various defaults and current settings - just in case.

Two Setups - Two Operations

Two *operations* usually mean machining one end of the part first, *for all parts in the batch*, then changing the setup and completing the same part machined at the other end. In this type of setup, each part will be completely finished only *after* the second operation is completed. Programming for two operations rather than two chuckings requires additional setup, but generally produces a part that is more accurate, particularly if concentricity is an issue. This method also requires two (or more) part programs that are stored separately, loaded separately, and processed separately. CNC operator should be aware of any job that is split over two or more programs.

Multi Cut Facing

Multi Cut facing is a term used for programming more than one facing cut to achieve a specific part length. If the supplied material is too long to program only one facing cut, the amount of material has to be known and separated into two, three, four or even more manageable facing cuts, as necessary. Such situation results in wasted material, but it is a reality any CNC programmer may have to face.

Two main keys to successful programming of multiple facing cuts are based on knowing the *longest material stock* in the batch, and making a decision as to the *maximum width* allowed for facing. In the example, the maximum stock to be faced off is 7 mm. Based on a previous experience with the same material, the programmer determines that the maximum facing depth can be only 2 mm. Using a simple math, three facing cuts are not enough, and four are too much. Well, that's not exactly true. Since three face cuts are not enough, four cuts *will be* necessary, even if one of them will be a bit narrower. It follows that the next step must involve the actual *distribution* of the four face cuts.

Width of Cut Distribution

With four facing cuts required, the distribution of cuts - in the order of machining - may be 1 mm plus three times 2 mm - *or* - three times 2 mm plus 1mm; in both cases disposing of the 7 mm extra face stock. Which method is better - 1+3×2 *or* 3×2+1? It seems to be a minor point, and probably it is; however, the programmer *does have* a clear choice and only one method can be selected. Finishing the final face cut with a *smaller* width will most likely produce a somewhat better finish, so the 3×2+1 method seems a better choice. If other method produces better results, use it.

```
N1 T0100 G50 S2500
N2 G96 S125 M03
N3 G00 X56.0 Z5.0 T0101 M08
N4 G01 X-1.8 F0.1          (FACE 1)
N5 G00 W2.5
N6 X56.0
N7 Z3.0
N8 G01 X-1.8               (FACE 2)
N9 G00 W2.5
N10 X56.0
N11 Z1.0
N12 G01 X-1.8              (FACE 3)
N13 G00 W2.5
N14 X56.0
N15 Z0
N16 G01 X-1.8              (FACE 4)
N17 G00 X100.0 Z200.0 T0100
N18 M01
```

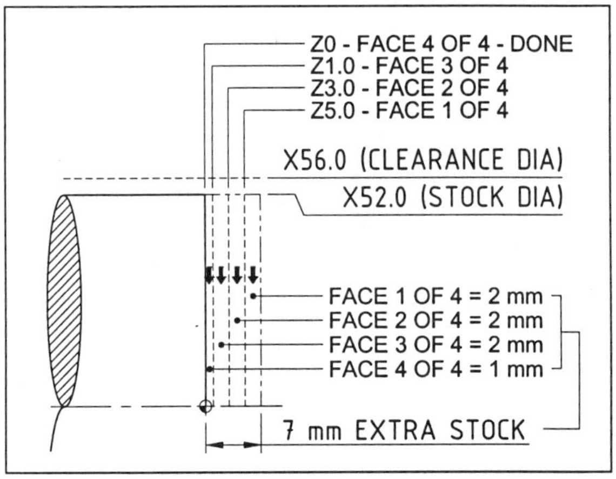

Note the cutting below centerline (for flatness) and the incremental Z-axis retract (convenience).

Breaking Corners

Machining corners between two major elements of a contour is called *corner-breaking*. Breaking sharp corners between two elements of a contour typically applies to a 90° corner between a face (shoulder) and a diameter, or between a diameter and a face. There are two general categories:

- **Defined corners**
- **Undefined corners**

Into the first category belong chamfers and radiuses that have been defined in the drawing. For example, a chamfer may be defined as 1×45° mm, or a fillet radius as R0.5 mm. The second category covers all sharp corners that have to be broken, without actually specifying the size. Drawings often identify such breaks with a comment *'Break all sharp corners unless otherwise specified'*. Undefined corners need a specific size in the part program, and the CNC programmer has to make that decision. Typically, such corners are programmed as 'minimum breaks', usually in the 0.05-0.25 mm range (0.002-0.01 inches). Although such breaks often appear to be only cosmetic in nature, they still may have a practical application, such ease of handling, ease of assembly or both.

Regardless of their actual definition or purpose, corner breaks can be automated, providing certain conditions are present. Fanuc control systems offer a special programming feature called *'Automatic Corner Breaking'* for 90° corners.

The illustration shows a typical application of a part that will benefit from the automatic corner breaking feature. Variety of chamfers and radiuses that have been used on external and internal contour may seem a bit busy, but provide a good example of this time saving programming tool. There are some rules and restrictions for programming of this feature:

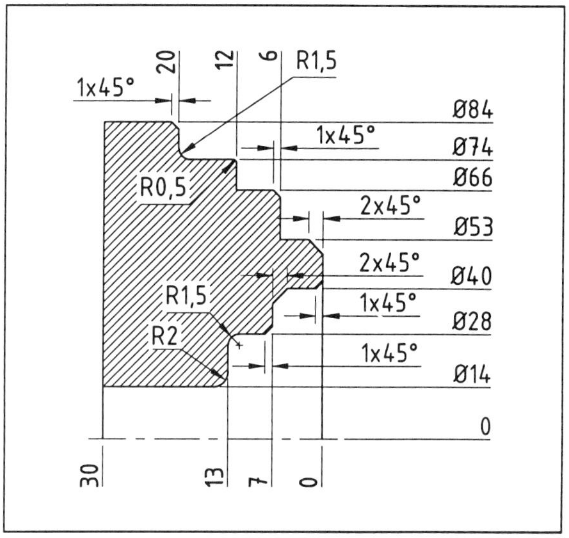

1. **Corner can only be between a shoulder and a diameter (no tapers or angles)**
2. **The motion *before* and *after* the chamfer or radius must be 90°**

Based on these restrictions, some chamfers in the drawing cannot use the automatic corner breaking method and must be calculated normally (manually).

In the program, both the turning tool and the boring bar will start with a 2.5 mm clearance off the front face. External chucking is assumed, so the ∅84 will *not* be machined. Also assumed is a completed face, roughing operation, and a hole through the part.

```
...
N21 T0500                               (T05 = TURNING TOOL)
N22 G96 S175 M03
N23 G42 G00 X44.0 Z2.5 T0505 M08        (START OF FRONT CHAMFER - OF)
N24 G01 X53.0 Z-2.0 F0.15               (MANUAL CHAMFER)
N25 Z-6.0
N26 X66.0 C-1.0                         (AUTOMATIC CHAMFER)
N27 Z-12.0
N28 X74.0 R-0.5                         (AUTOMATIC RADIUS)
N29 Z-20.0 R1.5                         (AUTOMATIC RADIUS)
```

```
N30 X82.0
N31 X85.0 Z-21.5                       (MANUAL CHAMFER)
N32 G00 U5.0                           (RETRACT FROM CONTOUR)
N33 G40 X100.0 Z200.0 T0500            (TOOL CHANGE POSITION)
N34 M01                                (OPTIONAL STOP)

N35 T0700                              (T07 = BORING BAR)
N36 G96 S150 M03
N37 G41 G00 X47.0 Z2.5 T0707 M08       (START OF FRONT CHAMFER - ID)
N38 G01 X40.0 Z-1.0 F0.125             (MANUAL CHAMFER)
N39 Z-7.0 C-2.0                        (AUTOMATIC CHAMFER)
N40 X28.0 C-1.0                        (AUTOMATIC CHAMFER)
N41 Z-13.0 R-1.5                       (AUTOMATIC RADIUS)
N42 X14.0 R-2.0                        (AUTOMATIC RADIUS)
N43 Z-32.0                             (BORE PAST PART LENGTH)
N44 G00 U-5.0                          (RETRACT FROM CONTOUR)
N45 Z2.5                               (RETURN TO FRONT OF PART)
N46 G40 X100.0 Z100.0 T0700            (TOOL CHANGE POSITION)
N47 M30                                (END OF PROGRAM)
%
```

Direction Specification

Although all Fanuc controls support the R-radius into *all* directions, some older controls do not support the C-chamfer. Instead, the I and J direction must be programmed, using the direction along an axis. Regardless of the address used (chamfer or radius), the direction always points into the *next* motion. For example, if cutting in X-plus and the following motion is Z-minus, the C® will be negative. For reference, both C and R directions are shown in the illustration bellow, along with the I and K.

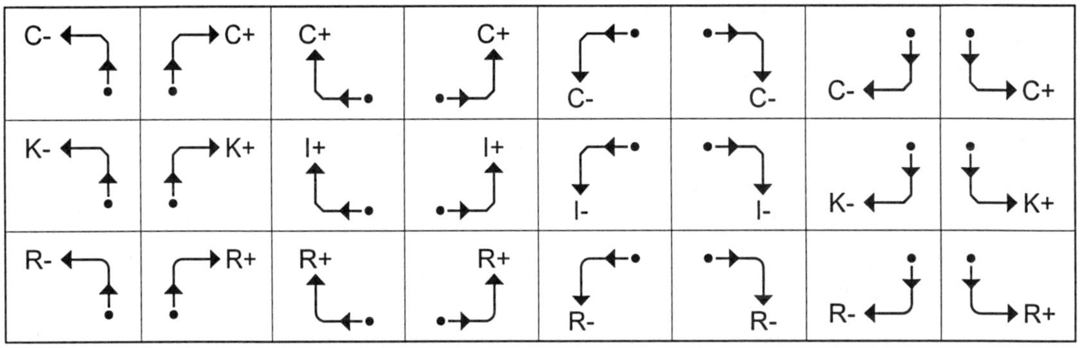

The heavy dot is origin of the cutting motion, and the arrows show the cutting direction.

Using Tailstock

When viewing a typical CNC lathe, the machine left side, where the chuck is located, is called the *headstock*. Opposite, at the far right side of the machine, is - the *tailstock*. Its design and purpose has not significantly changed for many decades, although it has acquired some modern features - tailstock is a *part supporting device*. In order to function properly, a small center-drilled hole has to be present on the front face of the part. This hole has a 60° angle that matches the supporting live or dead center, located in the tailstock.

Typical tailstock on CNC lathes consists of the main body that houses the internal mechanism, an extendable quill, and a tapered hole in the quill that can accommodate a live or dead center, depending on the tailstock design.

Types of Tailstock

From the programmer's perspective, there are two types of tailstock:

- **Fully programmable tailstock** ... **both body position and quill are programmable within a range**
- **Programmable quill only** ... **only the quill is programmable - body must be set manually**

If available, each operation requires a pair or miscellaneous functions (M-functions). For the fully programmable tailstock, there are four M-functions (examples only shown - check machine manual):

- **Tailstock body forward** ... **example: `M12`**
- **Tailstock body backward** ... **example: `M13`**
- **Quill forward** ... **example: `M21`**
- **Quill backward** ... **example: `M22`**

For both types, the CNC operator is responsible for proper setup of the tailstock (body and quill).

Programming a Tailstock with a Bar Stopper

In bar feeding operations, it is often necessary to extend the bar stock a certain distance from the chuck or collet face. For this purpose, a bar stopper is usually mounted in the turret and brought to the Z position to which the bar should be extended. If the bar extension is relatively short, for example, 2:1 or 3:1 ratio between the extended length and the bar diameter, the subsequent machining can usually take place without a tailstock support. For a bar that extends farther from the chuck/collet face, for example 5:1 ratio or higher, the tailstock support is required, which also means using a center drilling operation, often done on a CNC lathe as well.

The objective of this specially selected example is to *face and center drill* a bar stock that has been extended to a certain length from the chuck or collet face, then apply the tailstock *before* actual machining of the external contour. The contour has been selected as both simple in design and realistic for programming - see the drawing below. No M-codes have been used in the program in order to understand the individual steps.

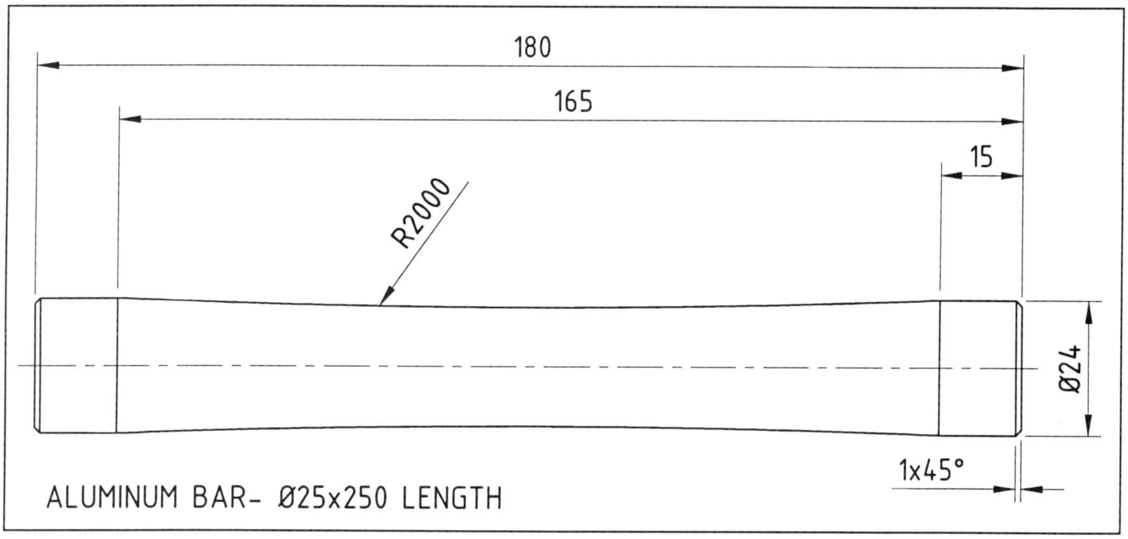

Once the drawing had been evaluated, the usual decisions have to be made.

To develop the part program, the following decisions will be reflected in the program:

1. **The bar stock is ∅25 mm - precut to 250 mm lengths, partially held in the chuck**
2. **Both the bar stock and the tailstock quill are handled manually *(no programmable functions are used)***
3. **T08 will be used as both the *bar stopper* and the *center drill* (see illustration below for dimensions)**
4. **T03 is a 35° turning tool (such as a VNMG type, with 0.8 nose radius) used for the contour**
5. **The selected bar extension for facing and center drilling is 50 mm from the chuck face**
6. **The selected bar extension for the OF turning is 200 mm from the chuck face**

The illustration shows relationship between the setting of offset 08 (tip of the center drill) and the tool holder corner that will be used as the bar stopper reference point. The two dimensions must be known and their difference included in the program. Wear offset cannot be used in this case, because the offset amount would exceed the allowable range.

Once these dimensions are known, the tool can be mounted permanently in the turret and used as necessary - programming will become that much easier for subsequent jobs.

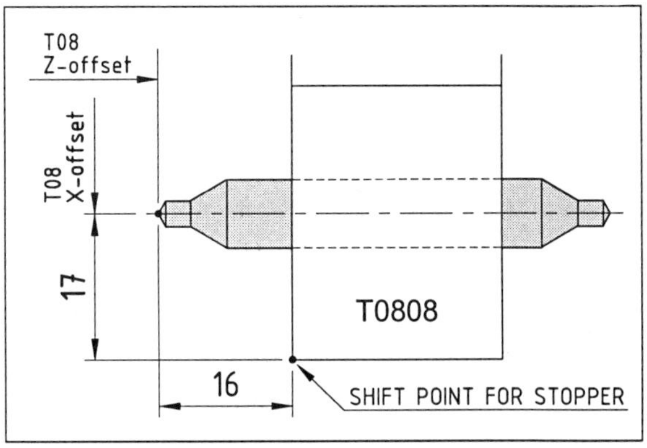

Solution to this project requires understanding of its initial requirements. For an accurate part setup, the Z-axis geometry offsets for T03 and T08 must be set for actual machining, which means the Z0 must be 200 mm from the chuck/collet face. For the stopper, 34 mm have to be shifted on diameter (17 mm per side) and 16 mm shift is necessary along the Z axis. The part can be finished to its final length in a subsequent operation, not important for this project.

```
(MACHINING WITH TAILSTOCK)
(25 MM DIA ALUMINUM BAR PRECUT TO 250 MM LENGTH)
(INITIAL BAR EXTENSION MUST BE LESS THAN 50 MM FROM JAWS FACE)

(T08 USED AS A STOPPER - MACHINE SETUP SHIFTED BY XZ AMOUNTS)
N1 G21 T0800                    (METRIC UNITS - TOOL 8 SELECTED)
N2 G00 Z-165.7 T0808            (0.3 FOR FACING - STOPPER SHIFT=16-0.3 MM)
N3 X55.0                        (STOPPER FACE AT 21.0 DIA - 21+2*17=55)
N4 M00                          (EXTEND BAR AGAINST STOPPER FOR CENTER DRL)
N5 G28 U0                       (RETURN TO MACHINE ZERO IN X-AXIS)
N6 G00 Z-100.0 T0800            (TOOL TO SAFE INDEXING POSITION FOR T03)
N7 M01                          (OPTIONAL STOP)

(T03 - FACE CUT - 35 DEG INSERT)
N8 T0300                        (TOOL 3 SELECTED)
N9 G97 S1800 M03                (1800 R/MIN FOR FACE CUT)
N10 G00 X28.0 Z-150.0 T0303 M08 (START POSITION FOR FACE CUT)
N11 G01 X-1.8 F0.1              (FACE CUT BELOW CENTER LINE FOR FLATNESS)
N12 G00 X28.0 Z-145.0           (SAFE POSITION OFF THE FACE)
N13 G28 U0                      (RETURN TO MACHINE ZERO IN X-AXIS)
N14 G00 Z-100.0 T0300 M09       (TOOL TO SAFE INDEXING POSITION FOR T08)
N15 M01                         (OPTIONAL STOP)

(T08 USED AS A CENTER DRILL - MACHINE SETUP USED)
N16 T0800                       (TOOL 8 SELECTED)
N17 G97 S1400 M03               (1400 R/MIN FRO CENTER DRILLING)
N18 G00 X0 Z-145.0 T0808 M08    (5 MM CLEARANCE OFF FACE)
```

```
N19 G01 W-11.0 F0.075           (6 MM DEPTH OF CENTER DRILL)
N20 G04 U0.5                    (0.5 SEC DWELL FOR BETTER FINISH)
N21 G00 W11.0 M09               (RETURN TO THE START POSITION)
N22 G28 U0                      (RETURN TO MACHINE ZERO IN X-AXIS)
N23 Z-11.0                      (ADJUST STOPPER POS. 5 MM OFF FACE 16-5=11)
N24 X55.0 Z-16.0                (STOPPER FACE AT 21.0 DIA - 21+2*17=55)
N25 M00                         (EXTEND BAR AGAINST STOPPER FOR TURNING)
N26 G28 U0                      (RETURN TO MACHINE ZERO IN X-AXIS)
N27 Z-11.25                     (TOOL TO SAVE INDEXING POSITION FOR T03)
N28 M00                         (APPLY TAILSTOCK)

(T03 - FACE CUT - 35 DEG INSERT)
N29 G50 S3000 T0300             (TOOL 3 SELECTED WITH 3000 R/MIN MAXIMUM)
N30 G96 S125 M03                (SURFACE SPEED SETTING OF 125 M/MIN)
N31 G00 Z2.5 T0303 M08          (START OF THE 1 MM CHAMFER IN Z-AXIS)
N32 G42 X17.0                   (START OF THE 1 MM CHAMFER IN X-AXIS)
N33 G01 X24.0 Z-1.0 F0.25       (FRONT CHAMFER CUTTING)
N34 Z-15.0                      (FRONT DIAMETER OF 24 MM)
N35 G02 Z-165.0 R2000.0         (CUTTING THE 2000 MM RADIUS)
N36 G01 Z-185.0                 (BACK DIAMETER OF 24 MM)
N37 U3.0                        (CLEAR ABOVE PART)
N38 G40 G28 U0                  (RETURN TO MACHINE ZERO IN X AXIS)
N39 G28 W0 T0300 M09            (RETURN TO MACHINE ZERO IN Z AXIS)
N40 M30                         (END OF PROGRAM)
%
```

This example shows the basic techniques used for programming a tailstock with a bar stopper. Adding any automated functions will nit change the basic approach, just makes the CNC operator's job somewhat easier.

Using 45-degree Tool

A turning tool that has the main cutting edge oriented at 45° is definitely considered a special tool for most machine shops, unless, of course, the standard type of work requires it on a daily basis, for example, in heavy turning. There are certain benefits to using this tool, as compared with a standard tool such as the CNMG type. Consider the following operation - a face and a diameter, with a chamfer between. Note that the part is large, but shown in the correct scale (critical dimensions only).

For convenience, the turning tool used is shown in the drawing, in the same scale as the part. Note the setting of the command point - it represents the easiest - *and safest* - setup at the CNC machine, using the touch-off method.

Both, the *inscribed circle* (IC) of the insert has to be known (from catalogue data) as well as its *corner radius* ®). IC of 12.7 mm indicates a one-half of an inch inscribed circle (insert width).

For machining, the left corner will cut the face, the bottom corner will cut the diameter, and the chamfer will be machined during transition between the face and the diameter. For safety, only one offset will be used, although it will require a few more calculations. Tool T08 is used in the program, with the wear offset 08 (T0808) for the face, chamfer and diameter.

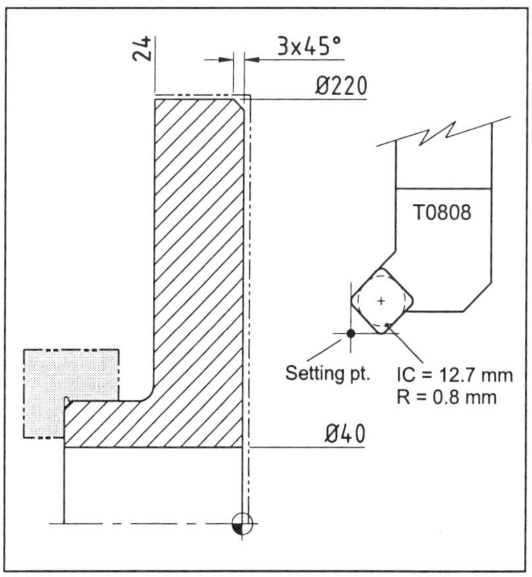

As the tool drawing shows, the command point setting is a combination of X-coordinate for the bottom tip (X-setting) and the Z-coordinate for the left tip (Z-setting). Because of its design, the command point is 'in the air'. the distance between each actual cutting point is shown on the drawing as the *d* distance, and can be calculated from the tooling data:

d = (12.7 - 2 x 0.8) x sin45 + 0.8
d = 8.64889
d = 8.649

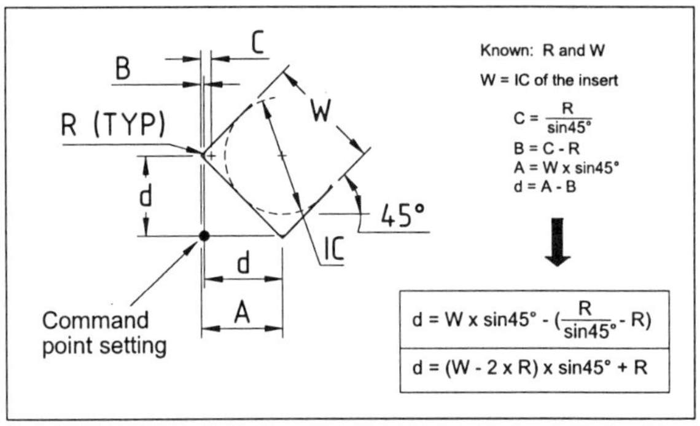

The whole programming process can be separated into seven major motions:

MOTION 1	**Rapid to start position for facing**
MOTION 2	**Machine the face all the way to the bore**
MOTION 3	**Rapid move off the part (clearance)**
MOTION 4	**Rapid to the chamfer start position**
MOTION 5	**Cut the chamfer**
MOTION 6	**Cut the diameter along its width**
MOTION 7	**Rapid to a tool change position**

In all cases, the X and Z coordinates in the program determine the position of the *command point*, as set on the machine (the imaginary point) and shown in the above illustration. The abbreviation *IC* is the insert *inscribed circle*, as defined in the tooling catalogue.

MOTION 1 - Rapid to start position for facing

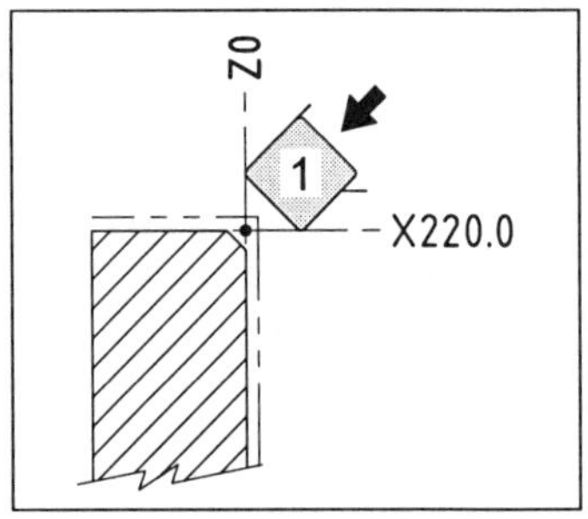

In the first motion towards the start position, the tool left tip must touch the face (Z0). At the same time, the bottom tip must be positioned in a safe location above the 220 mm diameter.

No calculations are necessary for stock up to about 4 mm. For larger stock, to tool should be positioned higher then 220 mm. Also keep in mind that the larger the stock, the heavier cut will be on both the face and the diameter, but particularly on the chamfer. In this case, additional cut (cuts) will be required.

MOTIONS 2 + 3 - Machine the face all the way to the bore / Rapid move off the part (clearance)

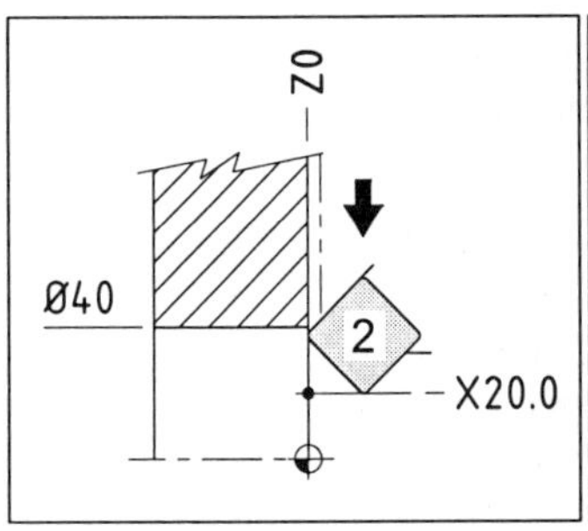

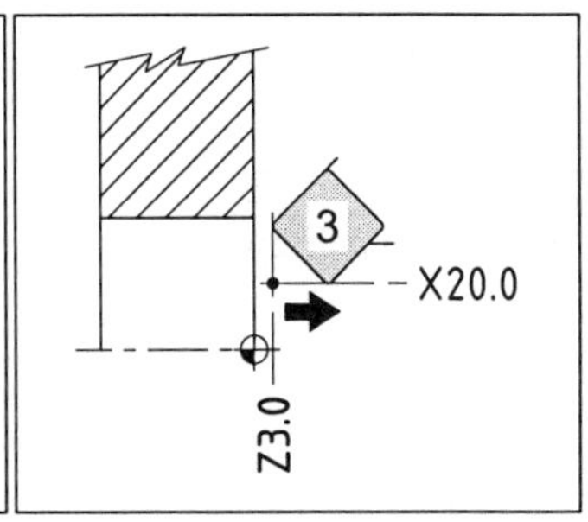

Only a small calculation is necessary for the facing cut. The insert left corner has to reach below the 40 mm bore diameter. As the distance *d* between the corners along an axis is 8.649 mm, 10 mm per side would be a sufficient clearance to complete the face cut.

When the facing cut is completed, the tool has to move off the face. The clearance used will be 3 mm, which is the Z-start position for chamfer cutting.

MOTION 4 - Rapid to the chamfer start position

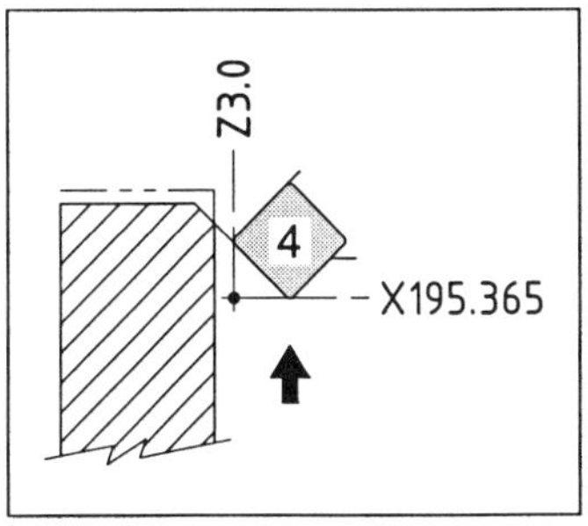

For precision work, the beginning and end of the chamfer position are the trickiest to calculate (particularly without a cutter radius offset). The two illustrations at right show the final position (left) and the calculations (right) used to find that position.

The sharp corner for chamfer start is at X208.0 Z3.0. The diameter of 208 mm is the result of:

Part diameter - 2 x chamfer - 2 x clearance
= 220 - 2 x 3 - 2 x 3 = 208.0

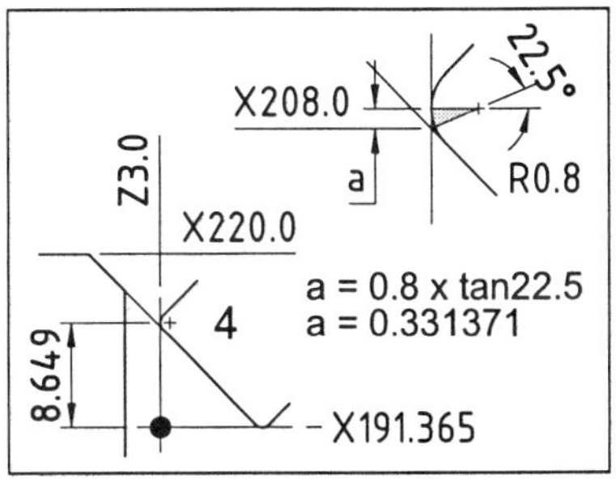

The double distance of 8.649 (*d*) from this point will make the diameter of the bottom tool point at:

208 - 2 x 8.649 = 190.702 mm

An adjustment *a* has to be made to position the tool edge in such a way that it matches the chamfer edge. If *a* is 0.331 mm, then the final diameter will be 190.702 + 2 x 0.331 = **X191.365**

MOTION 5 - Cut the chamfer

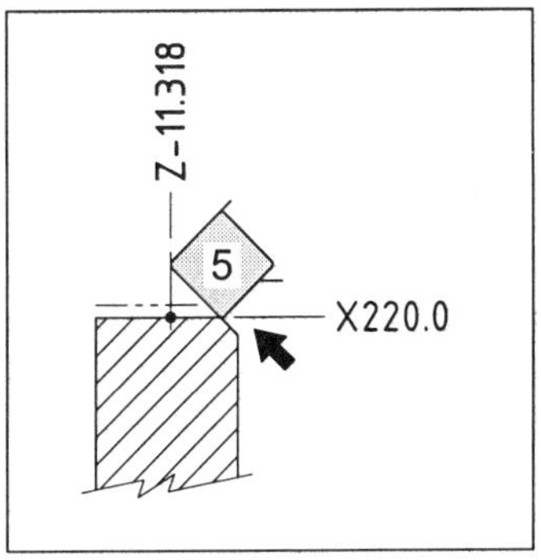

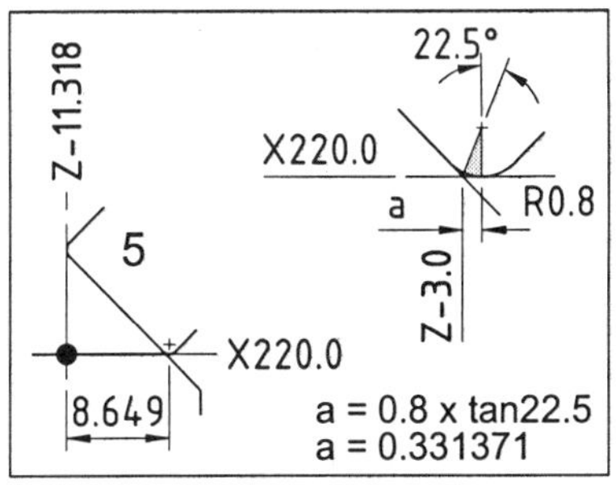

The same consideration that has been given to the chamfer X-start position must also be given to the chamfer Z-end position, at X220.0.

The chamfer is 3 mm, so the sharp corner at its end will be Z-3.0. Because the left corner is set 8.649 mm from the bottom corner, the 3 mm position becomes 11.649 position. As in the previous step, this position must be adjusted by the amount of *a* (0.331 mm), to 11.649 - 0.331 = **Z-11.318**

MOTION 6 - Cut the diameter along its width

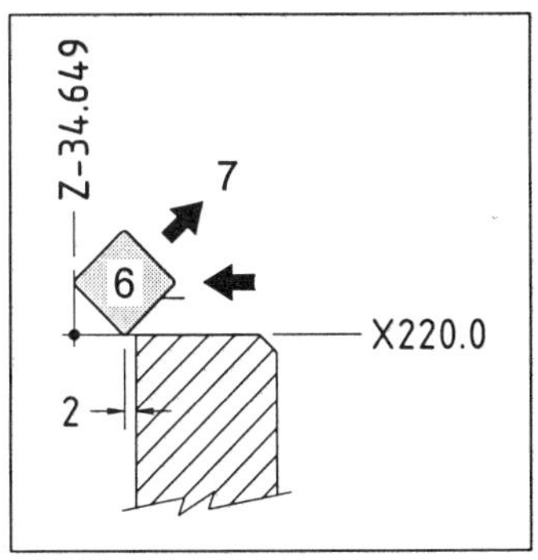

Overall part width is 24 mm. In order to cut the whole diameter length, the cutting motion will be physically extended by 2 mm, to Z-26.0. This position applies to the bottom tool corner only, and must be adjusted by the 8.649 mm (*d*) to **Z-34.649**

MOTION 7 - Rapid to a tool change position

Once the diameter has been completed, the tool can retract. This is the last tool motion, returning to a suitable tool change position.

The following program incorporates all decisions and calculations made to machine the three elements of the part, using a square insert, capable of machining with all exposed edges. The insert is oriented at 45° in an appropriate tool holder.

```
N1 G21
N2 T0100                          (90-DEG SQUARE INSERT IN 45-DEG TOOL)
N3 G96 S150 M03
N4 G00 X220.0 Z0 T0101            (POSITION 1 - START OF FACE)
N5 G01 X20.0 F0.35                (POSITION 2 - END OF FACE)
N6 G00 Z3.0                       (POSITION 3 - CLEARANCE OFF FACE)
N7 X195.365                       (POSITION 4 - START OF CHAMFER)
N8 G01 X220.0 Z-11.318 F0.2       (POSITION 5 - END OF CHAMFER)
N9 Z-34.649 F0.35                 (POSITION 6 - END OF DIAMETER)
N10 G00 X400.0 Z200.0             (POSITION 7 - TOOL CHANGE)
N11 M01 (M30)
```

In conclusion - this is an example of using a 45° square tool to machine a face and a diameter, with a chamfer between. However, it is also an example that can be used for similar applications of special tools, using the planning method and attention to detail. No cutter radius offset has been used, and only a single tool command point was selected. That does not mean they cannot be used.

Machining Thin Stock

Machining rings and various tubular material (pipes) requires fairly thin stock. As the material has to be clamped, the greatest problem with thin stock is its tendency to squeeze under pressure. There are at least three remedies that can be used - they relate mainly to the setup, but a good CNC programmer will be aware of them and takes them into consideration when developing the part program.

Adjusting Chuck Pressure

Most CNC lathes have a gage that shows the chuck pressure. Some may even have a programmable chuck pressure, using M-functions. In either case, the adjustment made to the chuck pressure does not always guarantee that the pressure will be weak enough to prevent part squeezing and strong enough to prevent the part dislocation from the chuck (a dangerous situation). Although decreasing chuck pressure may work in some cases, it is not a solution for the more demanding parts.

Using an Inner Plug / Outer Ring

If the chuck pressure adjustment does not provide optimal working conditions, some CNC machinists place a specially made cylindrical plug into the tube or pipe that fits in fairly tight, yet still can be removed when changing parts. Such a plug is easy to make (on a CNC lathe, of course) and can be made to any size desired. As a plug can only be used for outside chucking, a special ring can be fitted outside the tube diameter for inside chucking. Alternatives to this idea will depend on the actual job.

Using Special Split Jaws

For both external and internal machining, the three standard jaws will always produce a sort of an egg shape on thin stock, due to chuck pressure. Chuck pressure is not the only cause of the problem. The other reason - perhaps even greater one - is that the standard chuck/jaws combination has a large space between jaws. This is the space where the material deformities move into. The obvious solution is simple - if this empty space is not there, material has no place to move and stays in its own position. Many workholding companies offer a solution in forms of special *split soft jaws*.

Split jaws are special type of soft jaws that function on the same principle as a ring or a plug - they form almost a complete circle with only minor slots every 120°, for clearance. Split jaws are attached directly onto the chuck, in the same way as any soft jaws would be. This solution make the machining environment much more accurate and safer, even if the cost of such a setup is slightly higher.

G70/G71/G72 Cycle Methods

There are three *'bread-and-butter'* multiple repetitive cycles for turning and boring on Fanuc CNC lathe controls. A brief definition describes their general functions:

G70 - Finishing contour cycle - used after G71 or G72

G71 - Roughing contour cycle (external or internal) - main cutting direction is *horizontal*

G72 - Roughing contour cycle (external or internal) - main cutting direction is *vertical*

The first thing to remember when using the G70/G71/G72 method of material removal, is that there are *two* versions of these cycles - one that uses a single block to define the cutting data (Fanuc 10/11/15), the other method requires two blocks of cutting data (Fanuc 0/16/18/20/21+) and offer some additional programming options. These two types are *not* interchangeable - each can be used only on a particular control that supports it.

Programming Formats - G71

Older Fanuc control systems used a single block of program to define the cutting parameters for roughing (G71/G72). Later controls have some added features and require two blocks of program to perform the same cutting. The main difference is actually very small, almost negligible, but the program must reflect the cycle format that matches the control system of the CNC lathe used.

One-block format

One-block format is used for Fanuc models 10/11/15:

```
G71 P.. Q.. (I..) (K..) U.. W.. D.. F.. (S..)
```

☞ **where ...**

P	=	The first block number of the finishing profile
Q	=	The last block number of the finishing profile
I	=	Distance and direction of rough semifinishing in the X axis - per side
K	=	Distance and direction of rough semifinishing in the Z axis
U	=	Stock amount for finishing on the X axis diameter
W	=	Stock left for finishing on the Z axis
D	=	The depth of roughing cut (per side - no decimal point)
F	=	Cutting feedrate (in/rev or mm/rev) overrides feedrates between P and Q blocks
S	=	Spindle speed (ft/min or m/min) overrides spindle speeds between P and Q blocks

The *I* and *K* entries are commonly omitted, as is the spindle speed *S*. The actual finishing contour (profile) is defined between the P and Q blocks, where P-block represents the first point of the contour, and Q-block represents the last point of the contour.

Normally, the cycle executes Type I cutting, where the cut progresses in a uniform direction - that means a change in the cutting direction is not allowed. The other type, Type II allows for a change in one direction only.

Watch the amount of stock left on faces and/or shoulders. Because the tool cuts upwards, it is important to keep the amount to a bare minimum. The suggested amount of the stock W should not exceed 0.125 mm (0.005 inches).

Two-block format

Two-block format is used for Fanuc models 0/16/18/20/21+:

```
G71 U.. R..
G71 P.. Q.. U.. W.. F.. (S..)
```

☞ where ...

First block:

U = The depth of roughing cut - note the depth is U for G71
R = Amount of retract from each cut

Second block:

P = The first block number of the finishing profile
Q = The last block number of the finishing profile
U = Stock amount for finishing on the X axis diameter
W = Stock left for finishing on the Z axis
F = Cutting feedrate (in/rev or mm/rev) overrides feedrates between P and Q blocks
S = Spindle speed (ft/min or m/min) overrides spindle speeds between P and Q blocks

Only one addition distinguishes the features of the two-block method from the one-block method, and that is the inclusion of the R-address in the first block. While the one-block method uses an amount stored in system parameters, the two-block method allows the amount to be part of the cycle definition. Many CNC programmers do not consider this feature too important.

Comments relating to the Type I and Type II cutting described for the one-block method apply equally to the two-block method and so does the programming of the W-stock amount.

The actual finishing contour (profile) is defined between the P and Q blocks, where P-block represents the first point of the contour, and Q-block represents the last point of the contour.

Programming Formats - G72

In the G71 multiple repetitive cycle, the main cutting direction is horizontal. That cycle is generally used for parts that have a length significantly greater than their diameter. G72 multiple repetitive cycle is used for exactly the opposite type of parts - parts where the length is rather small but the diameter is comparatively large. In G72, the main cutting direction is vertical. At the machine, it is easy to see the effect of each cycle. While G71 cycle controls the machining direction horizontally (typically from *right to left*), the G72 cycle controls the machining direction vertically (typically *top to bottom*). In both cases, the P and Q are defined the same way, but the actual first and last positions are different. Examples in this section will illustrate those differences.

All above statements relating to G72 are correct in principle, but one has to take into consideration how the cutting tool design influences the cutting itself.

Note the machining forces are not favorable for cutting diameters in G72. Even with a very small amount of stock on diameter (U in the cycle), maintaining good quality surface finish may be a problem.

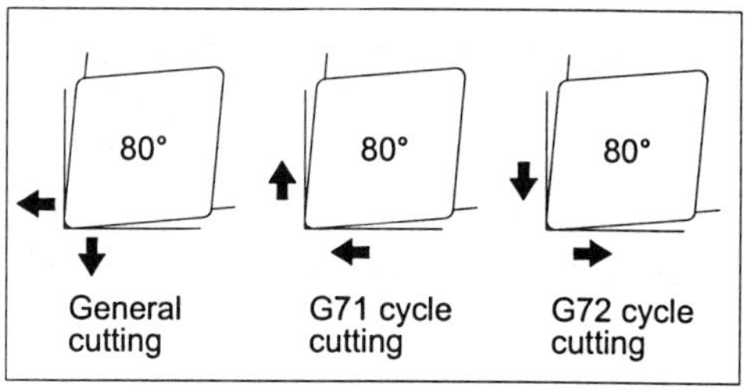

Programming formats for G72 are very similar to those of G71 - there are also two methods, depending on the control system available.

One-block format

One-block format is used for Fanuc models 10/11/15:

```
G72 P.. Q.. (I..) (K..) U.. W.. D.. F.. (S..)
```

☞ where ...

P	=	The first block number of the finishing profile
Q	=	The last block number of the finishing profile
I	=	Distance and direction of rough semifinishing in the X axis - per side
K	=	Distance and direction of rough semifinishing in the Z axis
U	=	Stock amount for finishing on the X axis diameter
W	=	Stock left for finishing on the Z axis
D	=	The depth of roughing cut (per side - no decimal point)
F	=	Cutting feedrate (in/rev or mm/rev) overrides feedrates between P and Q blocks
S	=	Spindle speed (ft/min or m/min) overrides spindle speeds between P and Q blocks

The *I* and *K* entries are seldom programmed, and so is the spindle speed *S*. The actual finishing contour (profile) is defined between the P and Q blocks, where P-block represents the first point of the contour, and Q-block represents the last point of the contour. because of the G72, the direction is different.

Two-block format

Two-block format is used for Fanuc models 0/16/18/20/21+:

```
G72 W.. R..
G72 P.. Q.. U.. W.. F.. (S..)
```

☞ where ...

First block:

W	=	The depth of roughing cut - note the depth is W for G72
R	=	Amount of retract from each cut

Second block:

P	=	The first block number of the finishing profile
Q	=	The last block number of the finishing profile
U	=	Stock amount for finishing on the X axis diameter
W	=	Stock left for finishing on the Z axis
F	=	Cutting feedrate (in/rev or mm/rev) overrides feedrates between P and Q blocks
S	=	Spindle speed (ft/min or m/min) overrides spindle speeds between P and Q blocks

There is only one major difference between a two-block G71 cycle and a two-block G72 cycle. In the first block, the depth is programmed using the U-address for G71, but the W-address for G72. Do not confuse U/W in the first block with the U and W in the second block.

G70 - Finishing Cycle

For either G71 or G72 roughing cycles there is only one finishing cycle - G70. Regardless of the control system, the format is the same:

```
G70 P.. Q.. (S..)
```

☞ where ...

P	=	The first block number of the finishing profile
Q	=	The last block number of the finishing profile
S	=	Spindle speed (ft/min or m/min) overrides spindle speeds between P and Q blocks

Finishing feedrate is usually programmed between the P and Q blocks, for better surface finish control of individual contour entities. P and Q blocks are typically the same as in the roughing cycle, but could be different, for example, if only a part of the contour needs finishing.

Make sure to use unique block numbers for G70-G71-G72 multiple repetitive cycles

G71 and G72 Compared

Two parts will be compared - one will use G71 cycle, the other will use G72 cycle. For a better comparison, both contours have the same basic shape, but in different orientation. Both drawings show the actual contour between P and Q points, including all necessary dimensions. The best way to study the program, is to match the program blocks to the illustration.

In the first example, G71 cycle is illustrated. The initial start point *SP*, the contour start point *P* and its end point *Q* are all identified. The *SP* point controls the physical diameter of the *first* cut, and may be adjusted as needed, by changing its X-coordinate in the CNC program. In this example, it could be a bit closer.

Clearances at the beginning and at the end of contour are also given, mainly for learning convenience, but also as good examples of a clearance selection.

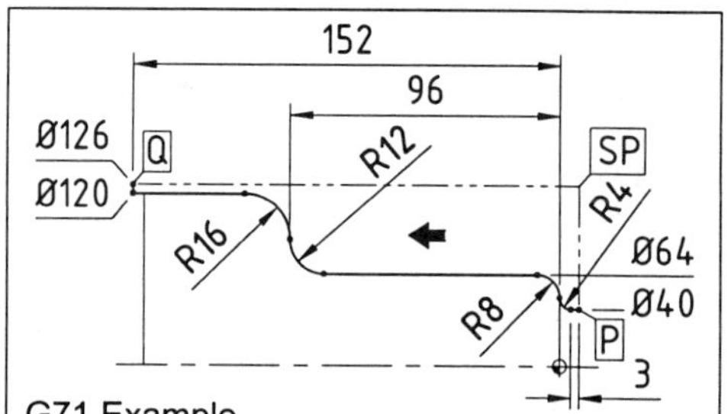

G71 Example

```
N1 G21 T0100                      (METRIC UNITS + TOOL 1 SELECTED)
N2 G96 S150 M03                   (150 M/MIN SURFACE SPEED)
N3 G00 G41 X126.0 Z7.0 T0101 M08  (START POINT SP + OFFSET + COOLANT)
N4 G71 U3.0 R0.75                 (3 MM DEPTH OF CUT - 0.75 RETRACT)
N5 G71 P6 Q14 U2.0 W0.125 F0.3    (CONTOUR IS BETWEEN BLOCK N6 AND N14)
N6 G00 X40.0                      (P - FIRST POINT OF THE CONTOUR)
N7 G01 Z3.0 F0.25                 (START OF LEADING ARC R4)
N8 G03 X48.0 Z0 R4.0 F0.2         (R4 LEADING ARC)
N9 G02 X64.0 Z-8.0 R8.0           (R8 CONTOUR ARC)
N10 G01 Z-84.0                    (DIAMETER 64 MM)
N11 G03 X88.0 Z-96.0 R12.0        (R12 BOTTOM CORNER RADIUS)
N12 G02 X120.0 Z-112.0 R16.0      (R16 TOP CORNER RADIUS)
N13 G01 Z-152.0                   (DIAMETER 120 MM PAST PART LENGTH)
N14 G00 X126.0                    (Q - LAST CONTOUR POINT = CLEARANCE)
N15 G70 P6 Q14 S200               (FINISHING CYCLE WITH THE SAME TOOL)
N16 G40 G00 X250.0 Z100.0 T0100   (TOOL CHANGE POSITION + OFFSET CANCEL)
N17 M01                           (OPTIONAL STOP)
...
```

In the example, the two block cycle is shown. The cutter radius offset will be ignored for the roughing cycle, but it will be applied to the G70 finishing cycle. Similar approach will be used for the G72 roughing cycle, using another example (see illustration).

In the second example, G72 cycle is illustrated.

The initial start point *SP*, the contour start point *P* and its end point *Q* are all identified - compare the positions of points *P* and *Q* to the G71 example - note that the P is above the largest diameter, because that is the place where the tool will start cutting from. Also note the main direction of the cut is vertical, from a large diameter to a small diameter. For internal use, the direction would be opposite, from a small diameter to a large diameter.

As before, clearances at the beginning and at the end of contour are also given, mainly for learning convenience, but also as good examples of a clearance selection.

The resulting program is not that much different from the previous example, at least in its structure.

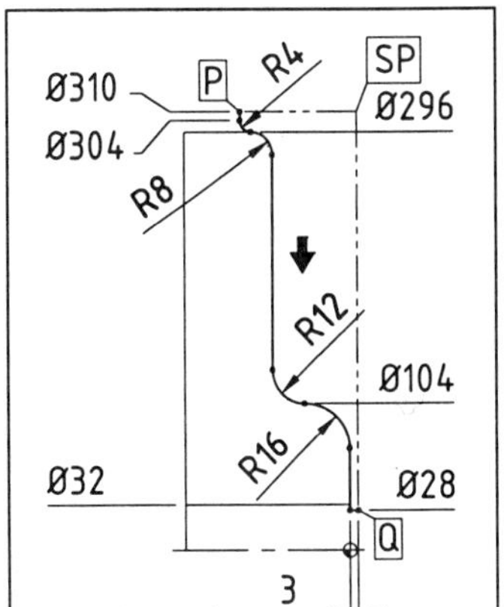

G72 Example

```
N1 G21 T0100                         (METRIC UNITS + TOOL 1 SELECTED)
N2 G96 S150 M03                      (150 M/MIN SURFACE SPEED)
N3 G00 G41 X310.0 Z3.0 T0101 M08     (START POINT SP + OFFSET + COOLANT)
N4 G72 W3.0 R0.75                    (3 MM WIDTH OF CUT - 0.75 RETRACT)
N5 G72 P6 Q14 U0.2 W2.0 F0.3         (CONTOUR IS BETWEEN BLOCK N6 AND N14)
N6 G00 Z-40.0                        (P - FIRST POINT OF THE CONTOUR)
N7 G01 X304.0 F0.25                  (START OF LEADING ARC R4)
N8 G03 X296.0 Z-36.0 R4.0 F0.2       (R4 LEADING ARC)
N9 G02 X280.0 Z-28.0 R8.0            (R8 CONTOUR ARC)
N10 G01 X128.0                       (FACE TO START OF R12 ARC)
N11 G03 X104.0 Z-16.0 R12.0          (R12 UPPER CORNER RADIUS)
N12 G02 X72.0 Z0 R16.0               (R16 LOWER CORNER RADIUS)
N13 G01 X28.0                        (DIAMETER 28 MM PAST BORE)
N14 G00 Z3.0                         (Q - LAST CONTOUR POINT = CLEARANCE)
N15 G70 P6 Q14 S200                  (FINISHING CYCLE WITH THE SAME TOOL)
N16 G40 G00 X250.0 Z100.0 T0100      (TOOL CHANGE POSITION + OFFSET CANCEL)
N17 M01                              (OPTIONAL STOP)
...
```

A few final comments - keep in mind that either cycle can be used for external or internal roughing. The control system will determine the cutting direction, based on the following criteria:

- **G71 - EXTERNAL** **X-coordinate (diameter) of the P-point is *larger* than X-coordinate of the Q-point**
- **G71 - INTERNAL** **X-coordinate (diameter) of the P-point is *smaller* than X-coordinate of the Q-point**
- **G72 - EXTERNAL** **X-coordinate (diameter) of the P-point is *larger* than X-coordinate of the Q-point**
- **G72 - INTERNAL** **X-coordinate (diameter) of the P-point is *smaller* than X-coordinate of the Q-point**

Multiple repetitive cycles are very powerful method of removing large areas of stock. The two roughing cycles described in this chapter - G71 and G72 - are the most commonly used cycles, with G71 being used much more than G72. There is a third cycle that also removes stock - G73. This cycle may have benefits for certain types of castings or forgings, but it does have several disadvantages as well. It is called the *pattern repeating cycle*, and - as the name suggests - the programmed contour between P and Q points is repeated several times, closer and closer to the final contour, based on programmed data. The main disadvantage of this cycle is that it is not always very efficient.

Programming Undercuts

Undercut - or a *recess* - is a special area of the finished part that serves as a clearance area. Undercut is often machined between a thread end and a shoulder, for example, to provide a suitable fit during parts assembly. The most common undercut at the end of a thread is a 45° chamfer, usually the same size as at the thread start. Its depth is typically a little greater than the thread depth. Roughing multiple repetitive cycles allow a direction change, if *Type II* is programmed. Other undercuts may have virtually any angle. In CNC lathe programming, the single most important consideration is the back angle clearance of the cutting tool.

The illustration shows general calculations for the three most common turning or boring inserts. The lead angle is normally 5° for the 80° insert, and 3° for the 55° and 35° inserts. Back angles can be easily calculated - they are show here as 5°, 32°, and 52° for 80°, 55°, and 35° inserts respectively.

Large (deep) undercuts may have to be programmed with one or more roughing passes, depending on the material and the tool nose radius.

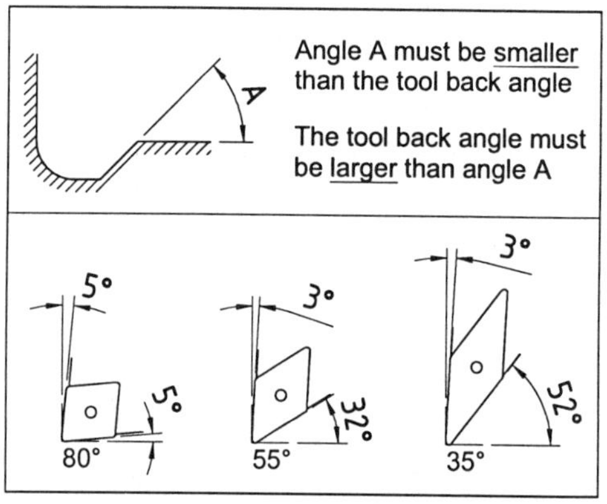

Hard Turning

As the word 'hard' in the title indicates, *hard turning* is a method of turning and boring parts that are made of material with a high degree of hardness, typically over 45 Rc (Rockwell hardness). Although the range of parts that can be hard-turned is large, typical applications include roller bearings, special gears, forging dies, axle shafts, injection valves and nozzles, and other parts that have been hardened *before* the final machining. The main reason many CNC machine shops adapt hard turning technology is to eliminate grinding operations. If that is not possible or practical, the least that is expected from hard turning is to *minimize* the grinding operations. In either case, it is the question of economics and productivity (as well as competition).

Hard turning does not require any special programming format or structure - in fact, the program looks just like for any other turning (or boring). What is different are the tools and the method of applying them to the cut. The most affected are the depth or width of cut, and the selection of spindle speeds and cutting feedrates.

14 PROGRAMMING TAPERS

Although tapers can be found in milling applications, they are mainly used for parts machined on CNC lathes. This chapter covers the mathematics behind tapers and presents a number of programming examples. Strictly speaking, every chamfer you program on a part is a taper.

What is a Taper?

First a common mathematical definition:

Taper is defined as a uniform change in the part diameter, measured along an axis

A taper is a conical shape between two diameters, which means a taper is the difference between two diameters at each end of a given part, measured along the centerline of the part. Taper can also be defined as the dimensional difference of two diameters over a given length.

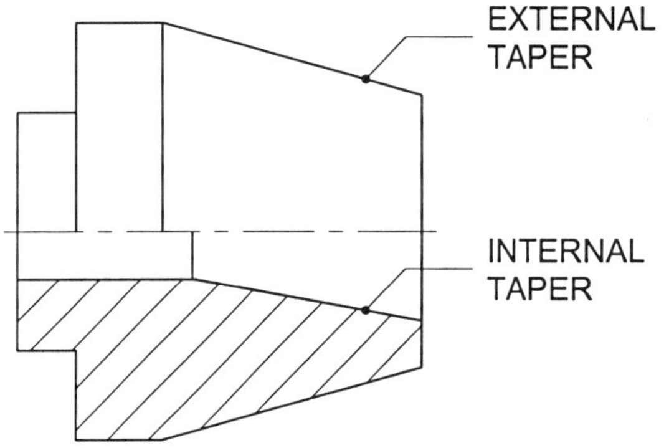

The illustration shows an external and an internal taper on a typical lathe part

Taper Definitions

Common tapers are expressed in ratios, typical in the metric system, but also used in the Imperial system, or as a taper per foot or a taper per inch. Tapers are defined in three ways:

1. **Taper-per-foot (TPF)** ... **Imperial units only (3 TPF)**
2. **Ratio of the taper length to diameter** ... **typically metric units (1 : 10)**
3. **Taper length and included angle in degrees** ... **Imperial or metric tapers (12×15°)**

In manufacturing, there are many tapers that are unique to a machine or cutting tools. For example, cutting tool holders identified as BT, CAT (V-flanges), HSK, etc., have a special taper to match the type of machine spindle. Other very common types of tapers are Morse™ and Brown & Sharpe™ self-holding tapers, typically used for taper shank drills, reamers, etc. Special tapers are also used on parts, for example, a pipe thread has a taper defined by established industrial standards.

To maintain consistency, all calculations in this chapter will use the abbreviations and symbols shown above, as required:

Symbol	Description	Units
D	Large diameter of the taper	millimeters or inches
d	Small diameter of the taper	millimeters or inches
L	Length of taper	millimeters or inches
A	Angle of taper per side	decimal angle
TPF	Taper per foot	inches
1 : X	Ratio	millimeters

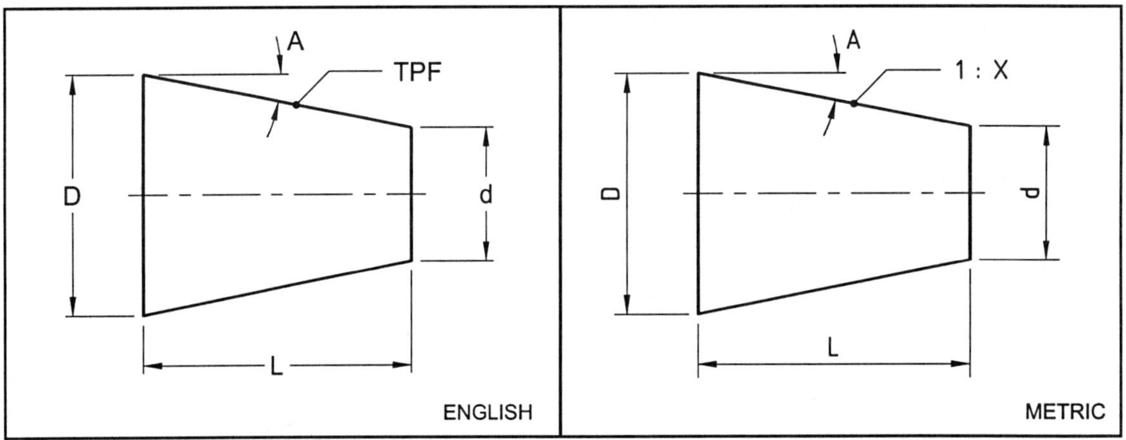

Note - the same schematic drawing will be used for all examples - they will be not-to-scale (NTS) !

Taper per Foot

Virtually all tapers of American or British origin use the designation *taper-per-foot*, abbreviated as *TPF*. For example, the original intent of Morse tapers in the 1860s was to have a uniform 0.625 taper per foot. That did not exactly happened, due to difficulties in grinding at the time, the but all Morse tapers are close to this original intent even today.

In order to make programming calculations correctly, it is important to understand the definition:

Taper per foot is the difference in diameter in inches over one foot of length

In a drawing, taper per foot will usually be identified by a leader with descriptive text, for example, 4 TPF or 4.0 TPF. As a programmer, you have to be able not only to interpret this description, but also define the required contour point on the part. In some drawings, the taper may be defined as *taper-per-inch*. This is not a traditional method, but easy to understand in both directions:

Taper-per-inch = Taper-per-foot / 12 *and* **Taper-per-foot = Taper-per-inch × 12**

The abbreviation *TPI* for taper-per-inch should not be used, as it can be confused with its more common meaning *threads-per-inch*.

EXAMPLE 1

If it becomes necessary to calculate the taper-per-foot *TPF*, both taper length and the two diameters must be known:

- **Formula:**

$$TPF = \frac{D - d}{L} \times 12$$

$$TPF = \frac{1.0 - 0.875}{6.0} \times 12 = 0.25$$

The answer is:
... the taper-per-foot *TPF* is 0.25

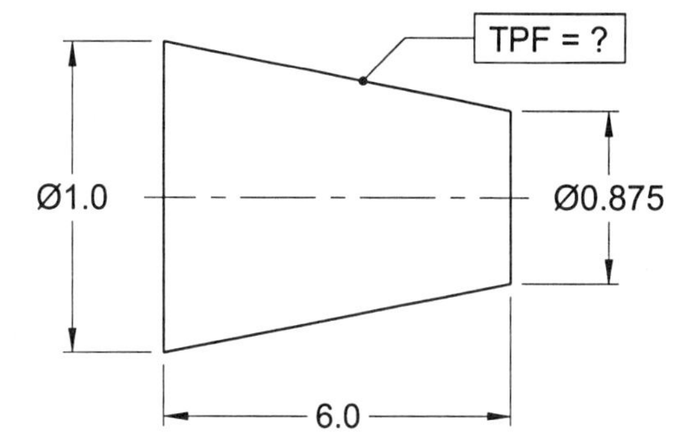

Example 1

This is not a calculation normally required in CNC programming, but it can be useful when confirming a result of another calculation, for example, one of the diameters or the taper length.

EXAMPLE 2

Another example that is useful but not too frequent in programming is the taper length calculation *L*, when the taper-per-foot and both diameters are known:

- **Formula:**

$$L = (D - d) \times \frac{12}{TPF}$$

$$L = (1.0 - 0.875) \times \frac{12}{0.25} = 6.0$$

The answer is:
... the taper length *L* is 6.0 inches

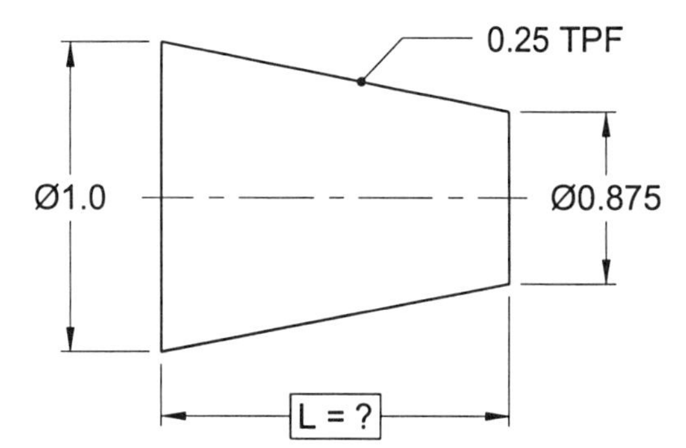

Example 2

Although the taper length calculation is not the most common application of solving tapers, it may come handy is several cases, for example, to check the accuracy of the drawing.

EXAMPLE 3

This is a very useful calculation for CNC programming - one of the two diameters is unknown - in this exercise, it will be the smaller diameter *d*:

- **Formula:**

$$d = D - \frac{L \times TPF}{12}$$

$$d = 1.0 - \frac{6 \times 0.25}{12} = 0.875$$

The answer is:
... the smaller diameter *d* is 0.875 inches

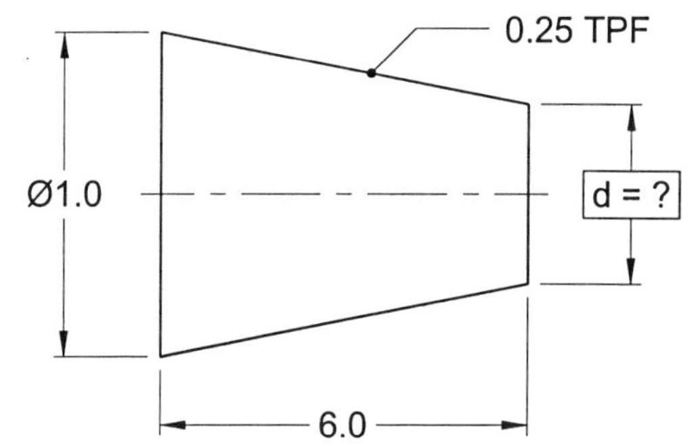

Example 3

EXAMPLE 4

Similar to the previous example, one of the two diameters is unknown - in this exercise, it will be the larger diameter *D*:

- **Formula:**

$$D = \frac{L \times TPF}{12} + d$$

$$D = \frac{6.0 \times 0.25}{12} + 0.875 = 1.0$$

The answer is:
... the larger diameter *D* is 1.0 inch

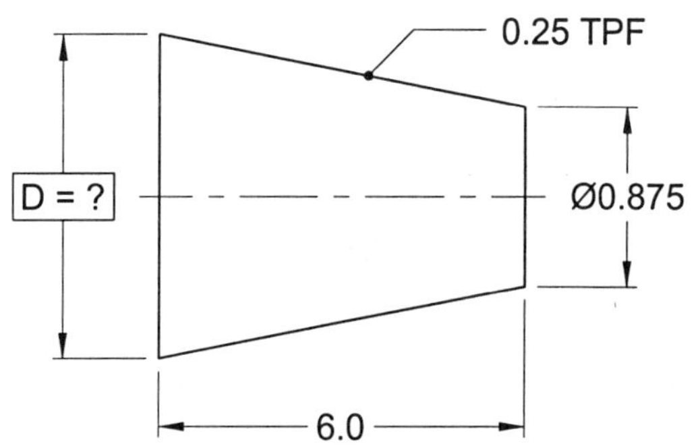

Example 4

The four formulas are indispensable to any CNC lathe programmer who works with Imperial engineering drawings using the *taper-per-foot* definitions. Keep in mind that these formulas provide only the basic calculations. Additional calculations will include clearances for the tool start position and chamfers or radii required at sharp corners.

In the next section, similar four formulas are provided to solve the unknown taper related dimensions in typical metric drawings.

Metric tapers are typically defined by the ratio of diameter to length.

Taper Ratios

More prevalent designation of metric tapers is a *taper ratio*:

Metric taper is defined as the ratio of difference between the large diameter and the small diameter over the given length of the cone

In simple terms, the metric designation of a taper can be presented as a mathematical ratio:

$$\frac{1}{X} = \frac{D - d}{L}$$

The ratio `1 : X` means that over the length of *X* mm, the diameter of the cone will change by one millimeter (1 mm). The change in diameter can either increase or decrease.

For example, a taper identified in the drawing as 1 : 10 will increase or decrease 1 mm on diameter every 10 mm of length. Similarly, a 1 : 5 taper will increase or decrease 1 mm on diameter every 5 mm of length.

EXAMPLE 5

If the taper ratio is unknown, the value *X* can be calculated from the following formula:

- **Formula:**

$$X = \frac{L}{D - d}$$

$$X = \frac{20}{14 - 10} = 5$$

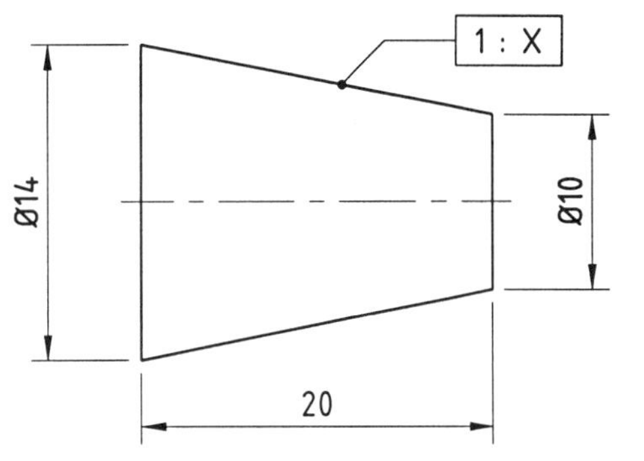

Example 5

The answer is:
... the taper ratio *X* is 5 (1 : 5)

Similar to the calculation of taper-per-foot *TPF*, this is not a calculation normally required in CNC programming, but it can be useful when confirming a result of another calculation, for example, one of the diameters or the taper length.

EXAMPLE 6

If the taper length is unknown, the value *L* can be calculated from the following formula:

- **Formula:**

$$L = (D - d) \times X$$

$$L = (14 - 10) \times 5 = 20$$

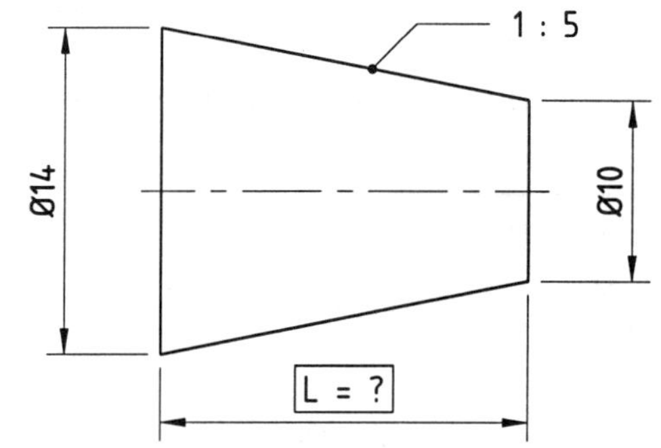

Example 6

The answer is:
... the taper length *L* is 20 mm

In practice, the taper length *L* is not a common calculation - the reason is that only enough information will be given on the drawing, so the programmer can easily figure out the rest.

EXAMPLE 7

If the small taper diameter is unknown, the value *d* can be calculated from the following formula:

- **Formula:**

$$d = D - \frac{L}{X}$$

$$d = 14 - \frac{20}{5} = 10$$

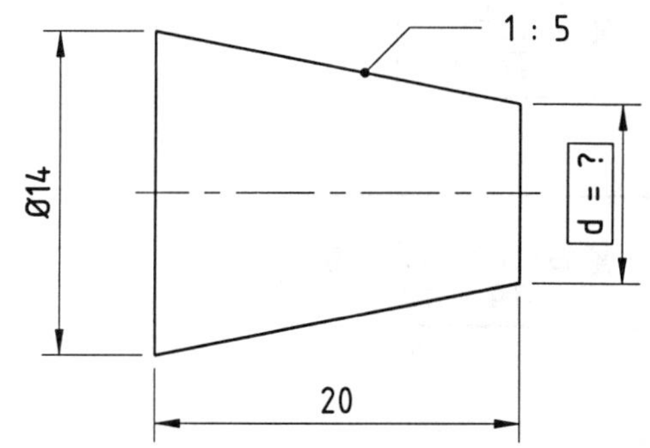

Example 7

The answer is:
... the smaller diameter *d* is 10 mm

This example can be easily adapted to a situation when the cutting tool physically starts in front of the small diameter *d*, for example, to provide clearance at the beginning of cut.

EXAMPLE 8

If the lager taper diameter is unknown, the value *D* can be calculated from the following formula:

- **Formula:**

$$D = \frac{L}{X} + d$$

$$D = \frac{20}{5} + 10 = 14$$

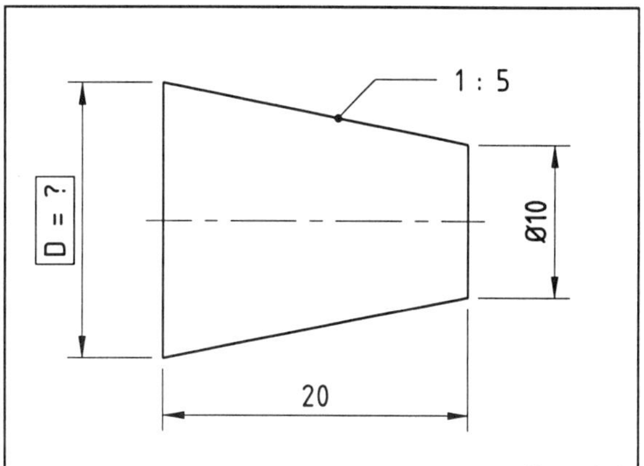

Example 8

The answer is:
... the larger diameter *D* is 14 mm

Taper Defined as Percentage

In some rare situations, a metric taper can also be defined in a drawing as a percentage of diameter change over given length, for example, a 20% taper. This is rather an uncommon method, used for special applications only. If you encounter a drawing with a taper defined as a percentage, it is easy to calculate the taper ratio. Once the ratio is known, the formulas listed elsewhere in this chapter can be used to calculate other dimensions of the part.

In order to convert percentage *P* to ratio *X* (1 : X), used the following formula:

$$X = \frac{100}{P}$$

The reverse method will work equally well, if you ever need to find out the percentage *P*:

$$P = 1 : X \times 100 = \frac{100}{X}$$

EXAMPLE - A taper defined as a ratio of 1 : 4 is equivalent to a 25% taper

EXAMPLE - A taper defined as a 2% taper is equivalent to the ratio of 1 : 50

Taper Angle Defined in D-M-S

D-M-S or *DMS* are common abbreviations that mean *Degrees-Minutes-Seconds*. Such a designation is considered obsolete today, and has been replaced by the decimal degrees *DD*. If you work with some older drawing, it may be necessary to convert the DMS to DD. Both abbreviations can also be found on scientific type calculators. If a calculator is not available, use the following formula:

$$DD = D + \frac{M}{60} + \frac{S}{3600}$$

☞ ... where:

DD = Decimal Degrees
D = Degrees
M = Minutes
S = Seconds

EXAMPLE - Convert 38° 30′ 00″ to decimal degrees DD:

DD = 38 + 30/60 + 00/3600 = 38.5°

EXAMPLE - Convert 26° 47′ 18″ to decimal degrees DD: (see illustration)

DD = 26 + 47/60 + 18/3600 = 26.78833°

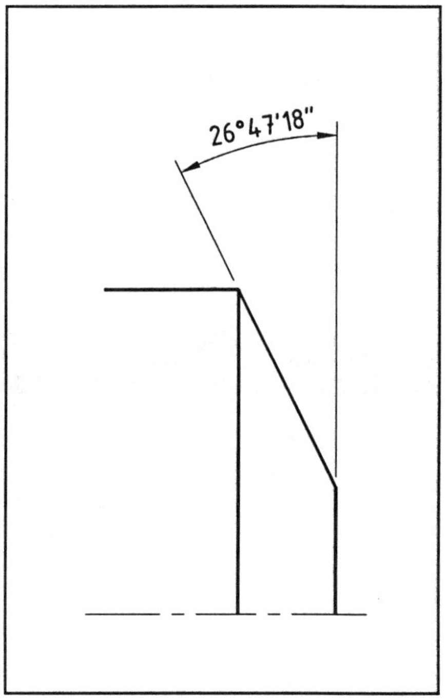

The next section further covers the subject of tapers when the *length and angle* are given.

Taper Length and Angle

Many non-standard tapers (cones) are defined in the drawing by their taper length and angle. The most common taper of this kind is a 45° chamfer, although in everyday life it is hardly referred to as a taper. The 'real' taper is often a visible cone, defined by its length an angle that has to be calculated mathematically, using trigonometric functions.

Chamfers

Chamfer is a term used for a beveled corner break. Virtually all chamfers have a 45° angle, which makes the necessary calculations much easier than with any other angle. Chamfers can be defined in the drawing as a horizontal dimension only, for example, 1.5x45 or as a separate dimension for the chamfer length and the angle. The following illustration shows both applications in a typical drawing.

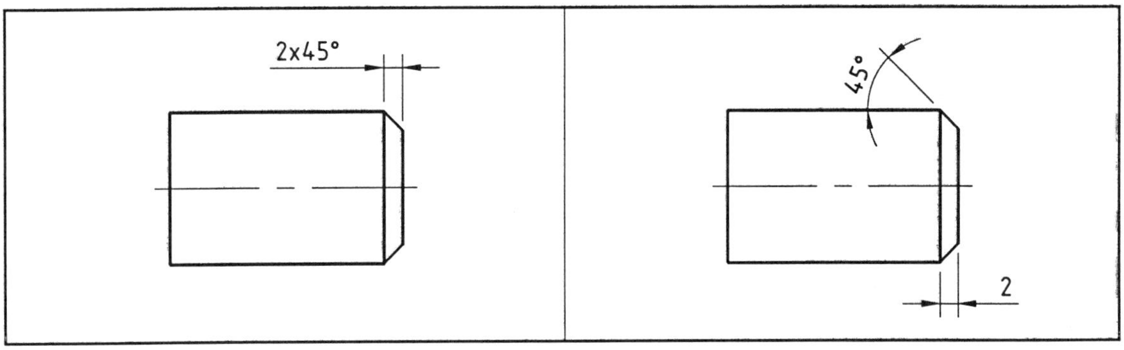

Both illustrations identify the same chamfer. In practice, the cutting tool will approach the part from a safe clearance distance, for example 3 mm off the face. The programmer has to *select* the Z-clearance but must *calculate* the X-position at this clearance.

45° Chamfer

In this practical example, three chamfers are used on the same part - for convenience, all will have the same size - 2×45°. The selection of three chamfers is intentional, as it illustrates all most common applications when chamfering on a CNC lathe. Although only external chamfers are shown, the approach applies equally to internal chamfers as well.

Why three chamfers? The three chamfers can be referred to as the *start* chamfer, the *middle* chamfer, and the *end* chamfer. There can be several chamfers in the 'middle', but only one each at the start and at the end. The start chamfer represents the tool approach towards the part, the end chamfer represents the tool exit from the part. Both require selection of the clearance position in one axis and calculation of the matching position in the other axis:

- **START chamfer - Z-axis clearance is *selected*, corresponding X-axis position must be *calculated***
- **MIDDLE chamfer - no clearances are required - XZ positions calculated as per drawing**
- **END chamfer - X-axis clearance is *selected*, corresponding Z-axis position must be *calculated***

All three chamfers are shown in the illustration:

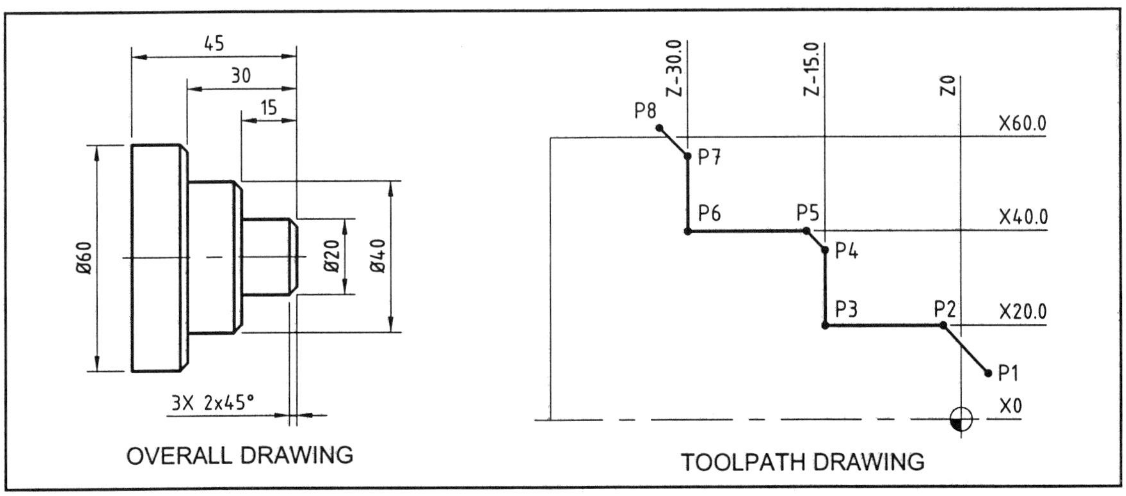

In terms of difficulty, the example drawing belongs to the category of super-easy ones. There are some calculations involved, so organizing the work well is essential. One proven method is the method of elimination. Eliminate all coordinate locations that are either known or can be calculated very easily. A *table form* is one successful method used to achieve this goal.

Table should include all toolpath points (coordinates), including those that are not part of the drawing, for example, temporary points needed for calculations, such as clearance points. The toolpath drawing in the illustration below shows eight points (coordinates) that have to be known to the CNC programmer. Those known points will be included in the table first:

Point	X-coordinate	Z-coordinate
P1	?	?
P2	X20.0	?
P3	X20.0	Z-15.0
P4	?	Z-15.0
P5	X40.0	?
P6	X40.0	Z-30.0
P7	?	Z-30.0
P8	?	?

The *middle* chamfers are the easiest to calculate, as there are no clearances involved - they are direct part of the toolpath - in the drawing, points *P4* and *P5* represent the middle chamfer. One coordinate of each point is already known, the other must be calculated and the point table updated:

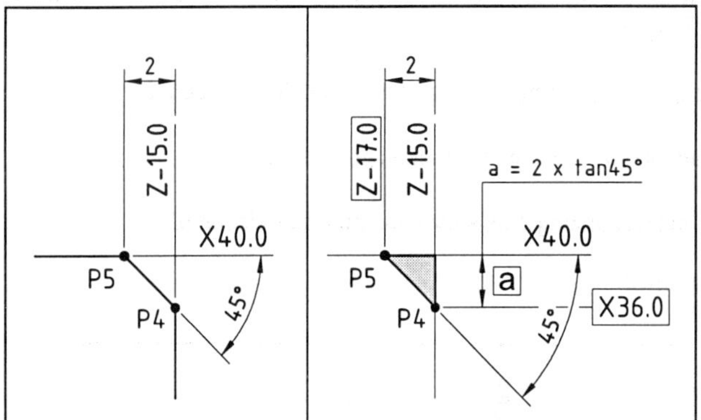

a = 2 x tan45° = 2.0

Provided dimensions | Calculated dimensions

Point	X-coordinate	Z-coordinate
P1	?	?
P2	X20.0	?
P3	X20.0	Z-15.0
P4	X36.0	Z-15.0
P5	X40.0	Z-17.0
P6	X40.0	Z-30.0
P7	?	Z-30.0
P8	?	?

Watch for a possible error! One programming error is quite common in cases like the one demonstrated. It is very easy to calculate (even without a calculator) the *a* dimension. After all, if the angle is 45° in one corner of the right triangle, it must also be 45° in the other corner. Logically, it follows, the length of one side of the triangle will be equal to the length of the other side. In our example, both sides of the triangle are 2 mm. So, where is likely error?

When programming the X-axis diameter! Keep in mind that the 2 mm length was calculated *per side of the part!* In practice, it means the calculated length has to be *doubled* in order to apply it to the diameter - that is why the X-coordinate of *P4* is **X36.0** and not **X34.0** - an easy error to make.

All calculations have been completed *between* the start and end points (*P1* and *P8*) - but not *at* the start and end points themselves. That will be the next step in order to find all coordinate points.

Start Chamfer with a Clearance

Calculation of *P1* - the start point of the contour - is the same as calculation of the previous points. The main difference is that the Z-coordinate (clear position) has to be *selected* by the programmer. Why the Z-clearance? Because the chamfer has to have 45° angle and end at *P2* (`X20.0 Z-2.0`), so the diameter of *P1* (X-coordinate) has to be *calculated* to match the conditions.

Here are the calculations with an updated table of *XZ* point coordinates:

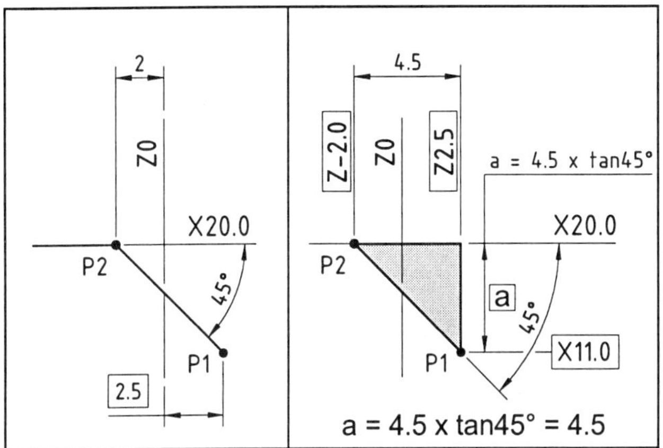

Provided dimensions | Calculated dimensions

Point	X-coordinate	Z-coordinate
P1	X11.0	Z2.5
P2	X20.0	Z-2.0
P3	X20.0	Z-15.0
P4	X36.0	Z-15.0
P5	X40.0	Z-17.0
P6	X40.0	Z-30.0
P7	?	Z-30.0
P8	?	?

The Z-clearance of 2.5 mm (highlighted in the table) could have easily been `Z2.0` or `Z3.0` or any other reasonable position. The key is that once you decide on the Z-clearance, the X-diameter *must be calculated* for that position.

End Chamfer with a Clearance

Following the start chamfer development carefully, it should not be difficult to develop the end chamfer. The difference will be the *selection* of the X-clearance and *calculation* of the Z-coordinate.

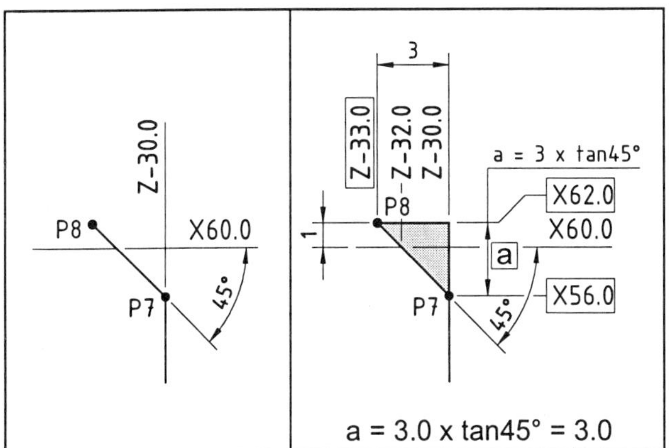

Provided dimensions | Calculated dimensions

Point	X-coordinate	Z-coordinate
P1	X11.0	Z2.5
P2	X20.0	Z-2.0
P3	X20.0	Z-15.0
P4	X36.0	Z-15.0
P5	X40.0	Z-17.0
P6	X40.0	Z-30.0
P7	X56.0	Z-30.0
P8	X62.0	Z-33.0

As the illustration shows, the physical X-clearance has been *selected* 1 mm above the 60 mm diameter, making it ∅62 (**X62.0**). From this location, the Z-position has to be *calculated*, which makes it 3 mm from the face at *P7*. The *finishing* contour has now been completed (T02 used for the example):

```
N31 (G21) T0200                     (METRIC MODE IN EFFECT - TOOL CHANGE)
N32 G96 S175 M03                    (CONSTANT SURFACE SPEED 175 M/MIN)
N33 G00 G42 X68.0 Z2.5 T0202 M08    (P1 POSITION - Z-AXIS)
N34 X11.0                           (P1 POSITION - X-AXIS)
N35 G01 X20.0 Z-2.0 F0.25           (P2 POSITION)
N36 Z-15.0                          (P3 POSITION)
N37 X36.0                           (P4 POSITION)
N38 X40.0 Z-17.0                    (P5 POSITION)
N39 Z-30.0                          (P6 POSITION)
N40 X56.0                           (P7 POSITION)
N41 X62.0 Z-33.0                    (P8 POSITION)
N42 G00 U3.0                        (RUN-OFF CLEARANCE)
N43 G40 X100.0 Z75.0 T0200          (TOOL CHANGE POSITION)
N44 M01                             (OPTIONAL STOP - COULD ALSO BE M30)
```

Face finishing may also be programmed for the same tool or it can be done by another tool earlier. Programming 45° chamfers is not demanding, but there are few calculations to watch for.

Other Chamfers

Not all chamfers used in a drawing are designed with a 45° angle. For example, the start of a shaft end may be designed as being 2.5 mm long at a 15° angle (see illustration). In this case, the general formulas used for calculations will remain the same, except the new data will be used instead.

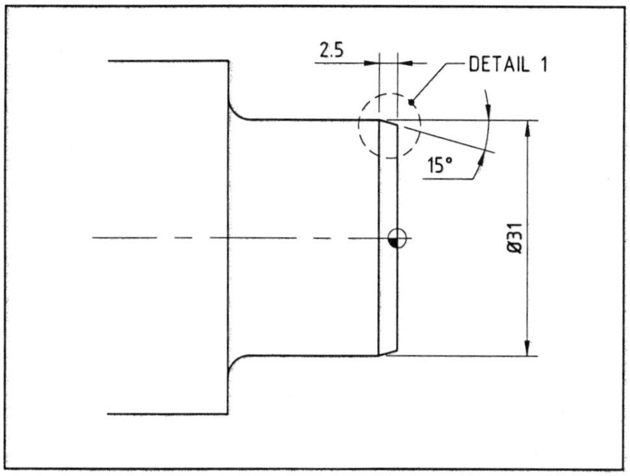

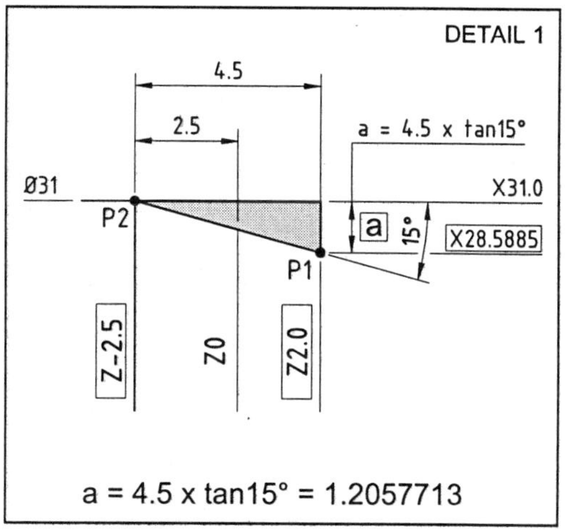

As the example shows, the calculation is not difficult. The Z-clearance has been *selected* as 2 mm and the X-diameter must be *calculated* (remember to double the *a* value). A common trigonometric function has been used to solve the problem.

There is another method of dimensioning tapers (including 45° chamfers) that can be found in engineering drawings. In this method, there is no angle given. Instead, the major and the minor chamfer diameters are given, along with the chamfer length. Clearance for these chamfers is usually very easy to solve because there is no need for trigonometric calculations. The illustration on the next page shows a typical drawing using this method. If the two diameters and the length between is known, so is the distance between the two diameters per side: **(31 - 28) / 2 = 1.5**. A mathematical method known as the *law of similar triangles* can be used very efficiently in virtually all such cases.

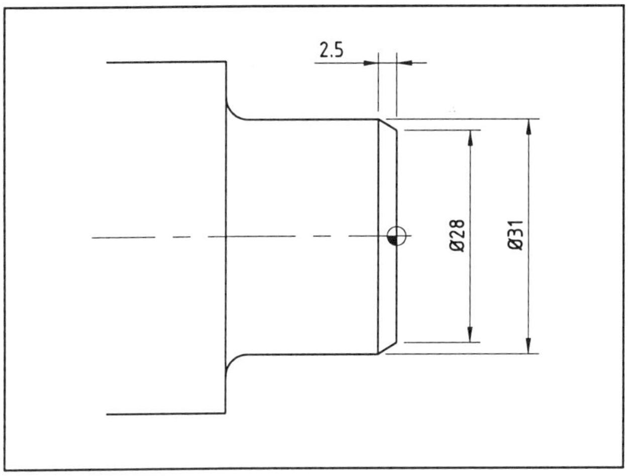

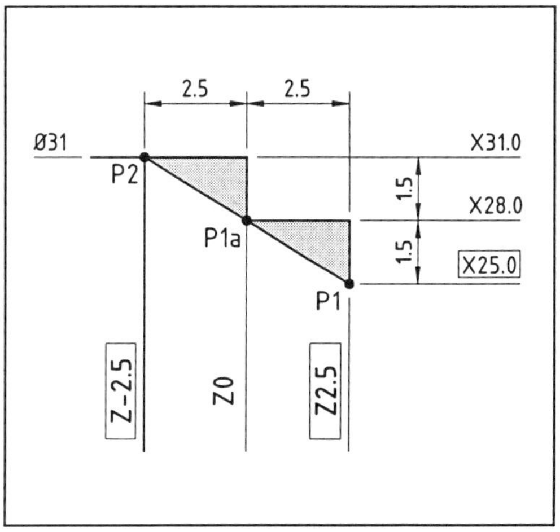

The illustration should require no explanation - the main key here is to select the Z-clearance carefully. This clearance can be the length of the chamfer (as is the case shown), it could be the double or half length of the chamfer or any other proportion. Based one common definition of similar triangles, the remaining side can be calculated:

Two triangles are similar, if their corresponding sides are proportional

For example, the ratio of *2.5 : 1.5* is the same as the ratio of *5: 3* or *1.25 : 0.75*. It is easy, but do not forget to double the calculated side to find the *X* diameter used in the CNC program.

In the next section, the focus will be on a different type of chamfers - or tapers - those that require *a blend* or a *lead* between the face and the taper. Such leads are usually designed to assist during the assembly of two parts.

Two common transitions - or *leads* - between the face and the taper are:

- **45° chamfer**
- **Blend radius (also known as a *fillet*)**

Tapers with Leads

Any tool approach towards the part that originates in the 'air' is an approach that includes *clearance* (provided for in the CNC program). The subject of clearance has been already discussed in its relation to chamfers. In this section, the issues relating to clearances will remain the same, but instead of working with a 90° chamfer between a face and a diameter, the angle and the taper length will vary between jobs.

In the two programming applications in this section, the taper that starts at the front face does not result in a sharp corner but is connected to the face either with a 45° chamfer or a blend radius (fillet). A chamfer or a fillet are common design features that are helpful during assembly, as they provides a useful lead.

Both of these applications will require a few steps, using trigonometric solutions. At first, the calculations may appear difficult, but once you get used to the procedure, the method is quite simple.

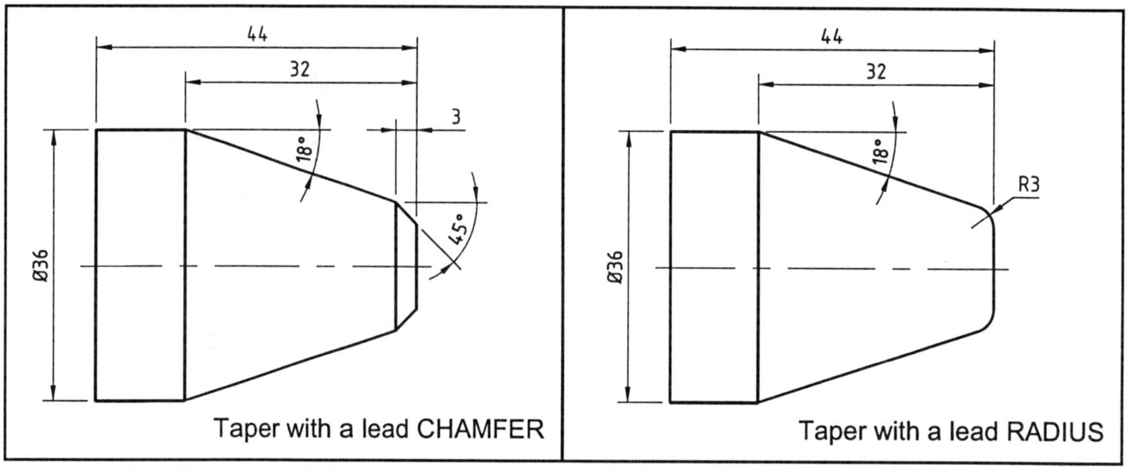

Taper with a lead CHAMFER

Taper with a lead RADIUS

Taper with a Lead Chamfer

A taper generally starts at a certain diameter and continues by its length and angle towards the front face, as defined in the drawing. The goal is to find *XZ* coordinates for the front chamfer (*P1* and *P2*). Both points have their Z-coordinate known from the drawing (`Z0` for *P1* and `Z-3.0` for *P2*). That leaves the calculation of two X-coordinates. A three-step calculation will be required:

Step 1 **- Calculate the *height* of taper a_1 without the chamfer - this will also calculate the front taper diameter X_a**

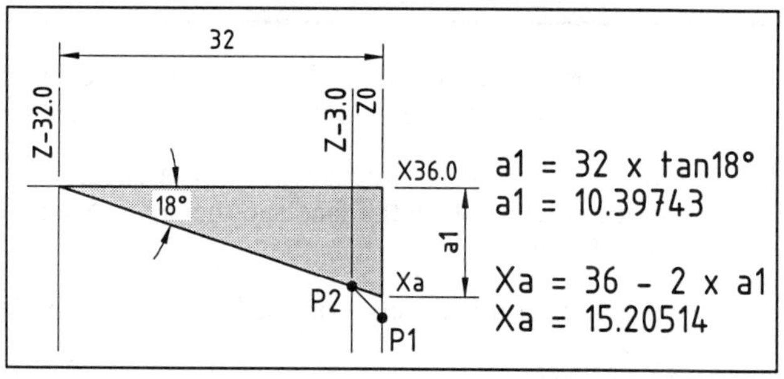

Step 2 **- Include the chamfer and calculate the opposite side a_2 of the small triangle - it will also calculate *X* coordinate of *P2* = *XP2***

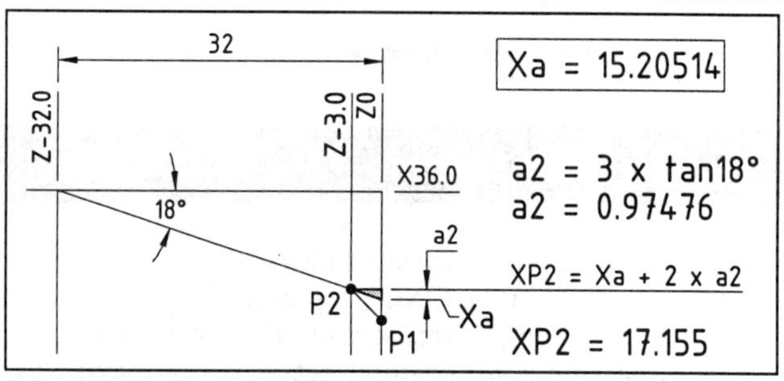

Step 3 **- From the calculated data available, the X-coordinate for *P1* can also be calculated easily, because the X-distance from P2 to P1 equals 3 mm (per side)**

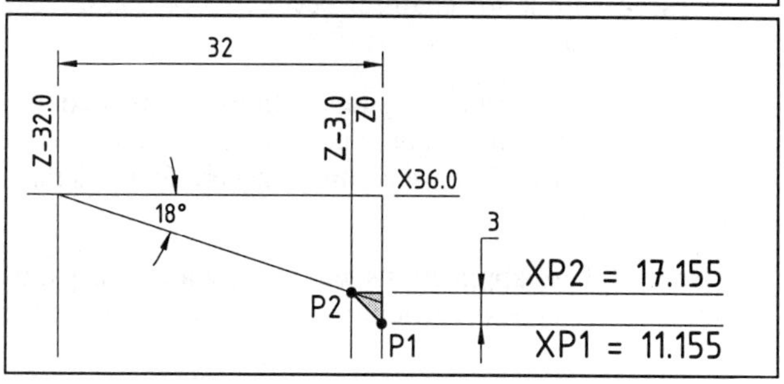

Once the physical chamfer coordinates are known, the actual clearance position calculation is simple - again, using the law of similar triangles.

The lead chamfer clearance is shown here as 2 mm (Z2.0). The start-up diameter can be calculated from the previous results

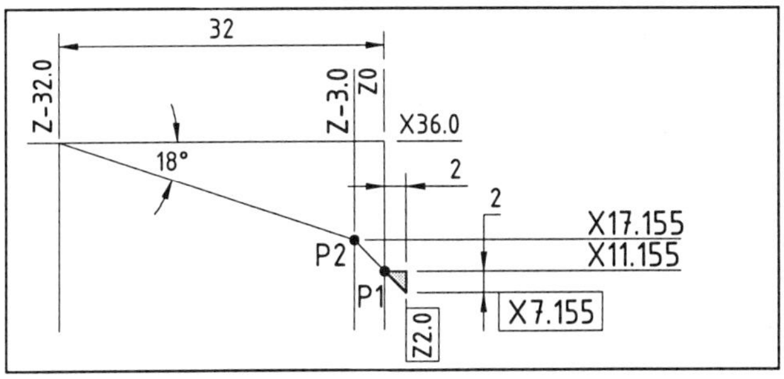

Taper with a Lead Fillet

Fillet is a common word used in machine shop (and CAD/CAM) to identify a blend radius. Blend radius is a radius that is tangent between two part features and always provides a smooth transition. There are no sharp corners or edges on a filleted corner. The fillet in the example is between the front face of the part and the 18° taper. Incidently, the lead radius is a somewhat easier to calculate than the lead chamfer. Note that the *P2* is *not* at `Z-3.0`.

The first step is the same as before - to calculate the *height* of taper *a*, without the radius - this will also calculate the front taper diameter X_a

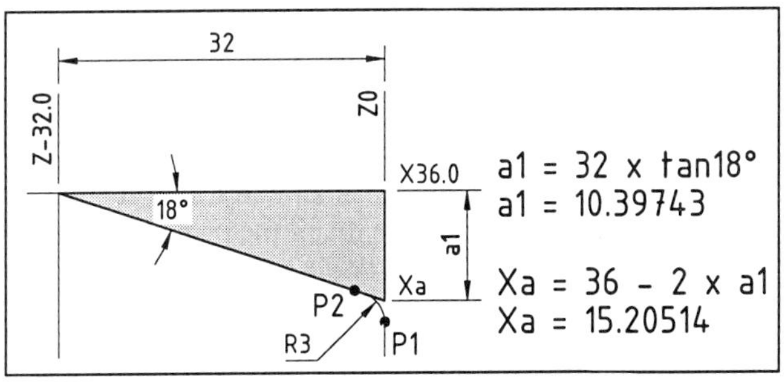

This type of calculation is so common in CNC programming, that for everyday work, a general formula can used:

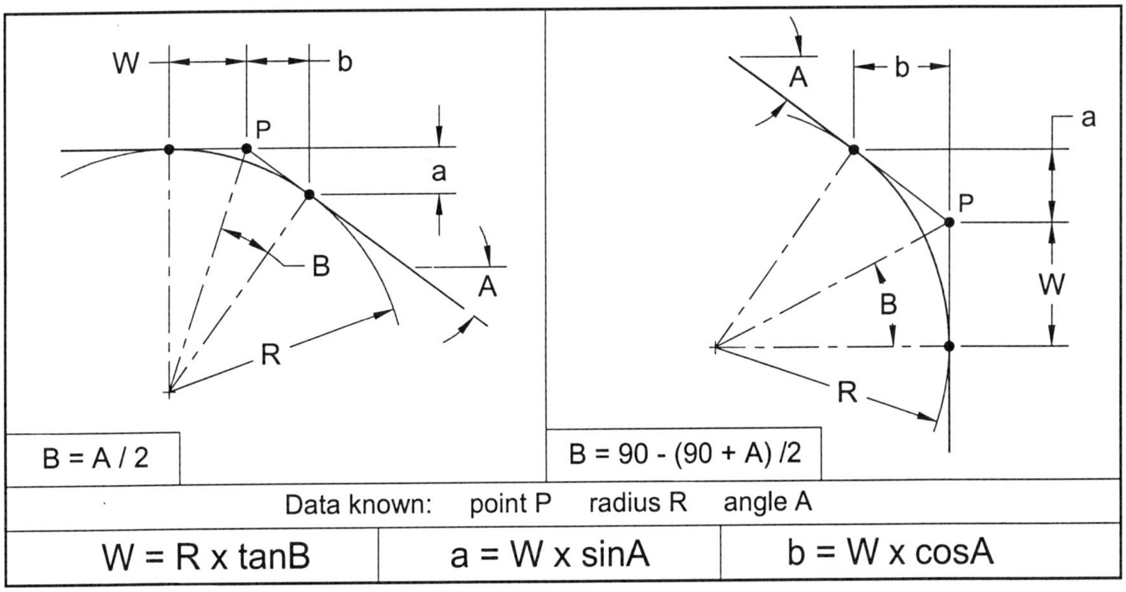

B = A / 2	B = 90 - (90 + A) /2	
Data known: point P radius R angle A		
W = R x tanB	a = W x sinA	b = W x cosA

As the lead radius starts on a face and continues into the taper, the formula at right will be used. At this time, the point *P* is known, from the previous calculation *(X_a)*. Its coordinates are **`X15.205 Z0`**. We need to calculate the dimension *W* to find the X-coordinate of point *P1*, the dimension *a* to find the X-coordinate of *P2*, and the dimension *b* to find the Z-coordinate of point *P2*. The formula can be applied in the detail:

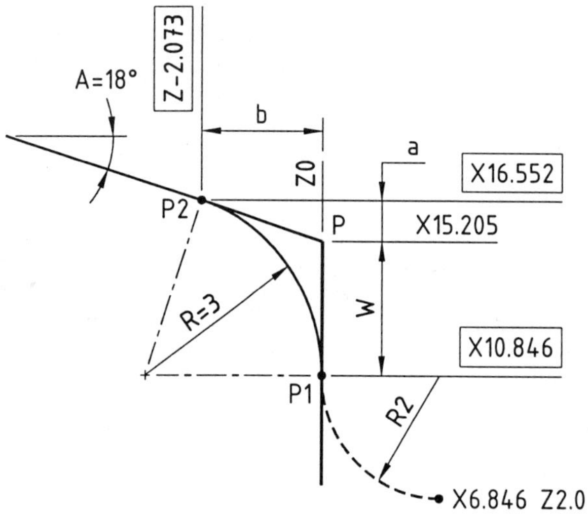

B = 90 - (90 + 18) / 2 = 36°

W = 3 x tan36°
W = 2.17963

a = 2.17963 x sin18°
a = 0.67354

b = 2.17963 x cos18°
b = 2.07295

XP1 = 15.205 - 2 x 2.17963 = X10.846
XP2 = 15.205 + 2 x 0.67354 = X16.552

ZP2 = b = Z-2.073

A 90° arc can be added as an extra lead for the tool approach - 2 mm arc shown next to *P1* with starting coordinates. Note that if the program requires cutter radius offset, there must be an additional linear motion, as *cutter radius offset cannot start on an arc.*

15 TECHNIQUES FOR GROOVING

As a machining operation, the process of machining grooves is typically performed on a CNC lathe. A general purpose program required to make only the simplest of grooves is very straightforward - it includes only two small operations of the grooving tool - the *plunge-in* as the first motion, then *rapid out* as the second toolpath motion. Grooves of this type are rather rare and seldom done on CNC lathes, as their cost would be too high as compared with alternate methods of production (such as manual machining). The majority of grooves that need to be programmed belong to the category of *precision* - or at least *semi-precision* - grooves. When planning a program for grooving operations, you have to consider the best selection out of many methods available. Once the groove machining selection is known, the rest of the programming process is rather simple.

Tooling for Grooves

Various tooling catalogues offer a multitude of grooving tools suitable for CNC lathe work. The two basic considerations regarding the tooling selection for grooving are:

- **Cutting width**
- **Cutting depth**

Of course, other features are also important, such as selecting the proper insert grade, insert coating, corner radius, chipbreaker, etc. Still, the width and the depth of grooving tool are the most important features for programming. For certain type of grooves, additional tooling considerations will also be very important, such as the radial clearance for face grooves.

Cutting Width

In an engineering drawing, the width of a square groove is typically given either as the actual width (wall-to-wall distance) or the absolute dimension to both walls. A combination of the two methods is also quite common. At the very least, groove width determines the grooving width insert. For precision groove machining, the insert selection is very important - the basic rule is that the insert width should always be smaller than the groove width. This will require more than one plunge, but the groove quality will improve. In order to minimize the number of plunge cuts, select the widest grooving tool suitable for the job.

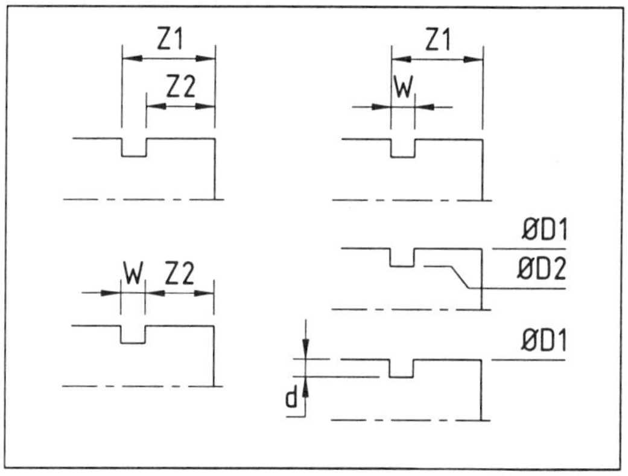

Cutting Depth

Any selection of grooving tool must also take into consideration the groove depth, as specified in the drawing. Catalogues specifications for grooving tools determine the maximum depth allowed for the selected tool. It is not unusual to *increase* the cutting depth by modifying the insert, usually by grinding off a small portion, providing the coating and the insert width are not compromised.

Groove Location

Groove location - or *position* - is normally measured as being relative to a particular face or a shoulder. The drawing designation has a very important impact on the selection of the grooving tool command point. As illustrated on the previous page, groove location can be defined in the drawing by *either* the Z1 position *or* the Z2 position *but not both.* For the objective of this subject, the dimensions are from the front face of the part, but could be defined from any other location. If both Z1 and Z2 are given, they also define the width of the groove.

Many grooves are classified as *precision* grooves rather than utility grooves, typically defined by adding tolerances to the groove dimensions. More details on grooving tolerances are described in the subject of *precision grooves*. Selection of the tool command point is important not only for the actual setup at the CNC machine, but also for the programming method.

Setting the Command Point

For programming purposes, the grooving tool should be treated with the same practical approach as any other tool in terms of the command point selection - *how it benefits the CNC operator*. As for all cutting tools, the command point of grooving inserts is the *reference* point of such tool, relative to part zero. This is the point that follows the XYZ (mill) or XZ (lathe) coordinates in the part program. Precision grooving programs typically require that the programmed command point is changed within the program, often combined with a change of offset.

Square Inserts

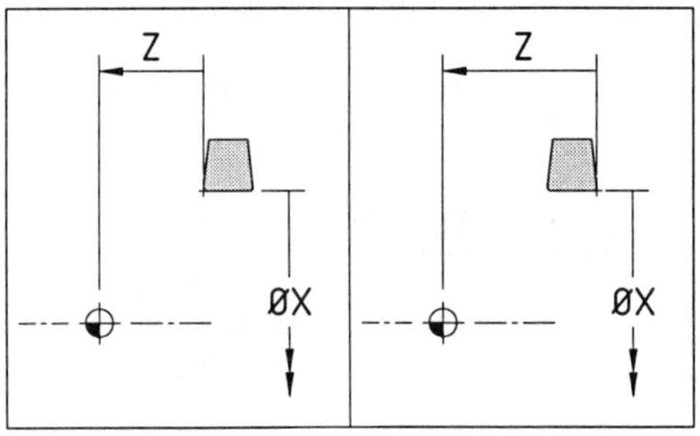

A command point for a typical square grooving insert used on CNC lathes is typically programmed either to the *left* or the *right* bottom corner of the insert, as viewed by the CNC operator. The same method applies for other square grooves, such as inserts with *specified* corner radiuses, also with a flat section at the bottom.

Note that all square inserts have ***some*** *corner radius, but not always specifically defined.*

Round Inserts

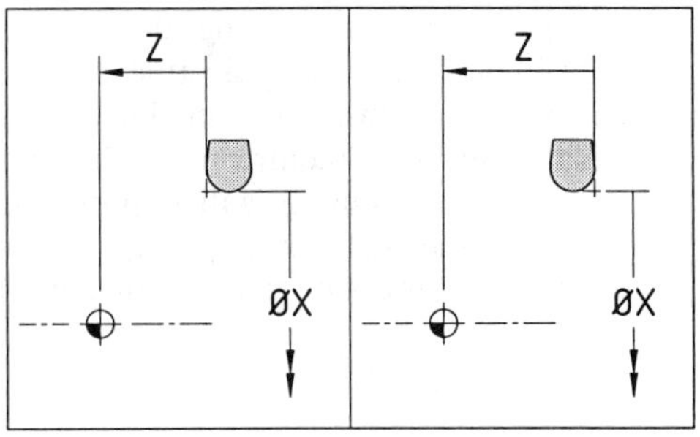

Many round inserts fit into the same tool holder as their square counterparts and are set the same way. Round inserts have no flat section at the bottom and are often used for recess grooving in shaft work, to increase the torque strength. In many cases, programming technique uses a simple *'plunge and retract'* method, without any contouring.

Although square inserts are often used for minor groove profiling (tapered walls, for example), round inserts have much larger field of applications for various groove contours.

Command point selection to the *left side* is more consistent with the general setup of other tools

Plunge and Retract Method

This is the simplest of all machining methods that is used in groove programming. The section title essentially defines the whole operation. The type of groove that belongs to this category is a rough groove, for example, one that will be completed later. Also in the same category may belong some utility grooves that are nothing more than open clearances. Edge burrs may or may not matter in such cases, but the general approach to programming such grooves is not precision oriented.

In the following example, the simplest groove possible is programmed *(any prior machining is assumed to have been completed)*. Note that there is no effort in the program to control groove tolerances and/or its surface finish, sharp corners or burrs included. This is a typical utility groove that is often machined on manual lathes, because its much cheaper than using a CNC lathe.

The groove is 5 mm wide (~0.20") and 3 mm deep (~0.12"), as the drawing shows. The command point for the selected tool is set at its lower left corner, and the program will use T02 for the example:

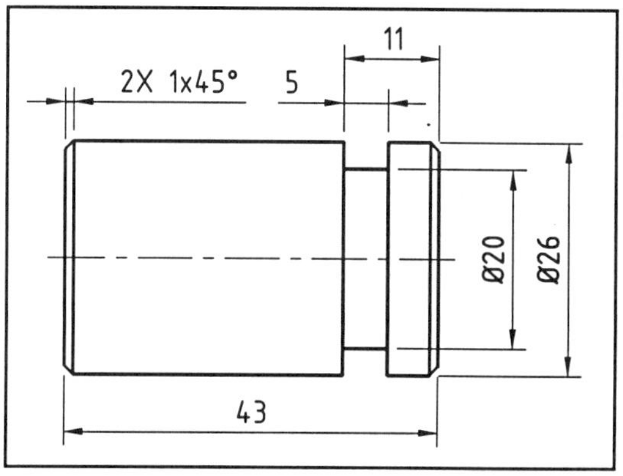


```
...
N11 T0100
N12 G96 S150 M03
N13 G00 X28.0 Z-11.0 T0202 M08
N14 G01 X20.0 F0.15
N15 G00 X28.0
N16 X100.0 Z200.0 T0200
N17 M01
...
```

A program with the same result can also use a special cycle G75, used for cutting along the X-axis.

G75 Cycle

Fanuc lathe controls offer a cycle listed under the multiple repetitive cycles section in the manuals. This cycle is designed for grooves that require an interrupted cut, for example, those machined in tough materials, deep grooves, or in any other type of machining where the chip breaking will improve the machining process. G75 cycle can also be used for cutting multiple grooves that share the same characteristics.

There are two formats for the cycle, depending on the control unit. For 10/11/15 Fanuc controls, use the one-block format:

```
G75 X.. Z.. I.. K.. D.. F..
```

☞ where ...

X	=	Final groove diameter
Z	=	Z-position of the groove (last groove for multiple grooves)
I	=	Depth of each cut (positive value)
K	=	Distance between grooves (positive value) - for multiple grooves only
D	=	Relief amount at the end of cut
F	=	Cutting feedrate in selected units

Spindle speed can also be programmed in the same block.

A two-block format is used for 0/16/18/20/21 Fanuc controls:

```
G75 R..
G75 X.. Z.. P.. Q.. R.. F..
```

☞ where ...

R	=	Clearance for each cut (return amount) (first block)
X	=	Final groove diameter
Z	=	Z-position of the groove (last groove for multiple grooves)
P	=	Depth of each cut (positive value)
Q	=	Distance between grooves (positive value) - for multiple grooves only
R	=	Relief amount at the end of cut (second block)
F	=	Cutting feedrate in selected units

The program on the previous page can be modified to use G75 cycle, even for a single groove:

```
...
N11 T0100
N12 G96 S150 M03
N13 G00 X28.0 Z-11.0 T0202 M08
N14 G75 X20.0 I1.0 F0.15                (EACH CUT WILL BE 1 MM DEEP)
N15 G00 X100.0 Z200.0 T0200
N16 M01
...
```

For a two-block format, the program will change only slightly:

```
...
N11 T0100
N12 G96 S150 M03
N13 G00 X28.0 Z-11.0 T0202 M08
N14 G75 R0.25
N15 G75 X20.0 P1.0 R0.25 F0.15          (EACH CUT WILL BE 1 MM DEEP)
N16 G00 X100.0 Z200.0 T0200
N17 M01
...
```

As these examples represent a single groove, note that in either program, there is no Z-location and there is no distance between grooves.

The main difference between the two formats is that the relief amount for each cut can only be programmed in the two-block format. In the one-block format, this amount is set by a system parameter. Both formats support programming of the relief at the bottom of the groove.

Grooving for Precision

The majority of programming CNC grooving operations falls into the category of *precision groove*. Although the term 'precision groove' may be interpreted in different ways, the common denominator of CNC programs developed for precision grooves is the focus on maintaining the precise *groove location* and *groove width*. This requirement is defined in the drawing by using various tolerances.

A typical precision groove requires good planning of the actual machining. The first requirement is the selection of a grooving insert, particularly its width. The width of grooving insert must always be *smaller* than the groove width, which means more tool motions than a simple plunge-retract method.

Once the suitable insert has been selected, the machining method has to be carefully planned. The technique can be itemized into steps. Once the steps are finalized, they can be transferred into the program itself. The steps suggested in this section apply to a groove that can be machined with only one roughing cut. For wider grooves, the technique is very similar, and an example will be provided later in this section. Command point of the grooving tool is set to the left for the following procedure.

Machining Procedure

MOTION 01	**From a tool change position, move the tool to a Z-position, where the middle of the insert width matches the vertical centerline of the groove**
MOTION 02	**In X-axis, move the tool close to the part diameter (1 mm or 0.020" per side suggested)**
	***** START POSITION OF THE GROOVING TOOL *****
MOTION 03	**Feed the tool into the material, leaving a very small amount at the groove bottom (0.125 mm or 0.005" per side suggested)**
MOTION 04	**In rapid mode, retract to the same diameter where the previous cut started from**
MOTION 05	**Still in rapid mode, use an incremental motion to the start of the *left* groove chamfer (or radius)**
MOTION 06	**In feedrate mode, cut the chamfer to size**
MOTION 07	**Continue the cut on the *left* wall of the groove, to the groove bottom - leave the same amount of material at the bottom, as in Motion 03**
MOTION 08	**In rapid mode, retract to the start position (as established in Motion 01 and 02)**
MOTION 09	**Still in rapid mode, use an incremental motion to the start of the *right* groove chamfer (or radius)**
MOTION 10	**In feedrate mode, cut the chamfer to size**
MOTION 11	**Continue the cut on the *right* wall of the groove, to the groove bottom (final diameter)**
MOTION 12	**Sweep the bottom of the groove, so the left corner of the insert touches the left groove wall**
MOTION 13	**In rapid mode, retract to the start position (as established in Motion 01 and 02)**

When described in such a step-by-step method, the process may seem a bit complicated. The best approach to take is to follow each step on its own, rather than all steps together.

Note - if properly calculated, the incremental *Motion 05* and *Motion 09* will have identical increments, although in the opposite directions. Programming these motions in incremental mode makes the programming much easier. This method allows a change of the command point from the left corner to the right corner (and vice versa), with only the simplest of calculations.

The groove shown at right is a typical groove suitable for the above described programming procedure.

If no tolerances are given or only the location tolerance is specified, a single offset in the program is sufficient. It will control the groove *location* relative to the front face, but it will *not* control the groove width. If groove width is to be controlled as well, an additional wear offset for the same tool must be programmed.

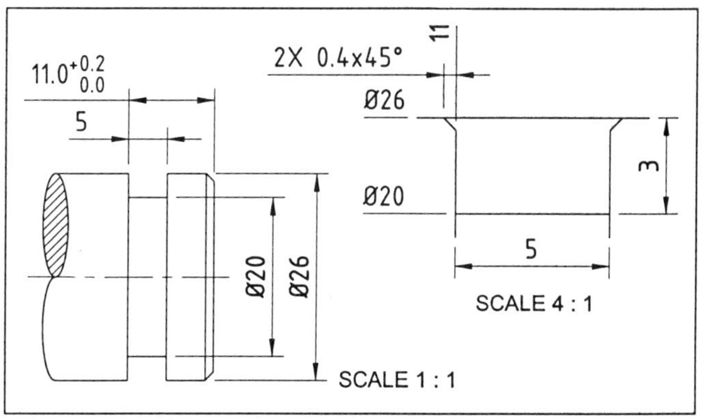

Programming Procedure

Once the machining procedure has been established, it can be applied to the part program. The programming procedure will follow all steps described above. Always study the drawing first! For the example, a 4 mm wide grooving tool will be used, with the command point set to its left corner. On the previous page is a drawing that shows a tolerance on the groove location, but not on its width. There will be only *one* wear offset required by the part program.

Now, the *machining* procedure can be converted to the *programming* procedure:

MOTION 01 **G00 Z-10.5 T0202 M08**

MOTION 02 **X28.0**

***** START POSITION OF THE GROOVING TOOL *****

MOTION 03 **G01 X20.25 F0.15**

MOTION 04 **G00 X28.0**

MOTION 05 **W-1.9**

MOTION 06 **G01 U-2.8 W1.4 F0.1**

MOTION 07 **X20.25 F0.15**

MOTION 08 **G00 X28.0 Z-10.5** **(START)**

MOTION 09 **W1.9**

MOTION 10 **G01 U-2.8 W-1.4 F0.1**

MOTION 11 **X20.0 F0.15**

MOTION 12 **Z-11.0**

MOTION 13 **G00 X28.0 Z-10.5** **(START)**

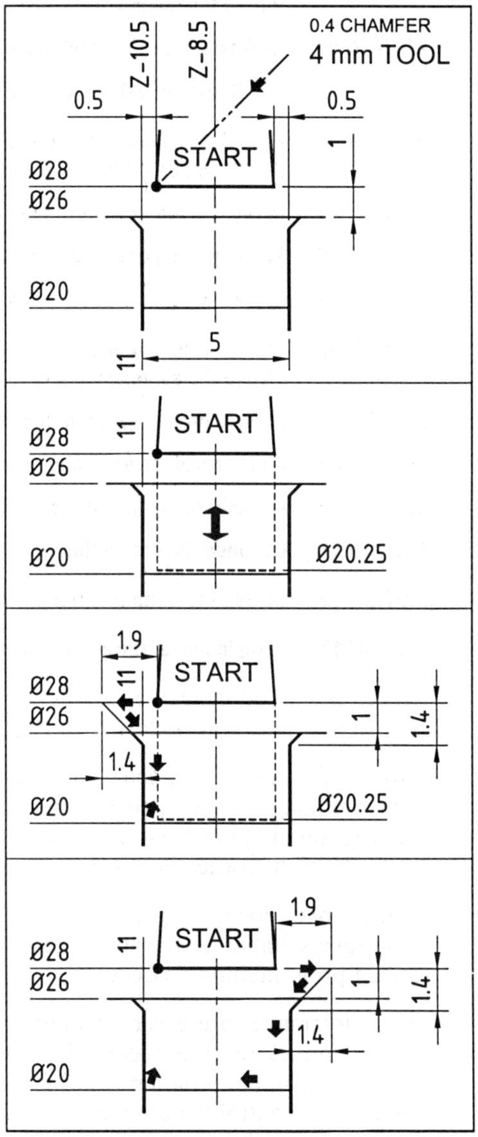

Actual program segment can now be developed:

```
...
N21 T0200
N22 G96 S120 M03
N23 G00 Z-10.5 T0202 M08      (MOTION 01)
N24 X28.0                     (MOTION 02)
N25 G01 X20.25 F0.15          (MOTION 03)
N26 G00 X28.0                 (MOTION 04)
N27 W-1.9                     (MOTION 05)
N28 G01 U-2.8 W1.4 F0.1       (MOTION 06)
N29 X20.25 F0.15              (MOTION 07)
N30 G00 X28.0 Z-10.5          (MOTION 08)
N31 W1.9                      (MOTION 09)
N32 G01 U-2.8 W-1.4 F0.1      (MOTION 10)
N33 X20.0 F0.15               (MOTION 11)
N34 Z-11.0                    (MOTION 12)
N35 G00 X28.0 Z-10.5          (MOTION 13)
N36 X100.0 Z200.00 T0200
N37 M01
...
```

This detailed example represents a typical programming method for most semi-precision and precision grooves machined on CNC lathes. Once you understand the programming technique used, it will be much easier to adapt it to any groove.

Deep Grooving

Majority of tooling catalogues define the term *'deep grooving'* as applicable to those grooves where the ratio of groove depth is 1.5 times greater that the groove width or higher. Selection of a deep grooving tool is determined not only by the groove width, but particularly by the groove *depth*. The grooving blade (insert) must be slightly longer than the groove itself, in order to provide all necessary clearances. Another very critical aspect of setting a deep grooving tool at the machine is the *tool orientation* - it must be *perpendicular* to the machine centerline, otherwise even a slight deviation may result in a distorted groove. Although correct tool setting is important for all grooves, it is much more important for deep grooves.

Deep grooves are often programmed using an interrupted cut, generally for the purpose of breaking the chips that may accumulate during machining. In the provided example, the deep groove is 18 mm deep. The grooving motion has to be interrupted every 6 mm, in order to clear the chips from the groove area. After each motion, the tool will retract 0.5 mm, before cutting the next 6 mm depth. If chip clogging is a problem (for some materials it can be), retract the grooving tool all the way above the part diameter.

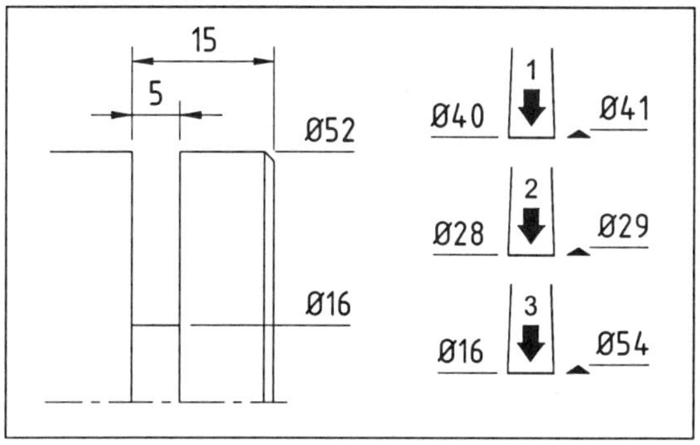

The grooving insert selected for this job is 5 mm and can cut into the full depth of the groove. Based on the actual material and selected cutting conditions, the CNC programmer has to select the *maximum depth of each cut*. For this example, the depth of each has been selected as 6 mm, resulting in three 'peck' grooves (interrupted cuts) required to complete the groove to its full depth. Breaking chips on deep grooves provides two main benefits - increase of the tool life and better groove quality.

```
(PARTIAL RETRACT 0.5 MM PER SIDE)

N51 T0400
N52 G96 S125 M03
N53 G00 Z-15.0 T0404 M08
N54 X54.0
N55 G01 X40.0 F0.25
N56 G00 X41.0
N57 G01 X28.0
N58 G00 X29.0
N59 G01 X16.0
N60 G00 X54.0
N61 X100.0 Z200.0 T0400
N62 M01
...
```

```
(FULL RETRACT TO X54.0)

N51 T0400
N52 G96 S125 M03
N53 G00 Z-15.0 T0404 M08
N54 X54.0
N55 G01 X40.0 F0.25
N56 G00 X54.0
N57 X41.0
N58 G01 X28.0
N59 G00 X54.0
N60 X29.0
N61 G01 X16.0
N62 G00 X54.0
N63 X100.0 Z200.0 T0400
N64 M01
...
```

The programs above show two versions with the same result. The program on the left is more efficient as the retract after each cut is only 0.5 mm above the current diameter. This method is suitable if chipbreaking only is desired. The program on the right shows the tool retracting to the clearance diameter (∅54) after each cut, then returning to make another cut from 0.5mm above the last diameter.

G75 cycle could have been used for this example, but block-by-block method is more effective for understanding the program. If a corner break (chamfer) is required to eliminate burrs or the groove is wider than the insert, apply the method used in the *'Grooving for Precision'*.

Grooves with Tapers - O-Ring Grooves

Pulley grooves described in the next section are special purpose grooves with tapered walls, used for very unique applications. There are many 'special purpose' grooves that do not have their two walls parallel with the machine axes. The most common of these grooves is a particular group of grooves called 'O-ring grooves'. This group also includes grooves with tapered walls, but with specifications unique to a particular industry or application. The purpose of the taper is to guarantee a high quality seal, as O-ring grooves are mainly used for the purpose of sealing, using a rubber ring or a similar sealing device. In this section, a typical approach to programming an O-ring groove is explained. Keep in mind that this technique can be used for any groove with tapered walls and corner radiuses. This section is not intended to cover the subject of O-ring grooves, only their programming.

The drawing at right shows a typical groove that belongs to this category. Common characteristics of such grooves are tapered walls, usually up to 5° for each wall, specified groove radius at the bottom diameter, groove corner breaks at the top diameter, and the groove width, provided as the distance between sharp corners.

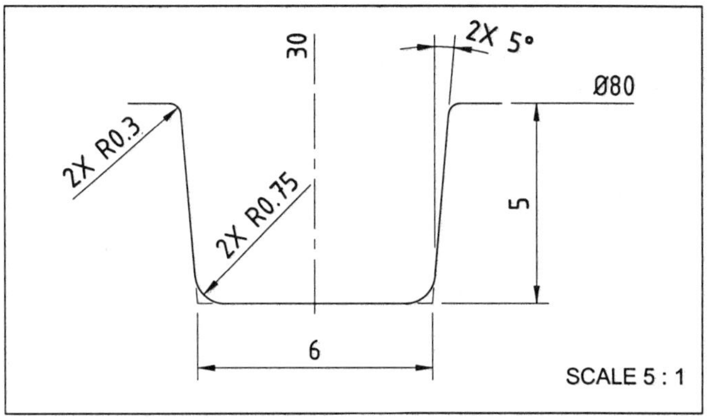

The first step in the process is to select a suitable grooving tool - note that a standard grooving tool is used.

As the groove width is 6 mm, it may be tempting to select a smaller width grooving tool just by guessing. This is not a professional approach, and may backfire - the proper approach is to calculate the exact length of the flat area at the groove bottom. Besides, once this length is known, the calculation can be used to find the coordinate of the few points required in the program.

In the detail shown, only one half of the groove is required, as the other half is symmetrical by the groove centerline.

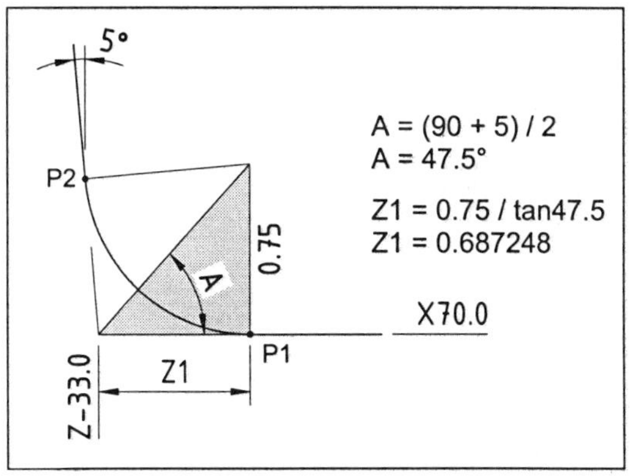

The first calculation is for the Z-value of point P1. Find the angle A, which is (90+5)/2=47.5°. As one side of the triangle is known, the other one can be calculated using trigonometric function.

Once the Z1 dimension is known (calculated as 0.687248 mm), the flat of the groove bottom may be calculated as well:

Width of groove - 2 x Z1 = 6 - 2 x 0.687248 = 4.626 mm

Grooving tool selected must be smaller than the flat width, and 4 mm wide grooving tool seems like the choice - it will be used for the example.

Z-coordinate of the point P1 can also be calculated by simple subtraction. If the groove center is at Z-30.0, the left sharp corner must be at Z-33.0. By subtraction of Z1, P1 can be calculated:

P1 (Z) = -(33 - Z1) = -(33 - 0.687248) = -32.313 .... **P1 =** `X70.0 Z-32.313`

Once all coordinates left of centerline are known, the corresponding coordinates at right are easy.

In order to calculate coordinates for P2, another calculation is necessary. The triangle to solve originates at the radius center, and the known hypotenuse of 0.75 is the radius.

Once the X2 and Z2 dimensions are known, the actual coordinates can be calculated as well.

P2 (X) = 70 + 2 x 0.75 - 2 x X2 = X71.369

From the previous calculation, Z1 is 0.687248:

P2 (Z) = -(33 - 0.687248 + Z2) = Z-33.060

P2 = `X71.369 Z-33.06`

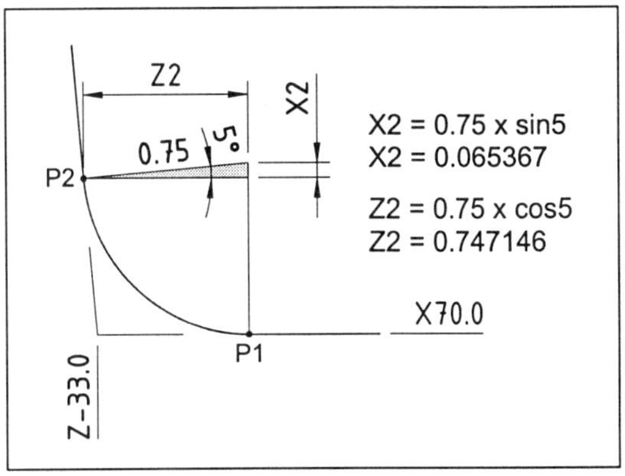

The last calculations are for the corner break between the groove top diameter and the wall. The left corner break will be calculated next.

Both points P3 and P4 can only be calculated from a known point. Since the sharp point of the corner break is unknown, it must be calculated.

Disregarding both radiuses, the dimension Z3 can be calculated. Adding this calculation to the known sharp corner at the bottom, the coordinate point for the sharp corner at the groove top is:

`X80.0 Z-33.437`

Now, points P3 and P4 can be calculated as well.

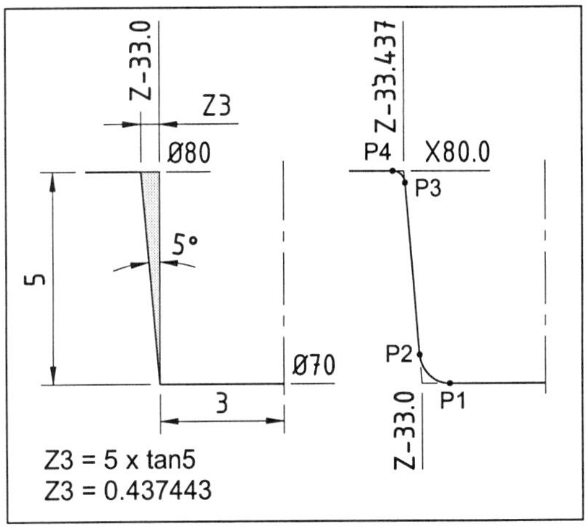

The illustration at right shows both calculations, starting with the angle A. Note that once the dimension Z4 has been calculated, it will be used in both calculations. Once as the opposite side, then as the hypotenuse.

P3 (X) = 80 - 2 x X5 = 80 - 2 x 0.273853 = X79.452

P3 (Z) = -(33.437 - Z5) = -(33.437 - 0.023959)
= Z-33.413

P4 (Z)= -(33.437 + Z4) = -(33.437 + 0.274899)
= Z-33.712

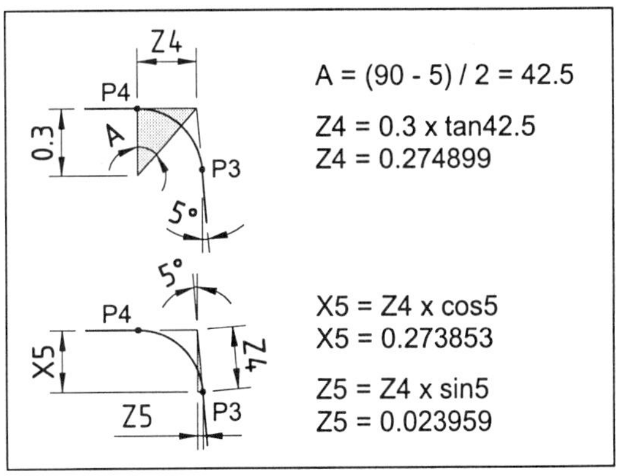

Program coordinates for P3 and P4 are:

P3 = `X79.452 Z-33.413`

P4 = `X80.0 Z-33.712`

Even if it appears like a lot of work, keep in mind that the radiuses are necessary for O-ring grooves, as they provide a smooth surface for the seal ring. Fortunately, the coordinate points to the right of centerline will use the same calculations - apart from some additions and subtractions, there are no new calculations.

For easier reference, P11 is related to P1, P12 to P2, P13 to P3, and P14 to P4. Note that the differences are only along the Z-axis, while the diameters remain the same. Here is the complete list in the order of machining!

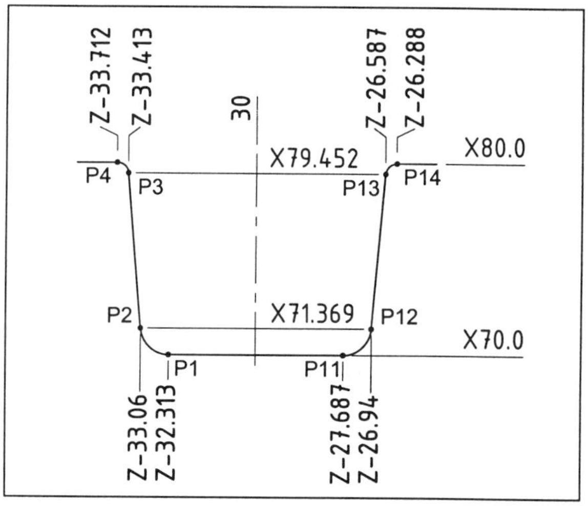


```
X80.0    Z-33.712    (P4)
X79.452  Z-33.413    (P3)
X71.369  Z-33.06     (P2)
X70.0    Z-32.313    (P1)
X80.0    Z-26.288    (P14)
X79.452  Z-26.587    (P13)
X71.369  Z-26.94     (P12)
X70.0    Z-27.687    (P11)
```

Complete program uses the points as defined, along with other program instructions:

```
(T06 = 4 MM WIDE GROOVING TOOL)
N1 G21
N2 T0600
N3 G96 S100 M03
N4 G00 Z-32.0 T0606 M08              (GROOVING MIDPOINT AT GROOVE CENTERLINE)
N5 X82.0                             (1 MM PER SIDE DIAMETER CLEARANCE)
N6 G01 X70.2 F0.25                   (0.1 MM LEFT ON GROOVE BOTTOM)
N7 G00 X82.0                         (RETURN TO ORIGINAL START POINT)
N8 Z-33.712                          (POSITION OF LEFT CORNER BREAK - P4)
N9 G01 X80.0 F0.15                   (START OF LEFT CORNER BREAK - P4)
N10 G02 X79.452 Z-33.413 R0.3 F0.1   (CUT LEFT CORNER BREAK - P3)
N11 G01 X71.369 Z-33.06 F0.15        (LEFT TAPERED WALL - P2)
N12 G03 X70.0 Z-32.313 R0.75         (LEFT GROOVE RADIUS - P1)
N13 G00 X82.0 Z-32.0                 (RETURN TO ORIGINAL START POINT)
N14 Z-30.288                         (* START OF RIGHT CORNER BREAK - P14)
N15 G01 X80.0                        (START OF RIGHT CORNER BREAK - P14)
N16 G03 X79.452 Z-30.587 R0.2 F0.1   (* CUT RIGHT CORNER BREAK - P13)
N17 G01 X71.369 Z-30.94 F0.15        (* RIGHT TAPERED WALL - P12)
N18 G02 X70.0 Z-31.687 R0.75         (* RIGHT GROOVE RADIUS - P11)
N19 G01 Z-32.313                     (SWEEP BOTTOM FLAT - P1)
N20 G00 X82.0 Z-32.0                 (RETURN TO ORIGINAL START POINT)
N21 X200.0 Z100.0 T0600              (TOOL CHANGE POSITION)
N22 M01
```

A few observations about the program - the middle of the groove is dimensioned as 30 mm from the front face (Z0). The selected insert is 4 mm wide. In order to position the insert exactly in the middle of the groove, the insert command point will be 2 mm further, at Z-32.0 (block N4). When the insert plunges into the material, it leaves a very small amount of stock (0.1 mm per side) at the groove bottom (block N6). The approach for the corner break is straight in the X-axis. Another option is to make this approach tangential, but that would involve another set of calculations that may be difficult to justify. Note the feedrate change for the corner breaks (blocks N10 and N16). The smaller the corner break, the slower the cutting feedrate should be - otherwise, there may not be a corner break at all or it may have a poor surface quality. The corner break method of programming described in this section can also be used for programming pulley grooves (corner breaks), described in the next section.

In this program, all Z-motions are in absolute mode. Note the blocks N14, N16, N17, and N18 (marked with an asterisk in the comment section). Each absolute Z-position has been increased by 4 mm, to compensate for the fact that it is the *right corner* of the insert that does the cutting. Grooves that are programmed symmetrically on both sides of their centerline may benefit from programming the Z-motions in incremental mode, to avoid this adjustment. In incremental mode, the same numbers will appear for both walls in the program, albeit with opposite signs.

Grooves with Tapers - V-Pulley Grooves

Pulley grooves are special - and specific - application grooves, quite common in many power transmission industries. Their sole purpose is to drive a belt. While most pulley grooves are cast with a high degree of precision, many others are machined from solid material. Pulleys made of steel or cast iron have more resistance to wear than their aluminum counterparts. Wear decreases the pulley efficiency, particularly for small diameters.

Production machining of pulley grooves on CNC lathes involves inserts of specific size and shape, designed exclusively for pulley grooves. Many cutting inserts for pulley grooves are made of ceramics and similar compounds and can be used at extremely high spindle speeds in aluminum, steel or cast iron (typically without coolant). The machining involves somewhat lower cutting feedrates, yet still yielding superior performance.

Pulley grooves are also grooves with an taper and are similar to taper grooves in many respects. Unlike the taper grooves described in the previous section, pulley grooves are *formed* to machine the groove in one plunge-in pass, using a dedicated insert. A standard square insert can also be used, along with multi-contour machining, but this method is highly inefficient and may be justified in one-time applications only. Groove angles of pulley grooves can be quite numerous, depending on the application, but 34°, 36°, and 38° angles are fairly common.

A typical pulley groove is shown at right, using Imperial dimensions. Note the dimensioning method used - the depth and the bottom width are *not* given, as they are not necessary. Top width and the groove angle determine these dimensions - for programming, they have to be calculated. Another dimension that is not given is the bottom groove radius - the *insert* determines its size, no special programming is necessary. Some older drawings specify the top corner with a note *'file break all sharp corners'*. This method is not acceptable on CNC lathes and corner breaks have to be calculated to be included in the part program.

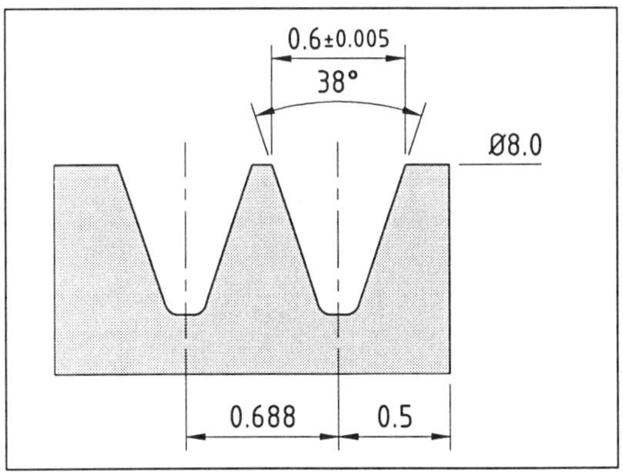

The most important part of programming V-pulley grooves (grooves in the shape of a letter V) is to guarantee the groove *width* dimension, specified at the top (0.6 in this example). By selecting the proper insert, all other dimensions fall into place, and the whole effort boils down to a simple plunge motion along the X-axis.

Insert Selection

Although numerous in numbers, pulley grooves represent rather a narrow area of practical applications. V-belts that will be used on these grooves are made according to certain standards, so it only makes sense to manufacture grooving inserts that can complement this standard. The selected insert illustrated at right is real - it has been taken from a tooling catalogue - only relevant dimensions are shown.

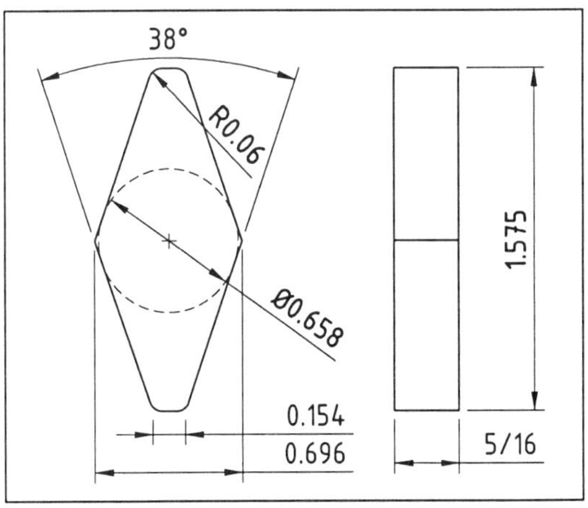

Depth Calculation

The program can be written once the groove depth is known. In order to calculate the groove depth, the insert selected must have an overall width *greater* than the groove width. In the example, groove width is 0.6 inches, overall insert width is 0.696 inches, and the bottom width is 0.154 inches (over sharp corners). Keep in mind that in any tapered groove, the groove depth influences the groove width (at the top) and vice versa. The goal of this calculation is to establish programmable grooving depth that will result in 0.600 groove width (top width).

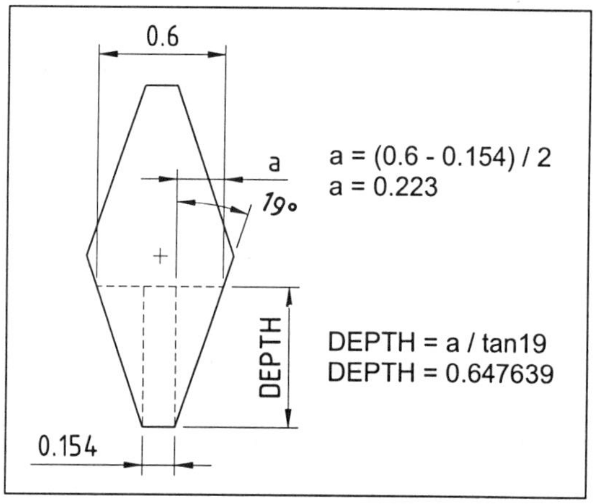

Groove corner radius at the bottom is not important for the calculation (not shown).

The actual grooving depth will be 0.6476 inches, programmed as 8.0 - 2 x 0.6476 = **X6.7048**.

Tool Setup and Program

Typical setup for this type of insert (angles at both sides) is to the middle point of its width. It may take a little more effort at the CNC lathe to complete the actual setup, but alternatives are even less practical. Having the middle of the insert as the command point, it is easy to position the tool, as pulley grooves are generally dimensioned to their center (middle).

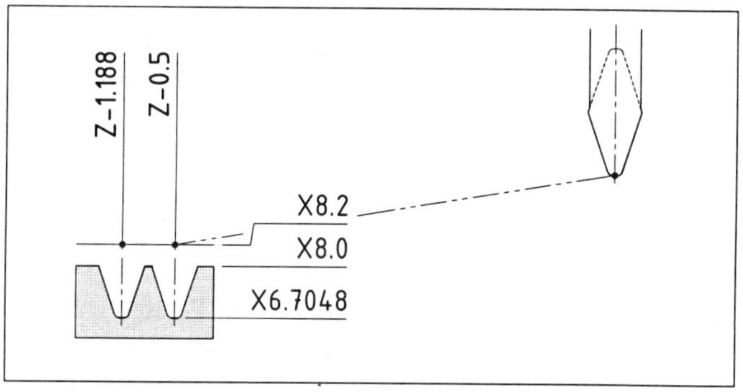

Writing the program is simple, as only in-out motions are necessary for each groove.

```
(T03 = SPECIAL CERAMIC PULLEY INSERT - MIDDLE SETTING)
N51 (G20) T0300
N52 G97 S4000 M03
N53 G00 X8.2 Z-0.5 T0303 M08      (START POSITION FOR THE RIGHT GROOVE)
N54 G01 X6.7048 F0.125            (HIGH SPEED - SLOW FEED PLUNGE CUT)
N55 G00 X8.2                      (RETURN TO ORIGINAL DIAMETER)
N56 Z-1.188                       (START POSITION FOR THE LEFT GROOVE)
N57 G01 X6.7048                   (HIGH SPEED - SLOW FEED PLUNGE CUT)
N58 G00 X8.2                      (RETURN TO ORIGINAL DIAMETER)
N59 X200.0 Z100.0 T0300           (TOOL CHANGE POSITION)
N60 M01
```

Pulley groove programming is one of the rare occasions where the insert dimensions determine certain aspects of the program data.

16 TECHNIQUES FOR THREADING

Many part programmers who have the opportunity to program single point threading operations for CNC lathes, have been accustomed to using the venerable G76 multiple repetitive cycle. Without a doubt, this is a modern and very flexible cycle that can be used for the majority of standard threading applications, external or internal. Threading operations can be very simple or they can be quite complex. There are many thread forms, many different specifications, many standards, and many different expectations. Some CNC programmers share a common view that threading on CNC lathe can be either *very smooth* or *very difficult*. The underlying message is that there is no *'in-between'* situation. Of course, this is not exactly the most accurate view, but it does illustrate certain frustrations that programmers and operators alike have to go through while programming or machining threads. Using a single arbitrary programming method for *all* threading applications is not always the best choice - in fact, it may even be the *wrong* choice.

Types of Thread Forms

Although the most common thread form is the 60° V-shape thread, used for both metric and imperial threading, there are many other thread forms that can be programmed - and machined - using a typical CNC lathe (turning center).

When it comes to machining, there is practically no subject larger than that of threads. Threads are everywhere, and they have so many different purposes, so many different forms, that even a seasoned CNC programmer may say find the subject of threading overwhelming. The reality of machining is that CNC programmers cannot pick and choose which threads to accept and which ones to reject - job is a job is a job.

Understanding threads in general is the first step towards understanding them in depth.

UN - Unified National and Metric

The most common thread form for imperial threads is the UN - *Unified National* - thread form. It shares its dimensions with the metric thread form. Both are 60° V-threads and both have their coarse and fine versions. UN threads have at least three versions, often identified on the drawing:

UN - UNC	**Unified National Coarse**	**4 - 8 TPI**
UNF	**Unified National Fine**	**12 - 18 TPI**
UNEF	**Unified National Extra Fine**	**20 - 32 TPI**

Metric threads are always specified with the *M* prefix, either with or without the pitch. Often the metric thread without the pitch specified indicates a *coarse* thread series, while a metric thread specified with a pitch indicates the *fine* series. For example:

M6	**Metric thread - coarse**	**... pitch is assumed 1 mm (same as M6 x 1)**
M6 x 0.75	**Metric thread - fine**	**... pitch specified**

If the pitch in metric thread is *not* provided always check the engineering intent

Other Thread Forms

Thread forms other than v-thread are often called *special* threads. The term 'special thread' has to be taken in the proper context. A thread form that is special to one machine shop, may be a standard thread for another machine shop. For the purposes of this chapter, the term 'special', as it relates to single point threading, will be used to describe several thread forms that are outside of the 60° thread angle. These forms include:

- **Whitworth thread**
- **ACME thread / Metric Trapezoidal thread**
- **Buttress Thread**
- **Round Thread**
- **Aero Thread / Round Thread**
- **API Threads**
- **Square Thread**
- **... and many more, including threads for a single purpose**

All general programming principles described in this chapter apply to all thread forms. The main difference is the cutting tool selection and thread parameters. Some unique forms are shown below.

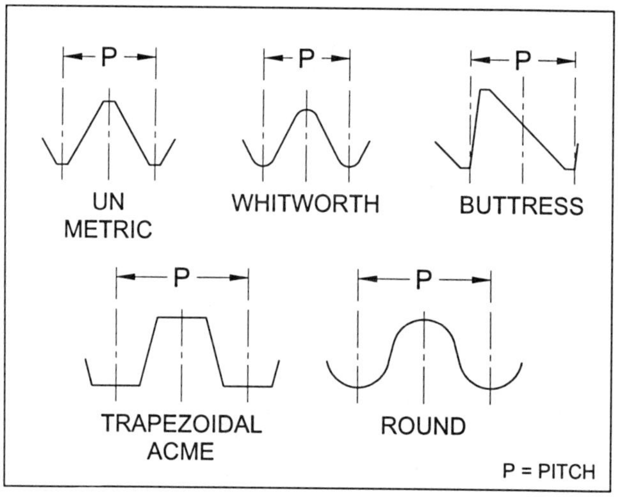

Without a question, the most common thread machined on CNC lathes is one with a symmetrical V-profile (V-thread). Regardless of the units selected for a particular part program, the V-profile thread (external or internal) has a 60° included angle. Both the ANSI (American) and the ISO (International) thread forms use this 60° profile as standard.

A symmetrical thread is a descriptive term that identifies a particular thread that has the same geometry on both sides of the thread centerline. Threads of this design are very common in manufacturing, as they present hardly any burden on the manufacturing process. Symmetrical threads are used to machine a variety of threads, particularly for a large variety of fasteners. Threads in the 'symmetrical' category include all 60° threads (metric and imperial), also the old British standard Whitworth threads (virtually obsolete now), and the two types of trapezoidal threads - the American ACME thread and the similar *Metric Trapezoidal* thread. Trapezoidal threads are not used as fasteners, but as threads that transfer motion. Their design provides a strong thread that can be used for a variety of applications. One of the more common application of a trapezoidal thread (ACME thread in North America) is a jack for replacing a car tire.

A thread that is somewhat similar to a trapezoidal thread form is a square thread. In fact, many applications that use an ACME (or its metric version) thread, used to be designed with a square thread. Both ACME and/or trapezoidal threads and square threads are threads strong enough to transfer a motion. However, square thread is not as easy to produce as a trapezoidal thread.

Thread Depth Calculation

Each thread form has its own set of formulas and calculations, often quite a few of them. For programming purposes, only a very few formulas are needed. For the two most common thread forms - *UN (Unified National)* and *Metric* - after the pitch calculation (if necessary), the most important calculation is the *thread depth.*

External and internal thread depths use different formulas, both based on the actual thread dimensions. The following two formulas provide at least theoretical depth for programming purposes:

TPI = Number of *threads per inch* (metric threads are defined by their pitch)

Pitch = P = 1 / TPI = distance between corresponding points of two adjacent threads

External thread depth (UN and Metric) = 0.613 × P = 0.613 / TPI

Internal thread depth (UN and Metric) = 0.541 × P = 0.541 / TPI

Both **0.613** and **0.541** are *constants*. ***Where did these constants come from?*** The UN or metric threads have specific characteristics - they have a 60° included angle and a known thread pitch. The constants have been calculated from actual thread data, based on established international standards.

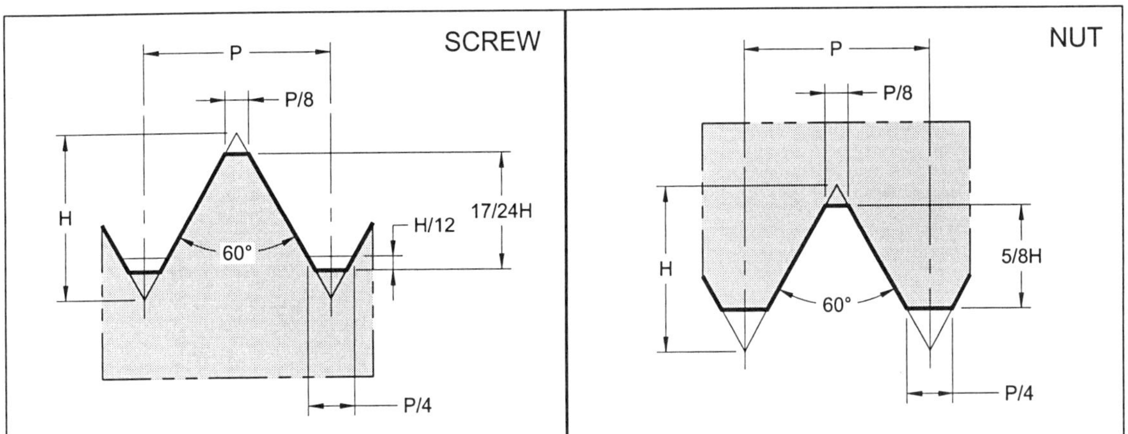

UN / METRIC THREAD - EXTERNAL

UN / METRIC THREAD - INTERNAL

The best way to understand the concept is to use unit of 1 for the pitch (P=1 inch *or* P=1 mm). That way, the constants calculated will be available for ***any*** pitch size, using the above formulas. Since the programmable thread depth is the most critical calculation, only one other dimension is important - the *dimension H* (thread height between sharp points).

For Pitch P = 1 ... the H = 1 / (2 × tan30) = 0.866025

☞ *therefore ...*

17/24H	=	**17/24 × 0.866025**	**= 0.708333 × 0.866025**	**= 0.613435**	**(EXTERNAL)**
5/8H	=	**5/8 × 0.866025**	**= 0.625 × 0.866025**	**= 0.541266**	**(INTERNAL)**

There is one other constant that may be used in various publications and thread specifications to calculate the thread depth. Instead of 0.613 constant for external thread depth, the constant is 0.649. What is the difference between them? While 0.613 is the standard constant specification used for *UN* and *metric* thread depths, 0.649 constant is used for calculating *American National* thread depth, a thread form similar to UN / metric form but not used too often. The constant for internal threads remains the same, at 0.541.

If pitch	**P = 1**
... then ...	**H = 1 / (2 × tan30) = 0.866025**
... and ...	*3/4H = 0.75 x 0.866025 = 0.649519*

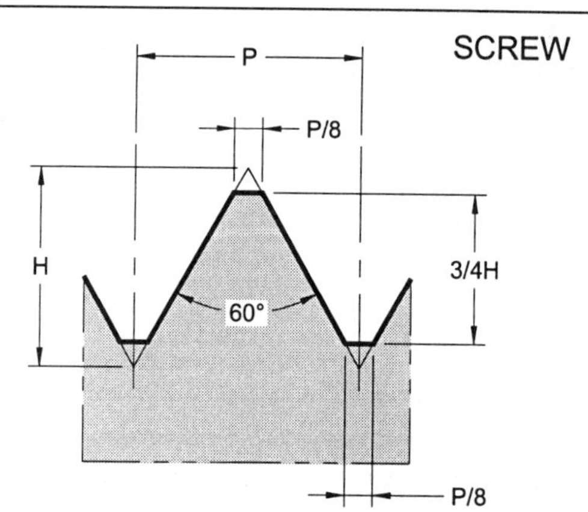

AMERICAN NATIONAL -EXTERNAL

Infeed Methods

Once the thread depth has been established, the next step is to select the thread machining method or - more accurately - the *infeed method* of the threading tool. There are four infeed methods that can be used to program single point threads - all are used with *decreasing* depth for each subsequent cut:

- **Radial infeed (also called *plunge* or *straight* infeed)**
- **Flank infeed (also called *compound* infeed)**
- **Modified flank infeed**
- **Alternate infeed**

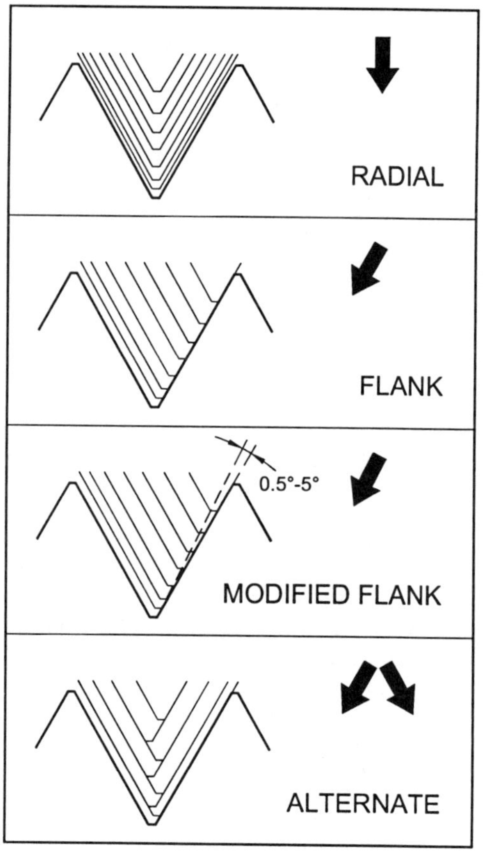

Radial infeed at 90° is the least desirable of the four. The problem is that the chip formation takes place at both cutting edges - chips form against each other and prevent coolant from reaching the cutting edge, causing heat and premature tool wear.

Flank infeed at 30° is more favorable, but its main problem is that one cutting edge rubs on one side of the thread, causing a tool wear and poor finish.

Modified flank infeed is much improved - instead of a 30° infeed, the cutting edge enters at 29°, for example. This method will cause very small step during each pass, but those same steps will provide clearance, so there is no tool edge rubbing against the thread wall. When the final depth is reached the small steps will be eliminated. ***This is the most preferred method !***

Alternate infeed is also a good choice, as each depth of thread uses opposite cutting edge, spreading the tool wear. It may require some experimenting.

Using the same depth for each threading pass is not recommended !

Tool Motions

Once the infeed method has been selected, the tool will have to make at least *four* motions to complete a single pass. The motions from a selected starting position are:

- **Motion 1 - Rapid to the thread diameter to cut**
- **Motion 2 - Feed towards the thread end**
- **Motion 3 - Rapid from the thread to the start position along the X-axis**
- **Motion 4 - Rapid to the start position along the Z-axis**

In some cases, there may be a fifth motion, programmed between the *Motion 2* and *Motion 3* as described above. This motion is sometimes called 'thread chamfering', which is not exactly the most suitable description. Its purpose is to leave the thread at an angle, over a short distance, usually the same length as the thread lead at 45°, still in feedrate mode. Programming for modern control systems does not require this step, although most control systems support it, particularly for compatibility with older programs.

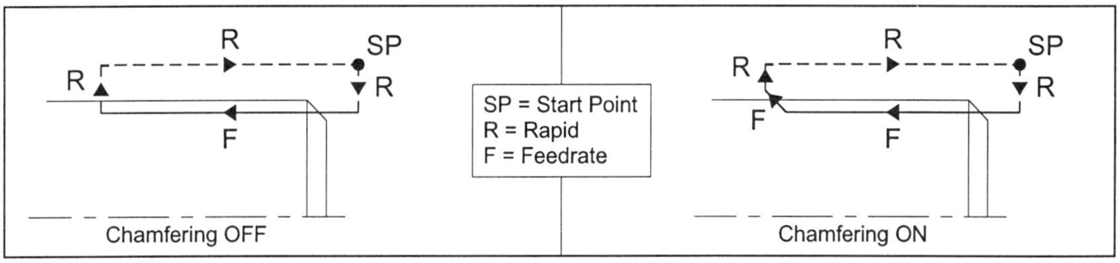

Later in this chapter, a sample thread will be programmed using different methods. Details relating to individual tool motions will be described in that section.

Cutting Conditions

Most programmers are aware that there is a direct relationship between the spindle speed and the feedrate. On CNC lathes, the spindle speed should always be programmed in the G97 mode (constant r/min). The feedrate is programmed in units per revolution - *mm/rev* or *in/rev*.

Acceleration and Deceleration

The time required to reach 100% of the programmed spindle speed and/or feedrate is called *acceleration*. The time required to reach a complete stop of the spindle and/or feedrate is called *deceleration*. It is important to account for both acceleration and deceleration in the thread program, by providing a sufficient clearance at the beginning of the thread (for acceleration) and at the end of the thread (for deceleration). Failure to provide clearance will result in *imperfect thread*.

Front clearance of ***three to four times lead*** should be sufficient for most threads. End clearance is often determined by the drawing and is less critical.

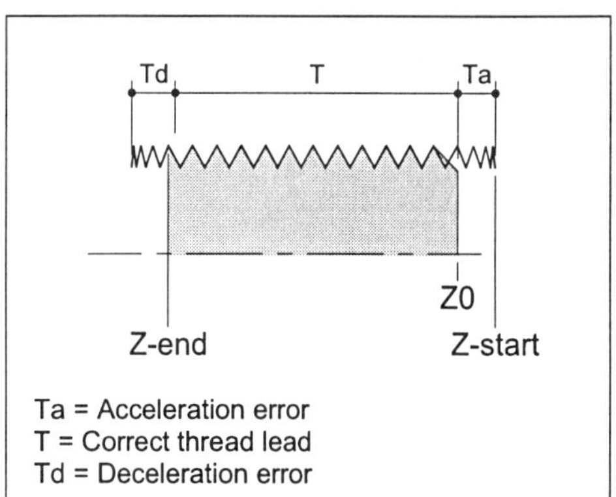

Cutting Depth

Previous calculations have defined the full depth of a given thread. It is not possible to machine a thread in a single pass - several passes have to be programmed in order to reach the required depth. As the tool moves deeper into the thread, it removes increasingly greater amount of material, making the tool work harder. As the previous schematic illustrations show, the depth of cut has to be *reduced* with each pass. Actual programming methods will be shown in the practical example.

Hand of Thread

Both external and internal threads can be machined in *right hand* orientation or *left hand* orientation. Three factors determine the hand of thread, and must be accounted for in the part program:

- Hand of the tool holder ... Right Hand *or* Left Hand (R/H or L/H)
- Direction of spindle rotation ... M03 => CW *or* M04 => CCW
- Direction of Z-axis motion ... Z-positive *or* Z-negative

Right hand thread is defined as a clockwise turn of a screw into a matching nut

In the following illustrations, the cutting insert is shown as *always* facing the CNC operator. For training purposes, it makes sense to show consistency within illustrations, to compare different configurations for external and internal threading. For a typical rear CNC lathe setting, the tool orientation shown may *not* match the actual tool orientation at the machine. In that case, all you have to do is to 'flip' the tool shown, by 180°. Keep in mind that no tool tip is *below* the spindle center line for rear lathes and no tool tip is *above* the spindle center line for front lathes. In order to 'convert' or 'flip' the shown view to the desired view, just imagine the selected tool rotated 180° in the X-axis. It is quite normal to see the 'back side' of the insert for right hand tools, when used on rear type CNC lathes.

The threading tool selection is even simpler with the four-line description shown in each illustration. All illustrations show the hand of thread (R/H or L/H), the tool holder selection (R/H or L/H), the programmed spindle rotation - M03 as CW or M04 as CCW as well as the Z-axis threading direction.

External Threading - Right Hand Thread

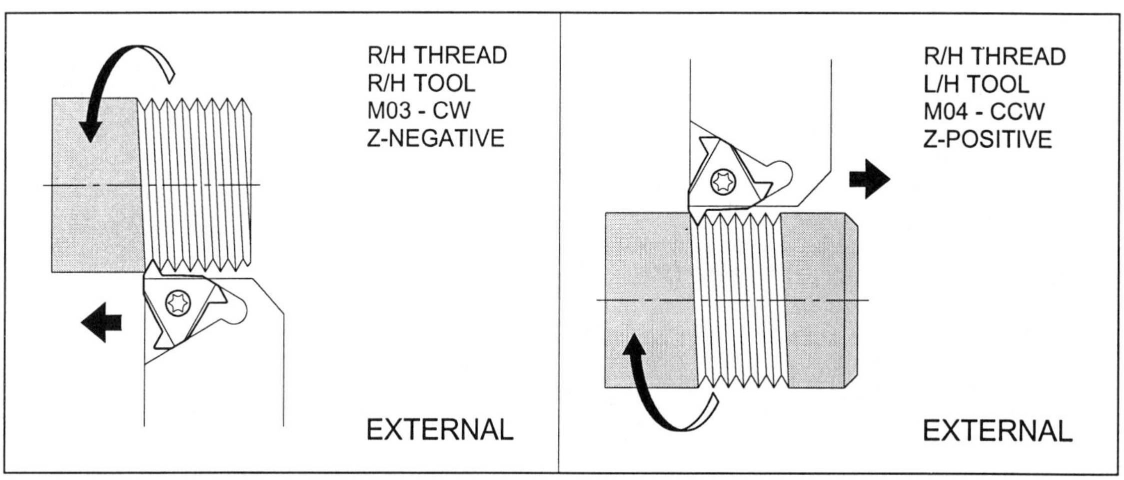

External Threading - Left Hand Thread

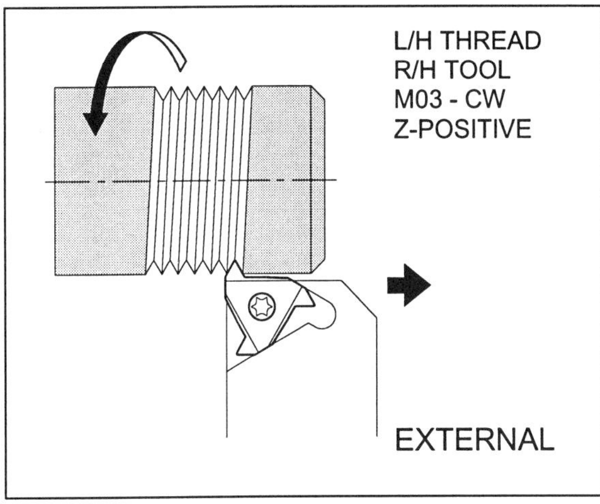

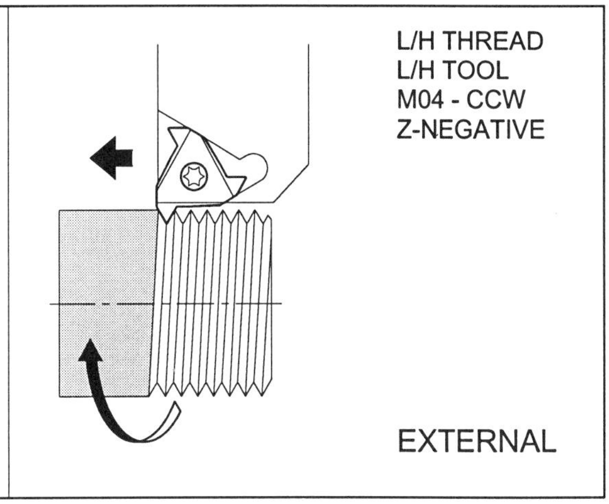

Internal Threading - Right Hand Thread

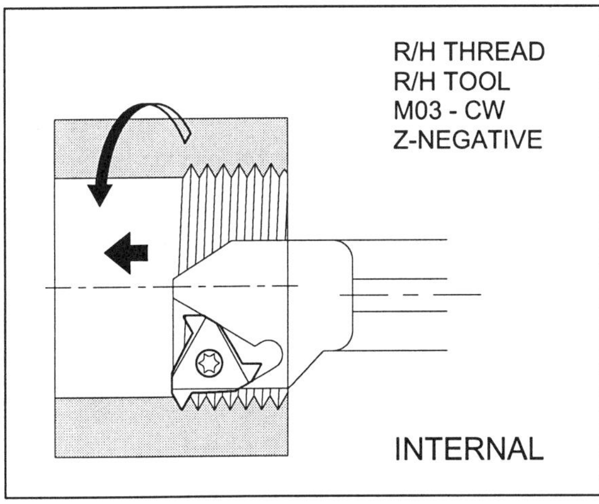

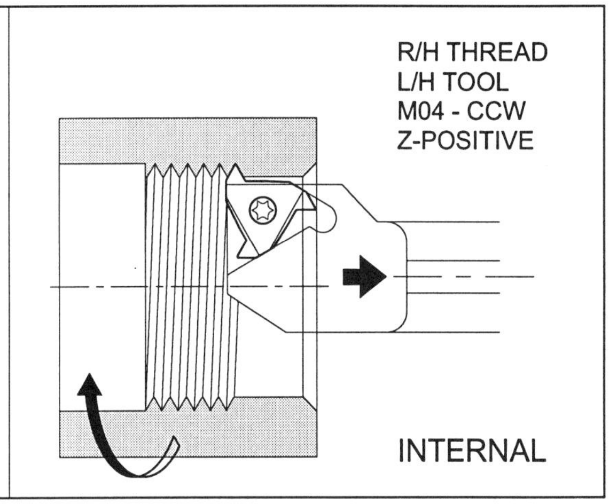

Internal Threading - Left Hand Thread

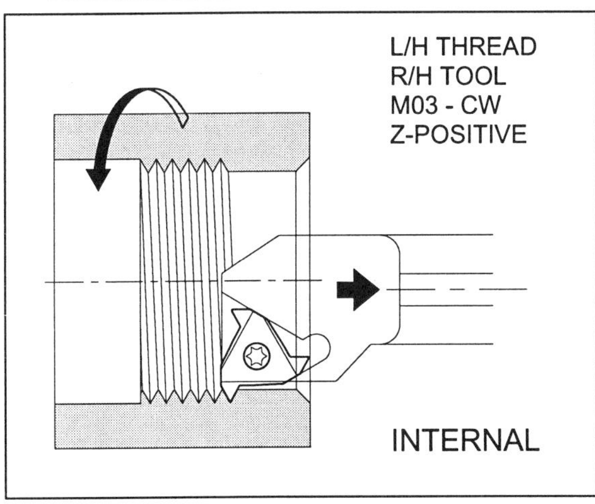

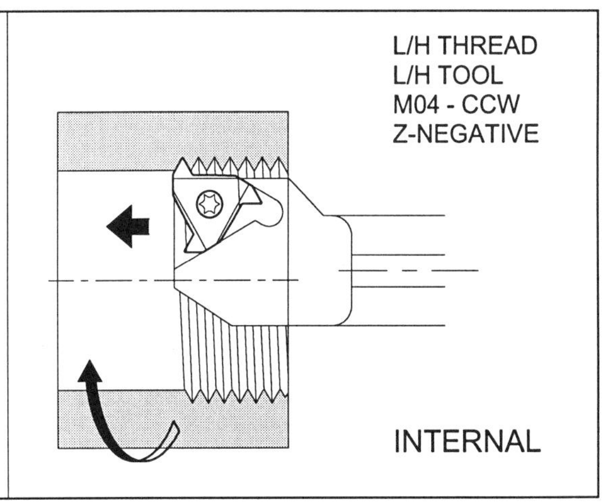

Sample Thread Evaluation

The external thread shown in the illustration is a common thread, specified in imperial measurements. It will be used to illustrate the different threading techniques, namely the long method using the G32 command, the original simple threading cycle G92 ('box' cycle), and the most modern multiple repetitive cycle G76. The G76 method will be shown in two versions - using one block and using two blocks of data input, depending on the control model. Only threading program is covered.

Initial Data

First, collect necessary drawing data and select the most suitable machining method:

- **Units:** **Imperial (inches)**
- **Size:** **2.25 external diameter - 18 TPI**
- **Pitch:** **1 / 18 = 0.055556 = 0.0556**
- **Length:** **1.75 inches**
- **Hand:** **Right hand thread - Z-negative**
- **Tool:** **Right hand (rear lathe mount)**
- **Spindle:** **M03 (CW)**

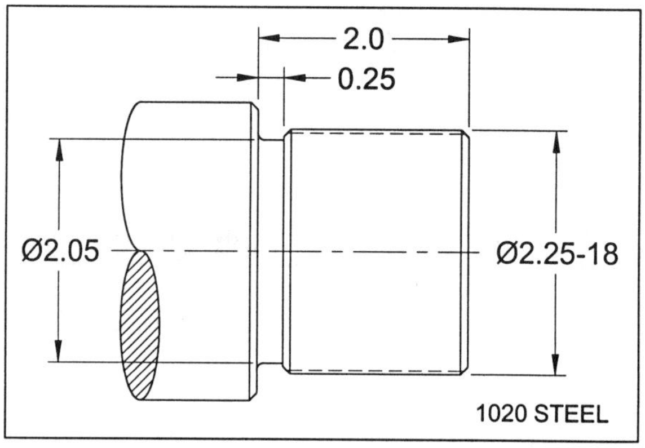

Cutting Conditions

Next, calculate the thread depth, using the constant of 0.613 introduced earlier:

Depth = 0.613 / 18 = 0.0341

In order to allow space for acceleration and deceleration, the front clearance will be 0.15 (over four times the pitch), the end clearance will be past the thread length, close to the shoulder. Z-1.95 will provide enough space for deceleration. In both cases, the calculations consider the command point of the insert as its left side at the bottom tip.

For mild steel (1020), suitable surface speed will be around 350 ft/min, used for spindle speed calculation *S* in the program:

$$r/min = \frac{ft/min \times 12}{\pi \times D}$$

r/min = 350 x 12 / (3.14 x 2.25) = 594 r/min = S594

Feedrate per revolution is the thread lead: 1 / 18 = 0.0556 = F0.0556

Although this particular thread does not present any problems, it may be necessary to check that the feedrate per minute does not exceed the machine capabilities. For example, if maximum cutting feedrate is given as 400 in/min, *594 × 0.0556* is equivalent to only 33 in/min, causing no problem.

Finally, the programmer selects five threading passes, using modified infeed at 29°. The number of passes is not required for G76 cycle, but it is required for G32 and G92 methods, also discussed in this chapter.

Number of Threading Passes

Determining the number of threading passes largely depends on the material machined, material of the threading tool, as well as cutting conditions and setup rigidity. Many cutting tool catalogues suggest the number of threading passes and even supply depth of each pass for different threads. For the sample thread, five passes should result in a high quality thread. These passes will be distributed over the thread depth (0.034 inches), which means between the top diameter 2.25 and bottom diameter of 2.182 (2.25 - 2 x 0.034 = 2.182).

Keep in mind that the depth of each pass has to decrease, in order to provide the best chip load with increasing thread depth. To calculate the depth of each pass, a special formula can be used, based on the number of passes calculated. One manufacturer provides these formulas, used with the sample thread data (several - *but not all* - letters correspond to cycle data used later in this chapter):

D = Thread diameter ... **2.25**
X = Final diameter ... **2.1819 (based on K calculation below)**
K = Depth of thread per side ... **0.0341 (0.613 / 18)**
Q = First pass depth ... **0.012**
R = Fixed finish allowance amount (after the last pass) ... **0.003**
N = Number of passes (round to integer if necessary) ... *to be calculated ...*

If both major and minor diameters are known, the depth of thread can be calculated:

K = (D - X) / 2

The depth *K* can also be calculated by using the constant ***0.613 / TPI*** or ***0.613 x Pitch***.

$$N = ((K - R) / Q)^2$$

N = ((0.0341 - 0.003) / 0.012)² = 6.7167 = 7 passes

Depth of each pass will use another formula - *n* is the pass number:

$$nth\ pass = Q \times \sqrt{n}$$

For the example, each single depth per pass will be measured from the major diameter:

Pass 1 = 0.012 × √1 = 0.0120	**... depth of cut = 0.0120**	**... cutting diameter = 2.25 - 2 x 0.0120 = 2.2260**
Pass 2 = 0.012 × √2 = 0.0170	**... depth of cut = 0.0050**	**... cutting diameter = 2.25 - 2 x 0.0120 = 2.2160**
Pass 3 = 0.012 × √3 = 0.0208	**... depth of cut = 0.0038**	**... cutting diameter = 2.25 - 2 x 0.0120 = 2.2084**
Pass 4 = 0.012 × √4 = 0.0240	**... depth of cut = 0.0032**	**... cutting diameter = 2.25 - 2 x 0.0120 = 2.2020**
Pass 5 = 0.012 × √5 = 0.0268	**... depth of cut = 0.0028**	**... cutting diameter = 2.25 - 2 x 0.0120 = 2.1964**
Pass 6 = 0.012 × √6 = 0.0294	**... depth of cut = 0.0026**	**... cutting diameter = 2.25 - 2 x 0.0120 = 2.1912**
Pass 7 = 0.012 × √7 = 0.0317	**... depth of cut = 0.0023**	**... cutting diameter = 2.25 - 2 x 0.0120 = 2.1866**
Final pass = 0.0317 + 0.003	**... depth of cut = 0.0030**	**... cutting diameter = 2.25 - 2 x 0.0120 = 2.1806**

There are a few problems using this - and similar - formulas. Although mathematically correct, one problem with such a formula is that it often generates more passes than practically needed. Compare the depth change between pass 1 and pass 2 (0.0070 difference) with passes 6 and 7 (0.0003 difference). In majority of material, the threading tool can thread deeper without a loss of quality. Another problem - a minor one - is that the last pass diameter does not match the initial calculation. This is due to initial rounding from the calculated 6.7167 passes to 7 passes. This error can be much higher. There is one more problem - the final pass has been selected as 0.003, whereby the depths of the last three passes are actually smaller than this amount. Use a formula as a guide, but with common sense.

If we discard the formula as impractical, what is the alternative calculation? The formula shown above depends on the *calculated* number of passes (7+1=8), not the number of passes preferred by the programmer, which is often more realistic. In the example, if only five passes are required, the formula will ***not*** work. As five passes will do the job, each depth has to be calculated separately. The technique is quite simple - use a *trial-and-error* method. Here is how it works:

First determine the single depth of thread (0.0341), then decide on the first pass depth (0.012), and the number of passes required (5). For passes 2 to 5, select progressively smaller cutting depth. The goal is to add all depth together and come up with the single depth of thread. Chances are that the first time you will not succeed - that is to be expected. After two or three trials, the results will match:

Pass 1 = 0.0120 ... cutting diameter = 2.25 - 2 x 0.0120 = 2.2260
Pass 2 = 0.0090 ... cutting diameter = 2.25 - 2 x 0.0090 = 2.2080
Pass 3 = 0.0070 ... cutting diameter = 2.25 - 2 x 0.0070 = 2.1940
Pass 4 = 0.0040 ... cutting diameter = 2.25 - 2 x 0.0040 = 2.1860
Pass 5 = 0.0021 ... cutting diameter = 2.25 - 2 x 0.0021 = 2.1818 *(0.0001 off because of rounding)*

This is not a scientific way - *it is a practical way* - it may not always work, but it should provide basis for similar approach, modified to suit a particular threading job.

Distribution of Depth Cuts

In the last section, the five passes were distributed as per desired depth of each pass. Only G32 command (*long hand* threading, also known as *block-by-block* threading method) and G92 cycle (*box type* threading cycle) require this type of calculations. Multiple repetitive threading cycle G76 determines the depth of each pass automatically - yes, using a software based formula.

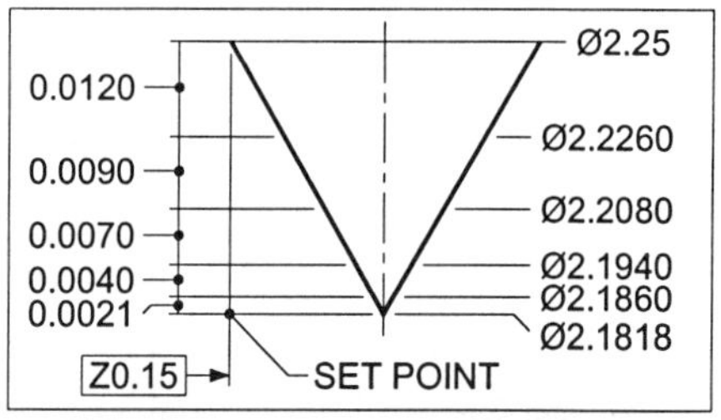

G32 Threading Method

The G32 is the oldest - and still *the most powerful* - of all single point threading commands. It is true that the programming can be *very* tedious; it is also true that any editing at the machine is virtually impossible. Having said that, it is equally true that the threading results can be absolutely outstanding, if the command is used correctly and for the right application.

The early control systems used the G33 command for threading that required one block entry for each threading motion. This command was later replaced with the G32 command, when Japanese controls became the standard. G33 is still used by some controls. In practical terms, both of these commands are identical in functionality and practical applications. If a particular control uses the G33 command instead of the standard G32 command used in the example, just replace G32 with G33 - other changes are not usually necessary.

Although the G32 command has been around much longer than any threading cycle, it still has advantages that no threading cycle can match. Here is only one, but it covers all advantages:

G32 offers absolute control over the thread, as each block is defined individually

Of course, there are also the inevitable disadvantages - the main disadvantages of the G32 command have already been addressed - long program output, often difficult calculations, and virtually zero convenience in program fine-tuning at the machine.

Radial Infeed Example

For the example shown, 18 TPI translates into *0.0556 pitch* and *0.0341 thread depth*. The first version will also be the simplest - not the best, just the simplest. It will use the G32 threading command to develop a threading program for the sample part, using five passes (as calculated previously), as well as using the *radial* infeed method - tool T05 will be used as an example within a program:

```
...                                 (==== G32 RADIAL INFEED METHOD ====)
N21 T0500                           (TOOL CHANGE - T05 ACTIVE)
N22 G97 S594 M03                    (CONSTANT SPINDLE SPEED - 594 R/MIN)
N23 G00 X2.5 Z0.15 T0505 M08        (START POINT OF THREAD)
N24 X2.226                          (*** PASS 1 - 0.0120 DEPTH ***)
N25 G32 Z-1.95 F0.0556              (2.2260 THREAD DIAMETER)
N26 G00 X2.5                        (RAPID RETRACT ABOVE THREAD)
N27 Z0.15                           (RAPID RETURN TO START POINT)
N28 X2.208                          (*** PASS 2 - 0.0090 DEPTH ***)
N29 G32 Z-1.95                      (2.2080 THREAD DIAMETER)
N30 G00 X2.5                        (RAPID RETRACT ABOVE THREAD)
N31 Z0.15                           (RAPID RETURN TO START POINT)
N32 X2.194                          (*** PASS 3 - 0.0070 DEPTH ***)
N33 G32 Z-1.95                      (2.1940 THREAD DIAMETER)
N34 G00 X2.5                        (RAPID RETRACT ABOVE THREAD)
N35 Z0.15                           (RAPID RETURN TO START POINT)
N36 X2.186                          (*** PASS 4 - 0.0040 DEPTH ***)
N37 G32 Z-1.95                      (2.1860 THREAD DIAMETER)
N38 G00 X2.5                        (RAPID RETRACT ABOVE THREAD)
N39 Z0.15                           (RAPID RETURN TO START POINT)
N40 X2.1818                         (*** PASS 5 - 0.0021 DEPTH ***)
N41 G32 Z-1.95                      (2.1818 THREAD DIAMETER - FINAL)
N42 G00 X2.5                        (RAPID RETRACT ABOVE THREAD)
N43 Z0.15                           (RAPID RETURN TO START POINT)
N44 X4.0 Z5.0 T0500                 (SAFE TOOL CHANGE POSITION)
N45 M01                             (==== END OF THREADING ====)
...
```

Although quite simple to program, the radial threading method will *not* yield quality results in most threads. In order to distribute the individual thread cuts for a *flank infeed* (compound infeed), some other considerations are necessary.

Flank Infeed Example

Flank infeed - or *compound infeed* - has been previously defined as the most desirable machining method for the majority of standard threads. This method uses the same calculations as the radial infeed method, but also requires a *shift* along the Z-axis for each depth pass. As five passes are required for the example, there will be five different starting positions in the Z-axis. Programmed thread diameters (X-axis) will not change.

Once the infeed angle has been selected (29° will be used for the example - angle *A*), all calculations will use trigonometric functions to find the Z-axis start position for each thread depth *(d_n)*.

Both the trigonometric formula and the Z-axis ***increments*** *(W_n)* are shown in the illustration.

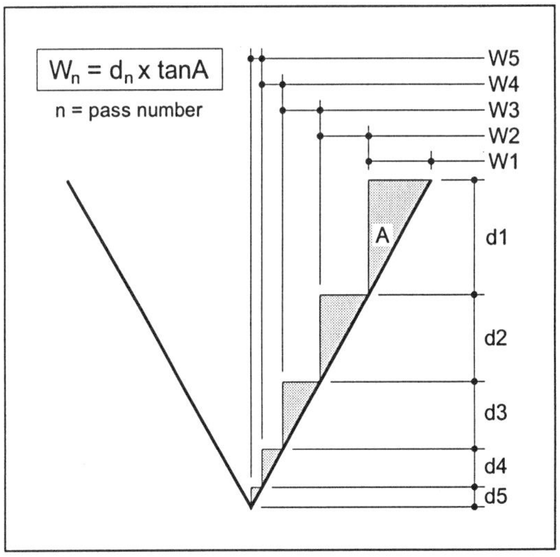

Note that the infeed angle is measured *per side*, making it equivalent to one half of the included angle, minus the clearance angle (60° / 2 - 1 = 29°). Using the formula above, the incremental shift *W* and the absolute *Z-position* for each threading pass can be calculated, with the following results:

Initial absolute Z-position = Z0.15			Infeed angle = 29°	
Thread pass number	**Thread pass diameter**	**Thread pass depth**	**Increment W**	**Z-position STARTS from ...**
Pass 1	2.2260	0.0120	0.0067	Z0.1500
Pass 2	2.2080	0.0090	0.0050	Z0.1567
Pass 3	2.1940	0.0070	0.0039	Z0.1617
Pass 4	2.1860	0.0040	0.0022	Z0.1656
Pass 5	2.1818	0.0021	0.0012	Z0.1678

Once these calculations are completed, the example using the radial method can be modified into the flank infeed method:

```
...                             (==== G32 FLANK INFEED METHOD ====)
N21 T0500                       (TOOL CHANGE - T05 ACTIVE)
N22 G97 S594 M03                (CONSTANT SPINDLE SPEED - 594 R/MIN)
N23 G00 X2.5 Z0.15 T0505 M08    (START POINT OF THREAD - PASS 1)
N24 X2.226                      (*** PASS 1 - 0.0120 DEPTH ***)
N25 G32 Z-1.95 F0.0556          (2.2260 THREAD DIAMETER)
N26 G00 X2.5                    (RAPID RETRACT ABOVE THREAD)
N27 Z0.1567                     (RAPID RETURN - START POINT PASS 2)
N28 X2.208                      (*** PASS 2 - 0.0090 DEPTH ***)
N29 G32 Z-1.95                  (2.2080 THREAD DIAMETER)
N30 G00 X2.5                    (RAPID RETRACT ABOVE THREAD)
N31 Z0.1617                     (RAPID RETURN - START POINT PASS 3)
N32 X2.194                      (*** PASS 3 - 0.0070 DEPTH ***)
N33 G32 Z-1.95                  (2.1940 THREAD DIAMETER)
N34 G00 X2.5                    (RAPID RETRACT ABOVE THREAD)
N35 Z0.1656                     (RAPID RETURN - START POINT PASS 4)
N36 X2.186                      (*** PASS 4 - 0.0040 DEPTH ***)
N37 G32 Z-1.95                  (2.1860 THREAD DIAMETER)
N38 G00 X2.5                    (RAPID RETRACT ABOVE THREAD)
N39 Z0.1678                     (RAPID RETURN - START POINT PASS 5)
N40 X2.1818                     (*** PASS 5 - 0.0021 DEPTH ***)
N41 G32 Z-1.95                  (2.1818 THREAD DIAMETER - FINAL)
N42 G00 X2.5                    (RAPID RETRACT ABOVE THREAD)
N43 Z0.15                       (RAPID RETURN TO START POINT)
N44 X4.0 Z5.0 T0500             (SAFE TOOL CHANGE POSITION)
N45 M01                         (==== END OF THREADING ====)
. . .
```

This method can be very time consuming but in certain cases, it may also be the only method available to program some special types of thread forms or machining applications.

Tapping with G32

Many control systems support a right-hand tapping cycle G84 and a left-hand tapping cycle G74. On CNC lathes, if tapping is necessary, the requirements are only for a single tapped hole, and the programming method reflects that. Rather than using a cycle (if available), CNC programmers use a regular programming approach, one instruction per block of program. There is nothing wrong with this approach, providing it is done correctly. The best way to illustrate this programming technique, is to show two identical methods - one wrong (very wrong, in fact), and the other one correct.

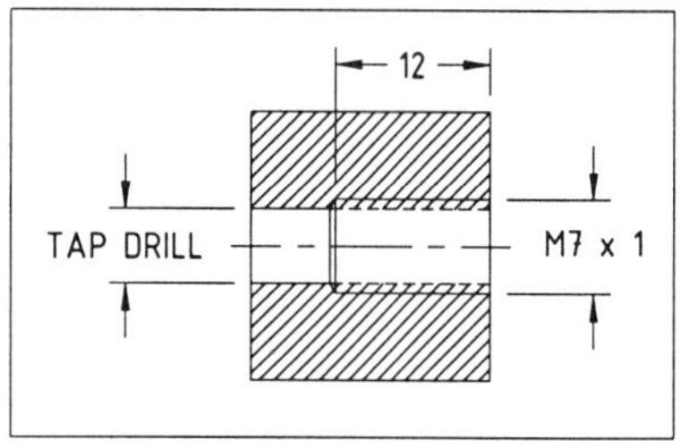

The illustration at right shows a typical metric thread - M7 x 1 - 12 mm deep. The tapping tool starts at Z3.0 and ends at Z-13.0, with all previous operations assumed to be completed.

In the table that follows, compare the program on the left with the program on the right. Both programs follow the same basic machining process:

Step 1 ...	**Start with normal spindle rotation**	**M03**
Step 2 ...	**Rapid to start position**	**X0 Z3.0**
Step 3 ...	**Feed-in to the thread depth**	**Z-13.0**
Step 4 ...	**Stop the spindle**	**M05**
Step 5 ...	**Reverse the spindle rotation**	**M04**
Step 6 ...	**Feed-out to the start position**	**Z3.0**
Step 7 ...	**Stop the spindle**	**M05**
Step 8 ...	**Resume normal spindle rotation (or end tool)**	**M03**

<table>
<tr><th>INCORRECT PROGRAM</th><th>Step #</th><th>CORRECT PROGRAM</th></tr>
<tr><td>N40 (G21) T0600</td><td>Tool change</td><td>N40 (G21) T0600</td></tr>
<tr><td>N41 G97 S600 M03</td><td>Step 1</td><td>N41 G97 S600 M03</td></tr>
<tr><td>N42 G00 X0 Z3.0 T0606 M08</td><td>Step 2</td><td>N42 G00 X0 Z3.0 T0606 M08</td></tr>
<tr><td>N43 <u>G01</u> Z-13.0 F1.0</td><td>Step 3</td><td rowspan="2">N43 <u>G32</u> Z-13.0 F1.0 M05</td></tr>
<tr><td>N44 M05</td><td>Step 4</td></tr>
<tr><td>N45 M04</td><td>Step 5</td><td rowspan="2">N44 Z3.0 M04</td></tr>
<tr><td>N46 Z3.0</td><td>Step 6</td></tr>
<tr><td>N47 M05</td><td>Step 7</td><td>N45 M05</td></tr>
<tr><td>N48 M03</td><td>Step 8</td><td>N46 M03</td></tr>
<tr><td>...</td><td></td><td>...</td></tr>
</table>

There are several major flaws in the left program. Although the contents look correct, and in a sense it *is* correct, it fails to address several realities:

- **Program will scrap the part when run in single block mode**
- **Program can be run at any feedrate override setting**
- **Program can be interrupted by pressing the feedhold button**

Any one of the three situations will cause a scrap part, stripped or broken tap, or both. The version on the right addresses all three issues:

- **Single block can be used safely, because M05 and M04 are programmed *together* with the tapping motion**
- **Feedrate override setting and feedhold will be ignored when G32 command is in effect**

This is just another example, where G32 can be used very efficiently in a part program.

G92 Threading Method

Look a few pages back, at the G32 programming method using a radial infeed. It does not take much to discover the great number of repetitions in that program. In fact, if there are significant repetitions for five passes, imagine how many repetitions would be in thread program with ten or more passes.

A typical threading pass has four individual parts, as described earlier. One of these parts is the pass diameter to be threaded. Once the new thread diameter has been programmed, it is followed by three block that never change, regardless of the number of passes. Of course, the more passes, the more repetitious blocks. A simple threading cycle G92 has been developed to address this issue.

In many ways, the G92 command is a very poor cousin of the G32 block-by-block method. It does shorten the threading program, and it can shorten the whole program quite significantly. On the other hand, it does not offer any extra features than those available in the G32 mode. In fact, you can do a lot *less* with the G92 than you can do with G32. On the other hand, for a very narrow field of jobs, the G92 threading command presents one benefit:

G92 'box' threading cycle is suitable only for *radial* infeed threads (also known as *straight* or *plunge* type infeed)

The main disadvantage of the G92 threading cycle is its virtually non-existing flexibility. Because of the four main thread motions of each pass create a rectangular area, this cycle is often called the 'box' threading cycle. The following program is nothing more than the G32 (radial) program modified.

In essence, G92 eliminates the repetitious motions inherent to each threading pass. Its programming format is quite simple:

```
G92 X.. Z.. F..
```

☞ where ...

X	=	Diameter of the current threading pass
Z	=	End position of the thread
F	=	Thread lead (feedrate)

```
...                                  (==== G92 RADIAL INFEED METHOD ====)
N21 T0500                            (TOOL CHANGE - T05 ACTIVE)
N22 G97 S594 M03                     (CONSTANT SPINDLE SPEED - 594 R/MIN)
N23 G00 X2.5 Z0.15 T0505 M08         (START POINT OF THREAD)
N24 G92 X2.226 Z-1.95 F0.0556        (*** PASS 1 - 0.0120 DEPTH ***)
N25 X2.208                           (*** PASS 2 - 0.0090 DEPTH ***)
N26 X2.194                           (*** PASS 3 - 0.0070 DEPTH ***)
N27 X2.186                           (*** PASS 4 - 0.0040 DEPTH ***)
N28 X2.1818                          (*** PASS 5 - 0.0021 DEPTH ***)
N29 X4.0 Z5.0 T0500                  (SAFE TOOL CHANGE POSITION)
N30 M01                              (==== END OF THREADING ====)
...
```

For those interested in getting more from this cycle, yes, there is a way to program the G92 threading cycle and get a flank infeed - with some drawbacks. As it is not possible to change the Z-start position within the cycle itself, it means the Z-position would still have to be calculated, and the cycle would have to be interrupted after each pass, defeating its rather short form.

One practical use of this cycle may be for roughing out a large portion coarse threads, followed by finishing thread operations (using the same tool or a different tool).

G76 Threading Method

By today's programming standards, the G76 multiple repetitive threading cycle is often considered the best in its class - and quite rightly. That does not mean the G76 command is a solution for *all* threading applications, far from it. At the same time, it can handle most of all threading jobs, external and internal, making it not only a very powerful feature of the CNC system, but also provides easy changes (fine-tuning) at the CNC machine. G76 threading command offers very fast and very efficient threading method and every programmer should know its capabilities. In current CNC programming environment, using the G76 cycle for most of the threads machined is virtually expected.

Programming Format

Multiple repetitive cycle G76 is one of several cycles that can be programmed in two input formats. One format is for older but still common control systems (10/11/15), using a single block program input. The other format is used for more recent control systems (0/16/18/20/21/+), where the G76 definitions are spread over two blocks of program input.

Regardless of the actual input format, G76 threading cycle offers unparallel flexibility at the machine, offering minor adjustments with major results.

One-block format is used for Fanuc models 10/11/15:

```
G76 X.. Z.. I.. K.. D.. A.. F.. (P..) (Q..)
```

☞ where ...

X	=	Diameter of the last threading pass
Z	=	Position indicating the thread end
I	=	Amount of taper over the total length
K	=	Single depth of the thread - positive
D	=	Depth of the first threading pass - positive
A	=	Included angle of the insert - positive
F	=	Cutting feedrate = thread lead
P	=	Infeed method (one of four) - positive
Q	=	Shift in degrees for multiple start threads (Fanuc 15)

Based on the program data entries, the threading program for the example is quite short. Unlike another short program - using G92 cycle - G76 offers quick changes to any programmed input data:

```
...                                        (==== G76 FLANK INFEED METHOD 1 ====)
N21 T0500                                  (TOOL CHANGE - T05 ACTIVE)
N22 G97 S594 M03                           (CONSTANT SPINDLE SPEED - 594 R/MIN)
N23 G00 X2.5 Z0.15 T0505 M08               (START POINT OF THREAD)
N24 G76 X2.1818 Z-1.95 I0 K0.0341
    D0120 A60 F0.0556                      (*** AUTOMATIC CALCULATIONS ***)
N25 G00 X4.0 Z5.0 T0500                    (SAFE TOOL CHANGE POSITION)
N26 M01                                    (==== END OF THREADING ====)
...
```

All thread related data is contained in a single block. If needed, the CNC operator will make some changes at the machine, to fine-tune the thread cutting process. Most likely, the most common change will be the *D* setting, which indirectly controls the number of threading passes.

Control systems that were developed after the 10/11/15 controls, use a two-block input format.

Two-block format is used for Fanuc models 0/16/18/20/21+:

```
G76 P.. Q.. R..
G76 X.. Z.. R.. P.. Q.. F..
```

☞ First block:

P	=	... is a six-digit data entry in three pairs: Digits 1 and 2 - number of finishing cuts (01-99) Digits 3 and 4 - number of leads for gradual pull-out (0.0-9.9 times lead), no decimal point used (00-99) Digits 5 and 6 - angle of thread (00, 29, 30, 55, 60, 80 degrees only)
Q	=	Minimum cutting depth (positive radial value - no decimal point)
R	=	Fixed amount for finish allowance (decimal point is allowed)

Second block:

X	=	(a) Last diameter of the thread (absolute diameter) *... or ...* (b) The distance from the start point to the last thread diameter (incremental)
Z	=	End of thread along the Z axis (can be an incremental distance W)
R	=	Radial difference between start and end positions of the thread at the final pass (R0 used for straight thread can be omitted)
P	=	Height of the thread (positive radial value - no decimal point)
Q	=	Depth of the first threading pass (positive radial value - no decimal point)
F	=	Feedrate of the thread (same as the thread lead)

Note that infeed method *P* or the multiple thread cutting shift *Q* are not available for the two-block cycles. Check the system parameters for mode switching (for example, FS21 to FS15). Also, make sure to understand the difference between P-Q-R words in the first block and P-Q-R words in the second block - they *do* have different meaning.

The previous one-block format can be modified to a two-block format, with a few extras:

```
...                                  (==== G76 FLANK INFEED METHOD 2 ====)
N21 T0500                            (TOOL CHANGE - T05 ACTIVE)
N22 G97 S594 M03                     (CONSTANT SPINDLE SPEED - 594 R/MIN)
N23 G00 X2.5 Z0.15 T0505 M08         (START POINT OF THREAD)
N24 G76 P010060 Q003 R0.002
N25 G76 X2.1818 Z-1.95 R0 P0341
    Q0120 F0.0556                    (*** AUTOMATIC CALCULATIONS ***)
N26 G00 X4.0 Z5.0 T0500              (SAFE TOOL CHANGE POSITION)
N27 M01                              (==== END OF THREADING ====)
...
```

The subjects in this chapter describe three methods (commands) that relate to single point threading. These three methods are different but share several features when used in the part program:

- Feedrate override will be ignored
- Spindle speed override will be ignored
- Feedhold function will be ineffective

The next chapter covers some restrictions that may be imposed in certain threading operations.

17 RESTRICTIONS IN THREADING

Single point threading can be a relatively easy job for both programming and machining, *if all goes well*. Fortunately, that is the case most of the time, but there are many situations that may present potential problems. At least two of them can be prevented:

- **Programming spindle speed incorrectly**
- **Programming cutting feedrate incorrectly**

The first item relates to spindle speeds that are either too high or too low for the given thread, the second item relates to certain threads being cut over a long length. Both potential problems can be eliminated mathematically, during the part programming process.

Thread Programming Basics

Every CNC machine tool is designed within certain parameters that relate to its various dimensions and capacities. On an every day basis, the part programmer must generally consider two of these parameters in most machining operations, but particularly in single point threading:

- **Range of spindle speeds**
- **Maximum cutting feedrate**

For typical turning and boring, the actual programmed spindle speed range and cutting feedrate fall comfortably within the machine specifications. Consider these features of medium to large size CNC lathes, as defined by the machine manufacturer (samples only):

Spindle Speed	Minimum	Maximum
	125 r/min	3000 r/min

Maximum Cutting Feedrate	Metric	Imperial
	6 350 mm/min - X-axis 12 700 mm/min - Z-axis	250 in/min - X-axis 500 in/min - Z-axis

Note that the imperial amounts are equivalent to the metric amounts. In the chart are only examples that will be used for further calculations of various threads. Also note, that some machines may have a lower maximum feedrate for the X-axis than for the Z-axis - *this may be a very significant factor*. Use the chart only as a reference and match it to the specifications of your CNC lathe.

In threading, these machine specifications are more important than in turning or boring, as they also define limits to what *sizes* of threads can be cut. For simplicity, there is only one spindle speed range used for the examples.

Spindle speed for single point threading (and for tapping) should always be programmed in the G97 mode - *constant spindle speed* (in revolutions per minute). Cutting feedrate needs some evaluation.

Threading Feedrate

Cutting feedrate for threading and tapping *is always the amount of the thread lead*, not the pitch. The confusion between the terms is not excusable, yet can easily become a cause of a major problem.

- **LEAD is defined as the distance the tool will advance along an axis in one spindle revolution**
- **PITCH is defined as the distance between corresponding points of two adjacent threads**
- **LEAD equals PITCH only for single start threads (including the majority of taps)**

As explained in the previous chapter, the feedrate for threading on CNC lathes is always programmed in *feedrate per revolution*, regardless of the machine specifications, which means we have to know the *lead of the thread*. In Imperial units, the thread is defined by the number of threads per inch *TPI* and the number of starts *n* (only two or more starts are specified in the drawing).

The formula for threading feedrate *F* is:

F = 1 / TPI × n = Lead = *Feedrate per revolution*

Since the majority of threads are threads with a single start, it is easy to understand the cause of a possible confusion between the terms *pitch* and *lead*.

The earlier chart for machine specifications lists the *feedrate per minute*, not the feedrate per revolution. The standard formula for calculating the feedrate per revolution is:

Feedrate per revolution = Feedrate per minute / Revolutions per minute

This is also the same formula applicable to calculations of the *maximum* feedrate per revolution, based on the *maximum* cutting feedrate of the machine and the *programmed* revolutions per minute:

MAX feedrate per revolution = MAX feedrate per minute / PROGRAMMED revolutions per minute

Standard Example

The drawing at right shows only the important dimensions of a simple standard thread to be cut in a material with excellent machinability. The spindle speed will be programmed at 1300 r/min.

Can the thread be cut on the machine specified in the chart? Such a question can be answered by a simple calculation. The result of this calculation *must be a lower value* than the maximum cutting feedrate of the machine, which is 250 in/min for the X-axis and 500 in/min for the Z-axis.

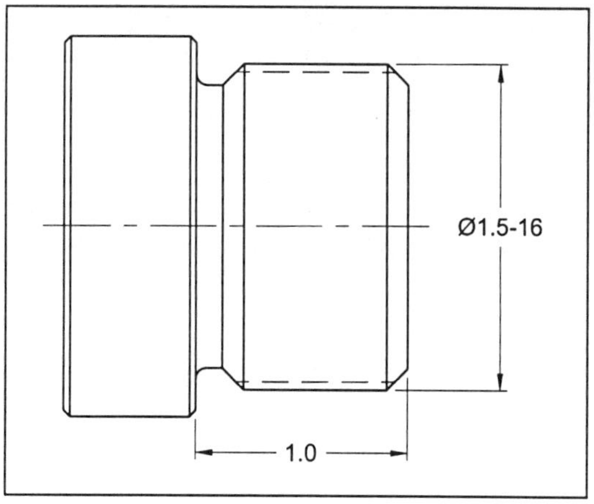

To guarantee thread integrity, always use the *lower* maximum feedrate !

Maximum machine feedrate is the lower value of 250 in/min, used for the calculations. The given thread has 16 threads per inch and a single start, so the threading feedrate (in/rev) is:

$$F = 1 / 16 \times 1 = 0.0625$$

Spindle speed has been selected at 1300 r/min. In order to find out if the machine can handle such data from the part program, we have to compare the respective feedrates per *minute*:

Maximum machine feedrate = 250 in/min

Effective cutting feedrate = 0.0625 × 1300 = 81.25 in/min

Providing the spindle speed has been selected correctly for the material used, the programmed data fall well within specified limits. Keep in mind that there is no need to consider any override settings at the machine, because they are all disabled in all available threading modes (G32, G92, G76).

Final Threading Program - Single Start

After all previous machining operations have been programmed (such as in blocks *N1* to *N30* - not shown), the threading section is quite short:

```
N31 T0300
N32 G97 S1300 M03                                  (SPINDLE SPEED 1300 RPM)
N33 G00 X1.7 Z0.2 T0303 M08                        (INITIAL CLEARANCE)
N34 G76 P011060 Q005 R0.002                        (G76 - LINE 1)
N35 G76 X1.4234 Z-0.95 R0 P0383 Q0100 F0.0625      (G76 - LINE 2 - 0.0625 IPR)
N36 G00 X5.0 Z3.0 T0300                            (TOOL CHANGE CLEARANCE)
N37 M01
```

The program reflects the selections relevant to the topic discussed, as well as other calculations.

Special Example

The drawing at right shows a small - but *very significant* - modification to the previous thread - the number of thread starts has changed from one to four, while other dimensions remain the same.

How does this change affect the program?

For multi start threads, the feedrate must change, because the thread lead will be different - also, each thread start will have a unique Z-axis location. There are several ways to program a multi start thread, this is just one of the methods.

The pitch of the thread does not change - *only its lead* - so the cutting feedrate must change as well:

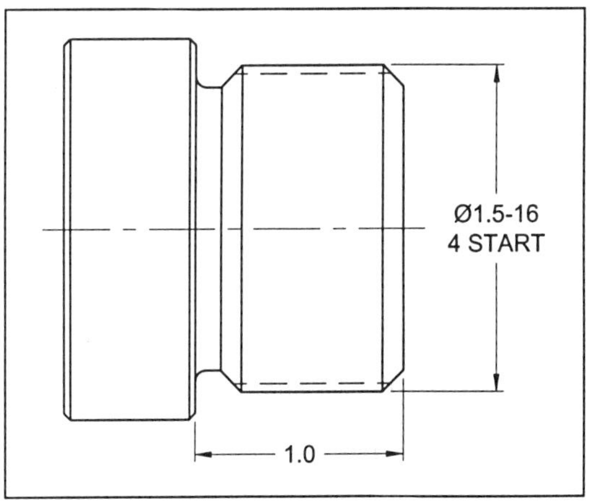

$$F = 1 / 16 \times 4 = 0.25$$

The machine spindle speed remains at 1300 r/min. In order to find out if the CNC lathe can handle cutting the thread at this spindle speed, we have to compare the respective *feedrates per minute*:

Maximum machine feedrate = 250 in/min

Effective cutting feedrate = 0.25 × 1300 = 325 in/min

For this thread, the 1300 r/min is *excessive*, even when the same spindle speed presented no problem for the similar thread with a single start. The thread lead is indirectly defined by the drawing and cannot be changed, which leaves the spindle speed alone to be decreased.

Rather than guessing, and possibly making an error, *always calculate* the maximum spindle speed for the programmed feedrate (lead) - the general formula is:

MAX revolutions per minute = MAX feedrate per minute / PROGRAMMED feedrate per revolution

In this example, the maximum spindle speed *S* will be:

S = 250 / 0.25 = 1000 r/min

Program the actual cutting speed *S* a little below any calculated maximum spindle speed, to avoid possible spindle fluctuation - 950 r/min in this case (`G97 S950` in the program).

Final Threading Program - Multi Start

After all previous operations have been programmed (such as in blocks *N1* to *N30* - not shown), the threading section will be a bit longer than that for a single start thread:

```
N31 T0300
N32 G97 S950 M03                                (SPINDLE SPEED 950 RPM)
(START NUMBER ONE)
N33 G00 X1.7 Z0.2 T0303 M08                     (INITIAL CLEARANCE)
N34 G76 P011060 Q005 R0.002                     (G76 - LINE 1)
N35 G76 X1.4234 Z-0.95 R0 P0383 Q0100 F0.25     (G76 - LINE 2 - 0.25 IPR)
(START NUMBER TWO)
N36 G00 W0.0625                                 (INCR. SHIFT TO START 2)
N37 G76 P011060 Q005 R0.002                     (G76 - LINE 1)
N38 G76 X1.4234 Z-0.95 R0 P0383 Q0100           (G76 - LINE 2 - 0.25 IPR)
(START NUMBER THREE)
N39 G00 W0.0625                                 (INCR. SHIFT TO START 3)
N40 G76 P011060 Q005 R0.002                     (G76 - LINE 1)
N41 G76 X1.4234 Z-0.95 R0 P0383 Q0100           (G76 - LINE 2 - 0.25 IPR)
(START NUMBER FOUR)
N42 G00 W0.0625                                 (INCR. SHIFT TO START 4)
N43 G76 P011060 Q005 R0.002                     (G76 - LINE 1)
N44 G76 X1.4234 Z-0.95 R0 P0383 Q0100           (G76 - LINE 2 - 0.25 IPR)
N45 G00 X5.0 Z3.0 T0300                         (TOOL CHANGE CLEARANCE)
N46 M01
```

From the actual machining perspective, this program will completely finish each thread from its start position, before completing the threads from their subsequent starting positions. This method may work well for machining where tool wear does not present a major problem. However, the technique can be easily adapted to other methods of programming a multi-start thread.

Slow Spindle Speed

In a much less common case, there may not be *enough* of the available spindle speed, rather than too much - which means the slowest spindle of the machine (125 r/min in the example) may be *too high*.

When is such a situation possible?

Consider the same threading example, but with a very rare - yet quite possible - definition of a thread with not one, not four starts, but with full *forty eight* starts!

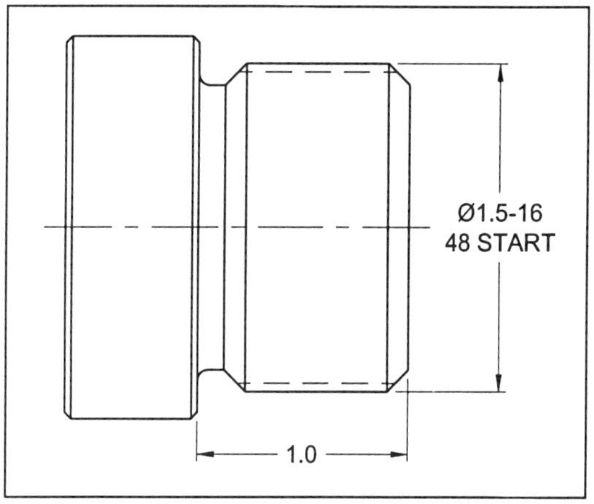

Multi start threads of this magnitude are not unusual in certain industries, for example, in optics and various instruments.

First, find the threading feedrate per revolution:

$$F = 1 / 16 \times 48 = 3.0$$

Next, calculate the *maximum* spindle speed *S* in revolutions per minute:

$$S = 250 / 3 = 83.333 = 83 \text{ r/min}$$

If the resulting spindle speed is *higher* than the minimum spindle speed of the CNC lathe, the thread can be machined without any difficulties. This is *not* the case for the example illustrated; the specifications of the CNC lathe described earlier are simply *not* suitable for precision machining of this particular thread. Even if possible to cut, a thread with unusually large number of starts will benefit from a more refined programming technique, one that minimizes a premature tool wear. Needless to say, expect a part program that will be significantly longer than more ordinary threading programs, regardless of the programming method used.

Metric Applications

A typical metric thread, similar to the one shown in imperial units, would be a thread identified as *M39×2* (with no spaces), for example. The letter *M* indicates a standard metric thread and is part of the designation, the number *39* is the thread nominal diameter (∅39 mm), the × symbol is a separator between the two parts, and the digit *2* is the thread *lead*. In a true metric standard, the metric designation will have a different form for multi start threads - for example:

- M39×2 thread with a single start: M39×2 ∅39 mm OD, 1 start, 2 mm lead, 2 mm pitch
- M39×2 thread with four starts: M39×8 P2 ∅39 mm OD, 4 starts, 8 mm lead, 2 mm pitch
- M39×2 thread with forty eight starts: M39×96 P2 ∅39 mm OD, 48 starts, 96 mm lead, 2 mm pitch

The maximum machine cutting feedrate is given in *mm/min* (for example, *6 350 mm/min* or *12 700 mm/min*). Programmed feedrate is given in *mm/rev*. All calculations will reflect these metric units.

Long Thread Programming

So far in this chapter, the focus was on controlling the spindle speed with a given feedrate. In part programming for CNC lathes, there is also a situation where the *feedrate* may have to be controlled in order to obtain desired results. The most common application that can benefit from such an adjustment is programming threads that are exceptionally long. CNC lathes capable of machining part lengths of between 40 and 160 inches are not uncommon. Suitable machine setup is assumed and is not discussed here. Note that this subject applies only to Imperial threads of any form, not metric threads. The reason is that metric threads are defined by their lead, whereby Imperial threads are defined by the number of threads per inch. Since the threading feedrate is always the thread lead, there is no rounding involved in metric threads, but some (not all) Imperial threads will use rounding when the threads per inch are converted to the thread lead (cutting feedrate).

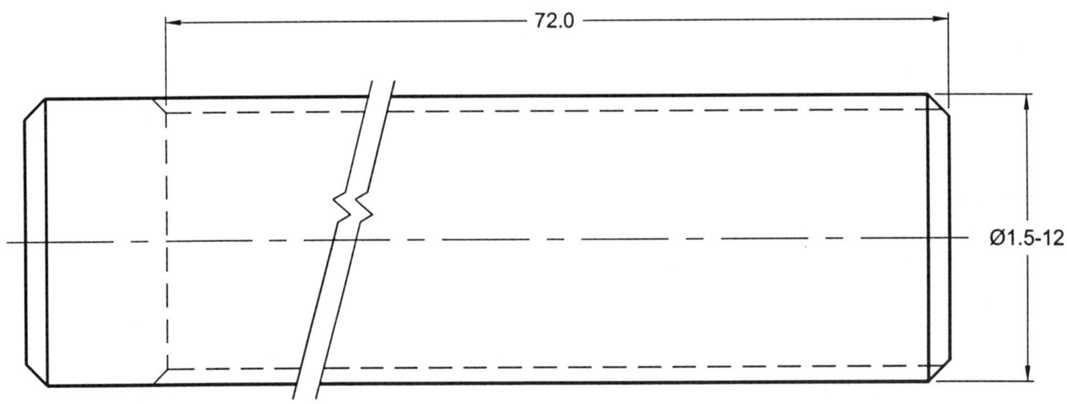

The simplified drawing above will serve as an example to illustrate the subject of this section. The thread could be a standard V-thread or a trapezoidal thread or any other form. The actual part program is not relevant, only the feedrate considerations in special single point threading.

Defining a Long Thread

In a machine shop, the term *long* presents a scenario of various interpretations in different situations. For setup purposes in lathe work, the length of a bar is measured relative to its diameter. In everyday threading operations, you will encounter most threads have a length relatively short to their diameter. These threads do not require any support. Long threads require support devices, such as a tailstock, a follower rest or a steady rest, in order to prevent deflection of the bar.

All CNC lathes provide a limited work area that is suitable for machining. Within this area, all threading, and other machining, takes place. A 0.25 bar can be considered long when its length exceeds 1-1.5 inches, depending on the material and setup used. The above illustration of the part provides all critical information:

Length of thread	Thread diameter	Threads per inch
72.0	1.5	12

With a suitable tooling and setup, it should not be difficult to make the threading program:

```
...
N31 T0300
N32 G97 S500 M03                                   (SPINDLE SPEED 500 RPM)
N33 G00 X1.7 Z0.2 T0303 M08                        (INITIAL CLEARANCE)
N34 G76 P011060 Q005 R0.002                        (G76 - LINE 1)
N35 G76 X1.4234 Z-72.0 R0 P0383 Q0100 F0.0833      (G76 - LINE 2 - 0.0833 IPR)
N36 G00 X5.0 Z3.0 T0300                            (TOOL CHANGE CLEARANCE)
N37 M01
```

Some minor improvements can be added or changes made, such as thread pullout and clearances, different infeed angle, etc. - apart from that, the program seems to be correct - *but is it?*

Lead Error

This is a single start thread *(n = 1)*, so the lead *L* of the thread will be equivalents to its pitch *P*:

$$P = 1 / TPI = 1 / 12 = 0.08333333$$

$$L = P \times n = 0.08333333 \times 1 = \text{Threading feedrate (in/rev)}$$

Neither calculation brings anything new to the subject - it is the same calculation that was used earlier in this chapter. The problem is not with the calculation itself but with its *rounding*.

Consider the programmed, expected and actual results in the form of a table:

Description	Calculation	Result
Programmed thread length	N/A	72.0 (*)
Programmed feedrate	N/A	0.0833
Expected number of threads	72.0 x 12 TPI	864.0
Actual number of threads	72.0 / 0.0833	864.345738
ERROR OVER THREAD LENGTH	864.345738 - 864.0	0.345738
ERROR OVER ONE INCH LENGTH	0.345738 / 72	0.004802

(*) = *front clearance in Z-axis is ignored*

The amount of error over the thread length is very large and must be eliminate (or severely minimized) for precision long threads. Look at two possible solutions (often incorrectly used):

- **Program the threading feedrate as F0.0832 ... in this case, the error over 1.0 inch is +0.019231**

- **Program the threading feedrate as F0.0834 ... in this case, the error over 1.0 inch is -0.009592**

Even without specific calculation details (use the above method), it is clear that both methods produce even worse results and neither one can be used to solve the problem. *The true solution is in changing the number of feedrate decimal places from four to six.*

Number of Decimal Places

Normally, the feedrate per revolution in Imperial mode (G20) is programmed with the maximum of *four* decimal places. This feature offers enough flexibility to the CNC programmer to fine-tune any feedrate as desired - except the feedrate for long threading. In this case, the *F* number in the program should contain up to *six* decimal places, for increased thread accuracy. Compare the table on the previous page with the modified table:

Description	Calculation	Result
Programmed thread length	N/A	72.0 (*)
Programmed feedrate	N/A	0.083333
Expected number of threads	72.0 x 12 TPI	864.0
Actual number of threads	72.0 / 0.083333	864.003456
ERROR OVER THREAD LENGTH	864.003456 - 864.0	0.003456
ERROR OVER ONE INCH LENGTH	0.003456 / 72	0.000048

(*) = *front clearance in Z-axis is ignored*

The amount of error over the thread length is virtually non-existent and the problem is solved.

In conclusion, only a few notes - the six decimal place accuracy is *not* required for threads whose lead can be divided into four or fewer decimal places. For example, *8 TPI = 1 / 8 = 0.1250* lead. Contrary to some beliefs, adding the extra zeros will *not* make the thread more accurate - this is a remnant of old methods defining tolerances by the number of decimal positions specified in the drawing. Today, this method of tolerancing is considered obsolete. Such trailing zeros should be omitted for clarity, for example, use `F0.125` instead of `F0.1250` or even `F0.125000`.

18 PRACTICAL THREAD MILLING

Cutting threads on CNC machines can take many forms. The typical methods include tapping, single point threading, even forming (threading die heads are not normally used in CNC work). If the conditions are favorable, one of the best methods of generating a high quality thread is - *to mill it*. Milling threads of many forms has proven to be one of the most effective and most cost-efficient methods of thread machining. Grinding threads is another available method of producing high precision threads, but it is time consuming and very costly. The area of thread milling can be fairly broad, yet the subject itself is not that difficult. First, some background and general considerations.

Thread Milling - General

The key to any type of thread milling is the availability of helical interpolation feature in the control system. This is the absolute must. It is not possible to program a thread milling toolpath without understanding the concept of helical interpolation. Even this knowledge cannot be used, unless the control system supports the helical interpolation feature in the first place. Helical interpolation is often an optional feature on most CNC systems - it may have to be purchased and installed on demand. A few controls provide this feature as standard, so it is always worthwhile to find out whether a particular machine (and its control system) supports - or does not support - this feature.

Note - it is actually possible to produce a thread milling cut by *simulation*. Instead of a smooth, continuous circular and linear motion, a series of very short line segments can be programmed instead. The shorter the lines, the better quality thread will be machined, but at the expense of a very long program. Some software provided by manufacturers of thread milling tools also supports simulation. If the control system has a feature called *macros* (typically *Fanuc User Macros B*), a fairly short program can be developed to perform such a simulation for virtually any thread. Using the macro feature is a realistic and practical alternative in cases where the helical interpolation option is not available.

Helical Interpolation

What exactly is *helical interpolation?* Experienced CNC programmers are already familiar with other forms of interpolation, namely *linear* and *circular* interpolation; so understanding helical interpolation should not be too difficult. In fact, helical interpolation is a *combination* of both - circular and linear - interpolations. It can be defined in a fairly simple statement:

> Helical interpolation is a simultaneous circular motion in two axes and one linear axis

It can also be added that the circular motion always takes place in the currently active plane; the linear axis is always perpendicular to the active plane. The rotating cutter is following the circular path (usually XY), while it moves up or down in the remaining axis (usually Z). Full XY-axes motion along 360° will produce one pitch along the Z-axis direction.

There are no special G-codes associated with helical interpolation - preparatory commands G02 and G03 are used in a similar fashion as in circular interpolation. The remaining axis (the third axis) defines the linear motion. The following table shows the relationships between the selected plane and helical interpolation commands as well as the programming format for each plane:

Plane	Circular motion axes	Linear motion axis	Program format
G17	in XY plane	along the Z-axis	G02 X.. Y.. Z.. I.. J.. F.. G03 X.. Y.. Z.. I.. J.. F..
G18	in ZX plane	along the Y-axis	G02 X.. Z.. Y.. I.. K.. F.. G03 X.. Z.. Y.. I.. K.. F..
G19	in YZ plane	along the X-axis	G02 Y.. Z.. X.. J.. K.. F.. G03 Y.. Z.. X.. J.. K.. F..

Similar to normal circular interpolation applications, instead of the arc vectors *IJK*, the direct radius address *R* can be used as well, providing the control system supports it.

Is Helical Interpolation Available ?

Without a doubt, thread milling is the most common application of helical interpolation. For most programmers, helical interpolation is not a daily routine, unless you are in the business that largely depends on milled threads. Only companies specializing in certain type of products will be using this feature frequently. For example, industries that manufacture fasteners, fittings, valves, pumps, and many others products will certainly benefit from this feature.

Thread milling may be the most common application of helical interpolation, but it certainly is not the only one. Helical interpolation can also be used for a number of non-thread-milling operations, for example, ramping into the material if plunging is either not possible or practical.

As helical interpolation as mostly an optional feature of the control system, how do we know the control supports this feature? Probably the easiest method to find out if the control has helical interpolation feature installed is to try one simple block of program in the MDI mode. Make sure the spindle is away from the machine zero in the X and Y axes, then enter these blocks:

```
G21   (or G20)
G91 G03 X-40.0 Y0 Z-2.0 I-20.0 J0                    ... or a similar statement
```

If the control accepts the command, you should see the actual motion by watching either the spindle motion or the position data on the control screen. If the helical interpolation option is not installed, the control will issue an alarm.

Benefits of Thread Milling

Thread milling is one of rather a few machining operations that offers a very extensive list of advantages and benefits, when compared with other methods of thread machining (or any machining).

The following list of benefits is fairly extensive, yet still far from being comprehensive:

- **External and internal threads can be cut with the same cutter**
- **Right hand and left hand threads can be cut with the same cutter**
- **Full 100% thread depth is possible with most thread milling cutters**
- **Reduced cycle time with higher cutting feedrates**
- **Smoother threads - minimizes or eliminates burrs**
- **Eliminates secondary operations - fewer tools used**

- **Increases precision and control when threading blind holes**
- **Increased accuracy and quality**
- **Spindle rotation does not change - spindle does not reverse**
- **No broken taps - hard material are easier to handle**
- **Better chip control - small chips eliminate jamming in the cutter flutes**
- **Only about 1/5 of power rating required**
- **Extended tool life**
- **Taper threads (pipe thread) can be machined**
- **Elimination of secondary operations**
- **Reduction of tooling and inventory costs**
- **Higher concentricity when compared with tapping**
- **... and many more**

When considering a thread milling operation, many users compare thread milling with tapping. Up to a certain level, the comparison is fair and objective - particularly when threading small holes that could also be tapped. The same comparison will not be objective for large holes or external threading, where tapping does not apply.

Selecting Tools

As with any other cutting tools, the first consideration when selecting a suitable thread mill is the material of the tool itself. The selection is between HSS (high-speed steel) tools and carbide tools. Without a doubt, carbide material is the preferred choice for production, as it benefits from the capabilities of modern machining centers - HSS is hardly ever used, but the current cutter development points to several alternatives to carbide. Overall, thread milling operations require a cutting tool that is already shaped in the same form as the thread to be cut. Thread milling is an operation that combines cutting and forming. The cutting action is generated by the tool motions in the program, the forming action is the result of the cutter geometry.

Initial Factors

Several factors relate to the thread design and the machining method - they have to be considered together, in order to achieve consistent results from one part to another. The group of primary importance contains four main features:

- **Material of the part to be machined**
- **Thread size of the part (specified thread diameter as per drawing)**
- **Lead (pitch) of the thread**
- **Full depth thread length**

Secondary factors are also important, as they include various tooling and machining decisions. Typical examples include programmed speeds and feedrates of the thread mill, the diameter of the cutter as it relates to the diameter of the thread, the number and type of flutes of the cutter, and some others.

In the area less related to programming but very important for machining, many thread milling cutters are available with built-in coolant channels (holes). These channels aim the coolant through the tool to the cutting edge(-s), extending the tool life and enhancing the thread surface finish.

Types of Thread Milling Cutters

Two general types of cutters are used in thread milling:

- **Single form thread mill**
  ... single tooth cutters
- **Multi-tooth thread mill**
  ... multi-tooth cutters

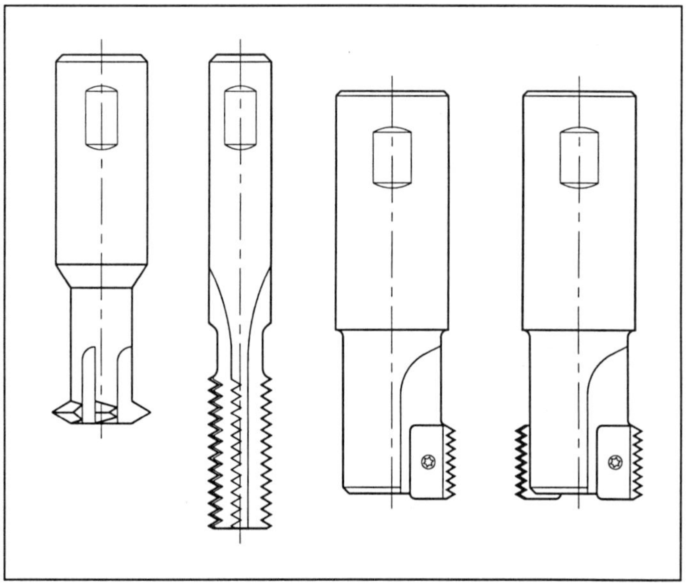

A single form tool can only cut one thread in a single pass, and must move around the part as many times as the number of threads required. This thread milling method is not very efficient for large number of parts. The greatest - and only - advantage of a single form thread mill is that it can cut any thread lead (pitch) that matches the angle of the tool (usually 60° V-threads). Because of a simple design, this type of tool is not very expensive.

In a multi-tooth cutter, the cutting teeth are arranged in a parallel design, rather than in a helical design that is typical to taps. The tool design reflects this requirement. Cost of multi-tooth cutter is much higher, for two reasons. One, the tool itself may cost 2-4 times more than a single form mill; two, each thread lead (pitch) requires a unique tool. Multi-tooth thread mills are designed to cut several threads at the same time - using increased feedrate is possible, as many cutting tips remove small amount of material each at the same time. The result is the ability of the tool to cut to the full depth in a single turn around the part. In spite of the initial higher costs, the efficiency of multi-tooth thread mills combined with their proper use will recover the tool investment very quickly.

Apart of the specific thread mill features mentioned, other considerations are the same as when selecting a standard end mill. Overall tool length, speeds and feedrates, flute design (spiral or straight), clearances and other machining considerations *do* influence the final thread quality. For example, spiral flutes allow the cutter to enter material gradually, decreasing the initial pressures on the tool.

Accuracy Issues

Even if taps can be used, thread milling will always produce better accuracy than the best taps. Part program controls the thread pitch, and when using cutter radius offset, the thread diameter can be adjusted at the machine with very high precision. Threading depth is uniform for each thread as well as over a large number of parts machined, and chip clogging is eliminated, also improving the thread surface quality. Surface finish on the threads is smoother than surface generated by tapping operations. If any of these benefits are required, particularly for a large volume machining, thread milling is certainly the machining method to consider.

Thread Mill Data

For practical CNC applications, the size and other data of a typical thread mill are important mainly in three areas - the cutter *diameter,* the maximum *length* of the cutter flutes, and the *pitch*. The pitch is the distance between threads. *Helix angle* is also an important part of the thread. A typical thread milling cutter is versatile, providing the thread profile and pitch of the thread and the cutter match. Both external and internal threads can be machined with the same cutter, regardless of their diameter. Common experience suggests that the machined thread should be about 20% larger then the diameter of the tool.

The flute length of the thread mill is determined by the tool manufacturer, so it is important to consider it, when purchasing the tool. There are many variations.

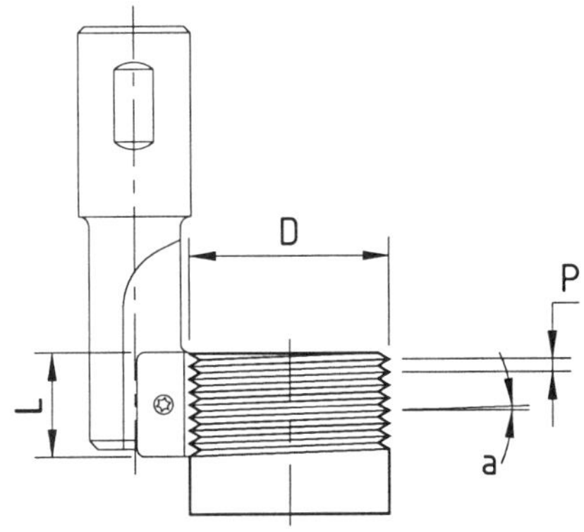

D = Thread diameter
P = Pitch (distance between threads)
L = Flute length
a = Helix angle

Cutting Direction

The title of this section does not seem to suggest any special treatment or any special approach to programming. Yet, the direction of cut is one of the most important subjects to consider when developing a thread milling toolpath. In addition to the tool rotation, cutting direction determines whether the thread will be milled in a climb milling mode or a conventional mode; it also determines the hand of the thread. These three subjects (with two choices each) are extremely critical for the part program development - even if they are listed under a rather unassuming heading of *'cutting direction'*.

External and Internal Thread Milling

The same thread milling cutter can be used for both external and internal thread milling, as long as the thread forms are identical. With the same cutter rotation (M03 for right hand tools), the direction of cutting (up or down) and the circular motion direction (G02 or G03) must be selected carefully.

Climb Milling and Conventional Thread Milling

Thread milling is just another form of milling - it shares exactly the same characteristics with contour milling - with focus on the cutting direction. The toolpath direction is always defined by following the tool motion in respect to the contour machined. Not surprisingly, the same considerations that are important for external and internal threading are also important for the type of cutting mode - climb or conventional.

Right Hand and Left Hand Thread Milling

Finally, whether a right hand thread or a left hand thread will be machined, also depends - yes - on the *same* three considerations described so far in this section.

External Thread Milling Illustrated

External thread milling in ***climb*** milling mode can produce either right hand or left hand threads by *changing the threading axis direction in G02 mode:*

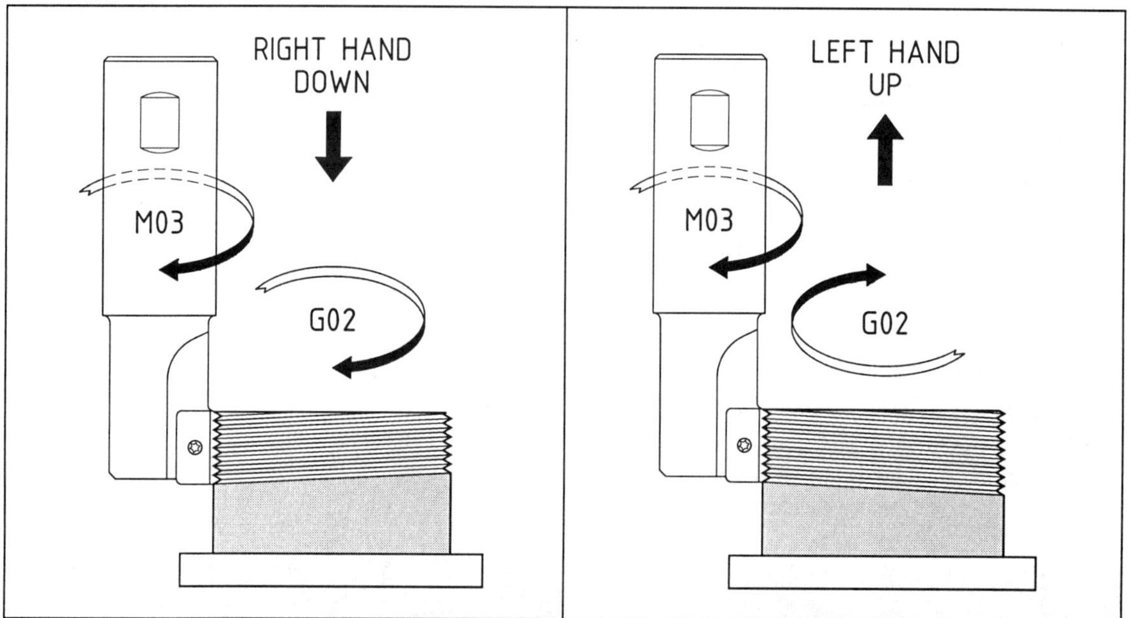

EXTERNAL THREAD - CLIMB MILLING METHODS

External thread milling in ***conventional*** milling mode can produce either right hand or left hand threads by *changing the threading axis direction in G03 mode:*

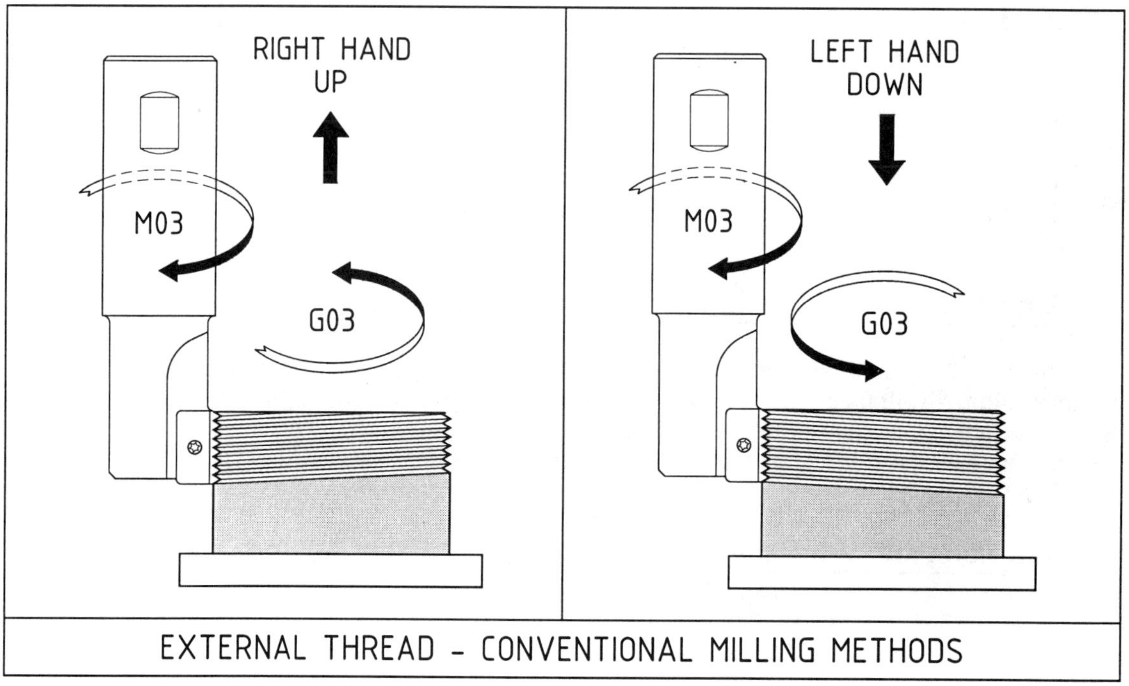

EXTERNAL THREAD - CONVENTIONAL MILLING METHODS

Internal Thread Milling Illustrated

Internal thread milling in ***climb*** milling mode can produce either right hand or left hand threads by *changing the threading axis direction in G03 mode:*

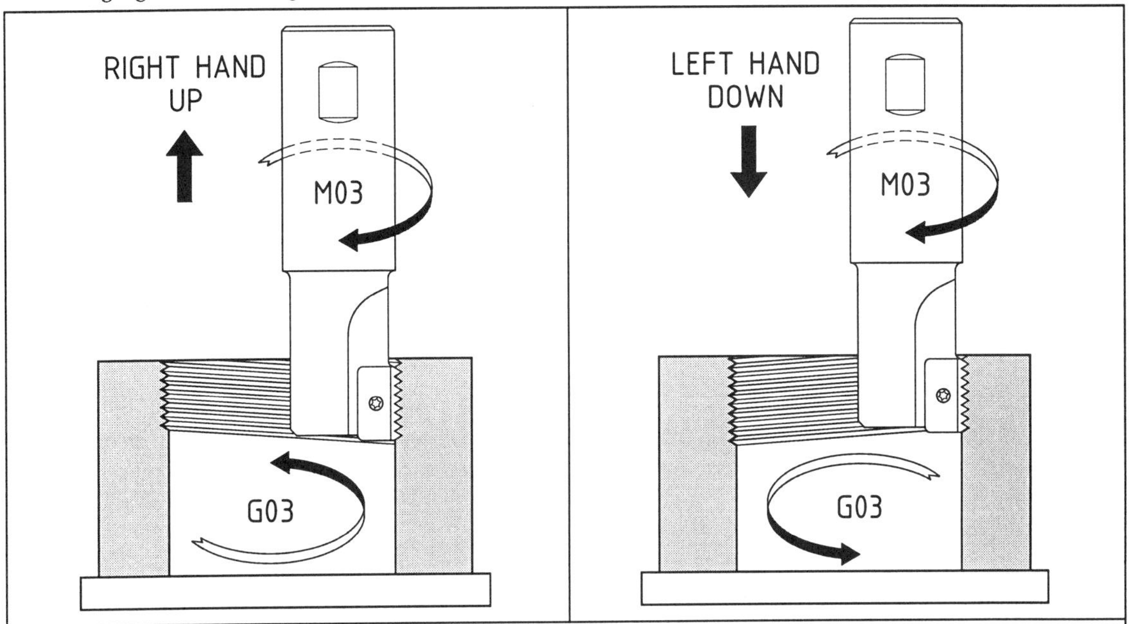

INTERNAL THREAD - CLIMB MILLING METHODS

Internal thread milling in ***conventional*** milling mode can produce either right hand or left hand threads by *changing the threading axis direction in G02 mode:*

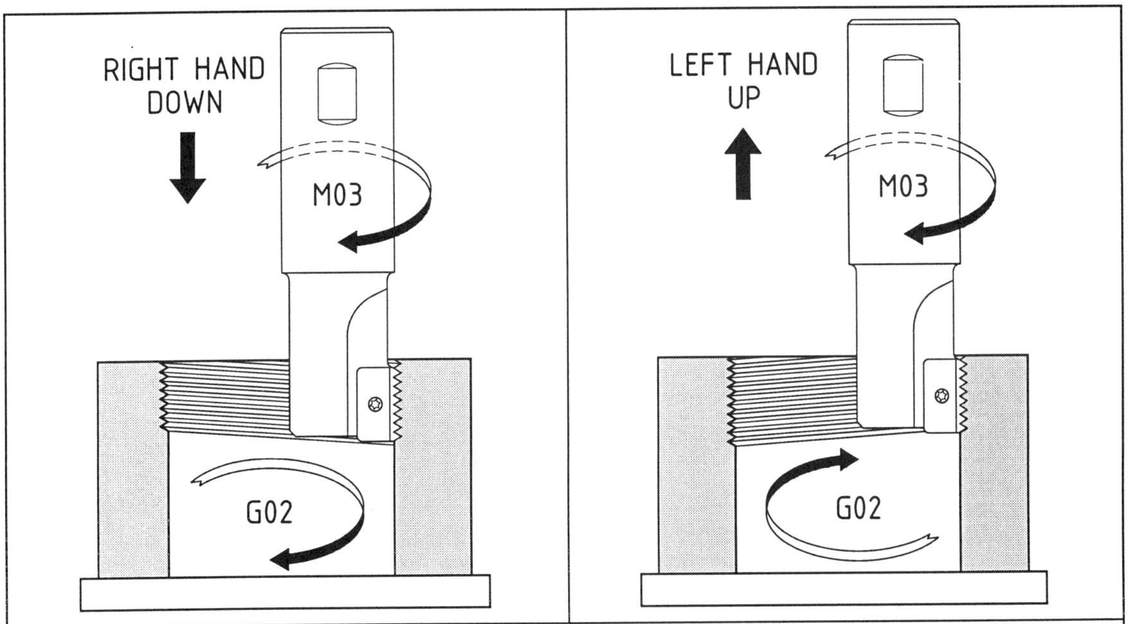

INTERNAL THREAD - CONVENTIONAL MILLING METHODS

All three related selections can be summed up in a simple table:

Standard *right hand* thread mill cutter used to cut ...	Spindle rotation	CLIMB MILLING		CONVENTIONAL MILLING	
		G-code / thread	Z-direction	G-code / thread	Z-direction
EXTERNAL RIGHT HAND thread	M03	G02	DOWN	G03	UP
EXTERNAL LEFT HAND thread	M03	G02	UP	G03	DOWN
INTERNAL RIGHT HAND thread	M03	G03	UP	G02	DOWN
INTERNAL LEFT HAND thread	M03	G03	DOWN	G02	UP

Both the illustrations and the summary table show cutting directions for all thread types.

Helix - Helical Curve

The helix is the single most important part of thread - any thread. By definition, a straight helix is a three dimensional curve that lies on a cylinder, created by a point path moving at a constant angle perpendicular to the plane axis. A tapered helix shares the same characteristics, but lies on a cone.

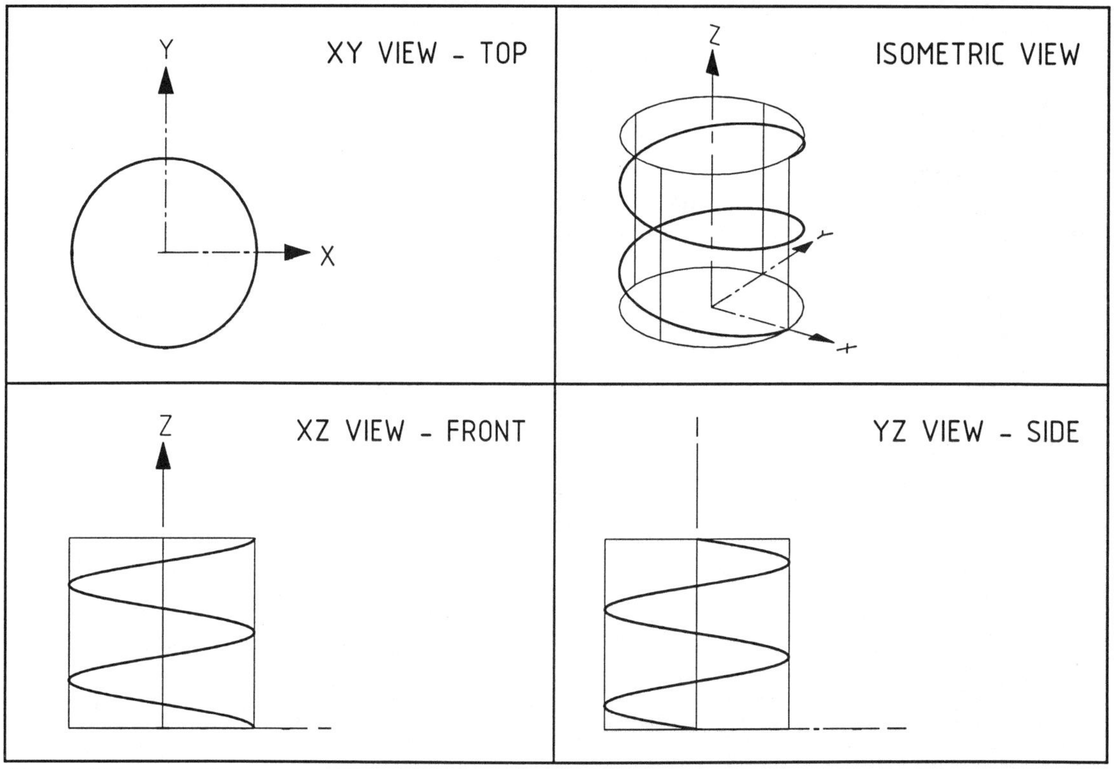

Milling a thread means following the circular curve in one plane that constantly changes along the axis that is perpendicular to this plane. Applied to helical interpolation, thread milling is a combination of a circular and linear motions.

The two illustrations on this page show views of a helical motion, applied at 360°.

In the illustration at right, the *top* and *side* views are shown with matching numbers. Each number represents a helical point at 30°. A spline representing the helix passes between through the points.

The illustration below is similar - it shows the same helix, but this time the view is *top* and *front*. Again, each helical point (there are twelve of them) represents a helical point at 30°. In addition, a flat view of the helix is also shown. A flat view is sometimes called a *rolled-out* view.

Regardless of the helix orientation, understanding the concepts of a helix as a curve will come very handy in helical interpolation, particularly if a partial helix is required (a helix that is less than 360°).

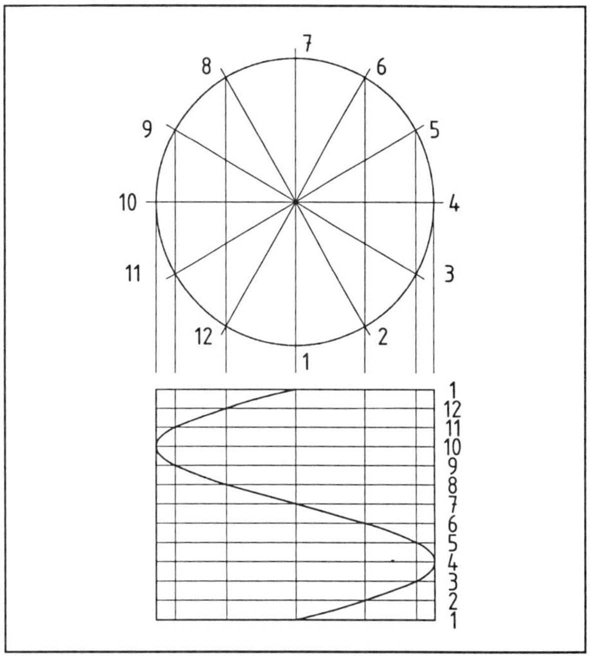

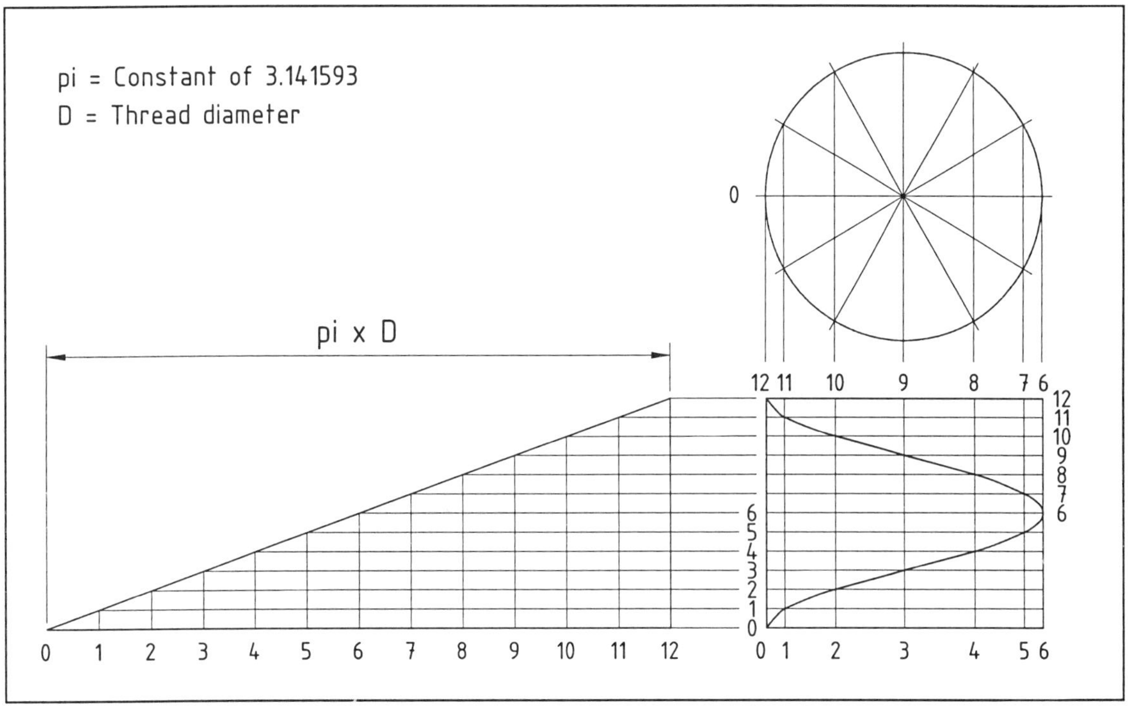

A partial helix is mainly applied to thread milling cuts used at the beginning or end of the full milling motions - they are so called *lead-in* and *lead-out* motions, already familiar from other chapters of this handbook. In the following sections, these lead-in and lead-out motions will be explained in detail, including practical examples.

Programming External Threads

In terms of thread milling, programming external threads and internal threads is very similar, with the major difference being the lead-in and lead-out motions. The four illustrations and the table presented in a previous section show all possibilities available for thread milling. A practical example could be developed for each variation, in Imperial and metric units, etc. One example of milling an external thread and one example of milling an internal thread should be sufficient to understand the concepts and the process.

This section covers the subject of ***external*** thread milling. For convenience, the drawing shown below will also be used for *internal* thread milling, described in the next section. Only relevant dimensions are shown. Two separate programs will be developed - one for external thread milling, the other for internal thread milling, both with suitable explanations.

The great majority of all thread milling applications takes place in the standard (default) G17 plane. This is also the plane used for both examples. For further details, see the chapter named *'Working with Planes'*.

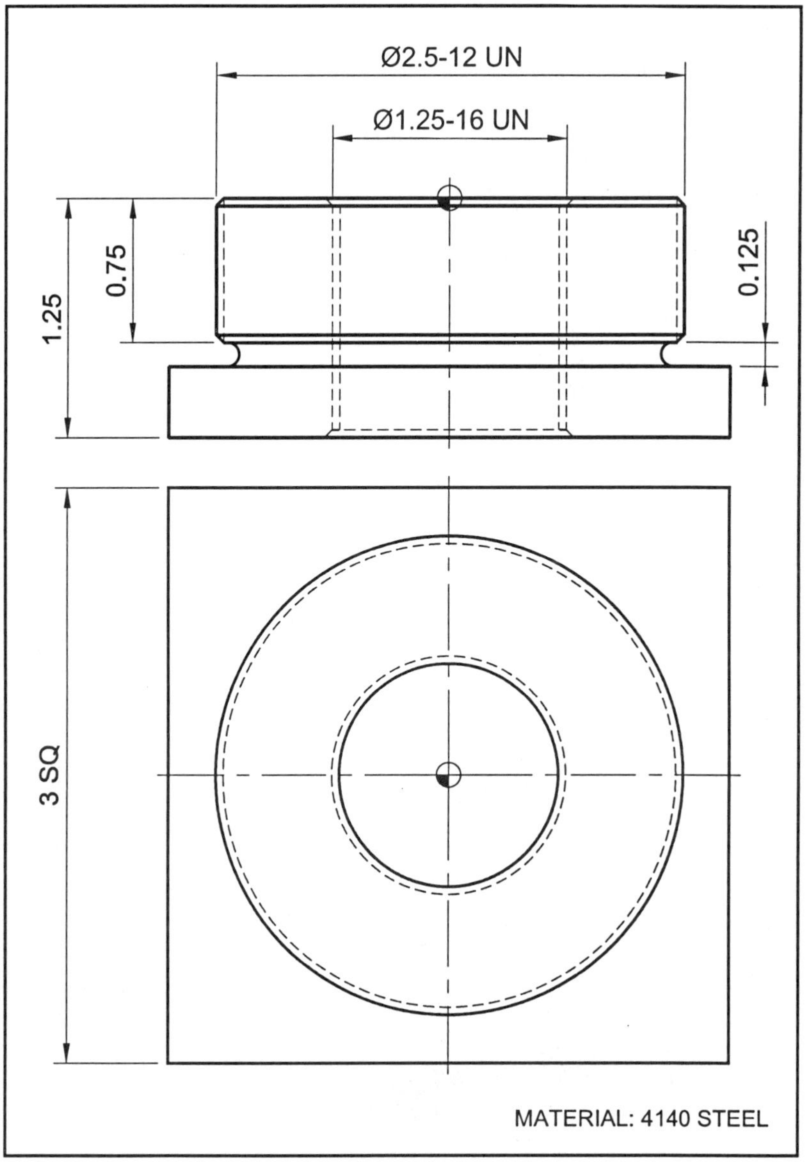

Tooling Selection

Selection of a thread mill cutter (also known as a *thread hob*), does not generally require any more effort than matching the thread (mainly its *pitch* and its *shape*) with the cutter specifications - they are the thread mill diameter and its flute length. For the *EXTERNAL* thread, a single *carbide* thread mill will be used in climb milling mode and cutter radius offset:

- **Right hand thread**
- **Carbide with four flutes**
- **Pitch = 0.08333 - 12 TPI**
- **Thread depth = 0.051**
- **Diameter = 0.625**
- **Thread length = 0.75**
- **Flute length = 1.125**

Cutting Conditions

Cutting conditions, namely spindle speed and cutting feedrate has to be selected by considering the material being cut and the material of the thread mill. In the example, a *four-flute* carbide cutter will be used to machine *4140 steel*. Depending on the exact conditions, the following selection may change, but for the example, the cutting speed used will be 325 feet/min with 0.003 feedrate per tooth. Based on the standard formula, the spindle speed will be:

$$\text{r/min} = \frac{325 \times 12}{\pi \times 0.625} = 1986 \text{ r/min}$$

With the feedrate per tooth of 0.003 and four flutes, the nominal cutting feedrate will be:

F_1 = 1986 × 0.003 × 4 = 23.8 inches per minute

This feedrate should be adjusted to the *center* of the thread mill, using another formula:

$$F = \frac{F_1 \times (D_t + D_c)}{D_t}$$

☞ *where ...*

F	=	Programmed feedrate
F_1	=	Nominal feedrate
D_t	=	Thread diameter
D_c	=	Cutter diameter

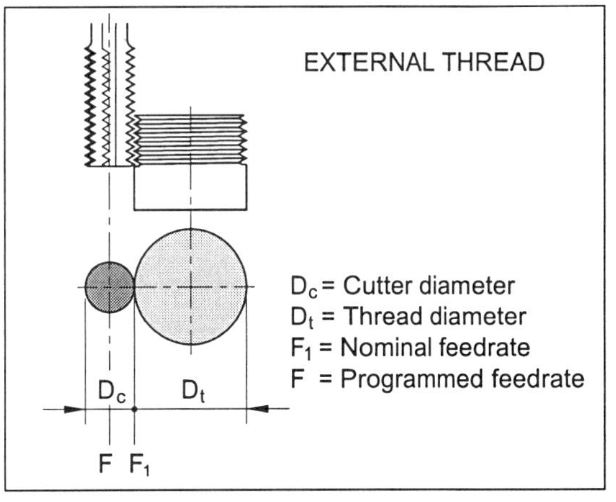

The programmed center-cutting feedrate is:

23.8 x (2.5 + 0.625) / 2.5 = 29.75 inches per minute (or **F29.8**).

Lead-In and Lead-Out

There are two types of lead motions towards a contour. The contour in the example is a thread and all rules of lead-in/out apply. In short, the thread milling cannot start *on* the contour, it must start *away from* the contour. Whether the cutter radius offset is applied to the actual cut or not, depends on the programmer's intent. However, the lead-in and lead-out motions do not change. There are two methods of programming the lead motions - one is a *straight* linear approach or departure (also called *radial* approach or departure), the other is a *tangential* approach or departure, using an arc or line-arc motion combination.

Tangential approach is the preferred choice and will be used in the program.

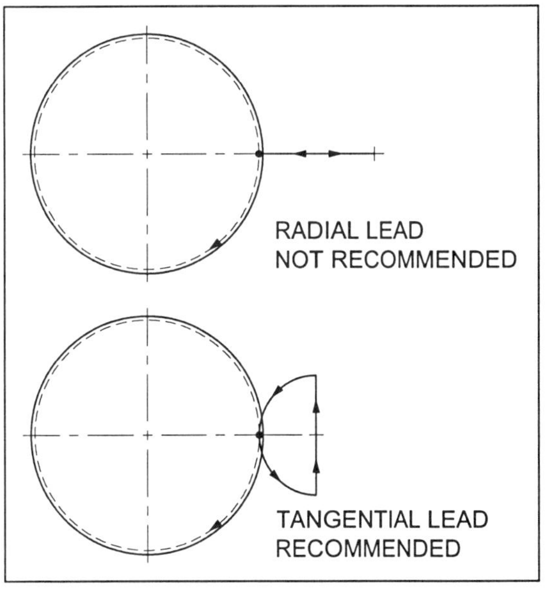

For external threads, the calculation of lead motions is not as critical as it is for internal threads. Providing there is enough space outside of the thread, the only consideration should be the radius of the lead arc. It should be greater than the cutter radius (thread mill radius). Lead-in and lead-out linear motions are necessary, as cutter radius offset cannot be started or canceled during arc motion.

In the example, the cutter diameter is 0.625 inches, so the radius is 0.3125 inches. For a smoother lead, the lead arc for the example will be selected with ***0.75 radius***.

Cutter Radius Offset

A thread can be milled without the use of cutter radius offset, by programming the toolpath of the cutter center. In this case, there will be no G41 or G42 in the program. This is not a recommended practice, as it offers no provision for thread size adjustment at the machine, during production. Using cutter radius offset is always the preferred method, even if cutter center toolpath is programmed.

Both G41 and G42 commands have to be programmed together with the D-address. The D-address specifies the offset number where the required offset amount will be stored:

- **Nominal offset setting will be *zero* for a threading toolpath representing the cutter center**
- **Nominal offset setting will be *cutter radius* for a threading toolpath representing the thread diameter**

In the example, the center of the cutter is programmed, and the nominal offset setting at the machine will be zero. By increasing this amount, the thread will become *larger* - by decreasing the amount, the thread will become *smaller*. In the program, climb milling mode of G41 will be used.

Program Development - External Thread

Before presenting the step-by-step evaluation of the program, here is its complete listing:

```
(2.5-12 EXTERNAL THREAD - 0.75 THD LENGTH - ONE TURN - Z-AXIS DOWN | Z-)
(X0 Y0 ............. CENTER OF PART)
(Z0 ................ TOP OF PART)
(LEAD ARC .......... 0.75 RADIUS)
(D51 SETTING ....... 0.0000 - INITIAL SUGGESTION)
(T01 ............... 0.625 DIA THREAD MILL = 0.3125 RADIUS - 12 TPI)
(PITCH ............. 0.08333)
(FLUTE LENGTH ...... 1.125)
(THREAD DEPTH ...... 0.613 / 12 TPI = 0.051)
N1 G20                                       (IMPERIAL UNITS)
N2 G17 G40 G80 T01                           (START UP BLOCK - T01 READY)
N3 M06                                       (TOOL 01 TO SPINDLE)
N4 G90 G54 G00 X1.949 Y0 S1986 M03 T02       (CENTER OF LEAD ARC)
N5 G43 Z1.0 H01 M08                          (INITIAL CLEARANCE ABOVE PART)
N6 Z-0.7251                                  (START OF THREAD IN THE CLEAR)
N7 G41 Y0.4375 D51                           (LEAD-IN RAPID MOTION + G41)
N8 G03 X1.5115 Y0 Z-0.7459 I0 J-0.4375 F29.8 (QUARTER TURN - LEAD-IN ARC)
N9 G02 Z-0.8292 I-1.5115 J0                  (FULL REVOLUTION TURN)
N10 G03 X1.949 Y-0.4375 Z-0.85 I0.4375 J0    (QUARTER TURN - LEAD-OUT ARC)
N11 G40 G00 Y0 M09                           (LEAD-OUT RAPID MOTION + G40)
N12 G28 Z1.0 M05                             (Z-AXIS HOME POSITION)
N13 M01                                      (OPTIONAL STOP)
```

The thread depth can be found in the tooling catalogue for each thread mill, or it can be calculated using the $0.613 \times P$ formula for external threads ($0.613 \times 0.0833 = 0.051$). Block N4 represents the start position for the tool, blocks N6 through N11 represent the actual thread milling, with the return to the original start position.

The following steps describe these important blocks in detail, along with illustrations of each step.

Z-depth Calculations

This part of the sample project covers thread milling of an external diameter, as specified in the drawing. In order to calculate the Z-depth for each tool motion, it is necessary to understand the concept of the linear motion in helical interpolation. Typical helical motion consists of a two axis motion along an arc (typically along the X and Y axes) and a simultaneous linear motion along the third axis, as defined by the current plane (typically along the Z-axis). In order to calculate correct Z-depths for the external thread milling example, it is necessary to evaluate individual (actual) motions, particularly how they affect the Z-axis programming.

First consideration is the cutter *start position* - its start depth. This is the maximum depth that cannot be exceeded. The drawing shows that there is a base of the part at Z-0.875. This is the maximum absolute depth that can be reached, at least theoretically. Theoretical depth does not include any clearances, so it is only good as a starting point for further decisions. *Practical* depth is selected by the CNC programmer, and includes a certain amount of clearance. In the example, the maximum depth the cutter can reach is **Z-0.85**, which provides 0.025 clearance above the top face of the base.

There will be four motions that affect not only the XY axes, but also the Z-axis:

Motion 1	**Start position - XY**	**Z-axis depth selected for suitable start, following Z-direction**
Motion 2	**Lead-in motions**	**Start cutter radius offset - Z-motion for 90° lead-in arc**
Motion 3	**Actual thread milling**	**Full 360° tool motion - starts and end at the same point**
Motion 4	**Lead-out motions to start**	**Cancel cutter radius offset - Z-motion for 90° lead-out arc**

The first concern should be prevention of any type of collision - in this case, there is a possible collision of the tool and the base top. This critical step has been taken care of by selecting ***Z-0.85* position** as the *maximum* programmed depth, making it 0.025 clearance above the base face.

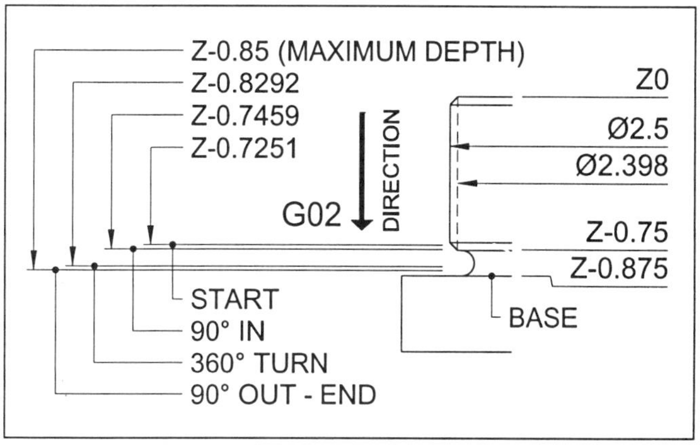

The next decision is the cutting direction. From the earlier charts, the thread milling direction for a climb milling mode (external thread) will be *DOWN* - along the Z-negative cutting motion.

The cutter flute length of the thread mill used is 1.125 - this is a larger dimension than the actual thread length (0.75), which requires only a single turn (360°) to complete the thread using the helical interpolation feature. This single turn is equivalent to the thread pitch (0.0833). For partial turns, the corresponding Z-motion has to be calculated. The most important part of Z-depth calculations is to identify each helical motion, based on the actual sweep angle:

360° turn => **Z-axis linear motion = Pitch of thread = 0.0833 inches = Z-axis change per 360°**

90° turn => **Z-axis linear motion (lead-in *and/or* lead-out arcs)**
¼ pitch of the thread = 0.0833 / 4 = 0.02083333 = 0.0208 = Z-axis change per 90°

In the following individual steps, these dimensions will be applied. Keep in mind that in thread milling - or any motion using helical interpolation - the main focus is to synchronize the two-axis circular motion with the single axis linear motion.

◆ **Step 1 - Start Position**

Once the depth calculations are completed, the actual contour points can be calculated.

The first step is to establish XY starting position for the cutter center. Thread radius is 2.5 / 2 = 1.25, thread depth is 0.051, lead-in arc radius is 0.75 (at Y0):

X= 1.25 - 0.051 + 0.75 = X1.949

```
N4 G90 G54 G00 X1.949 Y0 ...
N5 G43 Z1.0 H01 M08
N6 Z-0.7251
...
```

The cutter will start from **Z-0.7251**

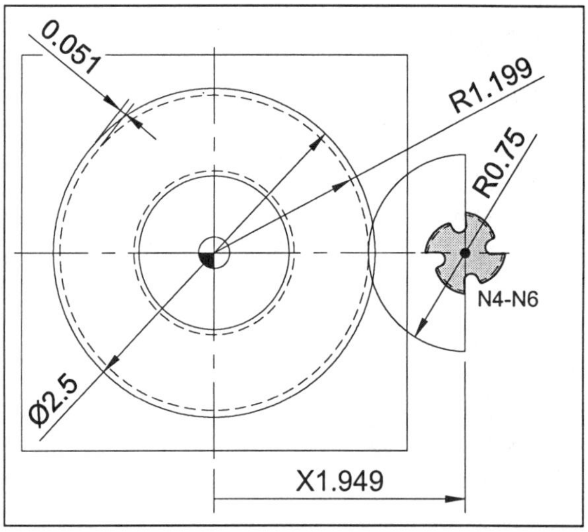

◆ **Step 2 - Linear Lead-in Motion**

Without changing the Z-depth, this linear motion is programmed for the purpose of turning on the G41 cutter radius offset.

Lead-in arc radius is 0.75, cutter radius is 0.3125, so the lead motion ends at:

0.75 - 0.3125 = Y0.4375

```
N7 G41 Y0.4375 D51
...
```

This motion takes place in the clear area, so it is programmed in G00 mode.

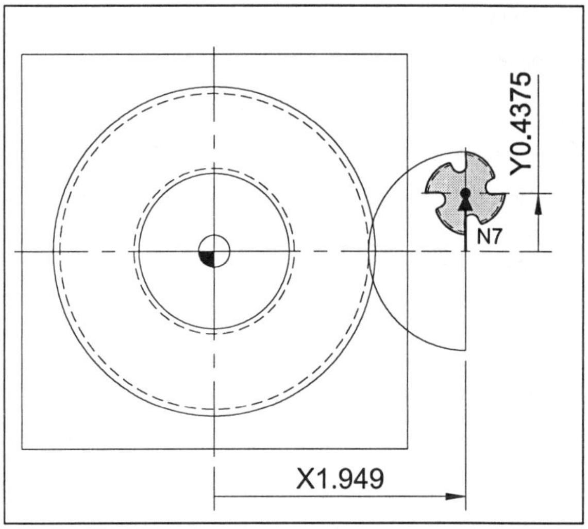

◆ **Step 3 - Arc Lead-in Motion**

In this motion, the cutter will make a physical contact with the material. In order to maintain integrity of the thread, the cutter has to enter the material along all three axes, in helical interpolation mode. A 90° lead-in is one-quarter of the turn, so the cutter must travel one-quarter of the pitch:

P / 4 = 0.08333 / 4 = 0.02083
Z = - (0.7251 + 0.0208) = Z-0.7459

```
N8 G03 X1.5115 Y0 Z-0.7459
       I0 J-0.4375 F29.8
...
```

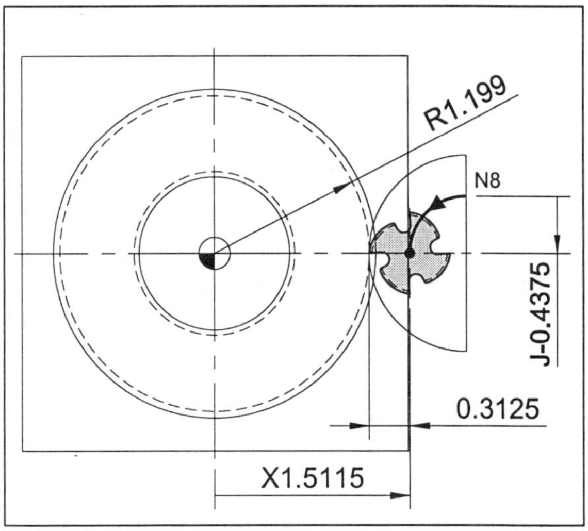

◆ **Step 4 - Full Turn - 1 x 360°**

Only one full turn of the cutter is required in helical interpolation to cut the thread. During this motion, the cutter has to move one full pitch (0.0833) along the Z-axis, from the previous position of Z-0.7459:

Z = - (0.7459 + 0.0833) = Z-0.8292

The start point and end point are identical, and no XY axes are necessary in the block:

```
N9 G02 Z-0.8292 I-1.5115 J0
...
```

In the next motion, the thread milling cutter will leave the material.

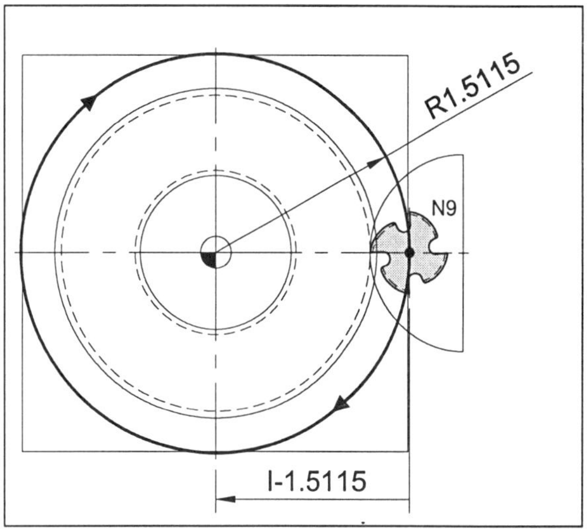

◆ **Step 5 - Arc Lead-out Motion**

As in the lead-in motion, the lead-out motion must also continue in helical interpolation mode - one quarter turn - in order to preserve the thread integrity:

Z = - (0.8292 + 0.0208) = Z-0.85

```
N10 G03 X1.949 Y-0.4375 Z-0.85
        I0.4375 J0
...
```

Once this motion is completed, the cutter has reached the maximum depth of Z-0.85 and is also off the part, in a clear location.

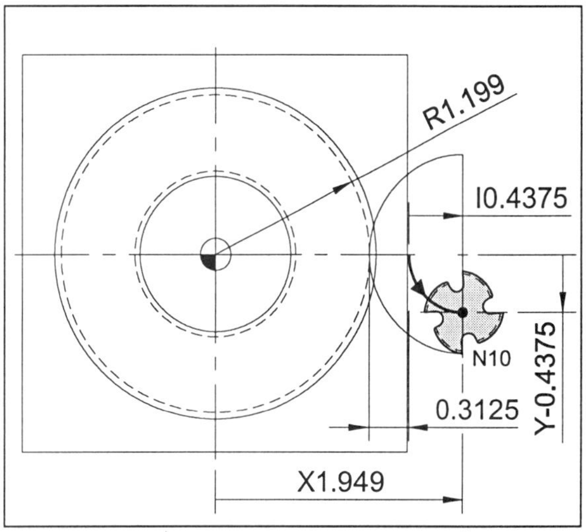

◆ **Step 6 - Linear Lead-out Motion**

The final motion is off the part, and rapid motion can be programmed for efficiency. This is also the motion where the cutter radius offset will be canceled.

The cutter returns to the same position it started from (X1.949 Y0):

```
N11 G40 G00 Y0 M09
...
```

With this motion, the detailed explanation is completed - study the complete program introduced earlier in this section.

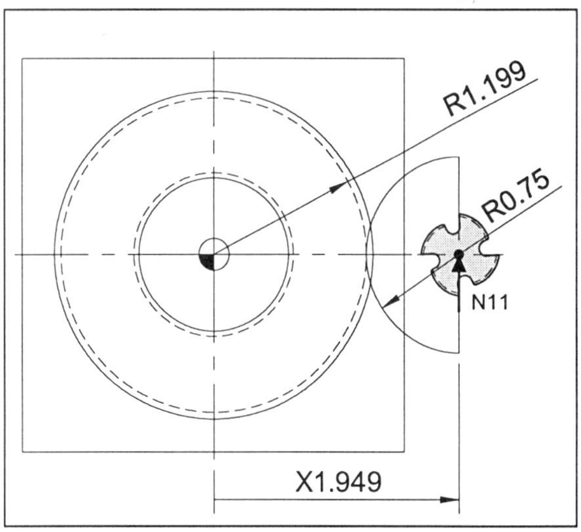

Programming Internal Threads

In the previous section, the main subject was *external* thread milling. The same careful attention given to external threads should be given to thread milling of internal threads. For both types, the direction of threading and circular motion have to be taken into account. Study the previous section carefully, to find out some details that are common to both types of threads, but not necessarily covered in this section with the same detail. Earlier illustrations in this chapter show all possibilities available for thread milling. The main difference between thread milling of external threads and internal threads is the method of lead-in and lead-out toolpath. While external thread milling offers much more available space to implement the lead motions, internal thread milling often takes place in some very confined environment. This section covers the subject of ***internal*** thread milling. A drawing that will be used for evaluation of the example is the same as introduced earlier. The program development for the internal thread will take off where the external threading ended.

Tooling Selection

One of the main benefits of thread milling is that the same thread mill cutter can be used for both external and internal cutting, providing the thread forms are the same. In the internal thread example, the thread form is *different* than for the external thread, so a different tool has to be used. For the example, the threading cutter will have the same diameter of 0.625 inches, but different pitch. Another consideration is the thread length versus the *usable* flute length of the cutter. For the *INTERNAL* thread, a single *carbide* thread mill will be used in climb milling mode and with cutter radius offset:

- **Right hand thread**
- **Carbide with four flutes**
- **Pitch = 0.0625 - 16 TPI**
- **Thread depth = 0.034**
- **Diameter = 0.625**
- **Thread length = 1.25**
- **Flute length = 1.15**

Although the changes are minimal, one change does stand out - the thread length is *greater* than the cutter flute length. This is a significant deviation from the external thread method and must be covered in the part program. When the full thread is machined, it will take more than one revolution to complete the thread. The number of threads will be calculated shortly.

Cutting Conditions

Cutting conditions - the spindle speed and cutting feedrate - have to be selected by considering the material being cut and the material of the thread mill. In the example, a *four-flute* carbide cutter will be used to machine *4140 steel*. Depending on the exact conditions, the following selection may change, but for the example, the cutting speed used will be 325 feet/min with 0.002 feedrate per tooth. Based on the standard formula, the spindle speed will be - the same as for the external thread:

$$\text{r/min} = \frac{325 \times 12}{\pi \times 0.625} = 1986 \text{ r/min}$$

With the feedrate per tooth of 0.002 and four flutes, the nominal cutting feedrate will be:

$F_1 = 1986 \times 0.002 \times 4 = 15.9$ **inches per minute**

The small changes in the cutting conditions have been made to illustrate calculation differences.

This feedrate of 15.9 must be adjusted to the *center* of the thread mill, using the previous formula, optimized for internal threading feedrate:

$$F = \frac{F_1 \times (D_t - D_c)}{D_t}$$

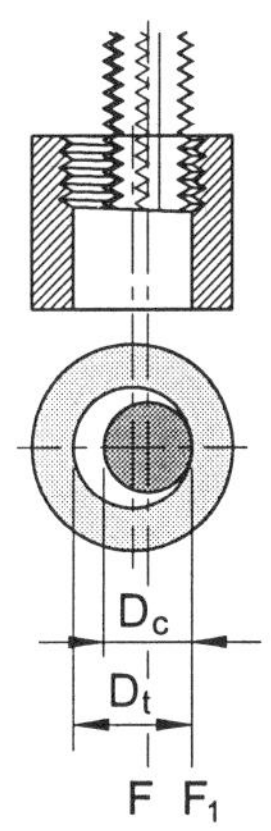

☞ *where ...*

- F = Programmed feedrate
- F_1 = Nominal feedrate
- D_t = Thread diameter
- D_c = Cutter diameter

The programmed center-cutting feedrate is much slower than that for external thread, although the size of the cutter and the material are both the same:

15.9 x (1.25 - 0.625) / 1.25 = 7.95 inches per minute (or **F8.0**) - ***this is very important part of the process***

Lead-In and Lead-Out

Lead-in and lead-out motions for internal threads present a greater challenge than those for external threads. The main reason is they have to be applied in a much smaller available work space. In order to calculate the leading lines and arcs, certain conditions have to be met:

$$R_t > R_l > R_c$$

☞ where ...

- R_t = Radius of the thread
- R_l = Radius of the lead arc
- R_c = Radius of the cutter

For the example, R_t = 1.25 / 2 = 0.625 and R_c = 0.625 / 2 = 0.3125.

In the example, the cutter diameter is 0.625 inches, so the radius is 0.3125 inches. The lead arc radius must be somewhere between these two amounts. For a smoother lead, the lead arc for the example will be selected with a ***0.5 radius***.

Internal thread milling presents a challenge that does not apply to an external thread - a confined space. The sweep angle of the lead arc should be 90 degrees whenever possible. A quarter of a circular motion is much easier to use in calculations than some odd angle. Lead-in and lead-out lines are necessary, as cutter radius offset cannot be started or canceled during an arc motion.

Cutter Radius Offset

The basic rules of cutter radius offset for external threads also apply to internal threads. The D-address used with G41 or G42 specifies the offset number where the required offset amount will be stored at the control. In the example, the cutter center is programmed, and the nominal offset setting at the machine will be zero. By increasing this amount, the thread will become *smaller* - by decreasing the amount, the thread will become *larger*. In the program, G41 will be used (climb milling).

Program Development - Internal Thread

Both threads (external and internal) are part of the same program. Once the external thread is completed (in block N13), the program continues with a tool change and the data for internal thread:

```
(1.25-16 INTERNAL THREAD - 1.25 THD LENGTH - THREE TURNS - Z-AXIS UP | Z+)
(LEAD ARC .......... 0.5 RADIUS)
(D52 SETTING ....... 0.0000 - INITIAL SUGGESTION)
(T02 ............... 0.625 DIA THREAD MILL = 0.3125 RADIUS - 16 TPI)
(PITCH ............. 0.0625)
(FLUTE LENGTH ...... 1.15)
(THREAD DEPTH ...... 0.541 / 16 TPI = 0.034)
N14 T02                                         (TOOL T02 READY)
N15 M06                                         (TOOL 02 TO SPINDLE)
N16 G90 G54 G00 X0 Y0 S1986 M03 T01             (CENTER OF LEAD ARC)
N17 G43 Z1.0 H02 M08                            (INITIAL CLEARANCE ABOVE PART)
N18 Z-1.35                                      (START OF THREAD IN THE CLEAR)
N19 G41 X0.125 Y-0.1875 D52                     (LEAD-IN RAPID MOTION + G41)
N20 G03 X0.3125 Y0 Z-1.3344 I0 J0.1875 F8.0     (QUARTER TURN - LEAD-IN ARC)
N21 Z-1.2719 I-0.3125 J0                        (FULL REVOLUTION TURN 1 OF 3)
N22 Z-1.2094 I-0.3125 J0                        (FULL REVOLUTION TURN 2 OF 3)
N23 Z-1.1469 I-0.3125 J0                        (FULL REVOLUTION TURN 3 OF 3)
N24 X0.125 Y0.1875 Z-1.1313 I-0.1875 J0         (QUARTER TURN - LEAD-OUT ARC)
N25 G40 G00 X0 Y0 M09                           (LEAD-OUT RAPID MOTION + G40)
N26 G28 Z1.0 M05                                (Z-AXIS HOME POSITION)
N27 M30                                         (END OF PROGRAM)
```

The thread depth can be found in the tooling catalogue for each thread mill, or it can be calculated using the $0.541 \times P$ formula for internal threads ($0.541 \times 0.0625 = 0.034$). Block N16 represents the start position for the tool, blocks N18 through N23 represent the actual thread milling, with the return to the original start position.

Z-depth Calculation

Z-depth calculation process is the same for internal as for external threads. To cut a right hand thread in climb milling mode, the cutter must start at the thread bottom and cut in Z-positive direction. There are no obstacles below the part, so the cutter can start almost at any depth, but 1 pitch below the part should be considered the minimum (0.0625). For a small additional clearance, the starting depth will be Z-1.35, 0.1 below the bottom. Calculations for a full or a quarter turn will be:

360° turn => Pitch = 1/16 = 0.0625
= Z-axis change per 360°

90° turn = Z-axis change for lead arcs
= ¼ pitch of thread = 0.0625 / 4
= 0.015625
= 0.0156 = Z-axis change per 90°

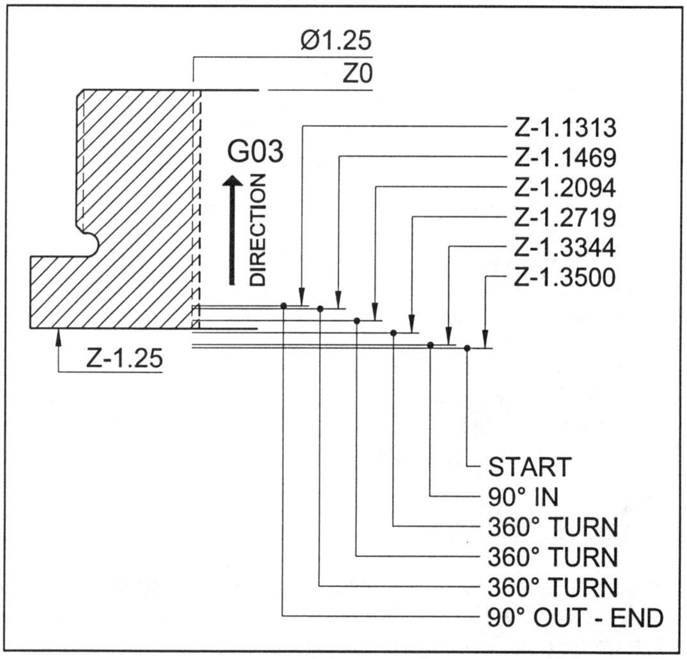

The following descriptions of individual steps use the above dimensions.

Number of Turns

Before the individual steps can be calculated, the number of full turns has to be calculated. A full turn is defined as a 360° helical motion. In this internal thread example, the effective flute length of the cutter is 1.15 inches, while the required thread length is 1.25 inches. That means a single full turn of 360° *cannot* complete the thread, and one or more full turns are required. In order to complete the thread as per drawing, the cutter top position has to be in the positive Z-axis area. Both the part program and its related illustration shown on the previous page, identify *three* full turns. The number of full turns was established on these dimensions - the three full turns are highlighted:

Motion Description	Programmed Z-depth	Cutter top position (for 1.15 flute length)
Start depth	Z-1.3500	Z-0.2000
90° lead-in arc	Z-1.3344	Z-0.1844
360° turn - 1 of 3	Z-1.2719	Z-0.1219
360° turn - 2 of 3	Z-1.2094	Z-0.0594
360° turn - 3 of 3	Z-1.1469	Z0.0031
90° lead-out arc	Z-1.1313	Z0.0187

As the table shows, each full turn (360° helical motion) brings the cutter top position closer to Z0. As long as this position is negative, an additional full turn will be required. In the example, the cutter top for the *second* turn is still in the negative area, which means at least one more full turn is required.

There is another way to look at the table above - once the absolute programmed Z-depth becomes *smaller than the cutter flute length*, the cutter can make a safe lead-out motion.

With all these prerequisites, the rest of the program should not be difficult to develop. The following steps describe these important blocks in detail, along with illustrations of each step.

◆ **Step 1 - Start Position**

For the internal thread, the most convenient start position is center of the circle that defines the thread. If its size is too large, the XY start position is usually programmed closer to the thread. In this example, the center has coordinates X0Y0, and the cutter starts at the pre-selected depth of Z-1.35 and will cut upwards, into the Z+ direction:

```
N16 G90 G54 G00 X0 Y0 ...
N17 G43 Z1.0 H02 M08
N18 Z-1.35
...
```

Additional dimensions are provided for reference. Subsequent steps show the XY cutter position for each depth, as calculated earlier - see illustration on the previous page.

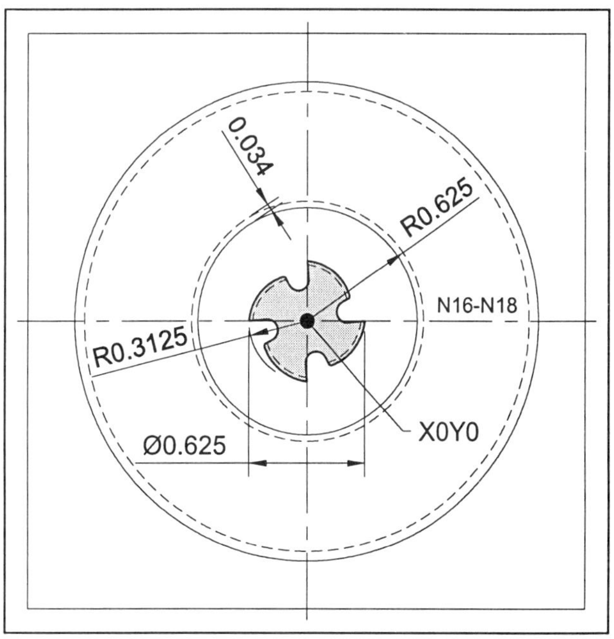

- **Step 2 - Linear Lead-in Motion**

There is no change to the Z-depth - this linear motion is programmed for the purpose of turning on the G41 cutter radius offset.

Lead-in arc radius is 0.5, cutter radius is 0.3125, so the lead motion ends at:

X = 0.625 - 0.5 = Y0.125
Y = - (0.5 - 0.3125) = Y-0.1875

```
N19 G41 X0.125 Y-0.1875 D52
...
```

This motion takes place in the clear area, so it is programmed in G00 mode.

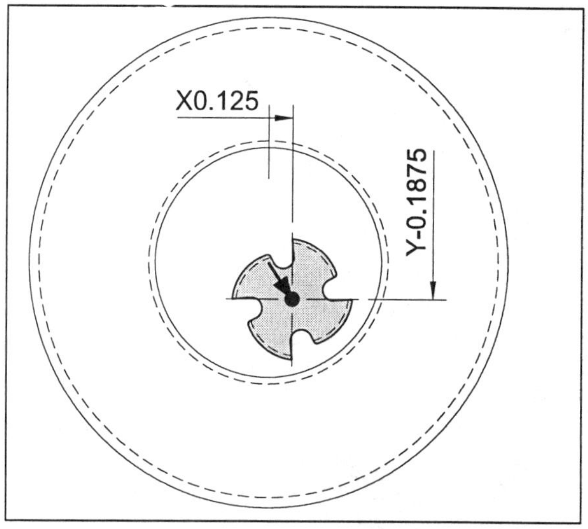

- **Step 3 - Arc Lead-in Motion**

In this motion, the cutter will make a physical contact with the material. In order to maintain integrity of the thread, the cutter has to enter the material along all three axes, in helical interpolation mode. A 90° lead-in is one-quarter of the turn, so the cutter must travel one-quarter of the pitch:

P / 4 = 0.0625 / 4 = 0.0156
Z = - (1.35 - 0.0156) = Z-1.3344
X = 0.625 - 0.3125 = X0.3125

```
N20 G03 X0.3125 Y0 Z-1.3344
        I0 J0.1875 F8.0
...
```

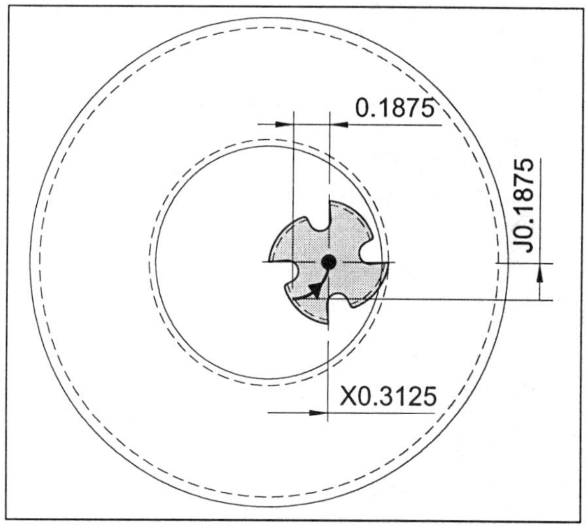

- **Step 4 - Full Turn - 3 x 360°**

There are three helical motions in this step - *three full turns* - completing the thread. Full circle cutting (I-0.3125 J0) and its direction (G03) do not change. The Z-depth changes - for each turn, the cutter moves up by one pitch (0.0625). The R-address for the radius cannot be used for a full circle.

Current position = Z-1.3344
Z-axis after turn 1 = - (1.3344 - 0.0625) = Z-1.2719
Z-axis after turn 2 = - (1.2719 - 0.0625) = Z-1.2094
Z-axis after turn 3 = - (1.2094 - 0.0625) = Z-1.1469

```
N21 Z-1.2719 I-0.3125 J0
N22 Z-1.2094 I-0.3125 J0
N23 Z-1.1469 I-0.3125 J0
...
```

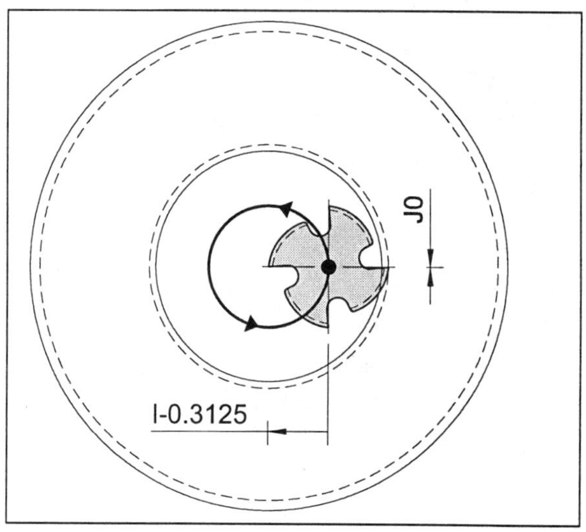

◆ **Step 5 - Arc Lead-out Motion**

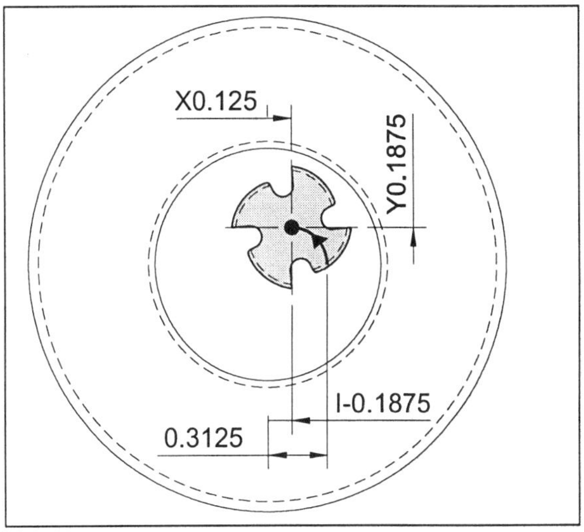

At this stage, the cutter top is just marginally above the part and the arc-out motion will complete the thread.

As in the lead-in motion, the lead-out motion must also continue in helical interpolation mode - one quarter turn - in order to preserve the thread integrity:

Current Z-position = Z-1.1469
Z = - (1.1469 - 0.0625 / 4) = Z-1.1313

```
N24 X0.125 Y0.1875 Z-1.1313
I-0.1875 J0
...
```

Once this motion is completed, the cutter has completed the thread and is also off the part, in a clear location.

◆ **Step 6 - Linear Lead-out Motion**

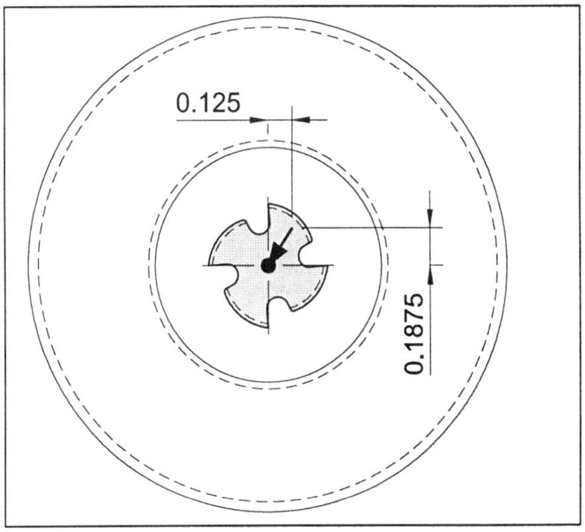

The final motion is off the part, and rapid motion can be programmed for efficiency. This is also the motion where the cutter radius offset will be canceled.

The cutter returns to the same position it started from (X0 Y0):

```
N25 G40 G00 Y0 M09
...
```

With this motion, the detailed explanations are completed - study the second part of the complete program introduced earlier in this section.

Pipe Thread Milling

Pipe thread is a specific form of a *conical* - or *tapered* - thread. Most related CNC programming utilities - as well as comprehensive thread milling dedicated software - offer pipe thread milling along a cone. Basic approach to program a pipe thread using helical interpolation is the same as described for either external or internal thread in this chapter.

The main difference is in compensating for the taper in each partial or full turn. This compensation will be reflected in the XY coordinates. A typical thread mill designed for pipe threads has the taper built-in, but following the general concept.

Details relating to pipe threads are beyond the scope of this handbook. As described in the next section, most of the companies that make thread mills also offer free programming software.

Thread Milling Software

Virtually all major manufacturers of thread milling tools provide their customers with a thread milling software. In most cases, there is no cost to such software, but keep in mind that the software is normally *not* generic - it is designed for particular cutters from each manufacturer's catalogue. That should not deter any experienced CNC programmer, as all tooling companies design their thread milling tools for standard threads. Most companies offer a free download from their web site, often providing valuable tips on machining and programming.

Typical thread milling software will let the user select all threading parameters and supply the tool data (usually from the catalogue of the particular manufacturer). Once all information has been supplied, the software produces a complete thread milling program segment that can be pasted into the normal part program. While this method provides a fairly reliable program, it is still up to the CNC programmer to fine-tune the output for best results.

Don't be surprised if some software will give you somewhat different results than those introduced in this chapter. Each manufacturer's software uses parameters and settings that relate to the thread mill cutters made by that company.

19 KNURLING ON CNC LATHES

When it comes to programming a knurling operation on CNC lathes, there is nothing simpler in terms of the actual tool motions:

- **Rapid to a start position**
- **Feed in for the length of knurl**
- **Retract from part diameter**
- **Return to the start point or a tool change position**

If it is not any special programming technique, what makes knurling different from other simple tool motions? This chapter will look at several issues important for successful knurling on CNC lathes.

Knurling Operations

Knurling is a special - usually uncommon - machining operation that can be produced on manual and CNC lathes. Its primary purpose is to develop depressions - or indentations - in the material surface by a rolling action of the knurling tool while it is moving along the horizontal machine axis. Depending on the knurling tool, these indentations can have either straight, angular or a diamond pattern. The diamond pattern is also called the 'diagonal pattern' or the 'cross pattern'. Angular pattern can be left hand or right hand. Knurling is not a metal cutting action but a metal forming action caused by force. A large variety of tool holders is available for different knurling requirements.

Tooling Selection

Several manufacturers have developed knurling tools especially for CNC lathe machining, including modular tools with interchangeable heads. These are the tools recommended for CNC knurling. Some designs are suitable for knurling to a square shoulder with minimum clearance. The most important of the tools is the knurling wheel itself.

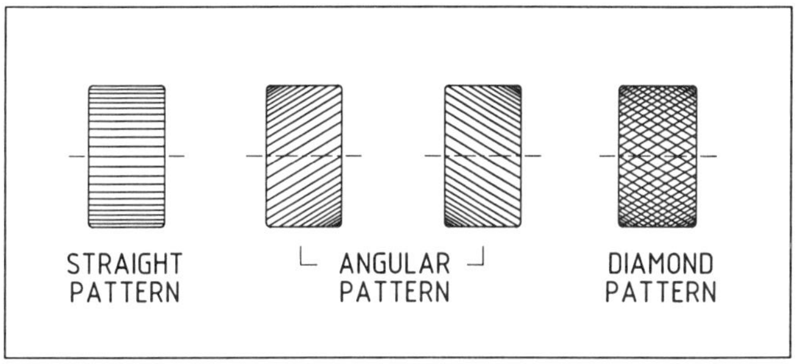

The three common pattern types are shown above. A typical engineering drawing will specify the type of knurl used on the part surface in one of three forms (patterns):

- **Straight knurl pattern** ... **parallel to work**
- **Angular knurl pattern** ... **typically 30° or 45°**
- **Diamond knurl pattern** ... **same angles as angular - also known as *diagonal* or *checkered* pattern**

Each of these three types has a specific purpose in practice. The straight knurl is used mainly for joining two cylindrical objects, for example, a pin and a hole. Typical usage of the diagonal knurl is to facilitate easier handling, for example, better hand grip on rotary knobs or handles.

Knurling tools can use one or two wheels. For example, using one left hand and one right hand wheel in the same holder and opposite to each other, a diamond pattern can be cut on the part. This configuration is illustrated at right.

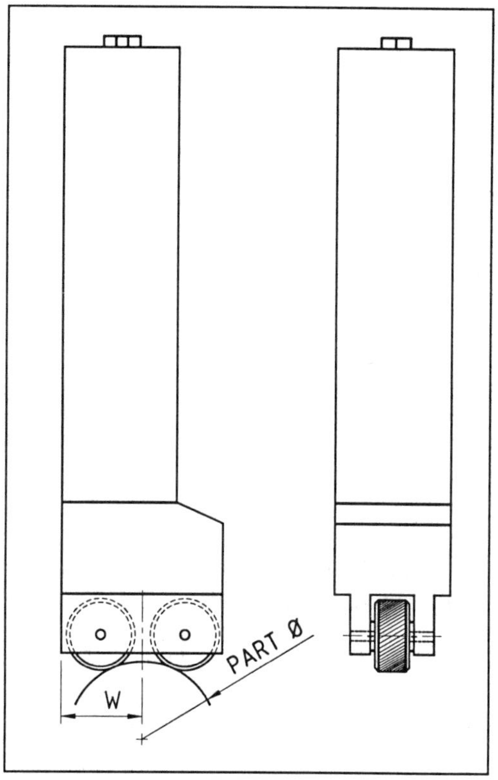

For easier setup, the dimension *W* is provided in the manufacturer's tooling catalogue and represents the center height of the tool.

Knurling tools designed for CNC lathes have floating (self-centering) positioning built-in, to provide an accurate alignment with the part to be knurled.

Depending on the type of knurling wheels mounted in the holder, these double-wheel holders can produce a diamond pattern or a straight pattern:

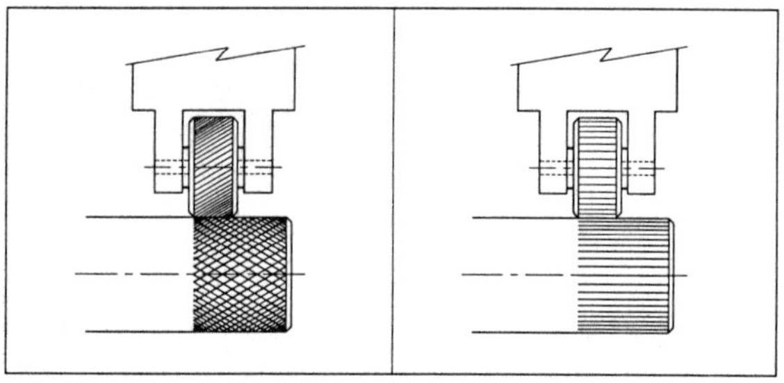

As the illustrations show, this tool design is suitable for open knurls only - it cannot be used for knurling towards a shoulder.

A different configuration of the tool end will allow such operations:

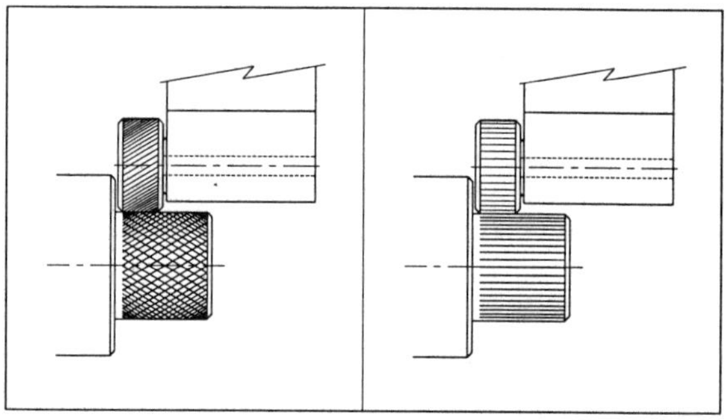

Many other designs are available, and they include single-wheel knurling tool for straight knurl, where the wheel is often mounted at an angle. Knurling wheels are available in a variety of pitch sizes.

For diamond threads, a knurling wheel can be chosen to produce a 30° knurl or a 45° knurl. The 45° pattern is often called a *square* knurl.

Knurling Pitch

A common designation of the pitch size of a knurling wheel is TPI - meaning *teeth per inch*. If such designation is defined in the drawing, do not confuse it with *threads per inch*, which also uses the same abbreviation. Pitch of a knurl is important not only for the proper wheel selection, but it plays a vital role in machining, particularly the feedrate selection.

In general terms, there are three groups of knurl pitch - *coarse*, *medium*, and *fine*.

Programming and Machining Techniques

Knurling is a final or finishing machining operation, usually on an external part diameter, although internal knurling is also possible. That means all other machining operations have to be done beforehand. As knurling is a rotary type of machining (lathe machining) the part has to be turned to the specified diameter before the knurling operation can be applied.

Tools for knurling have to be selected carefully, as they can only be used within certain range of work diameters. Some tools are designed to form the knurl, others are designed for cutting the knurl. Cutting knurls are faster than forming knurls, as they provide less pressure on the spindle and may be used in variety of materials.

The CNC programmer is responsible for programming the tool motions, selecting the spindle speed and the cutting feedrate. For small threads, an additional support may be required, such as a tailstock.

Tool Motions

Only four motions are required to machine a knurled surface. The *first motion* is to the start point - this point should be above the knurl, and about 0.100" (2.5 mm) into the part (Z-0.1 or Z-2.5 if the front face is Z0). The *second motion* has to be a feedrate motion, as the knurl will move to the cutting depth on the part diameter. Once the tool has reached the cutting diameter, it continues in the feedrate mode towards the end of the knurl, defined as length in the drawing - this is the *third motion*. Finally, once the knurling operation is completed, the tool can leave the part at rapid rate in the *fourth motion*. When the knurling tool leaves the diameter, it is recommended the motion is along the X-axis only.

Two other major requirement for successful knurling are sufficient supply of coolant and teeth free of any residual material, such as slivers of metal, burrs, or chips.

Lubrication is important for lowering the wheel temperature and smooth wheel turning

For some very short threads, there is no need for the third motion (normally along the Z-axis). If the knurling length is smaller than the cutting width of the knurling wheel, the tool can just plunge-in and leave the diameter immediately.

Depth and Feedrate

One of the benefits of CNC machines is a feature allowing the change of settings in the form of various offsets. Some programmers provide only an approximate depth in the program, and let the CNC operators to make offset adjustments at the machine. This method counts on the visual inspection of the knurl - if it looks good, both the depth and the feedrate are correct. Appearance of the knurl is important for exposed parts that require a good grip, such as handles.

Depth can also be calculated, based on circular pitch in terms of TPI (teeth per inch). This subject can get quite involved, but a typical example should illustrate the concept. Knurling tool manufacturers offer comprehensive data sheets for their tools. In the example, a medium knurl has ***50 TPI***. Circular pitch can be calculated by the formula ***1 / TPI = 0.0200***. This formula works for straight knurls and must be adjusted for angular knurls. For a 30° angle, the adjustment will be ***TPI × cos30 = 43.3*** in the example. For the diagonal depth, another formula is needed: ***Pitch / cos30 = 0.023***.

In typical examples of coarse, medium and fine type of knurls, the following table can be used as an example, including suggested feedrates in steel. All three calculations are based on the above data.

Type of knurl	Number of teeth TPI	Suggested depth		Approximate feedrate - in/rev
		STRAIGHT	ANGULAR	
COARSE	34	0.0294	0.0340	0.017
MEDIUM	50	0.0200	0.0231	0.011
FINE	80	0.0125	0.0144	0.007

The suggested feedrate may have to be adjusted for other materials, and some experimentation at the machine may also be necessary. Spindle speed is normally calculated for the material being machined, using roughing surface speeds (feet/min or m/min). The plunging feedrate should be pretty aggressive, and when knurl is completed, retract the tool right away.

Troubleshooting

If the knurl is not perfect, the chances are that either it is too deep or too shallow. A knurl that is too deep produces burrs and nicks on the part surface, while a knurl that is too shallow produces noticeable flats on the part surface. A short summary may provide some clues as to the causes:

- **Slivers, burrs and nicks - probable excessive cutting depth**
- **Lower than expected tool life - lubrication problems, hard material cutting**
- **Insufficient depth - wrong knurl or program data**

In certain ways, knurling is related to threading, and not for the common TPI abbreviation. Both machining operations are a combination of forming and cutting, both have a pitch that influences the feedrate, and both can produce very unwanted results or results of very high quality.

20 FOUR-AXIS LATHES

If a two axis CNC lathe programming presents enough challenges, the idea of a four axis lathe programming may seem a bit too much to handle, at least for the beginner in the CNC field. The actual situation is much simpler than that - and here is the good news - any four axis lathe is generally programmed as *two independent two axis lathe programs*. Of course, there are some expected differences, the greatest of them relates to certain several 'special functions' the control system has to support. These are special programming functions that are specific to most of four-axis lathe applications - these are functions unique to the particular 4-axis CNC lathe.

In principle, a four-axis lathe can be visualized as two two-axis lathes in one - along with two control systems - but both contained in a single machine. In terms of design, a four-axis lathe usually has one spindle and two independent turrets that must be synchronized to work simultaneously. The major advantage of using a four-axis CNC lathe is reduced setup and cycle time. That is not to say that all parts are suitable to machine on four-axis lathes. Although more tools can be mounted on two turrets than on one (usually up to 8 per turret), the main problem is often the lack of clearance, particularly for parts that are too small or too large.

Manual and CNC lathes with two turrets have been around for a long time, usually designed as flat-bed lathes, where the one turret is used mainly for external tools, the other for internal tools. These lathes cannot be considered four-axis lathes, because both turrets are mounted on a single cross slide (or an axis), and only one of them can work at any given time.

General Setup

Depending on the machine tool manufacturer, the design of four-axis CNC lathes may vary to some extent, but all lathes share the main characteristics. For the example used in this chapter, the four axis lathe will have the following layout:

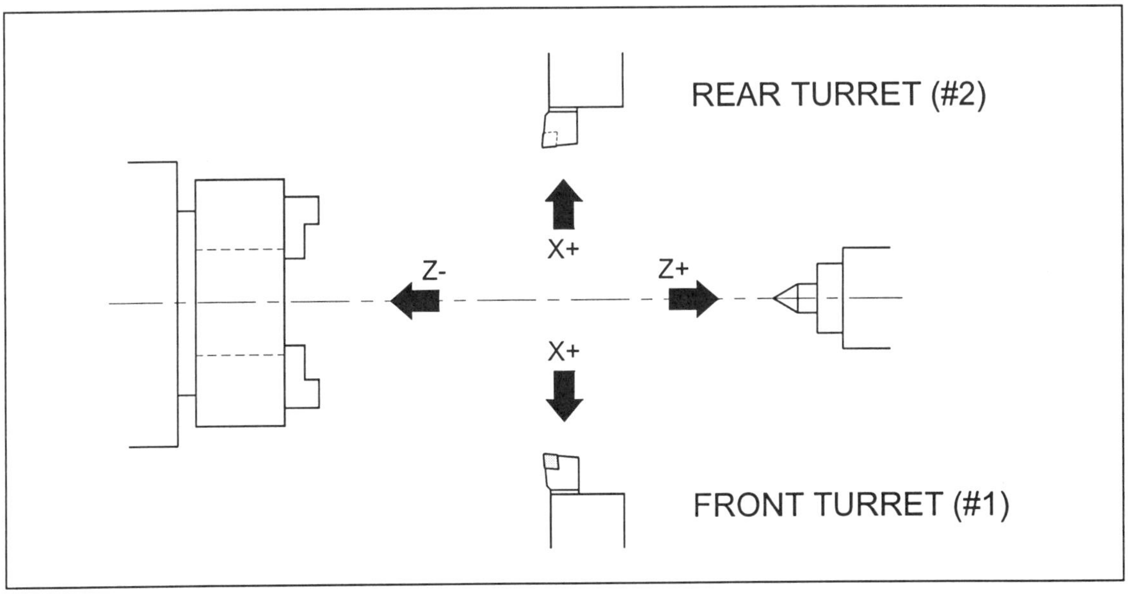

Note that the X+ is always the direction of increasing diameter. Also note, that the spindle rotation is the same for both turrets, which requires correct mounting of tools. The layout shown is only general, as each turret accepts external or internal tools, not just external, as illustrated.

Tool Tip Numbers

During setup, the CNC operator will have to input offsets data into the control register. One of the data that has to be input is the insert tip number, under the T-column of Fanuc controls.

Make sure to set the tool tip number for cutter radius offset correctly - they are different for each turret orientation, as shown in the illustration.

Both radius and the tip number must be positive when registered at the control system

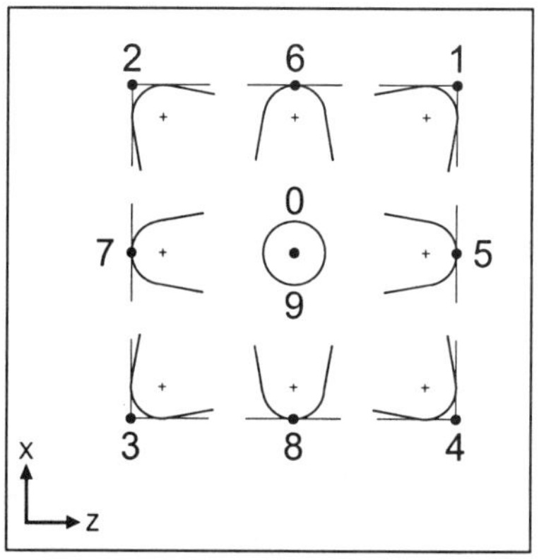

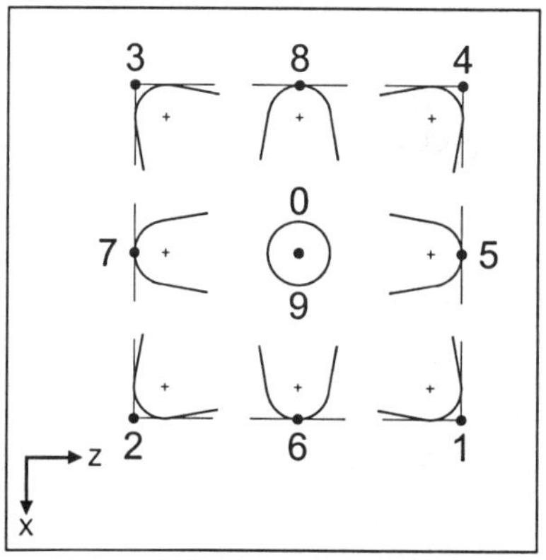

Programming Method

To program a four-axis CNC lathe requires not only a good knowledge of the machine and control system, but also the method of machining, particularly when distributed between two independent turrets. There are no serious or unexpected programming restrictions, but some controls require that the program number listed in the program be an odd number for the front turret (turret #1) and an even number for the rear turret (turret #2). Check your machine/control manual for any specific requirements.

Spindle Speed and Feedrate

With two independent turrets and only a single spindle, the question of spindle speed assignment in the program is very important. When machining two diameters simultaneously, typically an external and an internal diameter, chances are they will be different in size. If the difference is relatively small, the G96 command for *constant surface speed* (CSS) can be used in the program. If G96 is not a suitable selection, usually for diameters that deviate significantly, a compromise is necessary, in the form of programming G97 - *constant spindle speed*. Feedrate can be programmed independently, for each turret, as required - within general machine capabilities.

Typical rules may be different between CNC lathes and their corresponding control systems, but some general rules can be established:

If G96 command is programmed for one turret,
any G97 for the other turret will be ignored, until G97 replaces the G96

If both turrets use different G96 command,
the one specified for the first active turret becomes active

M-Functions

Similar restrictions related to spindle speeds also apply to certain M-functions, particularly to those associated with spindle rotation. Again, certain general rules apply, depending on the control:

- **If one turret is programmed with G96 in effect, spindle related functions M03, M04, and M05 become active, while the same functions specified for the other turret will be ignored**

- **M-functions that stop program processing - M00, M01, M02, and M30 - will only become effective when corresponding functions for the other turret have also been processed**

Note that all explanations for the spindle and miscellaneous functions mentioned in this chapter are correct in principle. This does not mean they will work on every four-axis CNC lathe. As is recommended for these special situations, use the provided explanations as guidelines only, and consult the manuals provided by the machine tool builder.

Synchronization Functions

Just by looking at the control system for a 4-axis CNC lathe, one can see a major difference from control systems designed for a two-axis lathe work. Several machine specific operation related functions, for example, feedrate override, are located there *twice*. It is important to be able to control machining operations for each turret independently. The program itself must contain special functions that provide a smooth transition between the actual work done by each turret.

The special and unique functions that are applied to four-axis lathes are called the *synchronization functions*. CNC programmer must use special program codes correctly. These are usually M-functions, designed to control activity of each turret, particularly when it comes to simultaneous start up and holding activity of one turret while machining with the other turret.

In just one typical machining application, one turret may be used for removing material from an external contour, while the other turret is used to remove material from an internal contour. For example, the synchronization functions may be three-digit M-codes, in the range of:

`M100` *to* `M999`

These special functions are called the *waiting M codes*. Each M-code must be programmed in pairs - it must exist in both programs. When a waiting code is programmed for one turret, the program will be stopped, until the same waiting code is encountered fo the other turret.

One last note - there may be times when the four-axis CNC lathe will be used with only a single turret being active. In this case, if there are any waiting codes in the program, they will be ignored.

The program structure described in the next section will illustrate the basic concept.

Program Structure

On the next page is an illustration of two generic, yet related, programs that exemplify the structure of a typical four-axis CNC program. Keep in mind that the illustration shows only the basic concepts. The actual length of each program can be as long or as short as necessary, and the waiting codes do not have to line-up optically. Also, the waiting codes do not have to be in increments of one, as long as they match in programs for both turrets.

The main purpose of waiting codes is to synchronize actual machining for both turrets

This structure will be applied to an actual program example, following this section. There are three tools for the rear turret, and two tools for the front turret. The control system will start the tool T01 for the rear turret, while the front turret is idle. Since the front turret machining starts with M100 waiting code, the control will wait until another M100 waiting code is found in the rear turret program. When that happens, both turrets will work simultaneously, until another waiting code is encountered. At that time, the program processing will stop for the turret affected only, until the corresponding waiting code is processed for the other turret.

The drawing below shows a drawing for a part suitable for a four-axis lathe - the actual program follows.

REAR TURRET	FRONT TURRET	
G20 T0100 G97 S1200 M03 G00 X.. Z.. T0101 M08 ... M01	IDLE	
M100	M100	WAITING
T0400 G97 S800 M03 G00 X.. Z.. T0404 M08 ... M01	T0100 G00 X.. Z.. T0101 M08 ... ... M01	
M101	M101	WAITING
T0600 G97 S1200 M03 G00 X..Z.. T0606 M08 ... M01	T0300 G00 X..Z.. T0303 M08 ... ... M01	
M102	M102	WAITING
M30 %	M30 %	

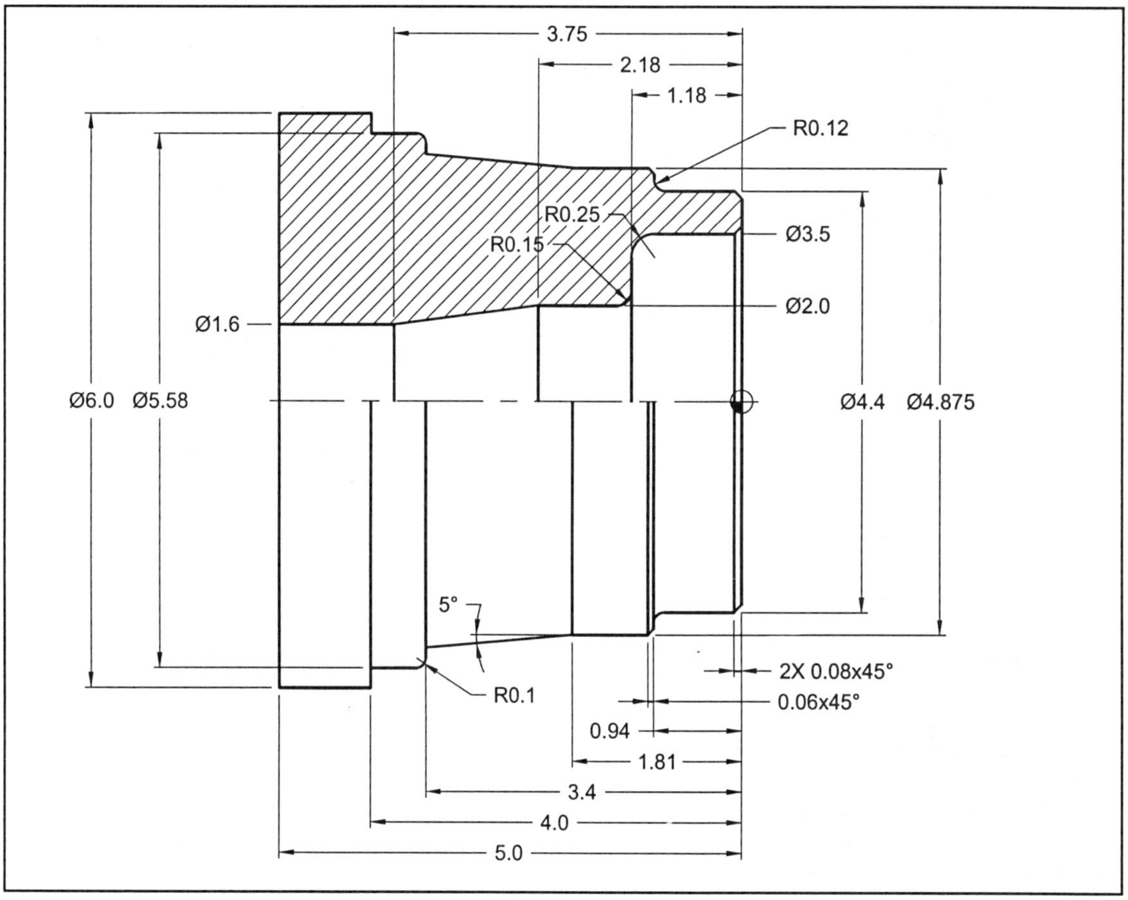

```
(REAR TURRET - INTERNAL CUTTING)
(T01 = 1.5 DIA INDEXABLE DRILL)
N1 G20                                    (IMPERIAL UNITS)
N2 T0100 M41                              (TOOL CHANGE + LOW GEAR)
N3 G97 S1200 M03                          (COMMON SPINDLE SPEED/ROTATION)
N4 G00 Z0.5 T0101 M08                     (SAFE CLEARANCE)
N5 X0                                     (CENTER LINE POSITION FOR DRILL)
N6 G01 Z0.1 F0.03                         (START OF DRILLING)
N7 Z-5.15 F0.01                           (DRILLING THROUGH THE PART)
N8 G00 Z0.5                               (RETURN FROM DRILLED HOLE)
N9 X10.0 Z3.0 T0100                       (TOOL CHANGE POSITION)
N10 M01                                   (OPTIONAL STOP)
N11 M100                                  (*** SYNCHRONIZATION M100 ***)

(T04 = ROUGH BORING = 80-DEG TOOL)
N12 T0400 M41                             (TOOL CHANGE + LOW GEAR)
N13 G97 S800 M03                          (COMMON SPINDLE SPEED/ROTATION)
N14 G00 X1.5 Z0.1 M08                     (START POINT FOR G71)
N15 G71 P16 Q23 U-0.06 W0.005 D2000 F0.025 (ROUGHING CYCLE - INTERNAL)
N16 G00 X3.86                             (START OF CHAMFER)
N17 G01 X3.5 Z-0.08 F0.012                (FRONT CHAMFER 0.08)
N18 Z-1.18 R-0.25                         (3.5 DIA + CORNER ROUNDING)
N19 X2.0 R-0.15                           (SHOULDER + CORNER ROUNDING)
N20 Z-2.18                                (2.0 DIAMETER)
N21 X1.6 Z-3.75                           (TAPER)
N22 Z-5.1                                 (1.6 DIAMETER)
N23 X1.4                                  (CLEAR OFF THE BORE)
N24 G00 G40 X10.0 Z3.0 T0400              (TOOL CHANGE POSITION)
N25 M01                                   (OPTIONAL STOP)
N26 M101                                  (*** SYNCHRONIZATION M101 ***)

(T06 = FINISH BORING = 55-DEG TOOL)
N27 T0600 M42                             (TOOL CHANGE + HIGH GEAR)
N28 G97 S950 M03                          (COMMON SPINDLE SPEED/ROTATION)
N29 G00 G42 X1.5 Z0.1 M08                 (START POINT FOR G70)
N30 G70 P16 Q23                           (FINISHING CYCLE - INTERNAL)
N31 G00 G40 X10.0 Z3.0 T0600              (TOOL CHANGE POSITION)
N32 M01                                   (OPTIONAL STOP)
N33 M102                                  (*** SYNCHRONIZATION M102 ***)
N34 M30                                   (PROGRAM END - REAR TURRET)
%
==========================================================================
```

```
(FRONT TURRET - EXTERNAL CUTTING)
(T01 = ROUGH TURNING - 80-DEG TOOL)
N1 M100                                   (*** SYNCHRONIZATION M100 ***)
N2 G20                                    (IMPERIAL UNITS)
N3 T0100                                  (TOOL CHANGE)
N4 G00 X4.2 Z0 T0101 M08                  (START POSITION FOR FACE CUT)
N5 G01 X1.4 F0.015                        (FACE CUT PAST THE BORE)
N6 G00 Z0.1                               (CLEAR OFF THE FACE)
N7 X6.1 Z0.1                              (START POINT FOR G71)
N8 G71 P9 Q17 U0.06 W0.005 D1800 F0.025   (ROUGHING CYCLE - EXTERNAL)
N9 G00 X4.04                              (START OF CHAMFER)
N10 G01 X4.4 Z-0.08 F0.012                (FRONT CHAMFER 0.08)
N11 Z-0.94 R0.12                          (4.4 DIA + CORNER ROUNDING)
N12 X4.875 C-0.06                         (SHOULDER + CORNER CHFR ?K-0.06)
N13 Z-1.81                                (4.875 DIAMETER)
N14 X5.1532 Z-3.4                         (5-DEGREE TAPER)
N15 X5.58 R-0.1                           (SHOULDER + CORNER ROUNDING)
N16 Z-4.0                                 (5.58 DIAMETER)
N17 X6.2                                  (CLEAR OFF THE DIAMETER)
N18 G00 G40 X10.0 Z3.0 T0100              (TOOL CHANGE POSITION)
N19 M01                                   (OPTIONAL STOP)
N20 M101                                  (*** SYNCHRONIZATION M101 ***)
```

```
(T03 = FINISH TURNING - 55-DEG TOOL)
N21 T0300                            (TOOL CHANGE)
N22 G00 G41 X6.1 Z0.1 T0303 M08      (START POINT FOR G70)
N23 G70 P9 Q17                       (FINISHING CYCLE - EXTERNAL)
N24 G00 G40 X10.0 Z3.0 T0300         (TOOL CHANGE POSITION)
N25 M01                              (OPTIONAL STOP)
N26 M102                             (*** SYNCHRONIZATION M102 ***)
N27 M30                              (PROGRAM END - FRONT TURRET)
%
```

Study the program carefully. It follows the structure presented earlier. There are three tools for the rear turret, and two tools for the front turret. The program starts with only the drill working on the rear turret, while the front turret is idle. When the drilling is completed, the waiting code M100 forces both turrets to start machining at the same time (rough boring and rough turning). There is a good chance that one of these operations completes earlier than the other. Waiting code M101 guarantees that no finishing takes place, until both roughing operations are completed. Finally, both finishing operations will be completed before program ends (waiting code M102 before M30).

Spindle speeds are programmed for only one turret in this case, at the programmer's discretion. G96 is an option, and so is programming the spindle speed for either turret.

From the example, it is apparent that the programming process for four-axis CNC is no different from the process for two-axis lathes. There are some additions and several considerations, but the main focus expected from the programmer is the most efficient use of both turrets.

21 PALLET CHANGERS

Pallet changers have been a part of CNC field for several decades. By grouping definition, pallet changers belong into the area of setup and work holding. Their purpose is quite simple - to minimize the idle time (non-profit time) between machining one part and machining the next part, particularly when the setup is complex or takes a long time. Idle time is always undesired side effect of any manufacturing. Machine tool builders have always made great efforts to minimize, and even *eliminate*, all or some idle time. One solution that has proven to be very effective in manufacturing is the concept of using special *pallets* as part of the machine setup. Pallets (or pallet tables) are special work tables that can be moved in and out of the working area. The only requirement is that the machine control system supports this feature, which means it has to be supported on both the machine side (hardware) and the control side (software).

The process of pallet changing is called *Automatic Pallet Changer*, abbreviated as APC

Types of Automatic Pallets

For horizontal and even vertical machining centers, manufacturers have developed a number of different pallet designs. Their design, purpose a general functions always share the same characteristics:

- **Design:** Two or more pallets, depending on the machine model
- **Function:** One pallet in the work area (active), the other pallet(-s) outside of the work area (inactive)
- **Utilization:** To machine part *A* on one pallet while part *B* is being set at the machine on another pallet

Each pallet can be loaded with the same part or a different part, and machining on both pallets can perform the same operation or another operation, depending on actual needs.

Regardless of the specific designs, the majority of pallets belong into one of two general types:

- **Rotary type APC**
- **Shuttle type APC**

The main differences between the two designs is the *method of transfer* - how the pallet table is loaded and unloaded.

Rotary Pallets

As the name of the design suggests, a *rotary* pallet is like a turntable with two halves. One half is the active area (used for machining), the other is the inactive area (used for setup). This design is not suitable for multi-part or multi-operation setup.

Shuttle Pallets

Shutlle type of pallets is also called *parallel* type. Shuttle type is designed with two sets of parallel sliders (or rails) that hold the pallet table. Multiple pallets (more than two) are also available.

Both types of pallet tables are illustrated on the next page.

Setup and Work Areas

A pallet table can be defined as a special purpose work table that can be moved between the *setup area* and the *work area* (machining area) by part program. Pallet table is similar to a regular fixed table, and may have T-slots and/or tapped holes built-in. For either setup, both pallet types are loaded from the machine front area (operator's area).

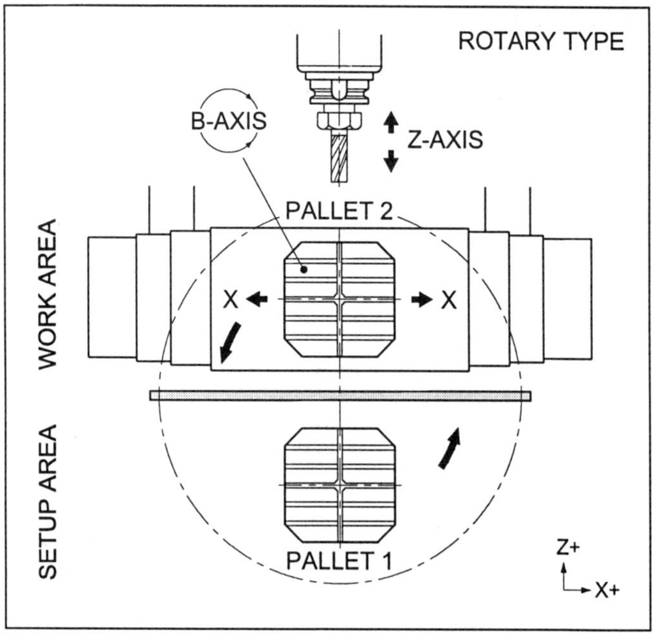

The illustrations show a typical layout for both pallet types, applied to a typical CNC horizontal machining center. The Y-axis is perpendicular to the view shown.

The work setup area is outside the machine, the work machining area is inside the machine.

In a typical operation, the current part is located on *Pallet 1*. When the setup is completed, the part is moved to the work area. While the machining takes place on part located on *Pallet 1* table, the operator sets up another part on *Pallet 2* table.

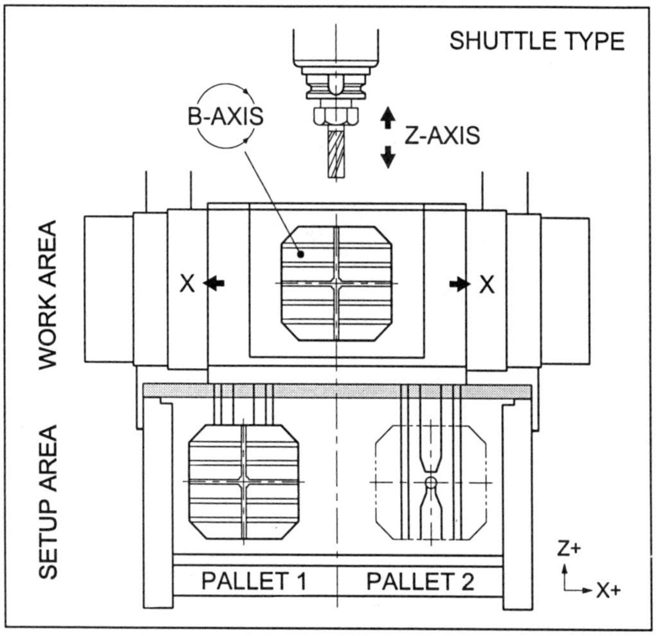

When describing the setup and work areas of a pallet system, there are two terms that manufacturers use to describe certain part of the equipment. One term is the *receiver* (also known as the *locator*), the other is the *loader* (also known as *pallet loader* or *transfer system*). The receiver holds the loaded pallet during machining and prevents any shifts. The pallet loader transfers either pallet between the two areas. *Loading* and *unloading* refers to moving the pallet into the machining area and back to the setup area respectively.

Understanding the concept of two pallet tables is important for understanding concept of multiple pallet tables. Some CNC machining centers can be supplied with up to twelve pallets in various configurations (round, rectangular, oval, etc.). The general concept of pallets is sometime grouped under the name *palletization* and it is as simple as it is ingenious.

In short, using pallets efficiently provides a greatly improved method of setup.

Programming Methods

In order to be able to write a part program using APC, the CNC programmer has to be fully knowledgeable about the actual pallet design and functions. There are significant differences between manufacturers, and the examples provided in this chapter should only serve as general guidelines.

First, there is the main pallet changer function M60. This is a miscellaneous function similar to the ATC (Automatic Tool Changer) function M06 (make sure to choose the right one). Second, for CNC horizontal machining centers, there will be the additional B-axis, as well as table clamp and unclamp functions, for example, M78 (clamp) and M79 (unclamp), if the design requires them.

Another important programming entry is the machine zero return command. It is quite common that machining centers equipped with pallets will have two machine zeros - yes, *two* machine reference positions. While the common G28 command is available to move the selected axes to the primary machine zero position, there is also the G30 command, available to move the selected axis (axes) to the secondary machine zero position. The main reason for the two reference points is to align the pallet table with the APC position.

Program output for the rotary type pallet changer is a little shorter than output for the shuttle type, as it involves fewer steps.

M60 Function

Virtually all machine tool builders support M60 as the common standard function for automatic pallet changing (APS). In order to use it successfully, one important condition must be met - the pallet table to be loaded or unloaded must be at machine zero position (primary or secondary). Usually only one axis is required at this position. The primary usage of M60 is shown in the general program format. As described already, M60 is only a part of a larger process.

General Format

The following program emphasizes the pallet changing on a typical shuttle pallet system. It can be easily adapted (simplified) to a rotary pallet system. In both cases, one pallet is in the machining area, the other is in the setup area.

For example, a typical shuttle type automatic pallet changer (APC) may have this structure:

```
G21 (G20)                                       (UNITS SELECTION)
G17 G40 G80                                     (START-UP BLOCK)
/ G91 G28 X0 Y0 Z0                              (XYZ AXES AT PRIMARY ZERO)
G28 B0                                          (B AXIS AT PRIMARY ZERO)
M60                                             (APC - LOAD PALLET 1)
  <... Machining part on Pallet 1 ...>
G91 G28 X0 Y0 Z0                                (XYZ AXES AT PRIMARY ZERO)
G28 B0                                          (B AXIS AT PRIMARY ZERO)
M60                                             (APC - UNLOAD PALLET 1)
G30 X0                                          (X AXIS AT SECONDARY ZERO)
M60                                             (APC - LOAD PALLET 2)
  <... Machining part on Pallet 2 ...>
G30 X0                                          (X AXIS AT SECONDARY ZERO)
M60                                             (APC - UNLOAD PALLET 2)
M30                                             (PROGRAM END)
%
```

Keep in mind that tool change position for horizontal machining centers is aligned at Y0.

Programming Example

This example is for a CNC vertical machining center, with a shuttle pallet located at the machine side, providing easy access to the machine work area. The objective of the example is to illustrate an actual contouring toolpath for two tools (roughing and finishing). The same machining will take place on both pallets - while one pallet is doing the work, the other part is prepared on the other pallet.

Vertical example has been chosen, as it provides a shorter program and reflects common programming methods for CNC vertical machining centers without a pallet changer.

Initial Conditions

Certain conditions have to be satisfied before the program can be developed and used. Some of these conditions are programmable, others have to be performed at the machine. These conditions apply to:

- **Position of machine axes**
- **Position of pallets**
- **Location of cutting tools**

Position of machine axes XYZ has to be at the primary machine reference point, established by the G28 command. Position confirmation lights at the control must be ON.

Position of pallets is on the side of the machine, and both pallets are unloaded (away from the machine area).

Location of cutting tools is the tool magazine, both tools are registered for random type *Automatic Tool Changer* (ATC), and the spindle is empty at the beginning of the program.

These same initial conditions must exist for each part - *i.e.,* the initial conditions before the work starts must be reestablished at the end of program.

Part Program

This is a general illustration how pallets work for CNC machines, nothing more. To maintain a realistic situation, the contouring program for the drawing (see illustration) will be a single roughing cut and a single finishing cut, using two tools:

- **T01 = ∅15 mm end mill for roughing**
- **T02 = ∅10 mm end mill for finishing**

The programmed contour is identical for both cutters and will be programmed in the form of a subprogram, for a shorter input and easier editing, if necessary.

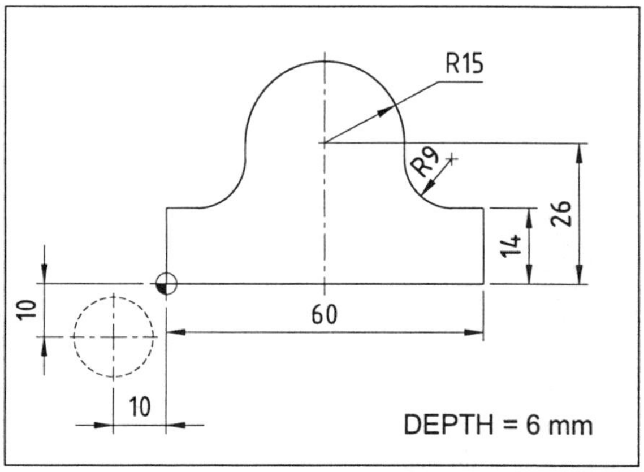

The drawing also shows the start and end position for both tools. This is a rather small object that can be portion of a much larger part.

In the next step, the machine layout will be shown - the contouring subprogram can be developed, to be used later by both tools at both pallets.

Subprogram

In the subprogram, only the XY coordinates will be programmed, with no depth, no cutter radius offset and no feedrate included - both will be supplied in the main program, as needed:

```
O7001 (SUBPROGRAM FOR THE COMMON CONTOUR)
N101 G41 G00 X0                          (MOVE TOWARDS THE CONTOUR)
N102 G01 Y14.0                           (--> )
N103 X6.0                                (  | )
N104 G03 X15.0 Y23.0 I0 J9.0             (  | )
N105 G01 Y26.0                           (  | )
N106 G02 X75.0 I15.0 J0                  (  ... CONTOURING ...)
N107 G01 Y23.0                           (  | )
N108 G03 X54.0 Y14.0 I9.0 J0             (  | )
N109 G01 X60.0                           (  | )
N110 Y0                                  (  | )
N111 X-10.0                              (<-- )
N112 G40 G00 Y-10.0                      (MOVE TO STARTING POINT)
N113 M99                                 (END OF SUBPROGRAM)
%
```

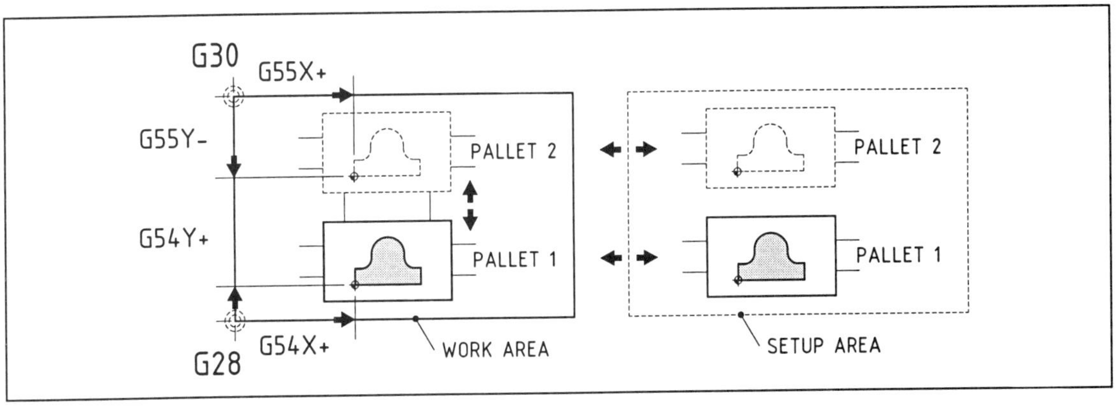

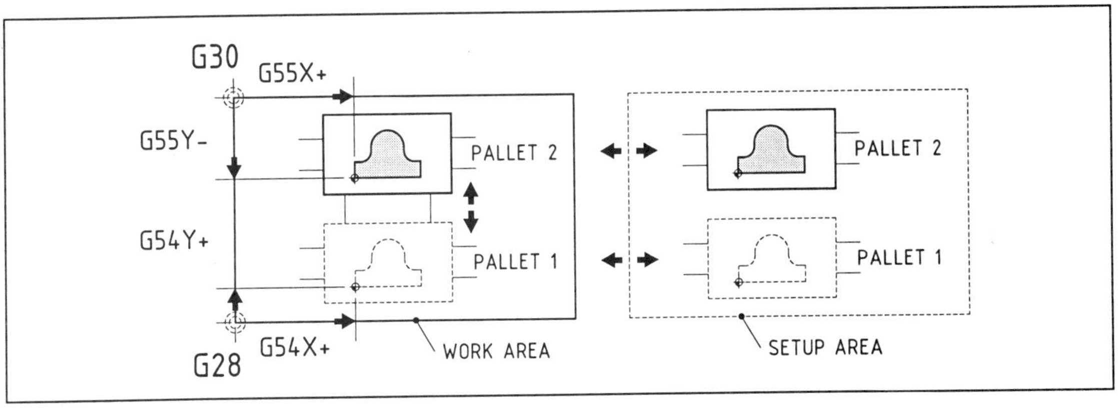

Once the subprogram is completed, the main program will focus mainly on the cutting tool and pallet selections, plus other data normally programmed.

Main Program

The main program has to include the initial conditions already established. Program starts from the *primary* machine zero position, with no tool in the spindle (T99 is an empty spindle).

```
(T01 = 15 MM ROUGHING END MILL - USES H01 AND H91)
(      D61 = CUTTER RADIUS FOR ROUGHING = 8.000 SUGGESTED FOR 0.5 MM STOCK)
(T02 = 10 MM FINISHING END MILL - USES H02 AND H92)
(      D51 = CUTTER RADIUS FOR FINISHING = 5.000 SUGGESTED)
N1 G21
N2 G17 G40 G80 T01
N3 M06
/ N4 G91 G28 X0 Y0 Z0                       (INITIAL POSITION)
N5 M60                                      (PALLET 1 LOAD)
N6 G90 G54 G00 X-10.0 Y-10.0 S800 M03 T02   (RAPID TO START POSITION)
N7 G43 Z25.0 H01 M08                        (H01 OFFSET FOR PALLET 1)
N8 G01 Z-6.0 F500.0                         (CUTTING DEPTH)
N9 M98 P7001 D61 F200.0                     (ROUGH PART ON PALLET 1)
N10 G28 Z25.0 M05                           (TOOL CHANGE POSITION)
N11 M01                                     (OPTIONAL STOP)

N12 T02
N13 M06
N14 G90 G54 G00 X-10.0 Y-10.0 S950 M03 T01  (RAPID TO START POSITION)
N15 G43 Z25.0 H02 M08                       (H02 OFFSET FOR PALLET 1)
N16 G01 Z-6.0 F500.0                        (CUTTING DEPTH)
N17 M98 P7001 D51 F175.0                    (FINISH PART ON PALLET 1)
N18 G28 Z25.0 M09                           (TOOL CHANGE POSITION)
N19 G28 X-10.0 Y-10.0 M05                   (XY TO PRIMARY HOME POS.)
N20 M60                                     (PALLET 1 UNLOAD)
N21 M01                                     (OPTIONAL STOP)

(==== PALLET 1 TO PALLET 2 CHANGE ====)

N22 T01
N23 M06
N24 G30 Y0
N25 M60                                     (PALLET 2 LOAD)
N26 G90 G55 G00 X-10.0 Y-10.0 S800 M03 T02  (RAPID TO START POSITION)
N27 G43 Z25.0 H91 M08                       (H91 OFFSET FOR PALLET 2)
N28 G01 Z-6.0 F500.0                        (CUTTING DEPTH)
N29 M98 P7001 D61 F200.0                    (ROUGH PART ON PALLET 2)
N30 G28 Z25.0 M05                           (TOOL CHANGE POSITION)
N31 M01                                     (OPTIONAL STOP)

N32 T02
N33 M06
N34 G90 G55 G00 X-10.0 Y-10.0 S950 M03 T99  (RAPID TO START POSITION)
N35 G43 Z25.0 H92 M08                       (H92 OFFSET FOR PALLET 2)
N36 G01 Z-6.0 F500.0                        (CUTTING DEPTH)
N37 M98 P7001 D51 F175.0                    (FINISH PART ON PALLET 2)
N38 G28 Z25.0 M09                           (TOOL CHANGE POSITION)
N39 G30 Y-10.0                              (Y-AXIS TO SECONDARY HOME POS.)
N40 M60                                     (PALLET 2 UNLOAD)
N41 G91 G28 X0 Y0                           (XY-AXES TO PRIMARY HOME POS.)
N42 M06                                     (T99 = EMPTY TOOL TO SPINDLE)
N43 M30                                     (PROGRAM END)
%
```

As you see, the programming APC is not much different than programming a single part layout. The most important differences have been underlined. For CNC horizontal machining centers, only the B-axis will be added, but the basic concept does not change.

22 WORKING WITH PLANES

By established standards, planes in CNC environment are related to mathematical planes. A plane is typically defined as a ***two-dimensional area bounded by two known lines or a single known arc***. A plane can also be defined by three points that do not lie on a single line. Any plane can be rotated or translated in space. In two-dimensional machining (2D), a plane is defined by any two standard axes (XYZ) and is programmable using preparatory commands (G-codes). In three-dimensional machining (3D), a plane can be set at any orientation and is not programmable. However, any plane can be defined in a CAM system for 3D work.

Mathematical Planes

As there are three axes named X, Y and Z, a mathematical plane is defined by any two axes combined, using the same letters. There is a standard international convention that dictates the definition of mathematical planes.

The best way to remember their exact definition, is to write down the three axes twice in a row without spaces, one after each other, and separate the resulting six letters into three pairs - *see illustration*.

Planes are defined by axes in their positive direction.

Axes XYZ written twice ...
... become XYZXYZ

When separated into three pairs, they become PLANES, defined as:

XY ZX YZ

*Understand the **mathematical** planes first - before defining the **machine** planes.*

Mathematical planes are always identified by positive axis direction, whereby the first letter of the definition represents the horizontal axis, and the second letter represents the vertical axis of the selected plane.

For example, in the YZ plane, the letter Y represent the horizontal axis, letter Z the vertical axis.

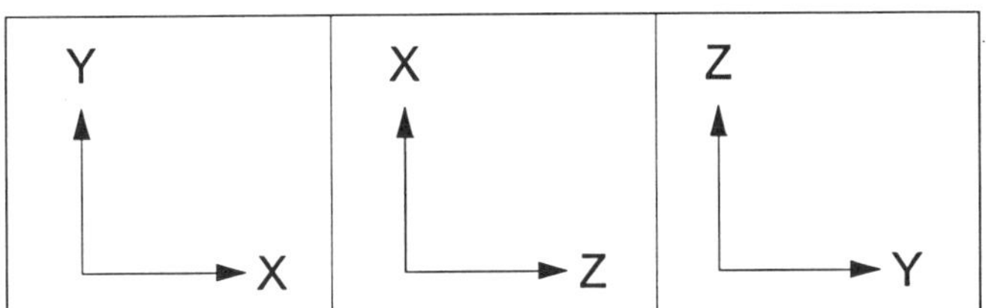

TOP - XY	FRONT - ZX	RIGHT - YZ

MATHEMATICAL DEFINITION OF PLANES

Planes of a CNC machine (machining center) are identical to mathematical planes, but one orientation appears different.

Compare the two illustrations and see the difference - the *front plane has a different orientation* on the machining center than its mathematical definition.

Let's look at the machine planes in more detail and see how they can be used in CNC programming.

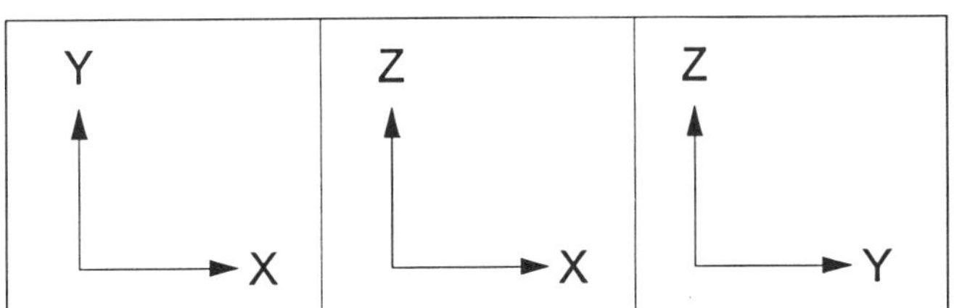

TOP - XY	FRONT - XZ	RIGHT - YZ

PLANES ON A VERTICAL MACHINING CENTER

Machine Planes

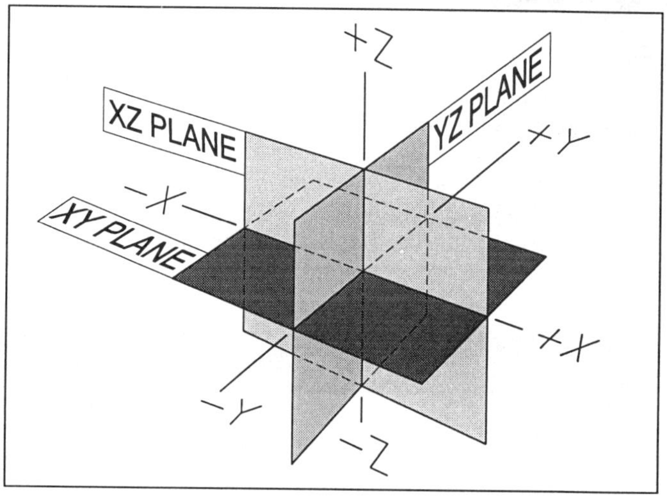

All three machine planes follow the standard mathematical definitions. In order to understand the concept of planes, it is important to look at each plane from the same relative direction. The viewing direction must be perpendicular to the plane, and the horizontal axis must point to the right, while the vertical axis of the plane must point to the left.

The illustration at right shows the three planes on a vertical machining center, where XZ machine plane is really the ZX mathematical plane, with the Z-axis being horizontal.

This fact requires some special consideration of the ZX/XZ plane, programmed with the G18 preparatory command. G17 (XY) and G19 (YZ) planes are normal and need no special considerations. Keep in mind that the ZX machine plane is not a different plane, it is the same mathematical plane, just viewed from a 'wrong' direction.

One general rule always applies:

Planes on CNC machines follow the rules of mathematical planes

Before getting into more detail, it is important to look at the commands available in CNC programming to define planes. There are three preparatory commands (G-codes) available.

G-codes for Plane Selection

From the list of G-codes, three commands define three planes:

G17	**XY plane**
G18	**ZX plane**
G19	**YZ plane**

None of the three commands require additional data (arguments). Many CNC programs include the preparatory command G17 (XY plane selection), usually at the beginning of the program in the so called 'safe block' or 'status block'. Including the G17 in the program is often done in an arbitrary way by part programmers, with no specific thinking of the reasons - and - yes, consequences. Justifications such as *'If other programmers do it, it must be right'* - *'If the manual says so, it must be right'* are rather feeble attempts to include the G17 command in the program without actually knowing why. Such programming approach does not follow the suggested professional method to make an important program selection - and the G17 command certainly falls into the category of important selections and should be treated as such. A professional method follows the systematic and logical approach to CNC programming - *every entry in the program must be justified.* This particular view is preferable - every program entry is there for a reason - it must be *justified*. This chapter describes the meaning of G17, as well as the other two G-codes relating to planes.

Effect of Planes in Programming

As the G17 command is the most common in programming, it makes sense to understand it first. Preparatory command G17 selects the XY working plane within the machine coordinate system. Perhaps the reason why plane selection G17 is so easily ignored is that most jobs do not require machining *in any other plane*. There are three planes available to the programmer - the required toolpath can be developed in any one of the three planes. Correct plane selection in the program is very important, as it influences the following programming functions and activities:

- **Circular interpolation movements**
- **Cutter radius offset**
- **Fixed cycles**
- **Coordinate system rotation**

On a typical daily basis, the most profound importance of planes can be found in circular motions.

Planes and Circular Motion

Consider the following program, using circular interpolation to machine a circular pocket:

The drawing shows the pocket to be machined, including the lead-in and lead-out linear and circular motions. For simplicity, the pocket center is X0Y0, Z0 is at the top of the part and tool used will be Ø16 mm center-cutting end mill in climb milling mode. This method reflects typical circular pocket machining with a cutter that can also clean-up the bottom in a continuous toolpath.

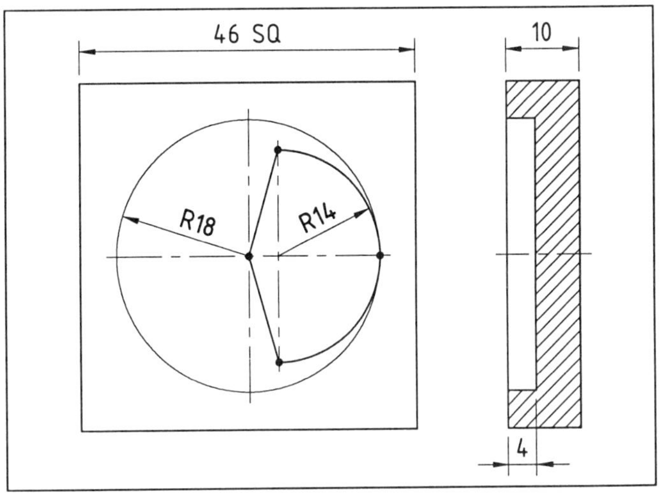

Here is a listing of such a program that will work equally well on both the vertical and horizontal machines:

```
N1 G21
N2 G17 G40 G80
N3 G90 G54 G00 X0 Y0 S950 M03
N4 G43 Z3.0 H01 M08
N5 G01 Z-4.0 F150.0
N6 G41 X4.0 Y-14.0 D51 F200.0
N7 G03 X18.0 Y0 I0 J14.0        (AXIS DATA IS INCLUDED IN BLOCK N7)
N8 I-18.0                       (AXIS DATA IS MISSING IN BLOCK N8)
N9 X4.0 Y14.0 I-14.0 J0         (AXIS DATA IS INCLUDED IN BLOCK N9)
N10 G40 G01 X0 Y0 M09
N11 G28 Z-4.0 M05
N12 M01
```

Note the three blocks - N7, N8, and N9 - and the attached comments. The G03 command is modal, and does not need repeating. Both blocks N7 and N9 include all data required to machine the particular arc. What about block N8? Apart from the G03, which is modal, and the unnecessary J0, which does not have be included - *there is no target point!* There are *no coordinates* specified for any axis.

If you have ever done or at least seen a program similar to this one, you also know that it will work perfectly well, with no problems whatsoever. The reason is not that the XY coordinates are modal and need no repeating - XY coordinates are *not* modal in the same sense as motion commands, for example. During linear motions, if a coordinate is missing, it will be ignored. That is not possible in a circular motion. What the control will do is to ***substitute*** the missing axis or axes. However, there are three main axes available and only two of them are needed to cut an arc or a circle. The question - which two axes - is answered by the *plane selection*.

Any arc or circle can only be machined in a single plane - there is no such a thing as a three-dimensional arc (helical motion is not a circular motion). In the example, both axes are missing and the control system has to substitute them. As G17 is programmed (or even selected by a default), the substitution will involve one or both axes from the active plane selection. G17 is the active plane for the program example, G17 is also defined as the XY plane, so X and/or Y axis will be substituted - or 'assumed' - as missing in the block N8.

```
N7 G03 X18.0 Y0 I0 J14.0          (AXIS DATA IS INCLUDED IN BLOCK N7)
N8 G03 X18.0 Y0 I-18.0 J0         (AXIS DATA IS INCLUDED IN BLOCK N8)
N9 G03 X4.0 Y14.0 I-14.0 J0       (AXIS DATA IS INCLUDED IN BLOCK N9)
```

Making a program to look like those three blocks is correct but cannot be justified for any production programs.

Comparing Arc Motion in Planes

At the end of this chapter, there is a complete example illustrating the arc motion in all three planes (including the use of cutter radius offset). As the current focus is on planes and arc motion within a plane, compare the two plane layouts introduced earlier, this time with included arc motions.

The three standard mathematical planes show the definition of an arc motion direction.

This direction is always relative to the direction of the ***positive*** *axes that define the plane.*

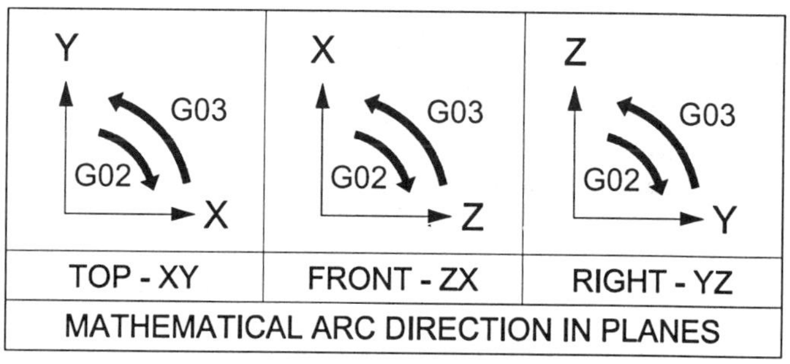

When the mathematical planes are applied to a CNC machining center, it becomes apparent that while the XY top plane and the YZ right side plane match in all respects, the front plane ZX/XZ does not match.

All machine planes adhere to the mathematical standard, so the front machine plane needs to match.

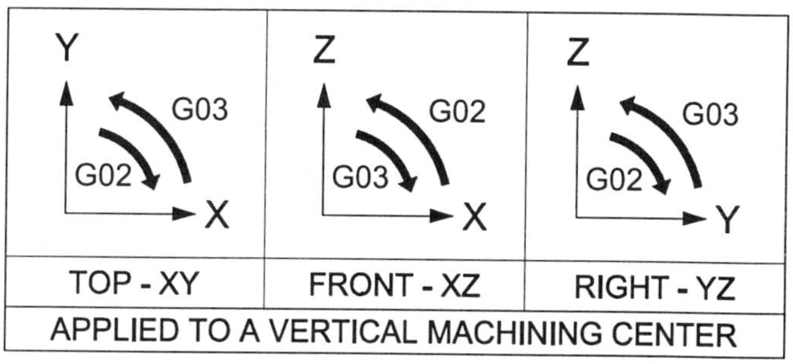

When standard planes are applied to machine planes, there is one *apparent* error. For the front plane, the arrow directions do not seem to match the standard G-codes for circular interpolation (G02 = clockwise, G03 = counterclockwise). The key word here is - *apparent*. In the part program, the front plane selection uses G18 command. ***Always exercise care when programming in G18 mode!***

When both axes are specified for a circular motion, they always take a priority over any active plane.

What exactly is involved in matching the plane orientation? The illustration shows an isometric view of the machining planes.

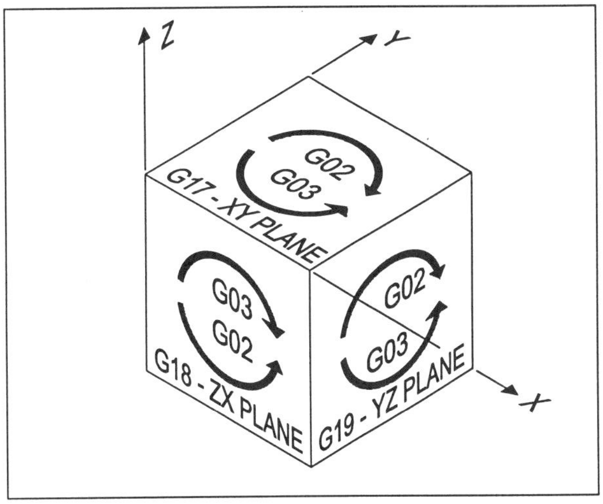

In order to understand the actual process of matching the machine plane to the standard plane, there are four steps involved:

- **Step 1 - Standard plane to match**
- **Step 2 - Standard plane mirrored**
- **Step 3 - Standard plane rotated**
- **Step 4 - Standard plane = Machine plane**

The illustration below show the four steps:

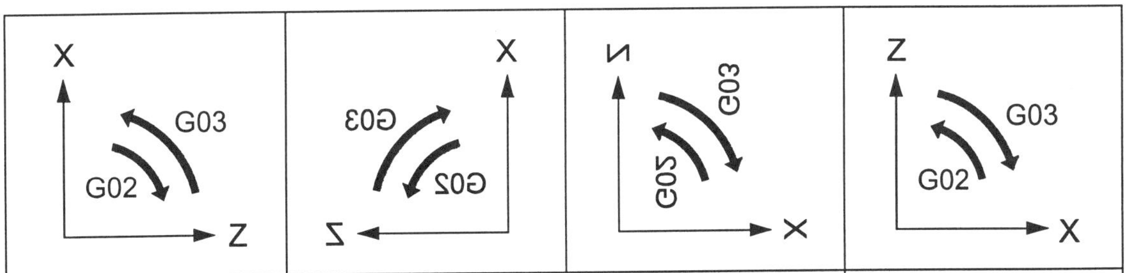

STANDARD MATHEMATICAL ZX plane	STANDARD ZX PLANE MIRRORED	ZX PLANE ROTATED AFTER MIRRORING	G18 PLANE FINALIZED

Equally important is the plane selection as it relates to the cutter radius offset.

Planes and Cutter Radius Offset

A two-dimensional contour programmed in one plane is no different than a similar contour programmed in another plane. In all cases, the current plane axes and their orientation always prevail. A fairly common knowledge of the cutter radius offset direction (to the *left* or to the *right*) is that is must be viewed along the direction of motion. A fact that is less known, is that this direction must be viewed in the selected standard plane. Direction of the cutter radius offset is programmed using the standard preparatory commands:

G41 - Cutter radius offset to the *left* of cutting direction *within the standard plane*

G42 - Cutter radius offset to the *right* of cutting direction *within the standard plane*

G40 - Cutter radius offset cancelled

Similar to all circular motions, the cutter radius offset direction matches the G17 (XY) plane and the G19 (YZ) plane. However, the G18 plane is different - see the comparative illustrations on the next page. Again, keep in mind that the difference is only apparent, not real.

This illustration shows the three standard mathematical planes. It also shows cutting direction and how the cutter radius offset is applied in individual planes.

Note that the toolpath includes both linear and circular motions. From left to right, the three standard planes are G17 - G18 - G19

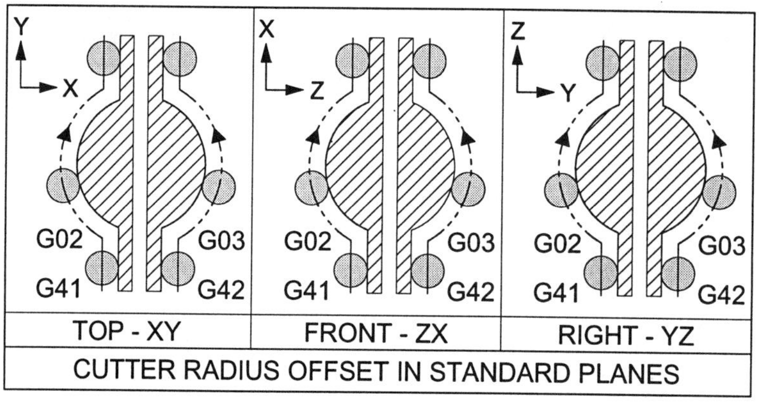

CUTTER RADIUS OFFSET IN STANDARD PLANES

The second illustration shows the standard planes applied to the CNC milling machines or machining centers. Note that the TOP and RIGHT planes match the standard planes. The middle illustration shows the *front* plane oriented to match the machine axes orientation. From left to right, the three planes are G17 - G18 - G19

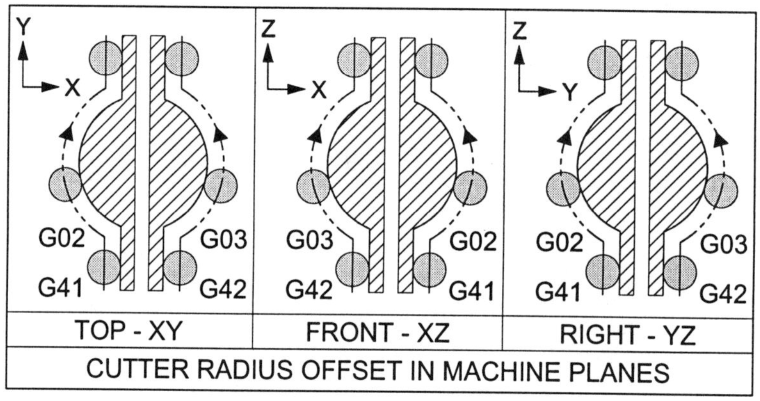

CUTTER RADIUS OFFSET IN MACHINE PLANES

The illustration below shows a combination of linear and circular motions. It will be used to program a single contour in all planes and both directions (1 to 6 and 6 to 1). Coordinates are provided.

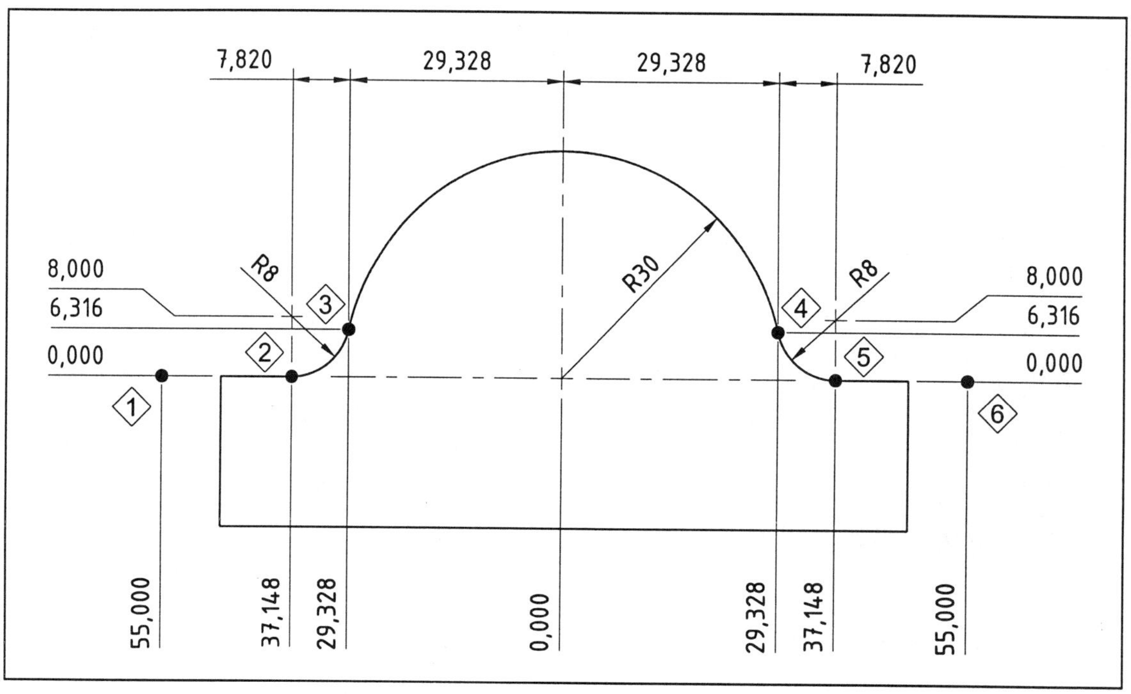

Working With Planes In Detail

The previous page illustration shows all coordinate points required to program a contour in any of the three planes. For the *top plane* (G17), a flat bottom end mill is used, for the *front plane* (G18) and the *side plane* (G19), a ball nose end mill is used. The examples are intended for educational purposes only, rather than actual machining. *The G-codes in the middle refers to the large radius (R30.0).*

G17 with G41 and G02

This method of programming will be the most common for the given contour. Machining will take place in the climb milling mode, and the direction of both the cutter radius offset and the circular motion matches the illustration. There is no depth of cut (change block N4 if needed).

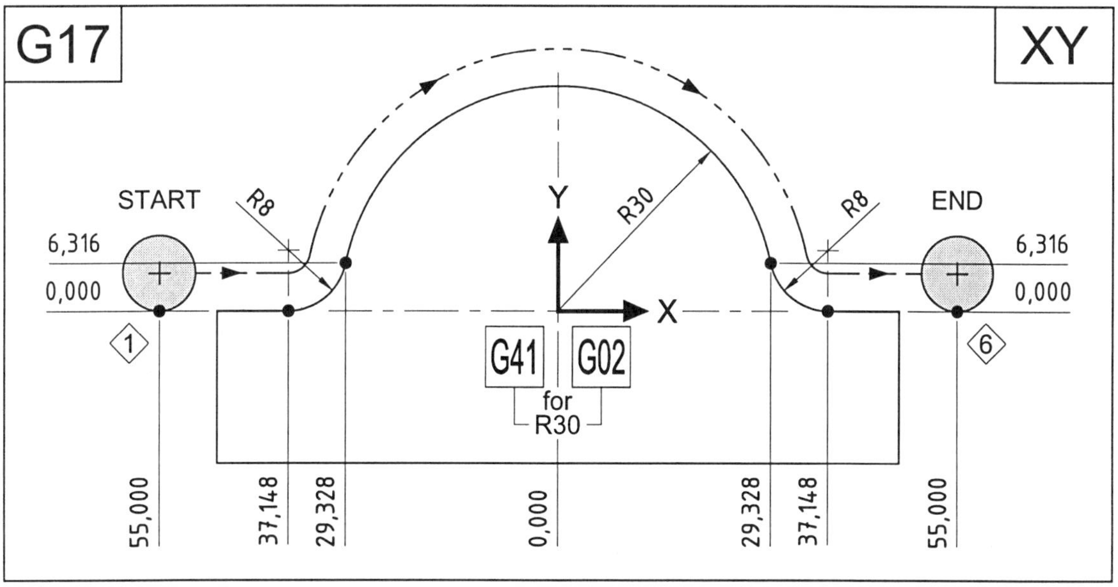


```
(*** G17 - XY PLANE - CUTTING DIRECTION FROM POINT 1 TO POINT 6 ***)
(*** USES I AND J ARC VECTORS ***)
N1 G21                                  (METRIC)
N2 G17 G40 G80                          (INITIALIZATION)
N3 G90 G54 G00 X-55.0 Y10.0 S1000 M03   (START POSITION)
N4 G43 Z0 H01 M08                       (TOOL LENGTH OFFSET)
N5 G41 G01 Y0 D51 F200.0                (CUTTER RADIUS OFFSET - START)
N6 X-37.148 F100.0                      (FIRST LINEAR MOTION)
N7 G03 X-29.328 Y6.316 I0 J8.0          (R8 ARC - CCW)
N8 G02 X29.328 I29.328 J-6.316          (R30 ARC - CW)
N9 G03 X37.148 Y0 I7.82 J1.684          (R8 ARC - CCW)
N10 G01 X55.0                           (LAST LINEAR MOTION)
N11 G40 G00 Y10.0 M09                   (CUTTER RADIUS OFFSET - END)
N12 G28 X55.0 Y10.0 Z0 M05              (MACHINE ZERO RETURN)
N13 M30                                 (PROGRAM END)
%
```

For this and all subsequent examples in this section, the D51 offset will be set at the control system for the actual cutter radius, which then can be modified for tool wear. Although the climb milling method of machining (G41) is preferable for most contouring on CNC mills and machining centers, the conventional mode (G42) will also be shown, for better understanding of the subject.

G17 with G42 and G03

This toolpath is similar to the previous one, but it represents the opposite direction of the previous example. Climb milling mode has changed to conventional milling mode and cutter radius offset to the left has changed to cutter radius offset to the right.

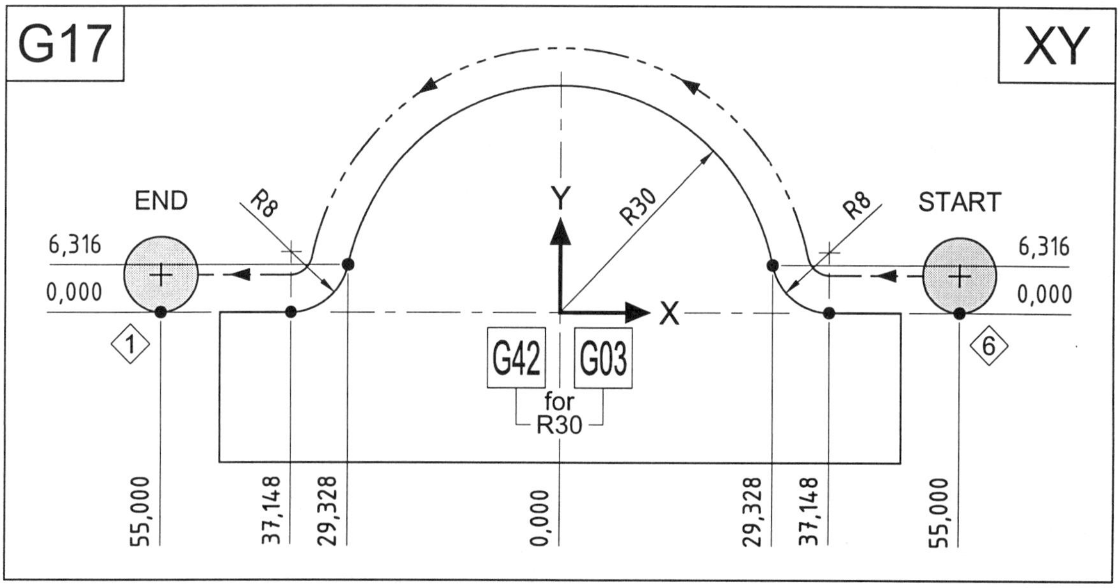


```
(*** G17 - XY PLANE - CUTTING DIRECTION FROM POINT 6 TO POINT 1 ***)
(*** USES I AND J ARC VECTORS ***)
N1 G21                                   (METRIC)
N2 G17 G40 G80                           (INITIALIZATION)
N3 G90 G54 G00 X55.0 Y10.0 S1000 M03     (START POSITION)
N4 G43 Z0 H01 M08                        (TOOL LENGTH OFFSET)
N5 G42 G01 Y0 D51 F200.0                 (CUTTER RADIUS OFFSET - START)
N6 X37.148 F100.0                        (FIRST LINEAR MOTION)
N7 G02 X29.328 Y6.316 I0 J8.0            (R8 ARC - CW)
N8 G03 X-29.328 I-29.328 J-6.316         (R30 ARC - CCW)
N9 G02 X-37.148 Y0 I-7.82 J1.684         (R8 ARC - CW)
N10 G01 X-55.0                           (LAST LINEAR MOTION)
N11 G40 G00 Y10.0 M09                    (CUTTER RADIUS OFFSET - END)
N12 G28 X-55.0 Y10.0 Z0 M05              (MACHINE ZERO RETURN)
N13 M30                                  (PROGRAM END)
%
```

G17 plane uses I and J vectors for circular motions

Toolpaths generated in the default plane G17 (XY) are most familiar to part programmers and operators alike. The direction of circular motion clockwise or counterclockwise - G02 or G03 - is easy to visualize, and so is the direction of the cutter radius offset to the left or to the right of the contour - G41 or G42. For the beginner, it is absolutely necessary to understand how the circular motions and cutter radius offset work in the standard (default) plane G17, before attempting to do any programming in the other two planes (G18 and G19).

Programming a contour toolpath in the XZ/ZX plane - G18 - requires the most care. The following two examples will show not only the direction of circular motion and cutter radius offset - they will also show two complete program examples.

Changing the active plane in the program is not exactly a daily part of CNC programming. Most jobs do not require the use of this feature. At the same time, when a job that requires selection of a different plane comes along, the programmer should be prepared. Machining in different planes requires a good knowledge of the actual applications - where the concept can be used.

The illustration at right shows a typical *two-dimensional* toolpath that can be programmed using the G18 (ZX) plane. Although the examples shown in this chapter relate only to a single toolpath across the cylinder, it is easy to put two of them together and develop a suitable subprogram for a similar toolpath.

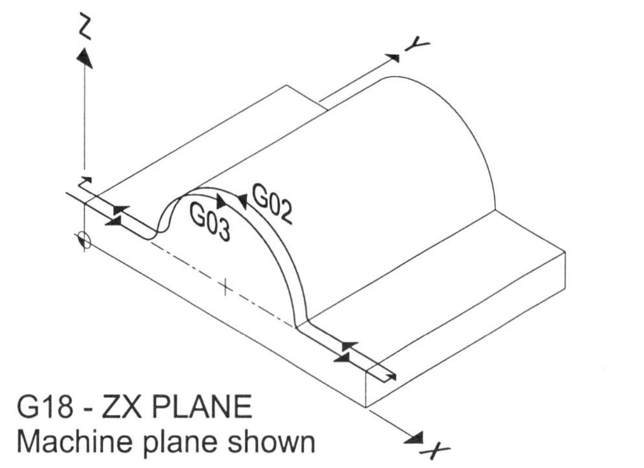

G18 - ZX PLANE
Machine plane shown

Machining in the three available planes is always along two axes at a time. Yet, with little ingenuity, a simple 3D part can be programmed. Cylinders, cones, spheres, etc. are good possibilities using this method.

G18 with G42 and G03

Climb milling applies equally to flat end mills as to *bull nose* or *ball nose* end mills. The difference between the standard ZX plane and the XZ machine plane has been explained already. This particular example shows the programming method for the same initial contour, but in a different plane.

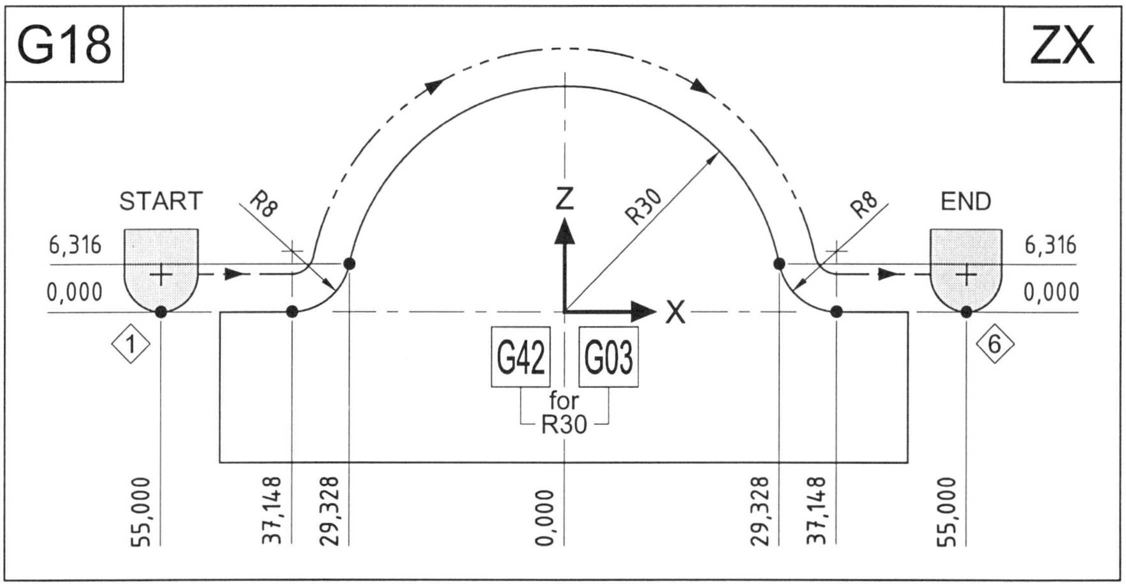

The illustration above (and the next one as well) shows the most misunderstood motion in CNC programming. Because of the differences between the standard mathematical plane orientation and the orientation of machine planes, the G18 command has to be programmed with full understanding of the concept of plane selection. Study the program on the next page carefully, and compare the toolpath with the similar toolpath in the previous G17 plane.

```
(*** G18 - ZX PLANE - CUTTING DIRECTION FROM POINT 1 TO POINT 6 ***)
(*** USES I AND K ARC VECTORS ***)
N1 G21                                    (METRIC)
N2 G17 G40 G80                            (INITIALIZATION)
N3 G90 G54 G00 X-55.0 Y0 S1000 M03        (START POSITION)
N4 G43 Z10.0 H01 M08                      (TOOL LENGTH OFFSET)
N5 G18 G42 G01 Z0 D51 F200.0              (CUTTER RADIUS OFFSET - START)
N6 X-37.148 F100.0                        (FIRST LINEAR MOTION)
N7 G02 X-29.328 Z6.316 I0 K8.0            (R8 ARC - CW)
N8 G03 X29.328 I29.328 K-6.316            (R30 ARC - CCW)
N9 G02 X37.148 Z0 I7.821 K1.684           (R8 ARC - CW)
N10 G01 X55.00                            (LAST LINEAR MOTION)
N11 G40 G00 Z10.0 M09                     (CUTTER RADIUS OFFSET - END)
N12 G17 G28 X55.0 Y0 Z10.0 M05            (MACHINE ZERO RETURN)
N13 M30                                   (PROGRAM END)
%
```

G18 with G41 and G02

The reverse order of cutting (point 6 to point 1) is shown next. Climb milling mode has changed to conventional milling mode and cutter radius offset left has changed to cutter radius offset right.

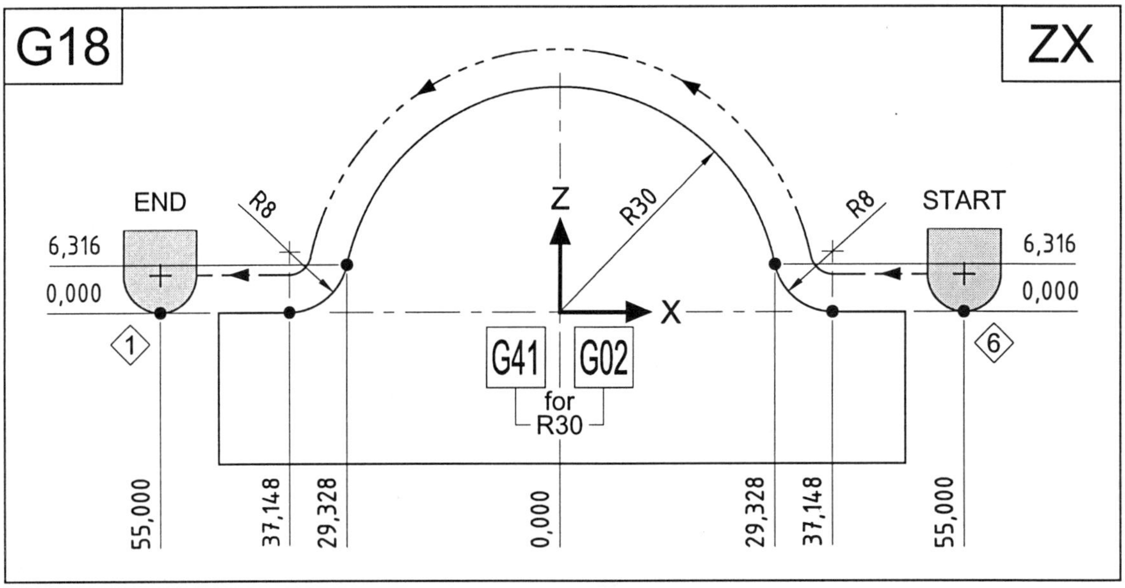


```
(*** G18 - ZX PLANE - CUTTING DIRECTION FROM POINT 6 TO POINT 1 ***)
(*** USES I AND K ARC VECTORS ***)
N1 G21                                    (METRIC)
N2 G17 G40 G80                            (INITIALIZATION)
N3 G90 G54 G00 X55.0 Y0 S1000 M03         (START POSITION)
N4 G43 Z10.0 H01 M08                      (TOOL LENGTH OFFSET)
N5 G18 G41 G01 Z0 F200.0                  (CUTTER RADIUS OFFSET - START)
N6 X37.148 F100.0                         (FIRST LINEAR MOTION)
N7 G03 X29.328 Z6.316 I0 K8.0             (R8 ARC - CCW)
N8 G02 X-29.328 I-29.328 K-6.316          (R30 ARC - CW)
N9 G03 X-37.148 Z0 I-7.821 K1.684         (R8 ARC - CCW)
N10 G01 X-55.0                            (LAST LINEAR MOTION)
N11 G40 G00 Z10.0 M09                     (CUTTER RADIUS OFFSET - END)
N12 G17 G28 X-55.0 Y0 Z10.0 M05           (MACHINE ZERO RETURN)
N13 M30                                   (PROGRAM END)
%
```

Each plane uses different arc vectors for circular interpolation, and G18 plane is no exception:

G18 plane uses I and K vectors for circular motions

When programming G18 ZX/XZ plane, think carefully about the motion direction. In your mind, align the part so it matches the standard mathematical planes. This way, the direction of the offset (G41 or G42), as well as the direction of the arc motion (G02 or G03) will become easy to see.

G19 with G41 and G02

Programming the G19 plane is as easy as programming the G17 plane - in both cases, the machine planes orientation matches the standard mathematical plane:

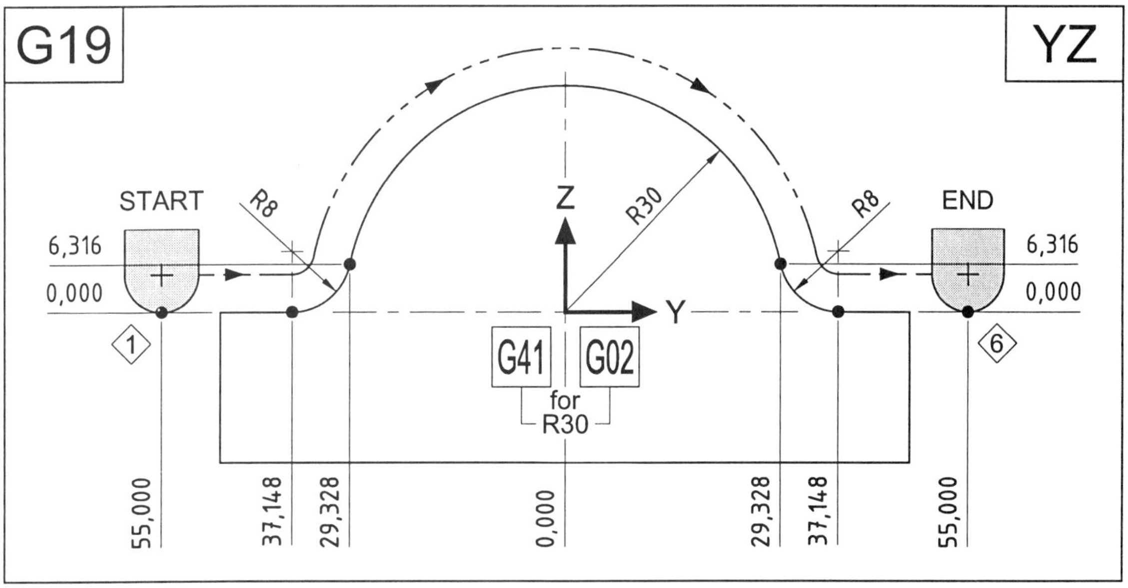

The complete program example reflects the conditions:

```
(*** G19 - YZ PLANE - CUTTING DIRECTION FROM POINT 1 TO POINT 6 ***)
(*** USES J AND K ARC VECTORS ***)
N1 G21                                   (METRIC)
N2 G17 G40 G80                           (INITIALIZATION)
N3 G90 G54 G00 X0 Y-55.0 S1000 M03       (START POSITION)
N4 G43 Z10.0 H01 M08                     (TOOL LENGTH OFFSET)
N5 G19 G41 G01 Z0 D51 F200.0             (CUTTER RADIUS OFFSET - START)
N6 Y-37.148 F100.0                       (FIRST LINEAR MOTION)
N7 G03 Y-29.328 Z6.316 J0 K8.0           (R8 ARC - CCW)
N8 G02 Y29.328 J29.328 K-6.316           (R30 ARC - CW)
N9 G03 Y37.148 Z0 J7.82 K1.684           (R8 ARC - CCW)
N10 G01 Y55.0                            (LAST LINEAR MOTION)
N11 G40 G00 Z10.0 M09                    (CUTTER RADIUS OFFSET - END)
N12 G17 G28 X0 Y55.0 Z10.0 M05           (MACHINE ZERO RETURN)
N13 M30                                  (PROGRAM END)
%
```

G19 with G42 and G03

This toolpath is the opposite direction of the previous example. Climb milling mode has changed to conventional milling mode and cutter radius offset left has changed to cutter radius offset right:

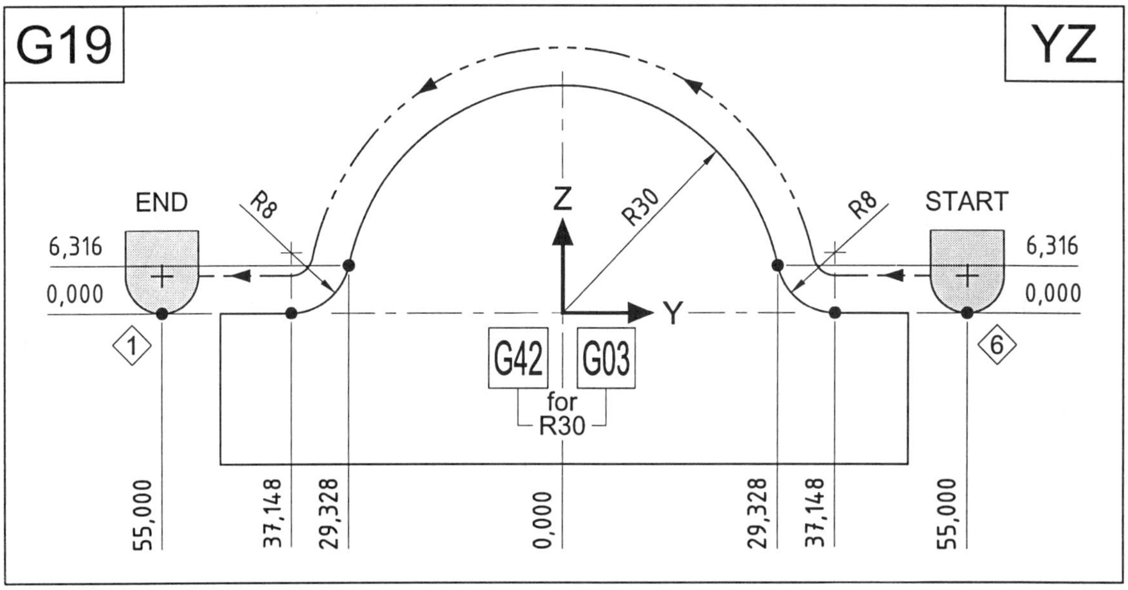


```
(*** G19 - YZ PLANE - CUTTING DIRECTION FROM POINT 6 TO POINT 1 ***)
(*** USES J AND K ARC VECTORS ***)
N1 G21                                   (METRIC)
N2 G17 G40 G80                           (INITIALIZATION)
N3 G90 G54 G00 X0 Y55.0 S1000 M03        (START POSITION)
N4 G43 Z10.0 H01 M08                     (TOOL LENGTH OFFSET)
N5 G19 G42 G01 Z0 D51 F200.0             (CUTTER RADIUS OFFSET - START)
N6 Y37.148 F100.0                        (FIRST LINEAR MOTION)
N7 G02 Y29.328 Z6.316 J0 K8.0            (R8 ARC - CW)
N8 G03 Y-29.328 J-29.328 K-6.316         (R30 ARC - CCW)
N9 G02 Y-37.148 Z0 J-7.82 K1.6849        (R8 ARC - CW)
N10 G01 Y-55.0                           (LAST LINEAR MOTION)
N11 G40 G00 Z10.0 M09                    (CUTTER RADIUS OFFSET - END)
N12 G17 G28 X0 Y-55.0 Z10.0 M05          (MACHINE ZERO RETURN)
N13 M30                                  (PROGRAM END)
%
```

G19 plane uses J and K vectors for circular motions

Note that for both planes where a ball nose end mill is typically used (G18 and G19), the tool length must be measured to the cutter tip. The program represents drawing dimensions, so the offset setting in the control will be the cutter radius. For example, if cutter diameter is 12 mm, the D51 will be set to 6.000. Of course, this nominal setting can be adjusted for tool wear or other fine tuning.

Hopefully, these examples have taken some mystery away from the subject seldom covered in training courses. With the proper application, programming a simple 3D part is only a matter of the suitable plane selection, cutter radius offset, and circular motions. Typically, such programs would heavily rely on subprograms or even macros, as a more flexible programming tool.

Using Right-Angle Attachment

Fixed cycles require neither circular motion nor cutter radius offset. Yet, they also have to be considered in relationship to planes. The application of a plane selection for fixed cycles is rather rare, as it requires a special machine attachment - a *90° angle head* (also known as *right-angle head*).

A right-angle head (see illustration) is a very flexible attachment that can be added to most CNC machining centers. It can be attached to a CNC vertical machining center, in order to allow machining on a side face, or it can be attached to a CNC horizontal machining centers for the same reason. Although this section is devoted to programming fixed cycles in different planes, keep in mind that other tools can also be mounted into a suitable toolholder. This very versatile head is used for many types of machining. The best way to explain this topic is to look at the basic concepts and show a practical example.

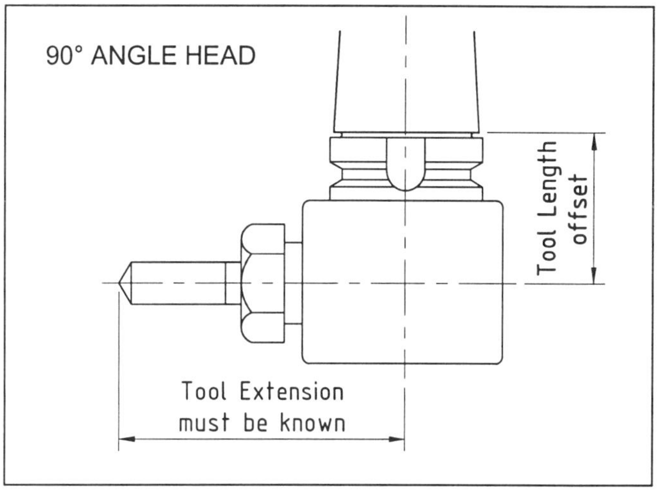

Basic Concepts

The most common machining of holes uses a particular fixed cycle in the XY plane, which is always perpendicular to the spindle center line. The XY coordinates specify center of the hole location, while other parameters (such as the R-level and the Z-depth) specify the drilling data. Machining in planes changes the dimensional cycle data to correspond to the active plane:

Active plane	Hole center coordinates	R-level clearance	Axis to program hole depth
G17	X.. Y..	Along Z-axis	Z..
G18	X.. Z..	Along Y-axis	Y..
G19	Y.. Z..	Along X-axis	X..

Employing a right angle head eliminates secondary operations and costly setup. The cost of the head can be recovered over a short period of time. Another option for machining on different sides is to use an indexing or rotary table. Regardless of whether a new setup or an indexing table is used, the machining plane does not change - the spindle center will still be perpendicular to the XY machining plane. When a 90° head is used in the spindle, it can perform various machining operations on another part face, without a separate setup. Depending on the head orientation, the face - the plane - can be either the front plane ZX (G18) or the right side plane YZ (G19). The above table shows how a different plane selection influences the program data - *hole position*, *clearance above face*, and the *cutting depth*. The *tool extension* shown in the above illustration must be known, as the work offset (G54, for example) is measured to the spindle center line. Using a presetter or a similar method, the extension can be established and applied for all X-motions (G18) or all Y-motions (G19). This section will explain how a 90° head is programmed for drilling operations, using fixed cycles.

Side Face Drilling

The drawing below will be used for a practical example. It shows two sets of holes in two planes. One set is a set of four ∅6 mm holes located in the top plane, programmed as G17 - XY plane. The other set is a set of five ∅4 mm holes, located in the right side plane, programmed as G19 - YZ plane.

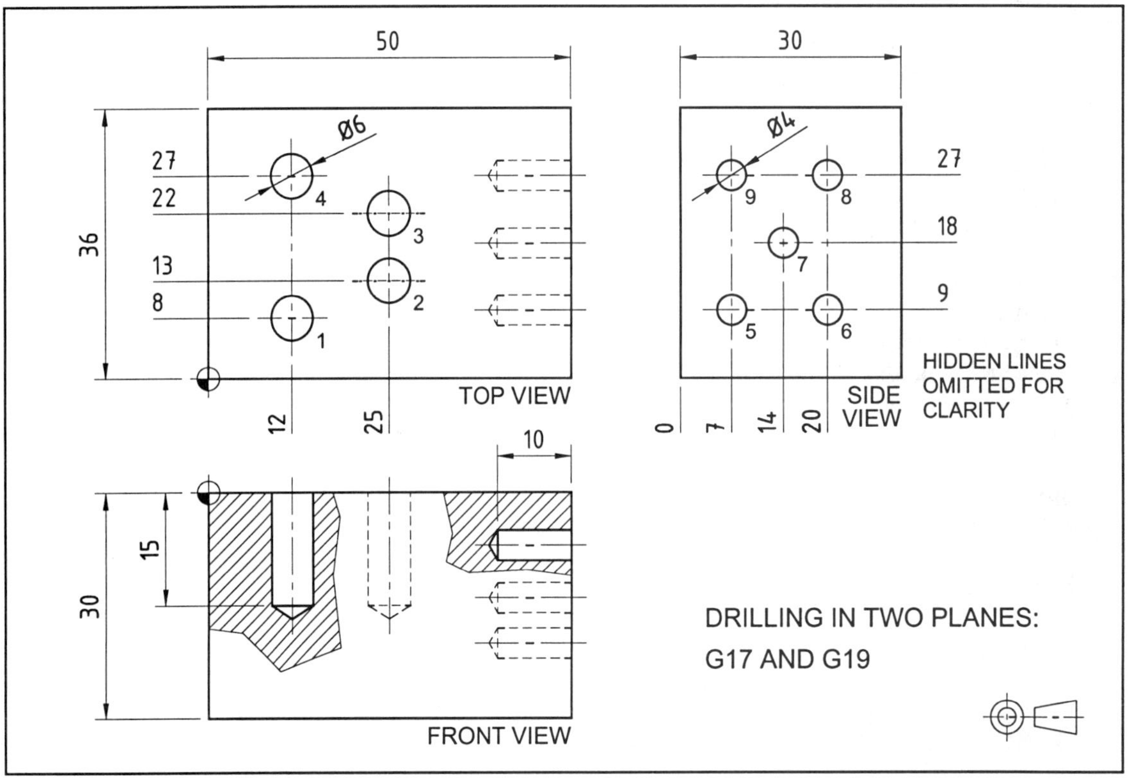

Drilling in the Top Plane - G17

For the purposes of the example, only the drilling operations will be shown. Holes are numbered in the order of machining for reference. Any other tool will follow the same logical approach. The program for the top is a standard drilling program, using the G81 fixed cycle. The hole depth will be 15 mm plus tool point length of ∅6 × 0.3: = 15 + 1.8 = Z-16.8.

```
N1 G21                                  (METRIC UNITS)
N2 G17 G40 G80 T01                      (G17 XY PLANE SELECTED)
N3 M06                                  (T01 = 6 MM DRILL)
N4 G90 G54 G00 X12.0 Y8.0 S1200 M03 T02 (POSITION TO HOLE 1)
N5 G43 Z10.0 H01 M08                    (INITIAL LEVEL 10 MM OFF TOP)
N6 G99 G81 X12.0 Y8.0 R2.0 Z-16.8 F150.0 (DRILL HOLE 1 - 2MM CLEAR)
N7 X25.0 Y13.0                          (DRILL HOLE 2)
N8 Y22.0                                (DRILL HOLE 3)
N9 X12.0 Y27.0                          (DRILL HOLE 4)
N10 G80 Z10.0 M09                       (CLEARANCE 10 MM OFF TOP)
N11 G28 Z10.0 M05                       (Z-AXIS ZERO RETURN)
N12 M01                                 (OPTIONAL BLOCK)
```

In addition to the G17 XY plane selection in block N2, note the initial level (Z10.0 in block N4), the R-level (R2.0 in block N6), and the Z-depth (Z-16.8, also in block N6). The same relative values will be used for the second tool.

Angular Head Settings

Before the second tool (angular head) can be used, is must be selected and its extension preset. The tool extension between the spindle center line and the tool tip must always be known. The example will use an amount of 150 mm as the extension length, as preset, usually off-machine. The illustration shows the front view of the ***INITIAL*** position of the angular head, 10 mm from the right side face, located at the lower left hole with the coordinates of Y9.0 Z-7.0 in G19 YZ plane. The initial position will be programmed as X210.0.

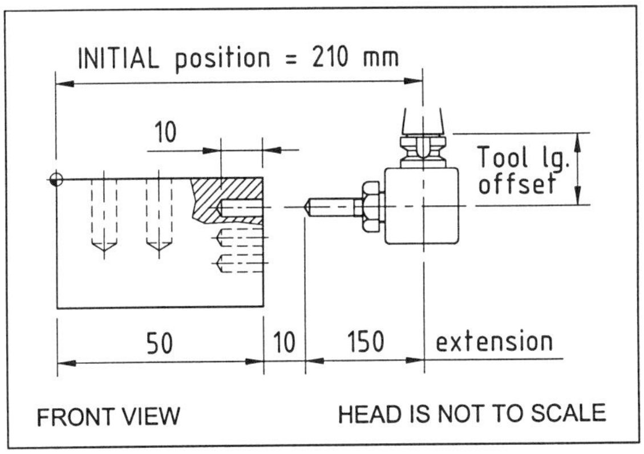

The next position for drilling will be the ***R-LEVEL***. In order to calculate the R-level, two methods may be used:

(a) Subtract the difference between the actual initial level clearance (10 mm) and the R-level clearance (2 mm), from the total initial position: 210 - 10 + 2 = **202.0** mm

(b) Total sum of 50 mm (part) + 2 mm (R-level clearance) + 150 mm (extension) = 202 mm programmed R-level.

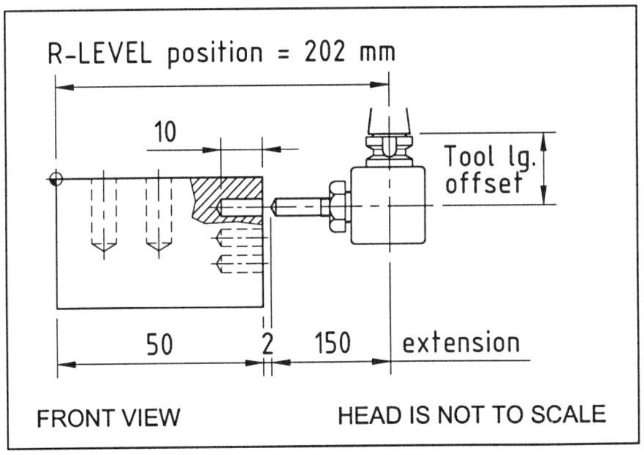

The final position is the ***DEPTH***. For a drill, the tool point length has to be accounted for. For a ∅4 mm drill, the tip length will be 4×0.3=1.2 mm. In this case, the full diameter is 10 mm, and the drill depth will be 10+1.2=11.2 mm. Subtract 11.2 mm depth from the total part width of 50 millimeters: 50-11.2=38.8 mm. The tool extension must be added, which results in the final depth: 38.8 + 150 = **188.8** mm programmed depth.

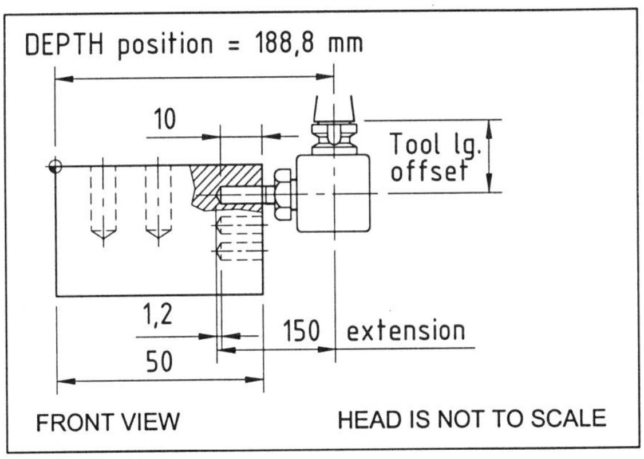

Programmed axis for the depth will always be the axis that is *perpendicular* to the active plane

Drilling in the Side Plane - G19

As the next tool will drill five holes located in the YZ plane, the above calculations have to be put into the program, still using the G81 fixed cycle. *In G19 plane (YZ), both the R-level and the depth are measured along the X-axis*. X-axis is the perpendicular axis of the YZ-plane (G19).

The second tool of the program can now be completed:

```
N13 T02                                        (TOO T02 READY - IF REQUIRED)
N14 M06                                        (T02 = 4 MM DRILL)
N15 G17                                        (G17 XY PLANE SELECTED)
N16 G90 G54 G00 X210.0 Y9.0 S1400 M03 T01     (POSITION TO HOLE 5)
N17 G43 Z-7.0 H02 M08                          (INITIAL LEVEL 10 MM OFF SIDE)
N18 G19                                        (G19 YZ PLANE SELECTED)
N19 G99 G81 Y9.0 Z-7.0 R202.0 X188.8 F140.0 (DRILL HOLE 5 - 2MM CLEAR)
N20 Z-20.0                                     (DRILL HOLE 6)
N21 Y18.0 Z-14.0                               (DRILL HOLE 7)
N22 Y27.0 Z-20.0                               (DRILL HOLE 8)
N23 Z-7.0                                      (DRILL HOLE 9)
N24 G80 Z210.0 M09
N25 G17
N26 G28 Z-7.0 M05
N27 G28 X210.0 Y27.0
N28 M30
%
```

All plane selections have been programmed as separate blocks, for better clarity of the program.

Side Face Milling

With all considerations applied to the fixed cycles in planes using an angular head, milling in planes with the same head is virtually identical. The cutting tool will be different, the toolpath motions will be different, but the calculations will follow the same pattern. The extension of the angular head must be known and added to the program. In this section, two independent examples will be shown - one for the angular head set perpendicular to the ZX plane, the other perpendicular to the YZ plane. In both cases, the angular head will be set to 120 mm extension, from the spindle center to the tip of the end mill.

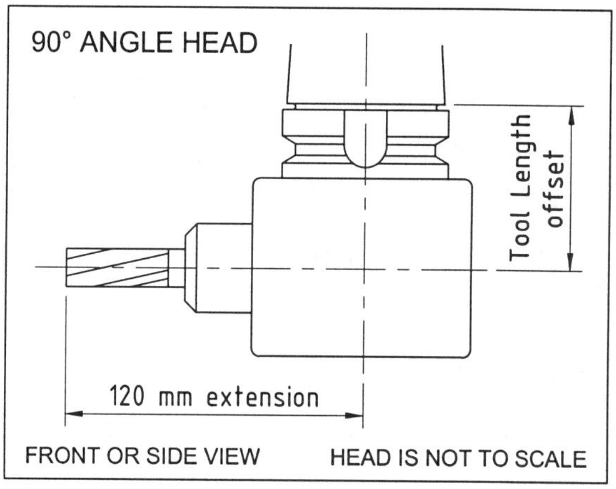

Milling in the Front Plane - G18 (ZX)

Only a single finishing toolpath will be programmed for this example, to illustrate the general concept only. The toolpath will show the necessary approach to *any* toolpath of a similar kind in G18 plane (ZX).

For the front plane machining, the head will be turned away from the CNC operator, requiring increased care during setup. The arc on the *front face* of the part is the arc orientation when the tool approaches the part.

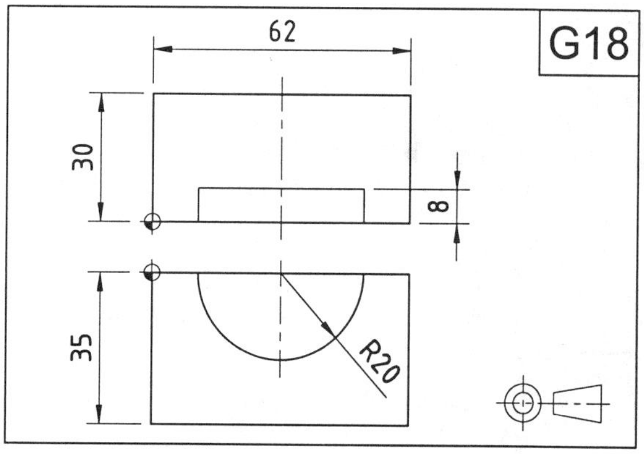

REMINDER:	**Watch for the arc direction and cutter radius offset selection in G18 plane**

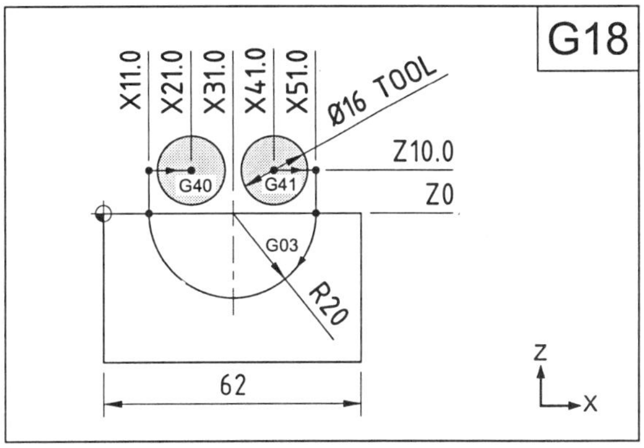

To develop the program, keep in mind that the front view at the machine does *not* correspond to the mathematical plane. The motions shown in the illustration at right show climb milling mode for the desired cut.

As the angular head orientation is perpendicular to the view shown, it means the 120 mm extension must be added to the Y-axis motions. Since the cutting depth is 8 mm, the Y-axis will be programmed at **112.0** mm at the cutting depth position:

120 - 8 = 112 mm = Y112.0

```
(*** MILLING IN G18 PLANE ***)
N1 G21
N2 G17 G40 G80 T01                      (G17 XY PLANE SELECTED)
N3 M06                                  (T01 = 16 MM END MILL)
N4 G90 G54 G00 X41.0 Y112.0 S1000 M03   (START POSITION)
N5 G43 Z10.0 H01 M08                    (TOOL CENTER 10 MM ABOVE PART)
N6 G18                                  (G18 ZX PLANE SELECTED)
N7 G41 G01 X51.0 D51 F150.0             (LEAD-IN MOTION)
N8 Z0                                   (ARC START POSITION)
N9 G03 X11.0 I20.0 K0                   (ARC END POSITION)
N10 G01 Z10.0                           (TOOL CENTER 10 MM ABOVE PART)
N11 G40 G00 X21.0 M09                   (LEAD-OUT MOTION-END POSITION)
N12 G17                                 (G17 XY PLANE SELECTED)
N13 G28 Z10.0 M05                       (Z-AXIS MACHINE ZERO RETURN)
N14 G28 X21.0 Y112.0                    (XY-AXES MACHINE ZERO RETURN)
N15 M30                                 (END OF PROGRAM)
%
```

Milling in the Side Plane - G19 (YZ)

This operation is very similar to the drilling example shown earlier, as well as the milling example in the previous section. As before, a single finishing toolpath will be shown, this time in the *side plane* G19 (YZ). When working with the side plane G19, the tool motions are clearly visible to the CNC operator, minimizing the possibility of a collision between the tool and the part.

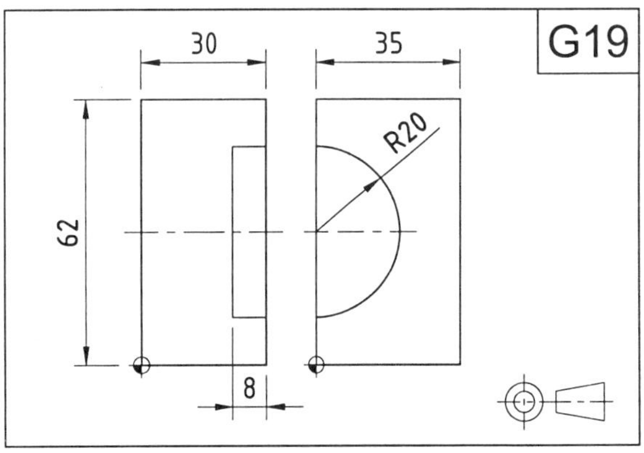

There are no special considerations for the side plane, except that the toll length extension of 120 mm will be added to the X-axis. As the part is 30 mm wide, and the depth of the cut is 8 mm, the X-axis position at cutting depth will be **142.0** mm:

30 - 8 + 120 = 142 mm = X142.0

The arc on the *side face* of the part is the arc orientation when the tool approaches the part. Both arc direction and cutter radius offset will be programmed normally, as viewed at the CNC machine.

The milling program (finishing only) for the side cutting can now be completed:

As before, all plane selections have been programmed as separate blocks (and underlined), for better clarity of the program.

The illustration shows the cutter start and end locations, cutter radius offset, and arc direction.

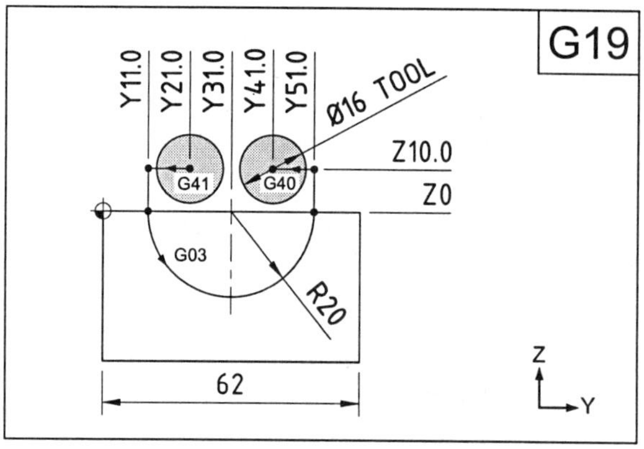


```
(*** MILLING IN G19 PLANE ***)
N1 G21
N2 G17 G40 G80 T01                      (G17 XY PLANE SELECTED)
N3 M06                                  (T01 = 16 MM END MILL)
N4 G90 G54 G00 X142.0 Y21.0 S1000 M03   (START POSITION)
N5 G43 Z10.0 H01 M08                    (TOOL CENTER 10 MM ABOVE PART)
N6 G19                                  (G19 YZ PLANE SELECTED)
N7 G41 G01 Y11.0 D51 F150.0             (LEAD-IN MOTION)
N8 Z0                                   (ARC START POSITION)
N9 G03 Y51.0 J20.0 K0                   (ARC END POSITION)
N10 G01 Z10.0                           (TOOL CENTER 10 MM ABOVE PART)
N11 G40 G00 Y41.0 M09                   (LEAD-OUT MOTION-END POSITION)
N12 G17                                 (G17 XY PLANE SELECTED)
N13 G28 Z10.0 M05                       (Z-AXIS MACHINE ZERO RETURN)
N14 G28 X142.0 Y41.0                    (XY-AXES MACHINE ZERO RETURN)
N15 M30                                 (END OF PROGRAM)
%
```

Programming an angular head is not much different than any similar milling operation - the tool length extension has to be known and added to the axis that machines the depth.

23 PROGRAMMING CAMS

First, a clarification - the word *cam* in this chapter refers to a specific type of design within the field of mechanical engineering - it does not have anything to do with the video type cams or even the CAM part of CAD/CAM. The focus will be on the developing a part program for a typical cam.

Overview of Cams

In geometrical terms, a typical cam can be described as a special contour profile that contains a combination of arcs and splines. In mechanical terms, the main purpose of a cam is to *transfer motion* - as a typical example, a rotary cam motion can be transferred into a linear motion. Common cam mechanism will have two moving parts attached to a single fixture (frame). The moving parts can be curved or grooved and they work in conjunction with a follower. The follower will cause a series of single motions (called *events*) to take place, while the cam is completing one cycle.

There are three events in a single cam cycle that the follower will provide:

- **Rise**
- **Dwell**
- **Fall (also called *return*)**

All three events are related to the following activity, relative to the cam center:

RISE	**... follower motion *away from* the cam center**	**(also called *positive* motion)**
DWELL	**... provides *no* change in follower motion**	**(follower is *idle*)**
FALL	**... follower motion *towards* the cam center**	**(also called *return* or *negative* motion)**

Cams are frequently used in many industries, particularly in automation machinery, gear cutting, screw machines, printing presses, packaging equipment, as well as in the automotive industry - all for the same goal - *to control motion by accurate timing*. As the cam is rotated, the follower moves closer to or farther away from a given cam surface - or - the direction does not change at all. One of the most important features of a cam is that it can control various motion profiles confined within small areas. For example, a typical application for cams in the automotive industry is opening and closing of engine cylinders. However, an engine cylinder does not just opens and closes, it also remains opened or closed for a period of time - there is no change in the motion direction and the dwell is employed.

Cams can be used for a large variety of applications and have many forms and shapes. The most common cam design a CNC programmer is likely to encounter is a *radial cam*. There are also two other cam designs, defined as a cylindrical cam and a spherical (ball) cam.

On the next page is a drawing that will be used in this chapter as an example. In the drawing, the groove represents the typical path of a cam follower. For the CNC programmer, it is very important to understand at least the basic concepts of cams.

Please note that the drawing used in this chapter contains only those dimensions and features that are directly related to the development of the grooved cam. Other dimensions are important for engineering reasons but not relevant to the CNC program development. The sample drawing is based on an actual cam drawing and is used as an example to illustrate the programming approach.

Cam Drawing Example

The drawing shown on this page represents a grooved cam. The groove has been designed to provide the path of a cam follower, which is 10 mm in diameter in this case. Like the vast majority of traditional drawings, a cam drawing also covers many expected features, such as the groove depth, which is shown, and many other features that have been omitted here. The CNC programmer's goal is to program the groove - actually, the center of the groove. Once the groove center toolpath is established, the issues related to fine tuning the toolpath can be resolved rather easily. *Study the drawing.*

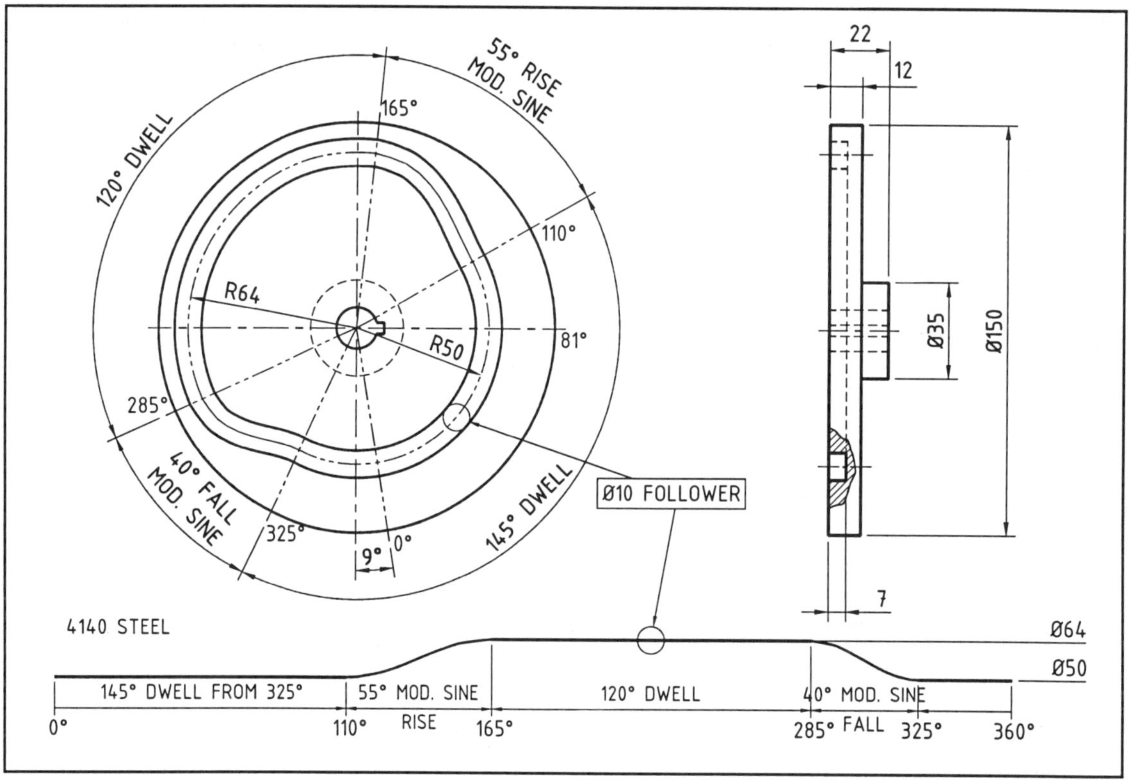

Yes, study the drawing first - particularly the cam groove that is the actual subject of program development. How is the groove dimensioned? The groove has *four* sections:

- **Arc radius of 50 mm**
- **Section between R50 and R64**
- **Arc radius of 64 mm**
- **Section between R64 and R50**

The word *'section'* here is worth of a special look. In one way, there are no dimensions provided to define the exact shape of each section. Second, implied dimensions *are* provided, in terms of descriptions - there are two such descriptions, one for each section. One refers to the cam *RISE* and the other to the *FALL* of the cam motion - both specified over a defined arc sweep angle (55° and 40° respectively). These two areas become a serious concern here, as they represent a serious challenge, a challenge that is not always a part of everyday CNC programming. ***There are no supplied dimensions!***

Cam RISE and FALL - Sections Evaluation

A typical cam drawing of this type will provide the rise and fall of the cam motion events, within a certain arc sweep angle. The job of the CNC programmer is to provide the *best curve* within this section, as defined in the drawing. If you are used to programming standard motions, namely lines and arcs, abandon this approach right now - just for this project. In the end, the final program blocks will still consist of simple motions, but the method of *obtaining the individual XY coordinates* will be much different than any other. The problem to be solved here is to create a toolpath along a curve that is neither linear nor radial. In other words, we *can not* cut the rise and fall sections of the cam using standard interpolations methods, either linear or circular. The special curve we want to develop the toolpath for is called a *spline*. Spline is a free-form curve, mathematically defined by a series of points, using established algorithms (mathematical rules). Mathematically, a spline curve stands at the same level as other curves, notably a line, arc and circle. The major problem is that the great majority of CNC systems *DO NOT* support any other interpolation except linear, circular and helical.

Any spline defined in the drawing can only be *approximated* in the CNC program. This process of approximation is done by connecting the spline points with lines and/or arcs, both types supported by the CNC system. Of course, any approximation is not *exactly* the same as the original calculations, but in machining, toolpath using the approximation calculations is generally accepted as a very accurate toolpath. As other methods are not available, this *'close enough'* toolpath is the only one to be programmed. Although arcs are possible to define a really fine cam contour, they are virtually impossible to calculate by using manual programming methods. Instead, short lines between points are acceptable alternatives. In a nutshell, the shorter the individual connecting lines, the more accurate the cam profile will be when machined. Most cams are defined by a modified sine curve. Sine curve has a standard mathematical definition, and modified sine curve is very close. The purpose of modified sine curve is to provide smoother start during acceleration of the cam. On the drawing, both sections identified as 55° rise and 40° fall are modified sine curves.

Cam Zero and Angle Orientation

When evaluating a cam drawing, the program zero has to be established, just as it has to be established for any other part program. This is the most important focus of CNC program development - to select one point, from which all absolute dimensions originate. For majority of cam profiles, all basic programming rules apply equally and the solution is often a very simple effort. Selecting the part (cam) zero (origin = X0Y0) is the first major step. Since the cam presented in the example is round (radial cam), the natural part zero selection will be the center point of the part.

The major difference in programming individual coordinate points of a cam profile is based on the cam angle orientation. For experienced CNC programmers, the discussion of angle orientation would seem futile and unnecessary. Even less experienced programmers learn early that zero degrees is normally to the right of the center point, at three o'clock (or east) position. Applying this knowledge to the follower cam in the example should seem as a breeze - after all, it's just another contour. This contour requires a rather special approach, in terms of developing the toolpath. Note that in the drawing area where we would normally expect the zero angle orientation, is in fact the angular orientation of 81°. The majority of cams do have a keyway (also shown in the drawing), and the orientation angle is usually related to the keyway.

Calculation of Coordinates

The greatest programming effort for even a simple cam will be the calculation of XY coordinates within the rise and fall sections. This will also be the main topic for the rest of this chapter. The main goal will be to establish rectangular coordinates, although the program can be somewhat simplified, if the polar coordinates option is available at the control system.

Regardless of which of the two sections is selected, there are two radiuses that connect the two angular sections. The first step is to find the radius change within the rise and fall sections:

R64 - R50 = 14 mm

The next step is to identify the range of angles:

55° angle between 110° and 165° => RISE between R50 and R64
40° angle between 285° and 325° => FALL between R64 and R50

Next, each section will be evaluated separately, although the development method is the same.

The RISE Section

As the coordinate points for both the rise and fall sections have to be calculated, it makes no difference which section is done first. The programming approach is logically identical to both rise and fall sections, with the goal of creating points on the cam contour that can later be connected as a contour.

The first step is to decide on how many segments the included angle will include. In the examples, the increment will be 10 segments within 55° rise. The smaller the angle, the more accurate approximation will be, but also more calculations will be needed:

55° rise / 10 spaces = 5.5° per segment

In the next step, a flat layout of the cam rise will be defined, where the horizontal axis represents the ten angular segments, and the vertical axis represents radius range between R50 and R64. The area will be divided into four quadrants of equal size, to accommodate the modified sine curve:

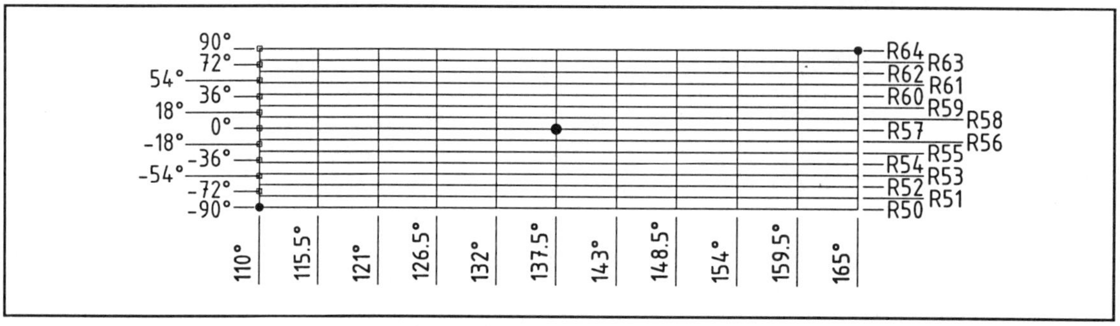

In the flat layout above, the heavy dot in the middle represents the exact halfway position between the range of angles as well as the range or radiuses - the sine curve must pass through this point:

110° + (165° - 110°) / 2 = 137.5°

50 + (64 - 50) / 2 = 57

On the left, the ten divisions represent total difference between R50 and R64, which is represented on a flat layout as 180° change. From the middle up, the degrees are positive, from the middle down they are negative. The increment of 18° is the result of dividing the 180° by the number of spaces:

180° / 10 = 18°

This concept may get a bit of time to get used to, but it represents a typical approach used in similar flat layout developments.

Calculating Radius Length

In this step, the first set of calculations can be made - this time, the purpose is to develop the *length* of each line, using the flat layout 18° incremental angle for each space. Total of eleven calculations will be required (10 spaces = 11 lines). Each line will originate at the cam center, and will end at its respective cam point - in effect, this line is a radius that will later become the hypotenuse of a triangle needed to calculate the actual XY coordinates.

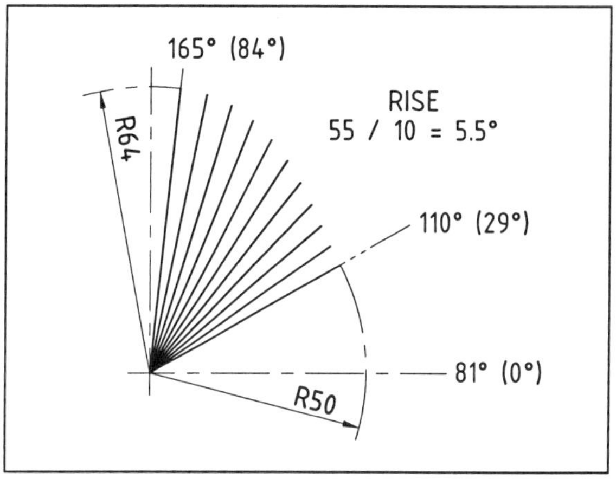

Because all calculations are based of the flat layout and its middle point, the middle radius of 7 mm will be used, along with the middle radius of 57 mm:

(64 - 50) / 2 = 7 and **(64 + 50) / 2 = 57**

The illustration shows all eleven lines with their lengths and angles as per the table below:

Flat layout angle (18°increment)	Calculation	Line length in mm (radius)	Angle orientation (5.5°)
-90°	7 × sin-90 + 57	50.000	110°
-72°	7 × sin-72 + 57	50.343	115.5°
-54°	7 × sin-54 + 57	51.337	121°
-36°	7 × sin-36 + 57	52.886	126.5°
-18°	7 × sin-18 + 57	54.837	132°
0°	7 × sin0 + 57	57.000	137.5°
18°	7 × sin18 + 57	59.163	143°
36°	7 × sin36 + 57	61.114	148.5°
54°	7 × sin54 + 57	62.663	154°
72°	7 × sin72 + 57	63.657	159.5°
90°	7 × sin90 + 57	64.000	165°

Calculating XY Coordinates

The line length calculated will now be used for calculating the XY coordinates of the endpoint of each line. Since the line represents the hypotenuse of a right angle triangle, and individual angles are already known, making calculations easier. Note that the cam orientation shows 81° where 0° would normally be expected. Although possible to calculate the XY coordinates in this orientation, it is much easier to consider the 81° as 0°, by subtracting 81° from 110°:

110° - 81° = 29° **... THIS ANGLE CORRESPONDS TO THE STANDARD DEFINITION OF ANGLES**

The spacing between lines is 5.5° per segment, so each angle will increase by that angle:

Line length (hypotenuse)	X-calculation	Y-calculation	Final X-coordinate	Final Y-coordinate
50.000	50 × cos29	50 × sin29	X43.731	Y24.240
50.343	50.343 × cos34.5	50.343 × sin34.5	X41.489	Y28.515
51.337	51.337 × cos40	51.337 × sin40	X39.326	Y32.999
52.886	52.886 × cos45.5	52.886 × sin45.5	X37.068	Y37.721
54.837	54.837 × cos51	54.837 × sin51	X34.510	Y42.616
57.000	57 × cos56.5	57 × sin56.5	X31.460	Y47.531
59.163	59.163 × cos62	59.163 × sin62	X27.775	Y52.238
61.114	61.114 × cos67.5	61.114 × sin67.5	X23.387	Y56.462
62.663	62.663 × cos73	62.663 × sin73	X18.321	Y59.925
63.657	63.657 × cos78.5	63.657 × sin78.5	X12.691	Y62.379
64.000	64 × cos84	64 × sin84	X6.690	Y63.649

Once the coordinates are known for each endpoint, the rise section of the cam is completed.

The FALL Section

Once the rise section has been calculated, the fall section calculations are much easier. The reason is that both sections share the radius of 64 mm, radius of 50 mm and the difference between them of 14 mm. The above formulas can be used in the reversed order, but using a smaller angle increment, still maintaining 10 segments within 40° fall, the same amount but smaller angle than for the rise:

40° fall / 10 spaces = 4° per segment

In the next step, a flat layout of the cam fall will be defined, where the horizontal axis represents the ten angular segments, and the vertical axis represents radius range between R50 and R64. The area will be divided into four quadrants of equal size, to accommodate the modified sine curve:

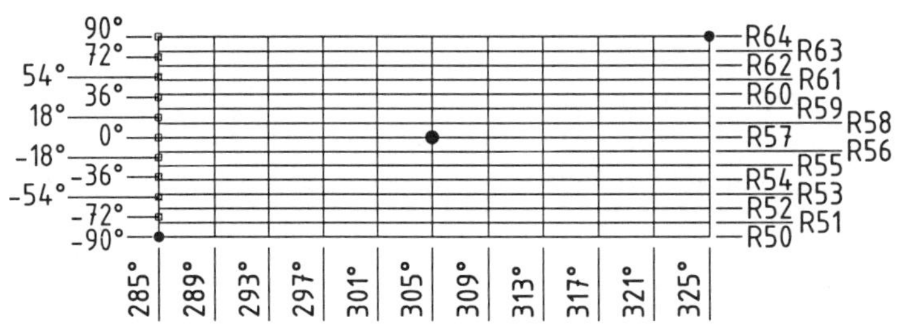

In the flat layout shown, the heavy dot in the middle represents the exact halfway position between the range of angles as well as the range or radiuses - the sine curve must pass through this point:

285° + (325° - 285°) / 2 = 305°

50 + (64 - 50) / 2 = 57

On the left, the ten divisions represent total difference between R50 and R64, which is represented on a flat layout as 180° change. From the middle up, the degrees are positive, from the middle down they are negative. The increment of 18° is the result of dividing the 180° by the number of spaces.

Calculating Radius Length

Again, the goal is to develop the *length* of each line, using the flat layout 18° incremental angle for each space. Total of eleven calculations will be required and each line will originate at the cam center and end at its respective cam point - it will become the hypotenuse of a triangle needed to calculate the actual XY coordinates.

Because all calculations are based of the flat layout and its middle point, the middle radius of 7 mm will be used, along with the middle radius of 57 mm:

(64 - 50) / 2 = 7 and **(64 + 50) / 2 = 57**

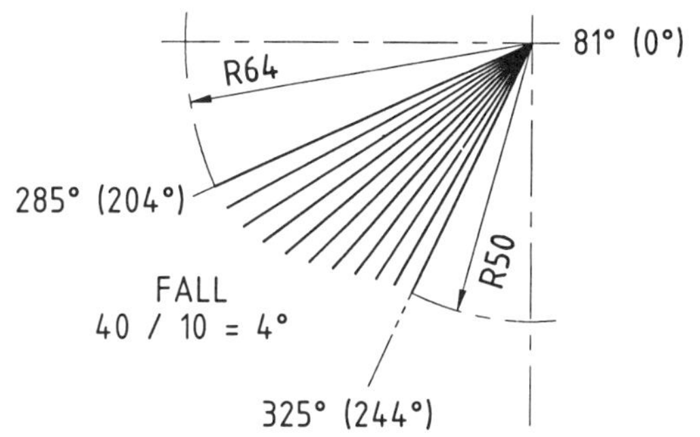

The illustration shows all eleven lines with their lengths and angles as per the table below:

Flat layout angle (18°increment)	Calculation	Line length in mm (radius)	Angle orientation (4°)
90°	7 × sin90 + 57	64.000	285°
72°	7 × sin72 + 57	63.657	289°
54°	7 × sin54 + 57	62.663	293°
36°	7 × sin36 + 57	61.114	297°
18°	7 × sin18 + 57	59.163	301°
0°	7 × sin0 + 57	57.000	305°
-18°	7 × sin-18 + 57	54.837	309°
-36°	7 × sin-36 + 57	52.886	313°
-54°	7 × sin-54 + 57	51.337	317°
-72°	7 × sin-72 + 57	50.343	321°
-90°	7 × sin-90 + 57	50.000	325°

Calculating XY Coordinates

The line length calculated will now be used for calculating the XY coordinates of the endpoint of each line, in the fall section of the cam. The calculation will continue from the rise calculation, and are also based on the standard angle orientation. The spacing between lines is 4° per segment, so each angle will increase by that angle:

Line length (hypotenuse)	X-calculation	Y-calculation	Final X-coordinate	Final Y-coordinate
64.000	64 × cos204	64 × sin204	X-58.467	Y-26.031
63.657	50.343 × cos208	50.343 × sin208	X-56.206	Y-29.885
62.663	51.337 × cos212	51.337 × sin212	X-53.141	Y-33.206
61.114	52.886 × cos216	52.886 × sin216	X-49.442	Y-35.922
59.163	54.837 × cos220	54.837 × sin220	X-45.321	Y-38.029
57.000	57 × cos224	57 × sin224	X-41.002	Y-39.596
54.837	59.163 × cos228	59.163 × sin228	X-36.693	Y-40.752
52.886	61.114 × cos232	61.114 × sin232	X-32.560	Y-41.675
51.337	62.663 × cos236	62.663 × sin236	X-28.707	Y-42.560
50.343	63.657 × cos240	63.657 × sin240	X-25.172	Y-43.598
50.000	50 × cos244	50 × sin244	X-21.919	Y-44.940

Summary

The two cam sections - its 55° *rise* and its 40° *fall* - are both presented at right. Each cam endpoint has been connected by short lines with the next endpoint. Connecting the endpoints with short lines to create the part profile is never as accurate as programming splines, but that is the limitation of programming cams using manual methods.

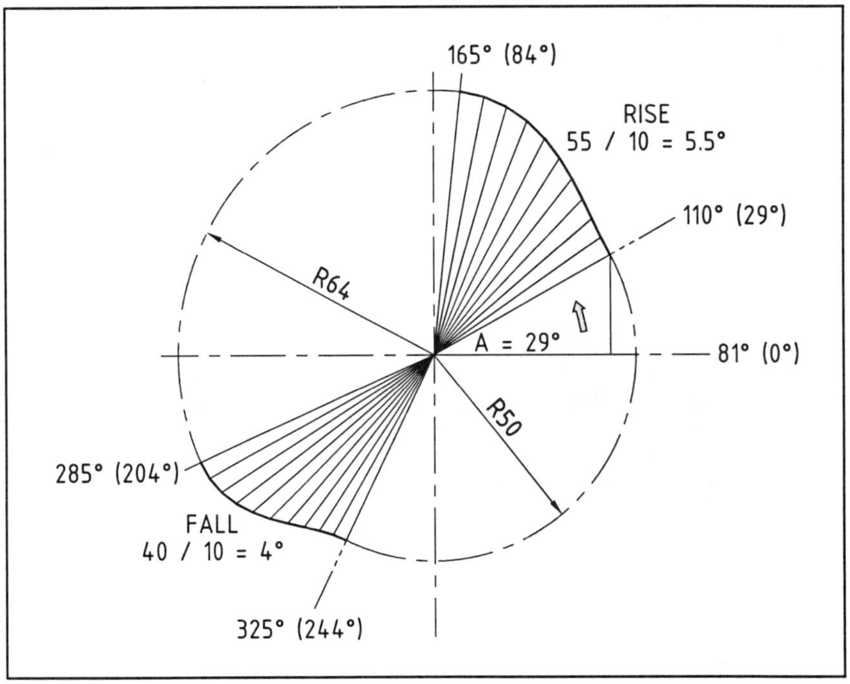

Finally, one more illustration - this time showing the flat layout for both sections. This layout is sometimes called the cam *time line*.

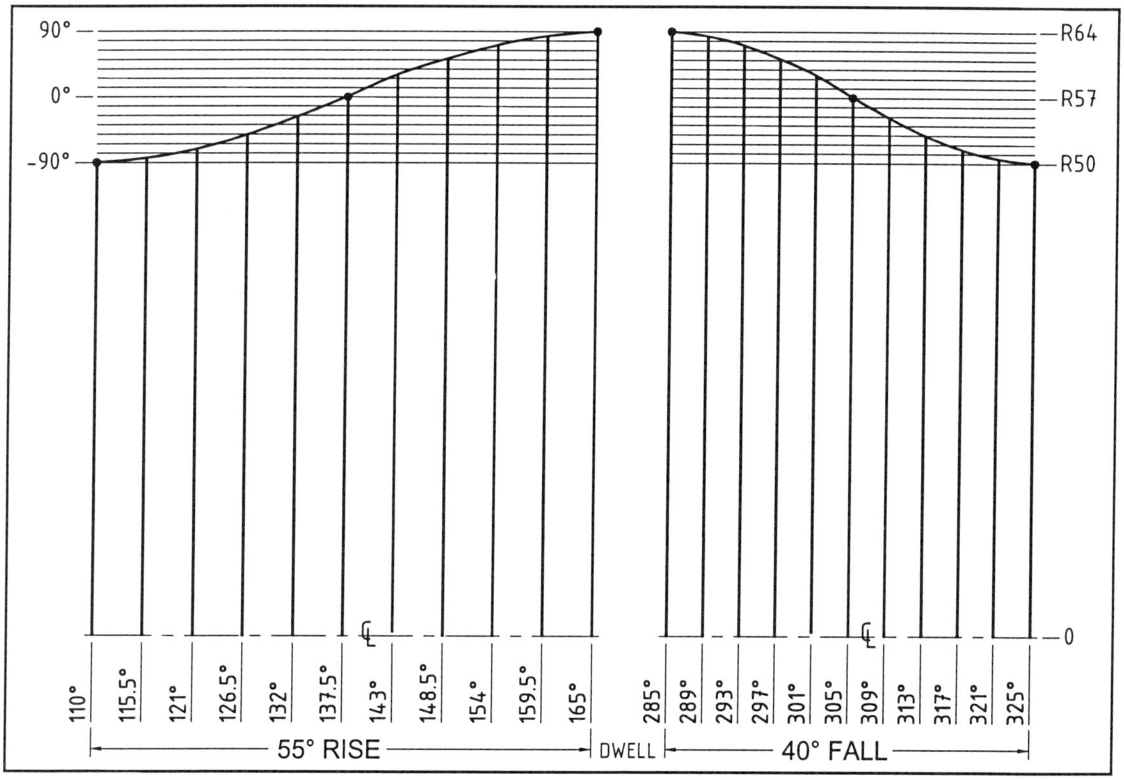

Writing the Program

Once the coordinates are known, the writing of the program itself is quite straightforward. The programmed toolpath will follow the layout of the drawing, starting at zero degrees (shown as 81° in the drawing). Just a simple grooving pass will be shown to illustrate the idea. You should have no problems modifying the version provided to a more complex version.

```
N1 G21
N2 G17 G40 G80
N3 G90 G54 G00 X50.0 Y0 S900 M03          (ENTER AT 81 DEGREES)
N4 G43 Z25.0 H01 M08
N5 Z2.0
N6 G01 Z-7.0 F100.0
N7 G03 X43.731 Y24.24 I-50.0 J0 F150.0    (START AT 110 DEGREES)
N8 G01 X41.489 Y28.515                    (END AT 115.5 DEGREES)
N9 X39.326 Y32.999                        (END AT 121 DEGREES)
N10 X37.068 Y37.721                       (END AT 126.5 DEGREES)
N11 X34.51 Y42.616                        (END AT 132 DEGREES)
N12 X31.46 Y47.531                        (END AT 137.5 DEGREES)
N13 X27.775 Y52.238                       (END AT 143 DEGREES)
N14 X23.387 Y56.562                       (END AT 148.5 DEGREES)
N15 X18.321 Y59.925                       (END AT 154 DEGREES)
N16 X12.691 Y62.379                       (END AT 159.5 DEGREES)
N17 X6.69 Y63.649                         (END AT 165 DEGREES)
N18 G03 X-58.467 Y-26.031 I-6.69 J-63.649 (START AT 285 DEGREES)
```

```
N19 G01 X-56.206 Y-29.885                  (END AT 289 DEGREES)
N20 X-53.141 Y-33.206                      (END AT 293 DEGREES)
N21 X-49.442 Y-35.922                      (END AT 297 DEGREES)
N22 X-45.163 Y-38.029                      (END AT 301 DEGREES)
N23 X-41.002 Y-39.596                      (END AT 305 DEGREES)
N24 X-36.693 Y-40.752                      (END AT 309 DEGREES)
N25 X-32.56 Y-41.675                       (END AT 313 DEGREES)
N26 X-28.707 Y-42.56                       (END AT 317 DEGREES)
N27 X-25.172 Y-43.598                      (END AT 321 DEGREES)
N28 X-21.919 Y-44.94                       (END AT 325 DEGREES)
N29 G03 X50.0 Y0 I21.919 J44.94            (EXIT AT 81 DEGREES)
N30 G00 Z2.0 M09
N31 G28 X50.0 Y0 Z2.0 M05
N32 M30
%
```

Program block N7 is the beginning of the 55° rise section, block N17 is the end of this section. Program block N18 is the beginning of the 40° fall section, block N28 is its end. Arc motions between the rise and fall represent the cam dwell (idle state).

The program also includes two calculations related to the I and J vectors of R64 and R50. The IJ vectors in blocks N18 and N29 are illustrated below (minus sign added):

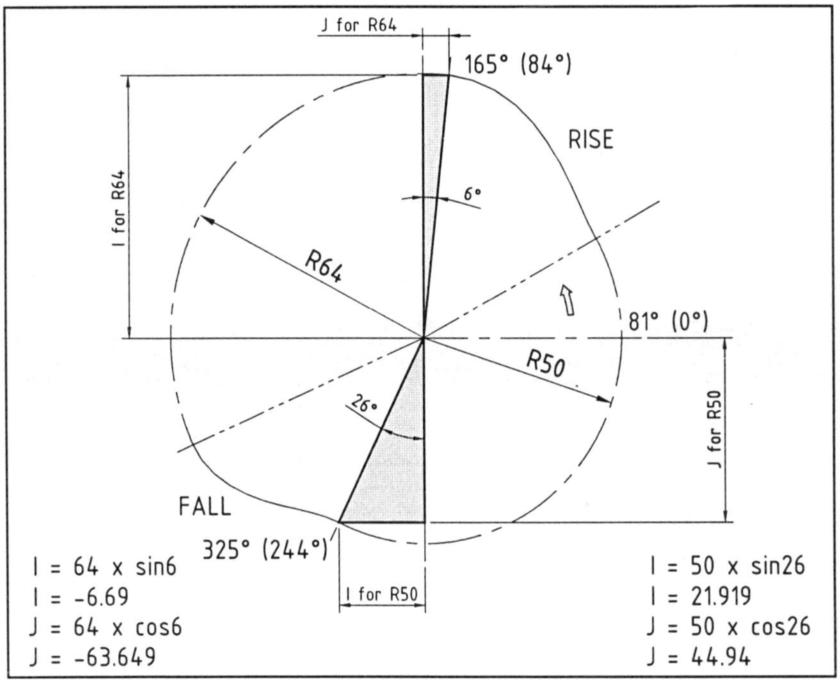

24 INTRODUCTION TO MACROS

Special Introduction

This chapter starts with somewhat unusual introduction. The publisher of this handbook, *Industrial Press, Inc., New York, NY, USA*, has also published my book called ***Fanuc CNC Custom Macros***, with the subtitle *Programming Resources for Fanuc Custom Macro B Users*. This handbook is a highly comprehensive look at macro applications in CNC machine environment for virtually all Fanuc control systems. It includes detailed explanations and many practical examples. In no way, the few page chapter presented here can replace a full size book and it is not my intention to do that.

The purpose of this chapter is to present an brief introduction to Fanuc macros to a CNC programmer who has no experience with this advanced method of programming. It is possible that this chapter will open up a new world of programming to many CNC users - a world that has many untapped possibilities. This chapter offers a short but concise introduction to Fanuc macros, using one practical example of actual development. Although the same subject is covered in the specialized publication as well, the approach used in this chapter is aimed at the programmer who is new to macros. When you feel that macros are useful to the type of your work, the ***Fanuc CNC Custom Macros*** handbook will help you to get into the depth of this fascinating subject.

Many will ask 'why macro, when there is CAM'. This is a very legitimate question and needs an answer. The majority of CAM software (*i.e.*, CNC programming using computers) is based on the development of a toolpath geometry and applying a suitable toolpath to this geometry. There is a lot more involved than that, but the purpose of such software is to provide solutions for *individual* projects, rather than *group* of projects. High level CAD software (design and drafting) offer *parametric* approach to design and/or drafting. In the CAM world, the parametric approach is rather rare. That is the reason why, in spite of the great advancement of CAD/CAM software, macros have their special place, *in addition to manual or computer assisted programming*. It is important to understand that macros do *not* replace manual programming, and certainly do *not* replace CAM programming. Macros offer the added opportunity to make parametric programs. This chapter looks at this subject.

Skills Required

Developing a macro program means developing a generic part program that can be used for many similar parts within a particular group. The parametric method that macros offer means a development of a single generic part program that is controlled by the supplied data. One such typical example is a bolt circle macro, where the data relating to the circle diameter, number of holes, etc., can be input as supplied data, and the macro will do the rest - calculate machining coordinates accordingly. Within the limitations of the macro, any bolt circle may be programmed within seconds - just by changing the supplied data. A typical bolt circle macro is used in this chapter to illustrate the macro development in detail.

Contrary to many beliefs, most macros are very focused in their objective. In fact, it is much better to develop a few specific macros than one that covers a lot. Developing any macro means learning a few more things - depending on the complexity of the macro to be developed, some basic knowledge and skills are required.

The following list identifies the most important points for successful macro development:

- **First and foremost - *THOROUGH KNOWLEDGE OF MANUAL PROGRAMMING***
- **Machining skills - how to machine a part**
- **CNC machines and controls - operation and programming (structure/flow, G-codes, M-codes, offsets, etc.)**
- **Basic mathematics skills - calculations, trigonometry, formulas**
- **Offsets and compensations applications skills - how offset adjustments work**
- **Program structure development skills - designed for both convenience and consistency**
- **Fixed cycles in depth - how they work in detail**
- **Subprograms in depth - including multi nesting applications**
- **Understanding the various methods of data settings (G10, three types of offsets, etc.)**
- **Control system parameters - their purpose, functions and methods of changing**

Most, or at great number of macros are written for a particular machine/control combination. This reality requires an in-depth knowledge of the machine tool and all control features available, and is very important during macro development. It is not uncommon to change a macro written for one control system to another control system, because of some small differences. Also, it is not uncommon to write a macro with the same objective for two different controls. Specialized applications will requires specialized knowledge. For example, if the macro is developed for *probing* or *in-process gauging*, the CNC programmer has to understand the complete gauging process as well as the objective of the gauging.

Macro is an Option

Fanuc Macro B is a special control option, which means it is not automatically available to the end user and must be activated - and paid for - first. The extra cost is easily justified by increased programming capabilities and should not be an issue. It is not unusual to find some users who do not even know whether they do or do not have the macro option installed. There is a simple way to find out if the macro feature is available. Just set the control system to *MDI* mode *(Manual Data Input)* and type in the following command:

```
#101=1
```

After pressing the *Cycle Start* button, the control will either accept the command or reject it. Make sure to enter the data as shown, including the # symbol that identifies the number that follows as a *variable* number 101 with an assigned value of 1. This statement uses the # symbol, which is unique to macros. If the control rejects it, or it is impossible to enter the symbols, that means the control does not support macro functions. In the other hand, if the control accepts the command, most likely you do have the macro feature installed and active. You may want to try some other tests to make sure.

Common Features and Applications

There are many features that Fanuc macros offer. Individually, these features offer many tools not available in standard programs or subprograms. When put together and used effectively, they provide a powerful programming environment.

The following list identifies the most important features of Fanuc macros:

- **Arithmetic and algebraic calculations (+ - * /)**
- **Rounding functions (ROUND, FIX, FUP**
- **Trigonometric calculations (SIN, COS, TAN, ATAN, ASIN, ACOS)**
- **Miscellaneous functions (SQRT ABS)**
- **Branching (IF, IF-THEN, GOTO)**
- **Logical operations (EQ, NE, GT, LT, GE, LE)**
- **Looping (WHILE)**
- **Conversion functions**
- **Variable data storage**
- **Error detection**
- **Alarm generation**
- **Input and Output**
- ***... and many other features***

In many ways, a typical macro program looks like a regular CNC program. Upon closer examination, the differences between a standard program and a macro become obvious. A macro program is structured as a subprogram - it is stored under its own program number (O....), and is typically called by the main program, using the G65 command. Macro is ended by the same function as a subprogram - M99. In its simplest form, some macro features can be used in the main program, without the actual macro call command (G65). The following list highlights some of the most common applications:

- **Groups of similar parts**
- **Offset control**
- **Custom cycles**
- **Non-standard tool motions (simulating curves)**
- **Special G-codes and M-codes**
- **Alarm and message generation**
- **Hiding and protecting macro programs**
- **Probing and gauging**
- **Various shortcuts and utilities**

Macro Structure

There are three basic areas to understand for successful macro development. These three areas offer many special functions that can be used *within* the macro:

- **Variables** ... *three types of data*
- **Functions and Constants** ... *mathematical calculations*
- **Logical Functions** ... *loops and branches*

Variables are the most important element of all macros. Variables provide the key element of any parametric programming - *flexibility* - variables store data that are subject to change. When a certain value is assigned to a variable, it is stored for some future use. All stored values are called either *defined variables* or *declared variables*.

Functions and ***Constants*** are also very important part of macros. Functions are used to *calculate* something - functions solve a mathematical calculation or a formula, for example, a + (plus) function will *sum* two or more values together. In addition to functions, *constants* can be defined in a macro as well, for example the π constant with the value of 3.14159265359...

Logical functions are used for *branching* and *looping* purposes. Looping and branching provide a *change in the program flow* that is based on a certain *condition* that has been previously defined. The given condition may be tested (evaluated), using the comparison functions, such as *'greater than'*, *'equal to'*, *'less than or equal to'*, and several others, used together with the IF function. They are called *Boolean* or *logical* operators (functions). The given condition can be tested only once, using the IF check, or it may be evaluated repeatedly, using the loop function WHILE - while the given condition is true, do the loop. The result of the evaluation will determine further flow of the program.

Macro Definition and Call

Macros are defined in a similar way as subprograms, but they are called by the G65 command:

```
O7654 (MACRO)
. . .
...
M99
%
```

G65 P7654 <*arguments*>

The G65 macro command accepts variable definitions, called *arguments*. They represent real program values that are required for a particular macro. Arguments are data *passed* to the macro body. Variable data in the macro are replaced with the supplied arguments. A typical program sample of a G65 macro using three arguments may have the following format:

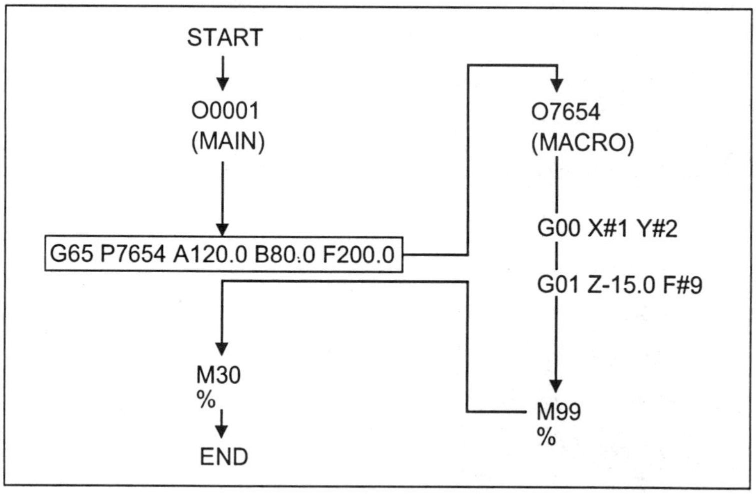


```
G65 P7654 A120.0 B80.0 F200.0
```

☞ **G65**	*Macro call command*
P7654	*Program number containing the macro - stored as O7654*
A120.0	*Assignment of local variable A (#1) argument to be passed to macro O7654*
B80.0	*Assignment of local variable B (#2) argument to be passed to macro O7654*
F200.0	*Assignment of local variable F (#9) argument to be passed to macro O7654*

The illustration on the previous page shows the flow between the main program and the macro, as well as the definition and use of the three variables. In the example, the two motion commands will be replaced with the values assigned (defined) as arguments:

`G00 X#1 Y#2`	*... will be interpreted as ...*	`G00 X120.0 Y80.0`
`G01 Z-15.0 F#9`	*... will be interpreted as ...*	`G01 Z-15.0 F200.0`

Assignments of macro variables as arguments is one of the most important subjects of developing any macro. Arguments are defined by a capital letter, followed by the required value. Each letter is associated with a local variable number that is used in the macro body.

Fanuc provides an arbitrary list of arguments and their associated variables, shown in the table:

Argument list 1 address	Local variable call in a macro
A	#1
B	#2
C	#3
D	#7
E	#8
F	#9
H	#11
I	#4
J	#5
K	#6
M	#13
Q	#17
R	#18
S	#19
T	#20
U	#21
V	#22
W	#23
X	#24
Y	#25
Z	#26

Fanuc also supports an *Assignment List 2*, but that method is very seldom used and is not important to general introduction to macros. In either list, these variables are called *LOCAL* variables and can be used in both arguments and the macro body. Although Imperial alphabet has twenty six letter, note that there are five letters missing - they are *G N O P L*. These letters are reserved and cannot be used as arguments in macro call G65. The letter *L* is used for the number of macro repetitions, if required. Local variables #10, #12, #14, #15, #16, #27, #28, #29, #30, #31, #32 and #33 can only be used inside of a macro.

There are four types of variables available for Fanuc macros:

- **Null variable**
- **Local variables**
- **Common variables**
- **System variables**

Variable number range		Variable type	Description
From	To		
#0		NULL variable	A *NULL* variable has no value. It is defined as #0 variable, it is an empty variable, often called a *vacant* variable. This variable can be read by the macro program, but it cannot be assigned a value, which means data cannot be assigned to it
#1	#33	LOCAL variables	*LOCAL* variables are only temporary - they are used in a macro body and hold certain data. When the macro is called, the local variables are set to their assigned values. When the user macro is completed and exits (using the miscellaneous function M99), or the control power is turned off, all local variables are set to null values - they cease to exist
#100 #500	#199 #599 (#999)	COMMON or Global variables	*COMMON* (also called *Global*) variables are still valid when a macro is completed. These variables are maintained by the system and they can be shared by several other macro programs. The higher level variables are normally cleared by a specially design macro program. Various options exist that define the actual range of common variables
#1000	...and up	SYSTEM variables	*SYSTEM* variables are used for setting and/or changing default conditions and can read and write different CNC data, for example, a current status of a G-code mode, the current work offset, etc. Their numbers are assigned by the Fanuc control

The above table shows the differences between the four types of variables. Null variable has very specific applications, and system variables cannot be declared.

Variable Declarations and Expressions

Variable data is not only defined in the argument section of the G65 macro call, variables can also be defined - or declared - in the macro body itself. Compare the two following declarations:

```
#19 = 1200
```

Value of 1200 is assigned to variable number 19 - possibly a spindle speed (rev/min)

```
#9 = 150.0
```

Value of 150.0 is assigned to variable number 9 - possibly a feedrate (mm/min, in/min, etc.)

Integers and real numbers have their special places in macros.

Both declarations store actual amounts or values - the value of 1200 is stored into the variable #19 and the value of 150.0 into the variable #9. Both values shown in the example are *numbers*, but they are two different *types* of a number. 1200 is an integer, whereby 150.0 is a real number. In macros, this distinction is important. Variables may also use special expressions - for example, the variable definition #9 = 150.0 can be redefined later, as #9 = #9 * 1.1, for example. If #9 defines cutting feedrate, the new feedrate will be F165.0. Expressions can also be nested, following the standard mathematical hierarchy of the calculation order:

```
#1 = 5.0
#2 = 8.5
#3 = 10.0
#4 = [#3*#2]/#1      Returns 17.0
```

Note that *brackets* are used for nesting, not parentheses. Parentheses indicate a program comment or a message:

```
#19 = 1200    (SPINDLE SPEED DEFINED)
#9 = 150.0    (CUTTING FEEDRATE DEFINED)
```

If macro feature is available, variables can be defined and used in the main program, for example:

```
#19 = 1200
...
N4 G90 G00 G54 X246.0 Y135.0 S#19 M03
...
```

Many variations and applications are possible.

Macro Functions

The available macro functions can be divided into six groups:

- **ARITHMETIC functions**
- **TRIGONOMETRIC functions**
- **ROUNDING functions**
- **MISCELLANEOUS functions**
- **LOGICAL functions and operations**
- **CONVERSION functions**

There are four ***arithmetic*** functions:

+ - * /	**(addition, subtraction, multiplication, division)**

#1 = 3.5	3.5	*Returned value of variable #1 is 3.5*
#2 = 4.25	4.25	*Returned value of variable #2 is 4.25*
#3 = 2.0 + 5.0	7.0	*Returned value of variable #3 is a sum of 2+5*
#4 = #3 + 1	8.0	*Add 1 to the result of variable #3*
#5 = #2 - 0.8	3.45	*Subtract 0.8 from the current value of variable #2*
#6 = #1 - #3	-3.5	*Subtract contents of #3 from the contents of #1*
#7 = #2 * 6	25.5	*Multiply contents of #2 by 6*
#8 = 7.0 / 8.0	0.875	*Divide 7 by 8 as real numbers*
#9 = 7 / 8	0.875	*Divide 7 by 8 as integer numbers*

Trigonometric functions are used for calculations of sides or angles of a right angle triangle.

The following trigonometric functions are generally available for macros:

SIN COS TAN ATAN ASIN ACOS

All input for **SIN**, **COS**, and **TAN** is in degrees and the output of the inverse functions **ATAN**, **ASIN** and **ACOS** is also in degrees. Inverse functions are usually marked as tan^{-1}, sin^{-1}, and cos^{-1} on a pocket calculator.

```
#1 = SIN[38]        0.6156615   Actual value must be in brackets
#2 = 23.7           23.7
#3 = COS[#2]        0.9156626   Reference to a variable must be in brackets
#4 = TAN[12.86]     0.2282959
```

The inverse trigonometric functions accept the length of two sides of a triangle, both enclosed in brackets, and separated by the slash symbol standing alone between them. The acceptable range is within 0 <= RESULT < 360:

```
#5 = ATAN[0.25]/[0.5]    26.5650512   Note the position of the slash !!!
```

ASIN and ACOS functions are not available on 0/16/18/21 model controls

Rounding functions control how decimal places are handled. There are there rounding functions:

ROUND FIX FUP

The **ROUND** function rounds off a value to the nearest whole number. The function disregards fractions that are *less than 0.5*. For fractions that are *equal to* or *greater than 0.5*, the next whole number is the rounded value:

```
ROUND[0.00001]          Returns 0.0
ROUND[0.5]              Returns 1.0
ROUND[0.99999]          Returns 1.0
ROUND[1.0]              Returns 1.0
```

There is one very important difference between using the **ROUND** function directly in a program statement or in a variable definition. Consider the following two blocks:

```
#101 = 27/32
G91 G00 X[ROUND[#101]]     The incremental motion will be X0.8438 (imperial) or X0.844 (metric)
```

However, if the **ROUND** function is used in a variable definition, the result will be different:

```
#101 = [ROUND[27/32]]      Returns 1.0
G91 G00 X[#101]            The incremental motion will be X1.0
```

Rounding can also be performed to a given number of decimal places, but it is a two step process. In the first step, the given value must be *multiplied by the factor* of:

10	**... to round off to *one* decimal place**	
100	**... to round off to *two* decimal places**	
1000	**... to round off to *three* decimal places**	*typical for metric system*
10000	**... to round off to *four* decimal places**	*typical for Imperial system*

The second step is to use the **ROUND** function together with the multiplier. For example, to round 123.56789 to three decimal places:

```
#101 = 123.456789
#101 = #101*1000          Returns 123456.789
#101 = ROUND[#101]        Returns 123457.0 (next whole number)
#101 = #101/1000          Returns 123.457
```

Note that only a single variable has been used for all four calculations. If the original value is not required later, this is a very efficient method of defining variables in the macro body.

The remaining two rounding functions **FIX** and **FUP** are used to round a given value *up* or *down* only, regardless of whether the decimal portion is over or under 0.5.

The **FUP** function is designed to *round up* the given value (raise fractions less than 1.0).

```
FUP[0.00001]              Returns 1.0
FUP[0.5]                  Returns 1.0
FUP[0.99999]              Returns 1.0
FUP[1.0]                  Returns 1.0
```

The **FIX** function is designed to *round down* the given value (discard fractions less than 1.0) - *i.e.,* strip all values after decimal point.

```
FIX[0.00001]              Returns 0.0
FIX[0.5]                  Returns 0.0
FIX[0.99999]              Returns 0.0
FIX[1.0]                  Returns 1.0
```

The **FUP** and **FIX** function are commonly used for counting rather than calculations.

Miscellaneous functions - there are five of them, but only two will be described in the course, as the other three are not available on all control models:

SQRT	**ABS**

The **SQRT** function calculates the *square root* of a number supplied between brackets:

```
SQRT[25.0]                Returns 5.0
```

The **ABS** function (absolute function) always returns a *positive* value of a given number:

```
ABS[-19.3]                Returns 19.3
ABS[19.3]                 Returns 19.3
```

Using the **ABS** function is very useful in many applications, for example, to guarantee a positive or a negative value:

```
#26 = 19.3                Positive input
G01 Z-[ABS[#26]]          The motion will be Z-19.3

#26 = -19.3               Negative input
G01 Z-[ABS[#26]]          The motion will be Z-19.3
```

As the example shows, the **ABS** application is often used to force a negative value of Z-depth.

Logical functions include Boolean functions and binary functions:

EQ NE GT LT GE LE

Boolean functions (operators) compare two values and return a *true* or *false* condition:

- EQ = Equal to
- NE = Not equal to
- GT = Greater than
- LT = Less than
- GE = Greater than or equal to
- LE = Less than or equal to

Binary functions are:

AND OR XOR

The **AND** and **OR** functions compare two defined conditions. The compared conditions are evaluated, and return either *TRUE* value or *FALSE* value. In plain English, it is easy to understand the difference between the **AND** and the **OR** functions, because they follow the basic logic of the Imperial language. Function **XOR** is seldom used in machining macros.

Conversion functions are used with programmable logical controllers (Fanuc version) and are not covered here.

Branching and Looping

Both *branches* and *loops* (iterations) are based on the return value of an evaluated condition. There are three macro functions available:

IF GOTOn WHILE	(also IF-THEN on some controls)

The **IF** function evaluates a condition and branches to a specified block number if the condition is true. If the condition is false, the macro processing continues in the next block. Effectively, the **IF** function bypasses one or more blocks if a certain condition is true. The programming format is:

```
IF [<this condition is true>] GOTOn
```

GOTOn function is an unconditional branch and is programmed by itself in the block. the *n* is replaced by a unique block number to branch to, for example GOTO78, will branch to a block N78. The programming format is a one word block entry:

```
GOTO78
```

WHILE is a looping function - also known as the *iteration* function - and also requires the specified condition to be true, in order for the loop to be processed. Consider the word *while* as meaning *as long as*. The most common mistake programmers make when using the WHILE function is not providing an exit method from the loop, creating an endless loop.

The programming format for the **WHILE** function is:

```
WHILE [<this condition is true>] DOn
...
. . .
ENDn
```

The ENDn is the end of loop identification. Up to three level loop nesting is possible, each nested loop will have is own WHILE-ENDn sequence. For a single loop, the ENDn is END1, matching DO1 in the WHILE block.

There is also **IF-THEN** method available on some higher level control systems. If available, the programming is much more streamlined. The programming format is:

```
IF [<this condition is true>] THEN [<argument>]
```

An argument is a one block statement, normally a definition based on the condition being true:

```
IF [#1 EQ 1] THEN #33=0
```

All branching and looping functions are decision making functions - they are very powerful and are fundamental part of effective use of Fanuc macros.

Macro Development - Bolt Circle

When it comes to learning opportunities, macro programming courses are not as easily available as learning standard CNC programming. Various courses that do offer macro programming also offer their own specific approach and contents. It should come as no surprise that one example that most courses offer is a macro to develop a bolt circle hole pattern. Macro development for a bolt hole pattern can be very simple or it can offer many additional features. This is where the courses differ from each other. The objective of this chapter is to present a *sample macro development process* - using a bolt circle hole pattern as an example - with many additional features that are not necessarily part of regular macro courses. There is also another reason - much more important - the presentation will focus on the generalities of developing *any* macro.

Knowing various macro functions is essential for any macro development. However, putting this knowledge into practice is much more involved. The bolt circle macro example in this section covers a common pattern of holes equally spaced on a circle. Developing the core of a macro for such pattern is quite simple, adding extra features will take more work, but the result will be that much more rewarding. The amount of flexibility a macro should have is the CNC programmer's decision. Programming a bolt hole circle pattern using the manual method is not difficult, but it could be time consuming and subject to errors. At the same time, a manually developed program for a bolt hole pattern is an excellent starting point for development of a similar program with parametric entries - a macro. There are several stand-alone utilities available for calculating bolt holes, running under the Windows operating system, and they all use the similar approach as the one used for the macro demonstrated in this chapter.

Developing any macro requires careful preparation. It is not good enough to take a single drawing and make a macro based on its data. Macros are parametric in nature - that means they are supposed to be used for other drawing with similar features. Even for a macro that is relatively easy to develop - such as the bolt hole macro - it is necessary to collect appropriate information from several similar drawings. If careful planning is very important for the development of standard CNC programs, it absolutely vital for the development of macros.

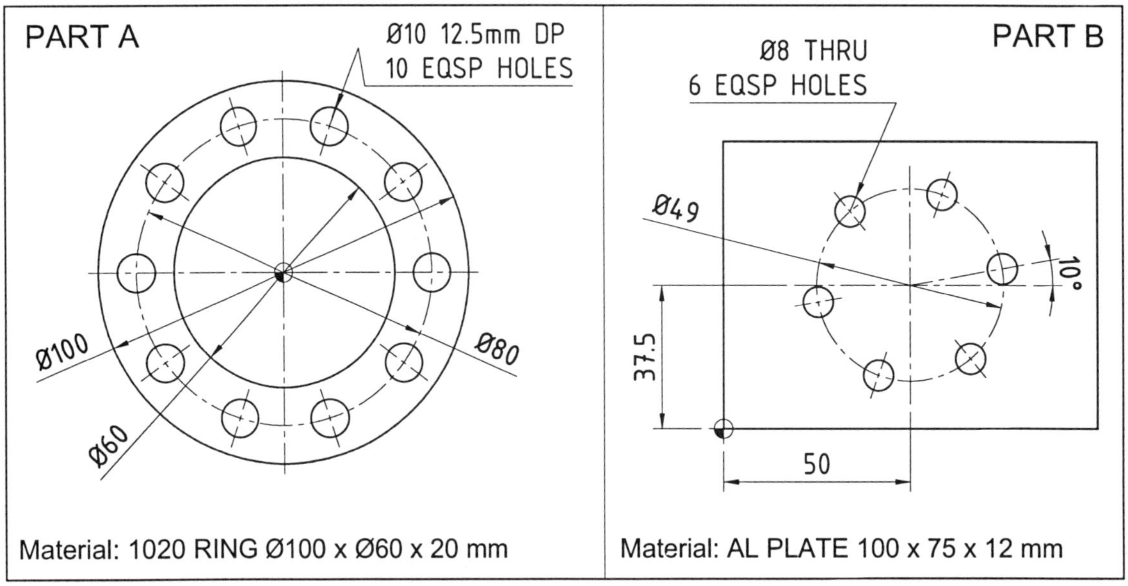

Two typical drawings of bolt hole patterns are shown above. These two drawings represent the selection of different types of bolt circles that should be covered by the same macro. All drawings in the selection must be *evaluated* first, to identify the common features as well as differences. Drawing are identified as *Part A* and *Part B* for easier reference.

Evaluation of Drawings

The result of a careful evaluation of several similar drawing will identify various features that may or may not be important. The following table lists both the common features and the differences between the two drawings - only the relevant items are shown:

Common Features	Differences
- Equally spaced bolt holes - typical to all bolt holes - Number of equally spaced holes - Bolt circle diameter	- Bolt circle center location - Angular orientation of the holes - Size, shape and type of the material - Diameter of holes - Depth of holes

The next step in the evaluation is to combine the data that matter to the bolt circle only - select the main objective first! Is the size or shape of the material important? Not for the macro. Its center location, on the other hand, is very important. Trying to 'squeeze' all kinds of data into a single macro is counterproductive. For example, using a fixed cycle within a macro is not recommended, because of the many variables that would have to be tracked. Fixed cycle should be outside of macro, which leaves only the bolt hole coordinates to be calculated. The above table can be now modified:

Features required to find XY coordinates	Features not used in the macro
- Equally spaced bolt holes - typical to all bolt holes - Number of equally spaced holes - Bolt circle diameter - Bolt circle center location - Angular orientation of the holes	- Size, shape and type of the material - Diameter of holes - *determined by the selected tool* - Depth of holes - *entered in the fixed cycle*

Once all important features have been identified, they can be listed, along with the drawing dimensions from the two sample drawings:

Feature	Part A	Part B
Bolt circle location (XY)	X0 Y0	X50.0 Y37.5
Bolt circle diameter	80.0	49.0
Number of equally spaced holes	10	6
Angular orientation from zero degrees	0	10

Bolt Hole Macro Features

All identified features have to be accounted for in the macro, to give it the necessary flexibility. A bolt hole pattern macro can be developed with the first hole location always located at zero degrees. Such a macro is easier to develop, but also offers only limited applications. Including the first hole angle in the macro call - even if it is at zero degrees - greatly enhances the macro usefulness and its applications. In fact, the macro can be structured in such a way, that if the first hole is at zero degrees, the angle does not have to be defined as an argument. This method is known as programming a *default* condition. Another feature that can be added without much effort, is the ability of the macro to start at any hole, skipping the hole or holes before. For example, on a 8-hole bolt pattern, starting at the third hole will skip the first two holes, and only six holes will be machined while still maintaining the bolt circle characteristics - no other changes will take place. Since the features represented by the last two examples are not going to be used very often, machining all holes can also be programmed as a default, so the programmer does not have to enter unnecessary arguments.

Another very important consideration is the selection of dimensional units. Can the macro be used for both metric and Imperial units? The answer is hidden in the above table. Both circle location and its diameter can be programmed in either units, with no special considerations. This question should be asked for any macro. A good case for considering the current units is the selection of a clearance amount. Macro features offer the option of programming Z2.5 (mm) or Z0.1 (inches) as a clearance, depending on which units are current. This does not apply to the bolt hole macro.

Decisions regarding the features and functionality of a macro must always be made *before* writing the macro. The bolt hole macro must cover the designs from a variety of drawings (such as the two representative samples shown earlier). The final macro should always be based on the conditions, restrictions and decisions made during the planning process. In the case of the bolt hole example, these are decisions and conditions under which the macro will operate:

- **Bolt hole diameter (pitch diameter) must always be known, along with any absolute center coordinates**
- **All bolt holes are spaced equally within the bolt circle pattern (angular measurement)**
- **Any number of holes is acceptable (within machine capabilities) - normally minimum of two, but one is allowed**
- **The angular location of the first hole can be anywhere, measured from zero degrees (3 o'clock position)**
- **Direction of machining is from the first hole into the CCW direction (arbitrary decision)**
- **Initial hole or holes can be skipped, resulting in an arc hole pattern - rather than a full circle hole pattern**
- **Macro must be available to any fixed cycle selection**

Careful planning is always very important; it may not be practical - or even possible - to add other features later to an existing macro.

Assignment of Variables

Once the initial planning has been completed, the macro development can begin. Typical process starts with definition of the input values - the arguments to be passed from the G65 macro call to the macro itself. The assignment of variable data will follow all earlier decisions:

Assignment letter	Associated variable #	Description & Comment		Part A Example	Part B Example
W	#23	Bolt circle diameter		80.0	49.0
X	#24	Bolt circle center - X-coordinate		0	50.5
Y	#25	Bolt circle center - Y-coordinate		0	37.5
H	#11	Number of equally spaced holes		10	6
(A)	#1	Angular orientation of the first hole	Default is 0	0	10.0
(S)	#19	Hole number to start at from 0 °	Default is 1	0	0

Although assignment of the letters can be any legitimate choice, it makes sense to choose letters that have some meaning, whenever possible.

Internal Calculations

Any programmer who has ever done a bolt circle program the traditional way knows that calculation of each hole coordinate requires trigonometric solution (polar coordinates are not considered here). This solution involves the *radius* of the bolt circle, the *increment angle*, combined with the *sine* and *cosine* functions. Looking at the table above, neither the radius nor the increment angle is listed in the definitions. There is no need for it. Both can be calculated in the macro body - the radius will be one half of the supplied diameter W (#23/2) and the angle between holes will be 360° divided by the number of holes H (360/#11). It makes sense to use the diameter in the G65 assignment rather than a radius, as the proper drafting technique is to supply *bolt circle diameter*. In order to save processing resources, the variable #23 will be redefined in the macro body, to mean *radius*, not diameter.

Other Calculations

Calculating the XY coordinates of each hole from the bolt circle center requires the radius (#23 redefined in the macro body), the current angle based on the original definition and increment, and the sine and cosine functions:

X-coordinate = #23 * cos<current angle>
Y-coordinate = #23 * sin<current angle>

In order to include the XY absolute center location, both X and Y have to be added to the formula:

X-coordinate = #23 * cos<current angle> + #24
Y-coordinate = #23 * sin<current angle> + #25

As each next hole location will be changing with each new angle increment, a looping technique (iteration) will be applied to the macro, using the WHILE function. To check other conditions required, the IF function will be used as necessary. Overall, macro development is not difficult - it just takes some serious planning - see the complete macro chart on the next page.

Final Considerations

For any macro to work properly, other important considerations have to be taken into account and maintained from one job to another. They include macro transportability - can the macro be used on different machines? They also include storing the program processing state before macro call and re-instating it before macros exits. Other considerations include error trapping routines, messages and comments, alarms, and other means to make the macro as reliable as possible.

Bolt hole circle pattern - declarations | MACRO O8500

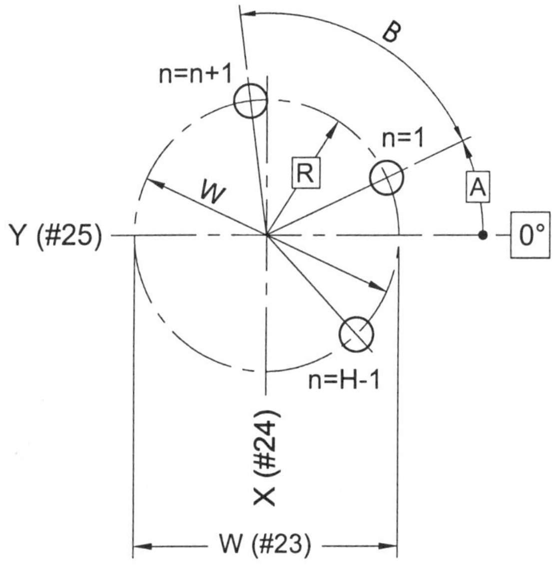

n = Hole counter (CCW from 0°)

VAR	#	Description
W	23	Bolt circle diameter
X	24	Absolute center location in X
Y	25	Absolute center location in Y
H	11	Number of EQSP holes
A	1	First hole angle from 0°
S	19	Hole number to start at
R	n/a	Bolt circle radius (#23/2)
B	n/a	Angle between holes (360/#11)
X and Y coordinates for any hole:		
X = cos((n-1) * B + A) * R + #24		
Y = sin((n-1) * B + A) * R + #25		

n=1 for Hole 1, n=2 for Hole 2, etc.

G65 P8500 W- X- Y- H- (A-) (S-)

To illustrate the complete application, the following eight blocks represent the main program that calls the macro O8500. In this example, the operation is standard drilling:

```
(=== MAIN PROGRAM FOR PART B ===)
N1 G21                                      Metric mode
N2 G90 G00 G54 X0 Y0 S1200 M03              First motion block + spindle speed
N3 G43 Z10.0 H01 M08                        Tool length offset + clearance above
N4 G99 G81 R2.0 Z-15.9 F225.0 L0            Fixed cycle call data - no machining (or K0)
N5 G65 P8500 X50.0 Y37.5 W49.0 H6 A10.0 S1  Macro call with all arguments - full circle
N6 G80 Z10.0 M09                            Retract above work
N7 G28 Z10.0 M05                            Return to machine zero
N8 M01                                      End of current tool - optional stop
...
%
```

The example is for *Part B*, as illustrated earlier in the chapter. Macro call for *Part A* (and *Part B*) are compared at the end of this section. For either drawing, the same macro O8500 will be used:

```
O8500 (=== BOLT HOLE CIRCLE MACRO ===)              Macro number and description
#10 = #4003                                         Store current setting of G90 or G91
IF[#23 LE 0] GOTO9101                               Bolt circle diameter to be greater than zero
IF[#11 NE FUP[#11]] GOTO9102                        No fractions allowed for number of holes
IF[#11 LE 0] GOTO9103                               Minimum number of holes is one
IF[#1 NE #0] GOTO101                                Branch to N101 if start angle is defined
 #1 = 0                                             Start hole angle = 0 (zero) if not defined
N101 IF[#19 NE #0] GOTO119                          Branch to N119 if start hole is defined
 #19 = 1                                            Start hole number = 1 (one) if not defined
N119 IF[#19 NE FUP[#19]] GOTO9102                   No fractions allowed for start hole number
IF[#19 LT 1] GOTO9104                               Start hole number must be one or higher
IF[#19 GT #11] GOTO9105                             Start hole number must be less than all holes
 #23 = #23/2                                        Change diameter of bolt circle to radius
WHILE[#19 LE #11] DO1                               Start loop for holes
 #30 = [#19-1]*360/#11+#1                           Calculate the current hole angle
 X[COS[#30]*#23+#24] Y[SIN[#30]*#23+#25]            Calculate current X and Y hole location
 #19 = #19+1                                        Update counter for the loop
END1                                                End of loop
GOTO9999                                            Bypass alarm messages
N9101 #3000=101 (DIA MUST BE GT 0)                  Alarm number 101 or 3101
N9102 #3000=102 (HOLES DATA MUST BE INTEGER)        Alarm number 102 or 3102
N9103 #3000=103 (ONLY POSITIVE NUM OF HOLES)        Alarm number 103 or 3103
N9104 #3000=104 (START HOLE IS 1 OR HIGHER)         Alarm number 104 or 3104
N9105 #3000=105 (START HOLE NUMBER TOO HIGH)        Alarm number 105 or 3105
N9999 G#10                                          Restore modal G-code
M99                                                 End of macro
%
```

There are two instances where the macro defines a default setting. One is for an angle of the first hole (#1), the other is for the start hole number (#19). This macro uses a rather inconvenient method with the GOTO statements. GOTO statements will work for all control systems, but the higher level Fanuc controls also support a shorter - and much friendlier mode - using the IF-THEN format:

```
IF[#1 EQ #0] THEN #1 = 0                            Start hole angle = 0 (zero) if not defined
IF[#19 EQ #0] THEN #19 = 1                          Start hole number = 1 (one) if not defined
```

Macro Call

In the main program, block N5 calls the macro. For the two sample programs, the macro call can be programmed as:

```
N5 G65 P8500 X0 Y0 W80.0 H6 A0 S1                   Macro call (Part A) with all arguments
N5 G65 P8500 X50.0 Y37.5 W49.0 H6                   Macro call (Part A) no A and S arguments

N5 G65 P8500 X50.0 Y37.5 W49.0 H6 A10.0 S1          Macro call (Part B) with all arguments
N5 G65 P8500 X50.0 Y37.5 W49.0 H6 A10.0             Macro call (Part B) no S argument
```

This macro offers a lot of flexibility. With some creative use of three variables (H, A, and S), any hole can be the first hole and any number of holes can be specified, as long as they are part of a bolt circle. Bolt hole patterns that have an even number of holes are generally more flexible that those with odd number of holes. Developing the basic core of a macro is usually not difficult. The real challenge is to add the 'extra power' to the macro. That takes some experience.

Please note that this chapter represents only a very minor portion of macro possibilities, and in no way is intended to be a comprehensive teaching tool

25 DID YOU KNOW THAT ... ?

This is not really a chapter in the traditional sense. These two pages contain bits and pieces I have collected in various notes, and all I have done is to summed them up into short questions, under the collective heading *"Did you know that ...?"*. Of course, there could have been several more pages added, but there are enough subjects covered here already. Use the check boxes to mark the answers you *did know*, and make an effort to learn those that you *did not know*.

There is no particular order in the listing.

- ☐ **G28 Z.. motion (vertical CNCs) will cancel the tool length offset ?**
- ☐ **Miscellaneous functions M00, M01, and several others become active only AFTER the programmed tool motion in the same block is completed ?**
- ☐ **Miscellaneous function M19 (M20) can be used to orient the spindle ?**
- ☐ **Spindle can be oriented at the machine (control) using MDI or a button ?**
- ☐ **G41 and/or G42 in a turning program will make a precise contour (providing other settings match) ?**
- ☐ **Cutter radius offset G41 or G42 must be programmed together with a D-address ? <mill only>**
- ☐ **G87 fixed cycle cannot be programmed in G98 mode ?**
- ☐ **G41 or G42 command cannot be started while in arc motion mode (G02 or G03) ?**
- ☐ **Start of cutter radius offset MUST be programmed only while a rapid or linear motion is active ?**
- ☐ **Miscellaneous function M99 is not only a subprogram end, but it can also make the main program run 'forever' ?**
- ☐ **Local coordinate offset G52 is only effective when a work offset G54-G59 is in effect ?**
- ☐ **Programming of G49 (tool length offset cancel) can be more harmful than ignoring it ?**
- ☐ **Program data T03 M06 is the same as M06 T03 ?**
- ☐ **Work offset G54 is NOT always the automatic default on most controls ?**
- ☐ **M00 will terminate all axis motion, coolant function, and spindle rotation ?**
- ☐ **On many CNC machines, the G00 rapid motion will not be the shortest distance between the two points ?**
- ☐ **Tool length offset G43 must be programmed together with an H-address ? <mill only>**

- ☐ Txx programmed for milling controls will only CALL the tool but not make a tool change ?
- ☐ The stored (registered) amount of cutter radius offset DOES NOT always mean the real cutter radius ?
- ☐ Spindle speed rotation and the actual spindle speed can be programmed in different blocks ?
- ☐ Programming the tool length offset number matching the tool number is easy way to remember it ?
- ☐ Wear offset is not used just for wear, but for any other necessary adjustments (fine tuning) ?
- ☐ Old G92 position register command is measured from part zero to the current tool position ?
- ☐ G54-G59 work offsets are measured from machine zero to the part zero ?
- ☐ M03 and M04 (Spindle rotation) when programmed with a tool motion will be activated simultaneously ?
- ☐ M05 (Spindle stop) when programmed with a tool motion will come into effect only after the motion is fully completed ?
- ☐ If both R- and I- J- are in the same block, the R- will take precedence ?
- ☐ G44 tool length offset negative is hardly ever used ?
- ☐ Programming G20 / G21 (Units selection) in a block by itself will never cause harm, but may be necessary for some controls ?
- ☐ The most suitable contouring clearance for lathe work should be 2.5 mm or 0.2 inches ?
- ☐ Tool number for a lathe tool is actually a combination of the geometry *and* offset numbers - two pairs ?
- ☐ G62 - Automatic Corner Override - can be used to eliminate chatter or poor surface finish when machining small inside corners with cutter radius offset active ?
- ☐ G91 G28 X0 Y0 Z0 (milling) or G28 U0 W0 (turning) is the fastest way to send machine to its zero position from any tool position (providing it is safe) ?
- ☐ Alarm lights on the operation panel can burn out and fail to warn the CNC operator ? Check them frequently.
- ☐ Programming M01 (Optional Stop function) at the end of each tool makes part inspection easier?
- ☐ Comment following the program number will show as a description on the control display screen ?

26 REFERENCES AND RESOURCES

Many readers will use this book in conjunction with the *CNC Programming Handbook*. Although both books take a different approach to their respective subjects and have been developed as completely separate and independent publications in their own right, the subject of CNC and common authorship connects them to a certain degree. To make these connections easier, this chapter lists the subjects in the *CNC Programming Handbook* relating to the subjects (chapters) in this publication.

For in-depth coverage of *Fanuc Macros B* subjects, refer to *Fanuc CNC Custom Macros* book.

*Chapter 1 - **PART PROGRAM DEVELOPMENT***
Check the reference chapters in CNC Programming Handbook:

Chapter 5 - Control System . . . . pages 19-30
Chapter 6 - Program Planning. . . . . pages 31-40
Chapter 7 - Part Program Structure . . . . pages 41-46
Chapter 47 -Writing a CNC Program . . . . pages 441-446

*Chapter 2 - **CALCULATING CONTOUR POINTS***
Check the reference chapters in CNC Programming Handbook:

Chapter 22 - Linear Interpolation . . . . pages 159-162
Chapter 29 - Circular Interpolation . . . . pages 235-246

*Chapter 3 - **FORMULAS FOR CONTOURING***
Check the reference chapters in CNC Programming Handbook:

Chapter 22 - Linear Interpolation . . . . pages 159-162
Chapter 29 - Circular Interpolation . . . . pages 235-246
Chapter 52 - Math in CNC Programming . . . . pages 471-482

*Chapter 4 - **USING CUTTER RADIUS OFFSET***
Check the reference chapters in CNC Programming Handbook:

Chapter 30 - Cutter Radius Offset. . . . . pages 247-268
Chapter 31 - Cutter Radius Offset in Planes . . . . pages 271-273

*Chapter 5 - **PART REVERSAL IN MILLING***
Check the reference chapters in CNC Programming Handbook:

Chapter 18 - Work Offsets . . . . pages 123-130
Chapter 19 - Tool Length Offset . . . . pages 131-142
Chapter 39 - Subprograms . . . . pages 367-380
Chapter 40 - Datum Shift . . . . pages 381-384 and 386-392

*Chapter 6 - **USING TAPERED END MILLS***

This subject is not covered in CNC Programming Handbook

*Chapter 7 - **SPECIAL PURPOSE G-CODES***
Check the reference chapters in CNC Programming Handbook:

Chapter 8 - Preparatory Commands . . . . pages 47-52

*Chapter 8 - **TOOL LENGTH OFFSET CHANGE***
Check the reference chapters in CNC Programming Handbook:

Chapter 19 - Tool Length Offset. . . . . pages 131-142

*Chapter 9 - **BLOCK SKIP APPLICATIONS***
Check the reference chapters in CNC Programming Handbook:

Chapter 23 - Block Skip Function . . . . page 163-170

*Chapter 10 - **STANDARD AND RIGID TAPPING***
Check the reference chapters in CNC Programming Handbook:

Chapter 25 - Fixed Cycles. . . . . pages 177-190
Chapter 25 - Tapping Cycles . . . . page 186
Chapter 26 - Machining Holes . . . . pages 191-216
Chapter 26 - Tapping . . . . pages 209-212
Chapter 27 - Pattern of Holes . . . . pages 217-226

*Chapter 11 - **POLAR COORDINATES***
Check the reference chapters in CNC Programming Handbook:

Chapter 27 - Pattern of Holes . . . . pages 217-226
Chapter 27 - Polar Coordinate System . . . . pages 225-226

*Chapter 12 - **SUBPROGRAM DEVELOPMENT***
Check the reference chapters in CNC Programming Handbook:

Chapter 39 - Subprograms . . . . pages 357 to 380

*Chapter 13 - **TURNING AND BORING IN DEPTH***
Check the reference chapters in CNC Programming Handbook:

Chapter 34 - Turning and Boring . . . . pages 293-306
Chapter 35 - Lathe Cycles. . . . . pages 307-320

Chapter 14 - ***PROGRAMMING TAPERS***
Check the reference chapters in CNC Programming Handbook:

Chapter 15 - ***TECHNIQUES FOR GROOVING***
Check the reference chapters in CNC Programming Handbook:

Chapter 16 - ***TECHNIQUES FOR THREADING***
Check the reference chapters in CNC Programming Handbook:

Chapter 17 - ***RESTRICTIONS IN THREADING***
Check the reference chapters in CNC Programming Handbook:

Chapter 18 - ***PRACTICAL THREAD MILLING***
Check the reference chapters in CNC Programming Handbook:

Chapter 19 - ***KNURLING***

This subject is not covered in CNC Programming Handbook

Chapter 20 - ***FOUR-AXIS LATHE***

This subject is not covered in CNC Programming Handbook

*Chapter 21 - **PALLET CHANGERS***
Check the reference chapters in CNC Programming Handbook:

Chapter 46 - Horizontal Machining . . . . pages 429-440
Chapter 46 - Automatic Pallet Changer. . . . . pages 437-439

*Chapter 22 - **WORKING WITH PLANES***
Check the reference chapters in CNC Programming Handbook:

Chapter 31 - Plane Selection . . . . pages 269-274

*Chapter 23 - **PROGRAMMING CAMS***

This subject is not covered in CNC Programming Handbook

*Chapter 24 - **INTRODUCTION TO MACROS***

This subject is not covered in CNC Programming Handbook

Another publication, Fanuc CNC Custom Macros, covers this subject in depth

*Chapter 25 - **DID YOU KNOW THAT ...?***

Various references throughout CNC Programming Handbook

INDEX

E

F

G

H

P

R

S

T

U

V

W